# Intermolecular Interactions: Physical Picture, Computational Methods and Model Potentials

# Intermolecular Interactions: Physical Picture, Computational Methods and Model Potentials

**Ilya G. Kaplan**
*Universidad Nacional Autónoma de México, Mexico*

John Wiley & Sons, Ltd

Telephone (+44) 1243 779777

Email (for orders and customer service enquiries): cs-books@wiley.co.uk
Visit our Home Page on www.wiley.com



***Other Wiley Editorial Offices***

John Wiley & Sons Inc., 111 River Street, Hoboken, NJ 07030, USA

Jossey-Bass, 989 Market Street, San Francisco, CA 94103-1741, USA

Wiley-VCH Verlag GmbH, Boschstr. 12, D-69469 Weinheim, Germany

John Wiley & Sons Australia Ltd, 42 McDougall Street, Milton, Queensland 4064, Australia

John Wiley & Sons (Asia) Pte Ltd, 2 Clementi Loop #02-01, Jin Xing Distripark, Singapore 129809

John Wiley & Sons Canada Ltd, 22 Worcester Road, Etobicoke, Ontario, Canada M9W 1L1

Wiley also publishes its books in a variety of electronic formats. Some content that appears in print may not be available in electronic books.

***Library of Congress Cataloging-in-Publication Data***

Kaplan, I. G. (Il'ëiìa Grigor'evich)
[Vvedenie v teoriëiìu mezhmolekulëiìarnykh vzaimodeæistviæi. English]
Intermolecular interactions : physical picture, computational methods and model potentials / Ilya G. Kaplan.
p. cm.
Includes bibliographical references and index.
ISBN-13: 978-0-470-86332-9 (cloth : alk. paper)
ISBN-10: 0-470-86332-3 (cloth : alk. paper)
1. Molecular dynamics. 2. Intermolecular forces. I. Title.
QD461.K29713 2006
541′.394 – dc22
2005021486

***British Library Cataloguing in Publication Data***

A catalogue record for this book is available from the British Library

ISBN-13 978-0-470-86332-9 (HB)
ISBN-10 0-470-86332-3 (HB)

Typeset in 10/12 Times by Laserwords Private Limited, Chennai, India

This book is printed on acid-free paper responsibly manufactured from sustainable forestry in which at least two trees are planted for each one used for paper production.

*To my sons:*
*Grigori and Vassili*

# Contents

# Preface

The subject of this book – intermolecular interactions – is as important in physics as in chemistry and molecular biology. Intermolecular interactions are responsible for the existence of liquids and solids in nature. They determine the physical and chemical properties of gases, liquids and crystals, the stability of chemical complexes and biological compounds. Such universality is due to the universality of the laws of quantum mechanics, on which the theory of intermolecular interactions is based. Atoms, molecules and crystals obey the same quantum-mechanical regularities.

Usually in quantum-mechanical textbooks, the theory of the intermolecular interactions is reduced to a discussion on the long-range behavior of atom–atom interactions and the introduction of dispersion forces. In many textbooks, even dispersion forces are not included. Physicists attribute intermolecular forces to chemical interactions that should be considered within the scope of quantum chemistry. However, in quantum-chemical textbooks also, intermolecular forces are not discussed in detail.

In this book the detailed qualitative descriptions of different types of intermolecular forces at large, intermediate and short-range distances are presented (Chapters 1 and 2). For the first time in monographic literature, the temperature dependence of the dispersion forces is analyzed. It is argued that at finite temperatures, the famous asymptotic Casimir–Polder formula for the dispersion energy is correct only within a narrow distance range. I have tried to make the presentation understandable to a broad range of readers without oversimplification. In Chapter 3, the methods for quantitative calculation of intermolecular interactions are discussed and modern achievements are presented. This chapter should be helpful for scientists performing computer calculations of many-electron systems.

The table of contents indicates other specific topics covered. Chapter 4 is devoted to the manifestations of nonadditive effects and to many-body forces responsible for that. The largest chapter (Chapter 5) collects more than 50 model potentials exploited for processing experimental data and for computer simulation in different fields of physics, chemistry and molecular biology. To the best of my knowledge, it is the most comprehensive account of model potentials published in the last 30 years. I have also included in Chapter 5 a special section where the widely used (at present) global optimization methods: simulated annealing, diffusion equation method, basin-hopping algorithm and genetic algorithm are described.

My previous book on this topic, I.G. Kaplan 'Theory of Molecular Interactions' (Elsevier, Amsterdam) was published in 1986. It is still often cited by scientists and used in many universities for teaching. But for the last twenty years many new methods have been elaborated and new theoretical and experimental data obtained. This resulted in a reassessment of some accepted concepts in the field. The new book is not an up-to-date edition of the old one – it is much more comprehensive and written in another methodological way.

Significant efforts were made to present this book so that it is self-sufficient for readers. All necessary mathematical apparatus, including vector and tensor calculus and the elements of the group theory, are given in Appendix 2. The main methods used for quantal calculation of many-electron systems are presented in Appendix 3. In this Appendix, I discuss not only the ideas of the computational methods but present a critical survey of modern achievements; in particular, it relates to the rapidly developing DFT approach. Appendix 3 can help readers who employ in their studies modern suites of programs to learn the basic principles of the methods they use.

I would like to acknowledge Isaak Bersuker, Grzegorz Chałasiński, Frans van Duijneveldt, Serguei Fomine, David Grier, Boris Ivlev, Wim Klopper, Eugene Kryachko, Karo Michaelyan, David Moule, Boris Plakhutin, Andrei Tchougreeff, Anatoly Titov and Victor Tuguchev for sending preprints and reprints of their publications and for useful discussions of different aspects of intermolecular interactions. Especially, I would like to thank my wife, Larisa, for her ongoing patience and support.

Mexico
June 2005

Ilya G. Kaplan

# 1 Background Knowledge

## 1.1 The Subject and its Specificity

The importance of intermolecular forces[1] in Nature is very difficult to overestimate. It is sufficient to say that the existence of liquids and solids is due to intermolecular interactions. In the absence of intermolecular interactions our world would be a uniform ideal gas.

A knowledge of the physics of intermolecular interactions is required to solve a wide class of problems in physics, chemistry and biology. The thermodynamic properties of gases and liquids and their kinetic characteristics (the coefficients of heat conductivity, diffusion etc.) are determined by the nature of intermolecular interactions. Intermolecular forces also determine to a large degree the properties of crystals, such as the equilibrium geometry, the binding energy, phonon spectra, etc.

Intermolecular interactions are involved in the formation of complicated chemical complexes, such as charge-transfer and hydrogen-bond complexes. Study of the mechanism of elementary chemical reactions is impossible without knowledge of the exchange processes between the translational and electron-vibration energies, which depend on the interaction of particles under collisions. Knowledge of the potential surface, characterizing the mutual trajectories of the reactants, is necessary to obtain the rates of chemical reactions.

A knowledge of intermolecular interactions is of great importance also in biology. It is enough to say that intermolecular forces account for the stability of such important compounds as DNA and RNA and also play an essential role in muscle contraction. The coagulation theory of colloidal solutions is based on a balance of the repulsive electrostatic forces and the attractive dispersion forces.

The development of modern technology required a knowledge of the macroscopic properties of gases under conditions that are almost inaccessible for experimental measurement (supersonic velocities, high temperatures $>1000$ K, hyper-high pressures in shock waves). The theoretical prediction of these properties requires a knowledge of the potential energy curves for a wide range of separations, which have to be found in independently.

[1] Here and in other chapters the term 'intermolecular forces' denotes also interatomic forces, which do not lead to the formation of chemical bond. Except for some special cases, no distinction will be made between intermolecular and interatomic interactions.

It should be emphasized that intermolecular forces are not measured directly in any experiment. It is other characteristics, such as the deviation angle under scattering, the transport coefficients, etc., which are connected functionally with the intermolecular forces, that are measured. The main source of experimental information about intermolecular interactions are:

- scattering experiments in atomic–molecular beams, which, in some cases, permit the potentials to be obtained directly from the experimental data;
- spectroscopic measurements (vibrational–rotational spectra, predissociation, broadening of lines by pressure, etc.);
- data on thermophysical properties of gases and liquids (virial coefficients, viscosity and transport coefficients, etc.);
- data on crystal properties (elastic constants, phonon spectra, sublimation energy, etc.);
- experiments on the formation of radioactive defects in solids (the focussing energy, the threshold displacement energy, etc.);
- nuclear magnetic resonance experiments in solids and liquids (the time of spin and spin-lattice relaxations).

To process the experimental data, different semiempirical model potentials with parameters obtained by fitting to the experimental data are usually used. The analytical form of these two-body potentials depends upon the system under study and the nature of the problem. Consider, as an example, the widely used (12–6) Lennard-Jones potential:

$$V^{LJ}(R) = \frac{a}{R^{12}} - \frac{b}{R^6} \tag{1.1}$$

This potential (see Section 5.1.2) has two parameters, $a$ and $b$. At small distances $V^{LJ}(R) > 0$, the repulsive forces dominate; at large distances $V^{LJ}(R) < 0$, the potential becomes attractive. So, at some point $R_0$, $V^{LJ}(R_0) = 0$. The point $R_0$ corresponds to the distance where the repulsive forces are compensated by the attractive ones. The potential function (Equation (1.1)) has a minimum, which is found from the condition $dV^{LJ}(R)\ /\ dR = 0$ (Figure 1.1) where $\varepsilon$ denotes the depth of the potential well. The potential can be expressed via two other independent parameters, $R_0$ and $\varepsilon$, having obvious physical and geometrical sense. These parameters are connected with the parameters $a$ and $b$ by the following relations:

$$R_0^6 = a/b \quad \text{and} \quad \varepsilon = b^2/4a$$

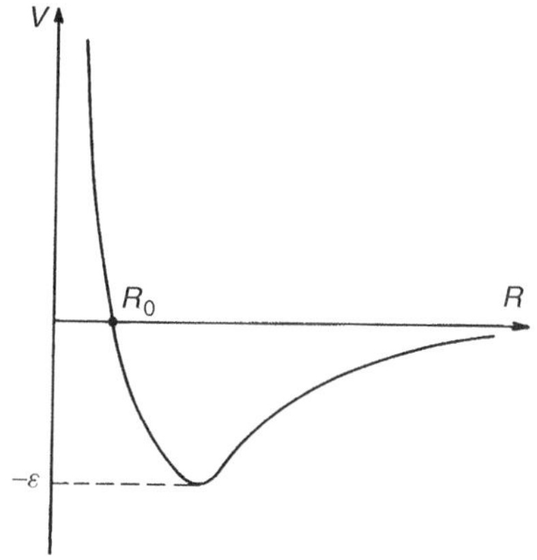


**Figure 1.1** The Lennard-Johns model potential (Equation (1.2))

and Equation (1.1) is transformed to:

$$V^{LJ}(R) = 4\varepsilon\left[\left(\frac{R_0}{R}\right)^{12} - \left(\frac{R_0}{R}\right)^{6}\right] \tag{1.2}$$

In general, a model potential can contain $n$ parameters ($p_1, p_2, \ldots, p_n$). To fit these parameters to experimental data, it is necessary to know a theoretical equation that expresses the measured property via a model potential, so that the theoretical equation for the measured property becomes a function of the parameters. Then the procedure that fits these parameters (see Section 5.2) is applied to obtain the best agreement of the calculated values of this function with the measured values.

It should be mentioned that semiempirical potentials cannot describe intermolecular potential adequately for a wide range of separations. A given potential with parameters calibrated for one property, often describes other properties inadequately, since different physical properties may be sensitive to different parts of the potential curve. Hence, to obtain more adequate potentials, the procedure of parameter matching must be carried out using as much of the experimental information as possible on the physical properties of the system under study. The use of high-speed computers brought about the possibility of the application of piecewise potentials that have different analytic form for different ranges of separations (Section 5.1.11).

Model potentials with different analytical forms can lead to the same observed relationship. In this case, experimental agreement with a given analytical dependence of model potential is not a sufficient indication of its correctness but merely a necessary condition. For example, the magnitude of the second virial coefficient is not sensitive to the form of the potential curve and its minimum position; it depends only on the ratio between the width and the depth of the potential well. In the same way, the viscosity coefficient is not sensitive to the dependence of the potential on the separation distance.

When estimating the reliability of the potential obtained, it is necessary to take into account not only the measurement errors but also the approximate nature of the formulae, which connect the measured characteristics with the molecular potential (an error of the theoretical approximation). It is also necessary to bear in mind that the optimized function is a nonlinear function of parameters. As a result, different sets of optimized parameters can give the same precision when fitted to experimental data.

In some cases, it is possible to solve the so-called inverse problem, i.e. to determine the potential directly from the experimental data without preliminary premises about its analytical form (as a rule, in a restricted range of distances). Different approaches to this complex problem are described in Section 5.3.

The discussion above emphasizes the importance of the theoretical determination of intermolecular potentials. The knowledge of the analytical dependence, which follows from the theory, permits more reliable model potentials to be constructed.

The basic concepts of the quantum-mechanical theory of intermolecular forces were formulated about 75 years ago. Nevertheless, it was only the last several decades that the number of studies on intermolecular interactions increased rapidly. This development has emerged for two reasons. Firstly, a general development of quantum-chemical methods of calculating the electronic structure of molecules has taken place, partly due to the availability of high-speed computers and the use of more refined mathematical methods. Secondly, more reliable experimental methods have appeared, which have allowed the theoretical predictions to be verified.

Before presenting the modern concepts about the nature of intermolecular forces, it is instructive to follow the evolution of these concepts. The discovery of the laws of intermolecular interactions, as any other discovery, was not a straightforward process. A great number of fallacies and wrong trends occurred along the way.

## 1.2 A Brief Historical Survey

Development of concepts about the nature of intermolecular forces is linked with the development of ideas about the atom. Concepts about the atomic structure of matter were formulated by the ancient philosophers. According to Democritus (fifth century BC) and to his semilegendary colleague Leukippos, all bodies consist of finest indivisible particles or *atoms*, separated by a vacuum (emptiness). A vacuum is required for the existence of motion. In the absence of a vacuum, the bodies would interfere under collisions, and motion would be impossible. Bodies differ from each other by the form of the atoms, their arrangement and their mutual orientation.

Introducing the concept of atoms required answering the question: In what manner are atoms linked to form different bodies? Since the interaction was considered to occur only by their direct contact, the problem was solved at the level of the simplest mechanical models, such as hooks, notches, and other devices, as described, for example, by Lucretius (first century BC) in his poem De Renum Natura [1].

In the Middle Ages, the problem of the atomic structure of matter did not attract any attention. An interest in science was completely lost and the scholastic mentality dominated. One of widely discussed 'actual' problems was: How many angels can be accommodated on the edge of a needle? The interest in the doctrine of the ancient atomistics arose only from the middle of the seventeenth century. This process occurred very gradually. For example, Galileo, who worked successfully in the field of the dynamics of bodies, became very abstract when was discussing the structure of these bodies. He reduced the particles, which compose the matter, to mathematical points separated by a vacuum and explained the strength of bodies as the disgust to vacuum, i.e. as the resistance of small emptinesses to their expansion.

The ideas of Descartes greatly influenced the development of concepts about the internal structure of matter. These ideas were presented in his treatise 'Les Passions de l'Ame', finished in 1633 but published only in 1664–1667 after his death. Descartes assumed that matter is equivalent to its extent but that, on the other hand, an extent does not exist without matter (cf. with the modern concept of space and matter!). In contrast to the ancient atomistics, Descartes suggested that matter is, in principle, infinitely divisible. This idea was also accepted by Leibniz, who assumed that there is no last small body because each body, even very small, is a whole world with an infinite number of creations.

According to Descartes, matter consists of particles, which differ from each other in form and size, and that they are also divisible. Descartes suggested that solids consist of fixed bodies, which are densely packed, and that liquids consist of particles, which move relative to each other. All motions are represented exclusively by mechanical displacements. Descartes assumed that 'hidden' interactive forces are absent.

Newton is known as the creator of Newton's laws in classical mechanics and Newton's gravitational law. But he considered also forces existing in material medium. A principal difference between Newton and his predecessors was in postulating, in addition to mechanical interaction, the existence of gravitational, magnetic and electric forces, as well as other attractions effective only at very small distances [2].

However, the concrete form of the dependence of interatomic forces upon the distance was not discussed neither by Newton nor by his predecessors. The law of the interaction between particles was introduced for the first time by the Croatian physicist Boscovich (1711–1787) in a work entitled 'Theory of Natural Science Reduced to the Single Law of Forces Existing in Nature' [3]. According to Boscovich, all bodies consist of point particles. The oscillating force between any two point particles increases infinitely as they approach each other and tends to the Newton gravitational forces $(\sim 1/R^2)$ at large distances (Figure 1.2). The existence of repulsive and attractive forces and their alternation is necessary, in Boscovichś opinion, in order to explain the physical properties of gases as well as the deformation of plastic bodies. The Boscovich interaction law was the first interatomic potential used to explain physical properties.

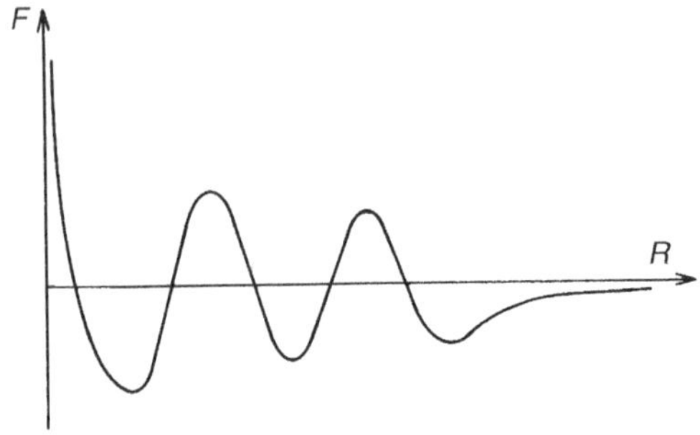


**Figure 1.2** The Boscovich universal potential

At approximately the same time, in 1743, the French physicist Clairault introduced the concept of forces interacting between molecules to explain the rising of liquids in capillaries [4]. The studies by Clairault in this field were developed further by Laplace [5] and by Gauss [6]. The intermolecular potential $V(R)$ was used in their papers in a most general functional form. Gauss studied the problem of liquids in capillaries using the principle of virtual work, requiring equilibrium of every mass point under acting forces. He considered three types of forces acting on molecules in liquid: 1. the force of gravity, 2. the mutual attractive forces between the molecules and 3. attractive forces between the molecules of the liquid and the molecules composing the walls of the capillary tube. Gauss showed that for the appearing integrals to be finite, the potential must behave at $R \to \infty$ as $1/R^n$ with $n$ not smaller than six.

The kinetic theory of gases was developed by Clausius [7], Maxwell [8, 9] and Boltzmann [10] in the second half of the nineteenth century. Clausius [7] accepted that molecules repel each other at small distances and attract each other at large distances (at present, this concept is well established). Maxwell presumed that the intermolecular forces are entirely repulsive at all distances. He used an analytical form for the interaction potential, $V(R) = A/R^n$, which corresponds to a repulsion, and obtained the expressions for the diffusion, heat capacity and elasticity coefficients. At that time it was known that the elasticity is proportional to the absolute temperature and does not depend on the gas density. The second property can be explained for any value of n, it had been derived previously on the assumption that the molecules are rigid spheres not interacting with each other except on direct contact. For an explanation of the first property, Maxwell concluded that $n = 4$. He assumed that the dependence $A/R^4$ is also valid at very small distances.

Today, it is well known that the repulsive part of potential is not described by the $A/R^4$ dependence. As Margenau and Kestner [11] remarked, Maxwell made a logical fallacy: from the fact that some premise led to the correct result, he concluded that the premise was valid, without testing the corollaries resulting from other premises. In the case considered, the proportionality of the elasticity to the temperature can be obtained from an infinite number of different potentials $V(R)$.

Although Maxwell's concept for a analytical form of model potential was erroneous, it represented a mathematically useful expression that allowed closed formulae for various kinetic characteristics of gases to be obtained and played a fundamental role in the subsequent development of the kinetic theory.

Boltzmann's work [10] was stimulated by his doubts on the simplicity of the interaction law proposed by Maxwell. Boltzmann repeated all the calculations using different attractive model potentials. He favored the existence of attractive forces since they provided the condensation of gases. All the models led to results which were similar to those of Maxwell and consistent with experimental data.

The attractive forces, which act between neutral atoms and molecules at large distances, were to be later named as the *van der Waals forces*. This name is not connected with the concrete studies by van der Waals on the nature of intermolecular forces, but with his well-known state equation [12], which accounts for the deviations from the ideal gas behavior:

$$\left(P + \frac{a}{V^2}\right)(V - b) = RT \tag{1.3}$$

where $a$ is a constant that accounts for the attraction between molecules of gas. That is, at constant volume and temperature, the pressure decreases with increasing $a$, in agreement with the fact that the attraction between the molecules must reduce the pressure on the vessel walls. The necessity of introducing a correction due to an attraction, in order to make the equation of state consistent with the experimental data, indicates the existence of the attractive forces between molecules.

In subsequent studies, different empirical potentials were used for the explanation of physical properties of gases and liquids. In a series of papers, Sutherland [13, 14] examined some analytical forms of the attractive potential between molecules in the gas phase (in particular, $V(R) \sim -A/R^3$) with parameters determined by fitting to experimental data. This kind of phenomenological approach became typical for subsequent studies in this field. The semiempirical procedure of fitting the potential parameters using experimental data has appeared to be very useful and was developed in the twentieth century, beginning with the paper by Lennard-Jones [15]. The well-known Lennard-Jones potential:

$$V(R) = \frac{\lambda}{R^n} - \frac{\mu}{R^m} \tag{1.4}$$

has been applied widely in studies on gases and condensed matter.

Towards the end of the nineteenth century and at the beginning of the twentieth century, in addition to the phenomenological approaches based on the application of empirical potentials, attempts were made to elucidate the physical nature of intermolecular forces. Some authors connected the intermolecular forces with gravitational ones. For instance, the Newton potential was corrected by a screening factor:

$$V(R) = -G\frac{m_1 m_2}{R} \exp\left(-\frac{R}{a}\right) \tag{1.5}$$

However, the small magnitude of the gravitational forces has resulted in a rejection of their use in the theory of intermolecular forces. Appearing at the same time, the data about the existence of electrical charges in atoms and molecules led to an assumption about the electromagnetic nature of intermolecular forces.

Reinganum [16] was the first to consider the interaction of two neutral molecules as the interaction of two permanent electric dipoles. According to the electrostatic theory, the interaction energy of two electric dipoles with moments $\mathbf{d}_1$ and $\mathbf{d}_2$, respectively, which are located at a distance $R$ from each other, depends on their mutual orientation and has the following form:

$$E_{dd} = \frac{1}{R^5}\left[R^2\left(\mathbf{d}_1 \cdot \mathbf{d}_2\right) - 3(\mathbf{d}_1 \cdot \mathbf{R})(\mathbf{d}_2 \cdot \mathbf{R})\right] \tag{1.6}$$

Introducing the spherical angles $\theta_1, \varphi_1, \theta_2, \varphi_2$, characterizing the directions of dipoles $\mathbf{d}_1$ and $\mathbf{d}_2$, with the z-axis along a line connecting their centers, Equation (1.6) becomes:

$$E_{dd} = -\frac{d_1 d_2}{R^3}\left[2\cos\theta_1\cos\theta_2 - \sin\theta_1\sin\theta_2\cos(\varphi_1 - \varphi_2)\right] \tag{1.7}$$

The energy $E_{dd}$ becomes positive (corresponds to the repulsive forces), if one of the dipole moments is perpendicular to the z-axis, and has a minimum corresponding to the attractive forces with the value:

$$(E_{dd})_{\min} = -\frac{2d_1 d_2}{R^3} \tag{1.8}$$

for the dipole orientation along the z-axis. Reinganum carried out a statistical calculation of dipole–dipole interactions, averaging all the orientations. Such an average, under the condition that all orientations are equally probable, leads to the zeroth interaction energy:

$$\langle E_{dd}\rangle = 0 \tag{1.9}$$

where the bracket notation denotes an average over orientations. But the probability of the dipole orientation, corresponding to the energy $E$, is determined by the Boltzmann factor $\exp(-E/kT)$. Taking this into account and $E_{dd}$ in the form of Equation (1.7), one obtains, at the condition $E_{dd} \ll kT$:

$$\left\langle E_{dd}\exp\left(-\frac{E_{dd}}{kT}\right)\right\rangle = \langle E_{dd}\rangle - \frac{1}{kT}\left\langle E_{dd}^2\right\rangle = -\frac{2}{3kT}\frac{d_1^2 d_2^2}{R^6} \tag{1.10}$$

So, the attraction forces between dipoles, obtained by Reinganum, decrease as the temperature increases and approach zero at high temperature. These forces were called *orientational forces*.

Intermolecular forces exist also at high temperatures. Therefore, the introduction of the orientational forces does not provide a complete answer to the question of the nature of intermolecular forces. Moreover, the existence of orientation forces requires the existence of permanent dipole moments in molecules. In fact, it was assumed at that time that all atoms and molecules possess dipole moments. This concept was strengthened by the early Debye studies on the theory of the dielectric permittivity [17]. According to that theory, the molecular dipoles align in an electric field. Debye verified his theory on five liquids, which accidentally occurred to be polar (alcohols), and arrived at the amusing (but not for his time!) conclusion that all molecules are polar.

In later studies it was found, however, that the simplest homoatomic molecules, such as hydrogen ($H_2$), nitrogen ($N_2$) and oxygen ($O_2$) molecules, have no dipole moments. In order to explain the interaction between non-dipole molecules, it was Debye [18] who made the next important step towards the understanding the nature of intermolecular forces. He assumed that the charges in a molecule are not fixed rigidly, but are able to move under the influence of the field produced by a permanent moment of the other molecule.

Since dipoles did not appear to be universal, Debye studied the induction of the dipole electric moment in one molecule by the permanent quadrupole moment of the other molecule. He evaluated the electrical field $D_A(R)$ created by the quadrupole moment $Q_A$ of molecule A at a distance R from it, $D_A = Q_A/R^4$. This field induces a dipole moment $\mathbf{d}_B$ in molecule B, $\mathbf{d}_B \sim \alpha_B \mathbf{D}_A$, where $\alpha_B$ is the polarizability of molecule B. The interaction energy of the induced dipole moment $\mathbf{d}_B$ with the electric field $\mathbf{D}_A$ is equal to $E_{qd} = -\mathbf{d}_B \bullet \mathbf{D}_A = -\alpha_B D_A^2$. Taking into account that a similar effect on molecule A is produced by the quadrupole moment of molecule B and carrying out an averaging over all the equally probable mutual orientations, Debye obtained the following expression for the interaction energy:

$$\langle E_{qq} \rangle = -\frac{2}{3}\left(\alpha_A \frac{Q_B^2}{R^8} + \alpha_B \frac{Q_A^2}{R^8}\right) \tag{1.11}$$

where the quantities $Q_A$ and $Q_B$ are components of the quadrupole moments of the molecules A and B, respectively. It is evident that Expression (1.11) does not depend on the temperature. Therefore, it provides an attractive interaction even at high temperatures when the effect of the orientational forces is almost zero. The induction interaction, induced by dipole moments, was studied later by Falkenhagen [19], who found it to be proportional to $1/R^6$. The forces of this type were then called the *Debye–Falkenhagen induction forces*.

While studying the induction interactions of molecules with quadrupole moments, Debye did not, however, consider the direct electrostatic interaction of quadrupole moments of molecules. This was done by Keesom [20] who generalized Reinganum's calculations by considering, in addition to the dipole–dipole interaction, the dipole-quadrupole and quadrupole-quadrupole ones. Sometimes the orientational forces are named the *Keesom forces*, although it is more correct to call them the *Reinganum–Keesom forces*.

Thus, classical physics has been able to explain, at least qualitatively, two types of interactions: the interactions between molecules with permanent multipole moments and the interactions between permanent and induced moments in molecules. This was achieved by introducing orientational forces that decrease with increasing temperature, and induction forces that do not depend on temperature.

In the case of some polar molecules, e.g. water, these forces make a considerable contribution to the intermolecular interaction. Although they constitute only a small part of the interaction for some other molecules, such as hydrogen chloride.

However, classical physics has completely failed to explain the origin of interaction among rare gas atoms. The electron shells of these atoms are spherically symmetric, which means that such atoms possess neither dipole nor other multipole moments, although the attractive forces in these systems are not negligible. At that time it was already known that at low temperatures rare gases undergo the transitions in liquid and solid states. In the framework of classical physics, one has been also unable to obtain the analytic form of a repulsion law at short distances.

The systematic and correct theory of intermolecular forces, which describes their behavior both at short and large distances, could be constructed only after the development of quantum mechanics between 1925 and 1927 (Bohr, Heisenberg, Schrödinger, Born, Dirac, Pauli). In 1927, Heitler and London [21] carried out the quantum-mechanical calculation of the potential curve for the simple system consisting of two hydrogen atoms. The Heitler–London calculation laid the foundation of the quantum theory of valence. As follows from their results, the repulsive behavior of the potential curve at short distances is determined by the antisymmetry of the wave function with respect to the permutations of electrons that stems from the Pauli exclusion principle. It causes the appearance of the specific *exchange interaction*. The repulsive exchange forces decrease exponentially with the distance.

In the same year, Wang [22] considered the quantum-mechanical attraction arising between two hydrogen atoms at large distances. He showed that it is proportional to $1/R^6$. These forces were called the *dispersion forces* and have a pure quantum-mechanical origin. The general theory of the dispersion forces was formulated in 1930 by Fritz London [23, 24]. The theory of dispersion forces resolved the problem of origin of the attraction between rare gas atoms. The leading term in the dispersion interactions for all systems falls down with distance as $1/R^6$. The dispersion forces are called also the *London forces,* or the *van der Waals forces*.

In 1948, while working on the problems of the coagulation theory of colloidal solutions, the Dutch theoretical physicists Casimir and Polder [25] took into account the retarding effects of the interaction between colloidal particles at large distances. They revealed that in this case, instead of the dependence $1/R^6$, there is a more rapid decrease of the dispersion forces with the distance, namely, the $1/R^7$–law (*the Casimir–Polder interaction*). In the same year, Casimir [26] found the analytical expression for the forces between two parallel metallic plates studying the zero-point energy of the vacuum electromagnetic field between plates. Finally, Lifshitz [27, 28] has created the general theory of the attractive van der Waals forces between macroscopic bodies with arbitrary permittivity.

At this point it is reasonable to conclude our brief historical survey of the evolution and main advances in the theory of intermolecular interactions. Their detailed account and the contemporary achievements are presented in the following parts of this book.

## 1.3 The Concept of Interatomic Potential and Adiabatic Approximation

A consistent theory of intermolecular forces can only be developed on the basis of quantum-mechanical principles. Because of the quantum nature of the electronic and nuclear motions, the solution of the intermolecular interaction problem reduces to solving the Schrödinger equation for a system of interacting molecules. Such a problem may be solved only after adopting some approximations. A substantial simplification may be achieved because of the possibility of separating the electronic and nuclear subsystems and introducing the concept of the *adiabatic potential.* This approach, denoted as the *adiabatic approximation,* is based on a large difference between the masses of electrons and nuclei (the detailed derivation is presented in Section A3.1).

Within the adiabatic approach, the electron subsystem is studied with fixed nuclei. In the Schrödinger equation, the nuclear kinetic energy operator is neglected and the nuclear coordinates are treated as fixed parameters. The Schrödinger equation splits into two: one for the electron motion with fixed nuclei and the other for the nuclear motion with the electron energy as the potential energy.

Below, we represent the basic equations for the system of two atoms A and B with $N_A$ and $N_B$ electrons, respectively, and the total number of electrons in the system $N = N_A + N_B$. Let us denote the set of $3N$ electron coordinates as $r$ and the distance between nuclei as $R$. The total wave function in the adiabatic approximation is written as a simple product:

$$\Psi_{m\nu}(r, R) = \chi_{m\nu}(R)\Psi_m(r, R) \tag{1.12}$$

where we denote the wave function describing the nuclear motion as $\chi_{m\nu}(R)$ and the wave function of the $N$-electron system in a quantum state $m$ as $\Psi_m(r, R)$; for each electronic quantum state $m$ there is a corresponding set of nuclear quantum states $\nu$.

The electronic wave function $\Psi_m(r, R)$ has to satisfy the Schrödinger equation for the electronic motion:

$$H_e\Psi_m(r, R) = E_m(R)\Psi_m(r, R) \tag{1.13}$$

with the Hamiltonian:

$$H_e = -\frac{\hbar^2}{2m}\sum_{i=1}^{N}\nabla_i^2 - \sum_{i=1}^{N}\left(\frac{Z_a e^2}{r_{ai}} + \frac{Z_b e^2}{r_{bi}}\right) + \sum_{i<j}\frac{e^2}{r_{ij}} + \frac{Z_a Z_b e^2}{R} \tag{1.14}$$

where $r_{ai}$ and $r_{bi}$ are the distances between electron $i$ and nuclei $a$ and $b$, having the charges $Z_a$ and $Z_b$, respectively, and $R$ is the fixed interatomic distance (see

Equation (A3.8)). The wave function $\Psi_m(r, R)$ describes the electronic motion at fixed nuclear coordinates or at an infinitely slow change of nuclear coordinates (adiabatic change). So, the eigenvalues $E_m$ of Equation (1.13) depend upon the value of parameter $R$. The solution of Equation (1.13) at different values of $R$ allows to obtain the function $E_m(R)$. This function plays the role of the potential energy in the Schrödinger equation for the nuclear motion, which in the so-called *Born–Oppenheimer approximation* [29] is presented as:

$$\left[-\frac{\hbar^2}{2\mu}\nabla_R^2 + E_m(R)\right]\chi_{m\nu}(R) = E_{m\nu}\chi_{m\nu}(R) \tag{1.15}$$

As was noted by Born [30, 31], the adiabatic potential energy $E_m\,(R)$ can be corrected by adding the diagonal contribution of the electron-nuclear interaction *(vibronic interaction)* (see Section A3.1). In this *Born adiabatic approximation*, the Schrödinger equation for the nuclear motion is written as:

$$\left[-\frac{\hbar^2}{2\mu}\nabla_R^2 + V_m\,(R)\right]\chi_{m\nu}\,(R) = E_m\chi_{m\nu}\,(R) \tag{1.16}$$

$$V_m\,(R) = E_m\,(R) - W_{mm}\,(R) \tag{1.17}$$

The adiabatic potential energy $V_m\,(R)$ is named the *interatomic potential*. The second term in Equation (1.17), $W_{mm}\,(R)$, is called *adiabatic correction*. The methods for calculating it are given in references [32–38]. Because of calculation difficulties, the adiabatic correction is often neglected and the interatomic potential is approximated by the energy $E_m\,(R)$ found in solution of the Schrödinger equation for the electronic motion at different values of the interatomic distance $R$ (Equation (1.13)).

In the vicinity of the minimum of the potential energy, the nuclear motion can be separated in the *vibration* motion of nuclei and the *rotation* motion of the system as a whole. The nuclear wave function is factorized as:

$$\chi_{m\nu}(R) = \Lambda_{m\nu}(Q)\Theta^J_{MK}(\vartheta) \tag{1.18}$$

where $\nu$ enumerates the vibrational states, $Q$ is the normal coordinate, $J$ is the rotational angular moment, $M$ and $K$ are its projections on z-axes of the laboratory frame and molecular frame, respectively, and $\vartheta$ is the set of the Euler angles. Vibrational energy levels $E_{m\nu}$ for some electronic quantum state $m$ are depicted in Figure 1.3.

The knowledge of interatomic potentials is necessary for studying the behavior of interacting atoms (molecules) at different distances and at different temperatures. But it must be kept in mind that the adiabatic approximation is not valid for degenerate electronic states, which are often the case for excited electronic states. It is also a bad approximation when the electronic terms are close to each other. In these cases the electron-nuclear interaction terms in the coupled equations (Equation (A3.13)) cannot be neglected. The adiabatic potential loses its physical meaning as the potential energy for the nuclear motion. The physical and chemical phenomena associated with the degeneracy and quasi-degeneracy of electronic

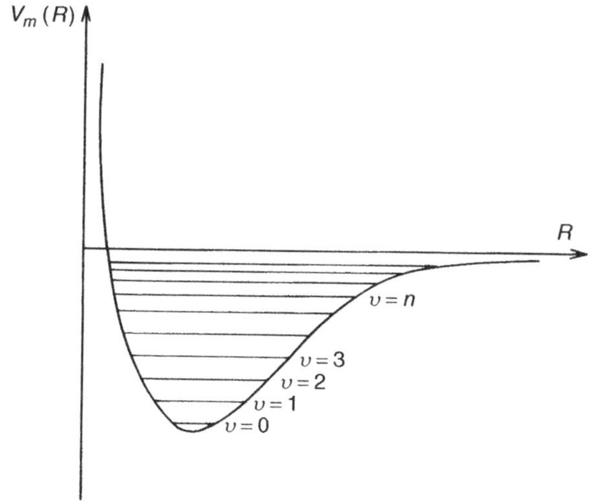


**Figure 1.3** Vibrational energy levels in a potential well formed in an electronic quantum state (m)

states are called the *Jahn–Teller effect* [39–43]. In Jahn–Teller systems, the adiabatical approach cannot be applied and the role of the vibronic interactions becomes crucial. Among such systems a very important role is played by molecular systems with potential surfaces characterized by so-called *conical intersections* [44], in which the nonadiabatic coupling terms are singular at isolated points. In the last decade, conical intersection effects have attracted great attention in theoretical and experimental studies as an essential aspect of various electronically nonadiabatic processes (see reviews by Baer [45] and Yarkony [46], and references therein).

For the nondegenerate ground electronic state, the adiabatic approximation works surprisingly well. The discussion of the accuracy of the adiabatic approximation is presented in reviews by Kołos [47] and Rychlewski [48]. For a good check, a most precisely calculated molecular system has to be chosen. The best choice is the hydrogen molecule. On one hand, it is the simplest molecular system for which the numerous precise theoretical and experimental studies have been performed. On the other hand, it contains the lightest nuclei; so, it can be expected that the obtained error of the adiabatic approximation will be an upper limit.

In Table 1.1, the most precise theoretical and experimental values of the dissociation energy and the ionization potential for the hydrogen molecule are presented. The discrepancy is smaller than the experimental error. It is a great success of computational chemistry based on quantum-mechanical concepts of chemical bonding. It should be stressed that this excellent agreement between theory and experiment is a result of more than six decades of research on both sides–theoretical and experimental. Thus, the calculations for hydrogen are the most trustworthy and can be used to check the adiabatic approximation.

**Table 1.1** Comparison of theoretical calculations with experimental values for the hydrogen molecule in (in $cm^{-1}$)

| | Theory | Experiment | Discrepancy |
|---|---|---|---|
| a) Dissociation energy | | | |
| Kołos–Rychlewski [49] | 36 118.049 | | $0.006 \pm 0.08$ |
| Wolniewicz [50] | 36 118.069 | | $0.004 \pm 0.008$ |
| | | $36\,118.11 \pm 0.08^a$ | |
| b) Ionization potential | | | |
| Kołos–Rychlewski [49] | 124 417.471 | | $0.017 \pm 0.017$ |
| Wolniewicz [50] | 124 417.491 | | $0.003 \pm 0.017$ |
| | | $124\,417.488 \pm 0.017^b$ | |

[a]Reference [51]
[b]Reference [52]

According to the calculation of the nonadiabatic corrections to the hydrogen ground state dissociation energy by Wolniewicz [50, 53], the nonadiabatic correction to the lowest vibrational level $\Delta D_0\,(v=0)^{nonad} = 0.50\ \mathrm{cm}^{-1}$ and the maximum nonadiabatic correction was found for the vibrational level $v = 9$, $\Delta D_0\,(v=9)^{nonad} = 5.20\ \mathrm{cm}^{-1}$. The precise value of $D_0$ is equal to $36\,118.069\ \mathrm{cm}^{-1}$ [50]. Hence, the relative error of the adiabatic approximation in the ground state of the hydrogen molecule is $\Delta D_0^{nonad}/D_0 = 10^{-5}$ to $10^{-4}$. From this follows that the accuracy of the adiabatic approximation is very high. It is considerably higher than the value of the small parameter $(m_e/M_p)^{\frac{1}{4}} = 0.15$, on which the Born–Oppenheimer adiabatic approximation [29] was based. For molecules with heavier atoms one can expect even larger precision. But it is difficult to check because of calculation problems.

For two atoms, the interatomic potential is represented by a potential curve $V(R)$. In the ground electronic state of two-atom systems, the interatomic potential $V_0(R)$ usually has a minimum. The depth of potential well, $E_0$, depends upon a system under study. In covalent bonded molecules, the well depth is equal to several electron-volts (eV) in the van der Waals dimers of noble gas atoms, it is equal to $10^{-2}$ to $10^{-3}$ eV. The ratio of the well depths in helium and hydrogen molecules (the two extreme cases) is about $10^{-4}$. The helium dimer was observed experimentally only recently [54]. It is the most weakly bound molecule known at present. Its dissociation energy ($D_0 = E_0 - (1/2)\hbar\omega_0$) is equal to 1.2 mK $\simeq 10^{-7}$ eV, while in hydrogen the dissociation energy $D_0 = 4.48$ eV. Also the great difference is in the optimal (equilibrium) distances: from 1.4 $a_0$ in the hydrogen molecule to 5.6 $a_0$ in the helium dimer.

The potentials describing the interaction between two molecules may be, as in the two-atom case, represented by a potential curve depending only on one variable, if the interaction is averaged over all molecular orientations in space. This potential, $V(R)$, where $R$ is the distance between the centers of masses of the molecules, is named the *potential of intermolecular interaction*, or briefly, the *intermolecular potential*. When the interacting molecules are fixed in space, the intermolecular potential, $V(R,\vartheta)$, depends on a set of Euler's angles, $\vartheta$, which determines the mutual orientation of the molecules. In addition, the interaction depends on the electronic state of the interacting systems. Therefore, for any pair of molecules there is a set of intermolecular potentials, $V_m(R, \vartheta)$, where $m$ labels the quantum states of the systems.

For three atoms taking part in a substitution reaction:

$$A + BC \longrightarrow AB + C \tag{1.19}$$

the adiabatic potential is represented by a potential surface depending on three interatomic separations:

$$V_m = V_m(R_{AB}, R_{BC}, R_{AC}) \tag{1.20}$$

If two of the interatomic separations remain constant, the resulting potential curve represents a cross section of this surface. Similarly to a two-atom case, in a three-atom system there is a set of potential surfaces that depends on the quantum states of the reacting atoms.

In a system with three (or more) atoms, there is the problem of nonadditivity. It is necessary to study the deviations from the additive approach, in which the potential of the system can be presented as a sum of pair potentials $V_{ab}$:

$$V(ABC\ldots) = \sum_{a<b} V_{ab} \tag{1.21}$$

This problem will be discussed in Chapter 4.

As shown in the preceding text, the concept of interatomic potential is based on the adiabatic approximation. Some authors studying the pairwise atom–atom potentials consider them as a classical concept. This is an incorrect statement. Even in the case where the motion of atoms is treated classically, as it is often done in the molecular dynamic studies, the interatomic potentials applied in these approaches have a quantum mechanical origin. The analytical form of numerous semiempirical model potentials (see Chapter 5) is based on the analytical expressions for different types of intermolecular forces (exchange, dispersion etc.) found by an approximate solution of the Schrödinger equation in the adiabatic approximation.

It should be noted that the adiabatic approximation may be invalid in high-energy collisions if the magnitude of energy is so high that the nuclear kinetic energy operator cannot be neglected. In this case, the electrons will not be able to adjust adiabatically to the nuclear rearrangement. In the theory of atomic collisions [55, 56] it is shown that the adiabatic approximation is valid only for large values of the so-called the Massey parameter $\zeta_{mn}$:

$$\zeta_{mn} = \omega_{mn} a / v \tag{1.22}$$

where $\omega_{mn}$ is the frequency of electronic transitions between states $m$ and $n$, $v$ is the velocity of the atomic motion, and $a$ is the distance at which the adiabatic electronic wave function undergoes essential changes. Parameter $a$ is defined as a characteristic distance such that a shift by $a$ results in an appreciable change of the matrix element with a semiclassical analogue of the nonadiabatic operator. The essence of the Massey criterion may be reduced to a requirement for the collision time, $\sim a/v$, to be much larger than the transition time, $\sim \omega_{mn}^{-1}$, between the adiabatic terms of the interacting system. When the equality (Equation (1.22)) breaks down, a large probability of nonadiabatic transitions exists, that is, the mixing of adiabatic terms of the system takes place.

In the early 1960s, the validity of applying the adiabatic approximation in elastic atomic collisions was discussed in connection with the high-energy helium (He) collision experiments by Armdur and Bertrand [57]. The experimental repulsive part of the He–He potential at distances $R \simeq 0.5$ Å was 9 eV lower than published theoretical values. The accurate quantum-mechanical calculation, with the inclusion of configuration interaction, carried out by Philipson [58], led to a lowering of the theoretical curve by 2 to 3 eV only. It was suggested that the adiabatic approximation breaks down in the experiment since the energy of the helium atoms was of order of 1000 eV. However, inclusion of nonadiabatic corrections resulted in an insignificant lowering of the theoretical curves, due to some compensating factors [59]. It is instructive that the reason for the discrepancy was found to be in the experimental measurements. After revising all the details of the experiment, Armdur [60] obtained a good agreement between the new experimental curve and the curve calculated by Philipson. Thus, the experimental data in such elastic atomic collisions, at least in the range of energies of a few keV, are described adequately within the frame of the adiabatic approximation. But it can be expected that the adiabatic approach will break down for higher collision energies.

In collision experiments with the helium beam [57], the helium atoms stayed in the ground electronic state. So, it was the elastic scattering process. The adiabatic approximation is not adequate for most of inelastic scattering processes (the charge exchange, predissociation etc.) even at low energy. For a non-negligible probability of transition from some electronic state to another, the adiabatic potential curves have to be located at a transition point $R$ sufficiently close to each other. Hence, the system will correspond to quasi-degenerate case and it is the vibronic interaction that causes such kind of inelastic transitions. This was realized in 1932 in three independent studies: Landau [61], Zener [62] and Stueckelberg [63].

Thus, the adiabatic representation, in which the electronic Hamiltonian matrix is diagonal, cannot be a satisfactory basis for study most inelastic scattering processes. At present in these studies, the so-called *diabatic representation* is used. It was introduced by Lichten [64] and rigorously formulated by Smith [65]. The diabatic electronic states are defined by an unitary transformation chosen in a such manner that the first derivative couplings in coupled equations for the nuclear motion, Equation (A3.13), become sufficiently small to be neglected. In the diabatic representation, the electronic Hamiltonian matrix is nondiagonal. Its nondiagonal elements are responsible for the transitions from one adiabatic electronic state to another.

Analyzing the atom–atom collisions, Smith [65] considered the case with a single internal coordinate. The multidimensional case was analyzed by Baer [66] and Mead and Truhlar [67]. One of the last derivations of the adiabatic-to-diabatic transformation can be found in Reference [45]. To study the topological effects in scattering processes and nonadiabatic phenomena in many-atomic molecules, the extended versions of the Born–Oppenheimer equations were developed [68–70]; see also Reference [45]. Vibronic dynamics is now a well elaborated and active field of investigation [46, 71, 72].

## 1.4 General Classification of Intermolecular Interactions

The classification of intermolecular interactions depends on the distance between interacting objects; although it must be kept in mind that all types of intermolecular interactions have the same physical nature: namely, the electromagnetic one. Various types of intermolecular interactions are presented in Figure 1.4. They are

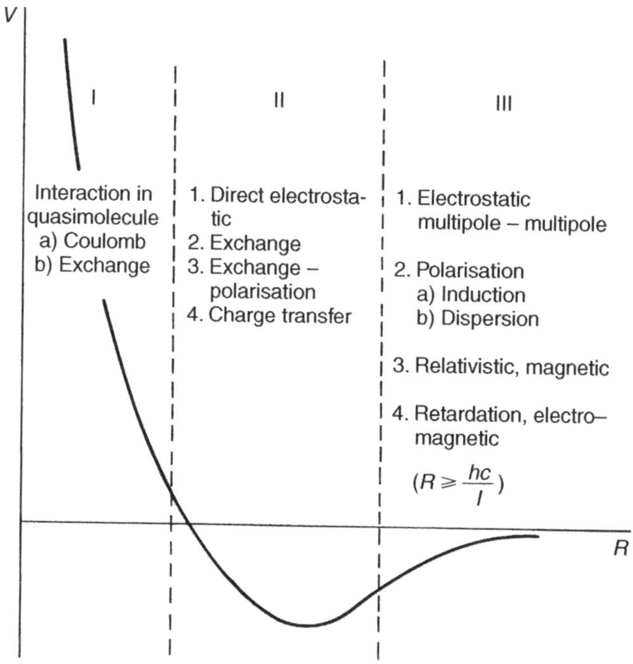


**Figure 1.4** Classification of intermolecular interactions

classified according to the three ranges of interatomic separation for a typical interatomic potential. These three ranges are:

I. A range of short distances at which the potential has a repulsive nature and the electronic exchange, due to the overlap of the molecular electronic shells, dominates.

II. A range of intermediate distances with the van der Waals minimum, which is a result of the balance of the repulsive and attractive forces.

III. A range of large distances at which the electronic exchange is negligible and the intermolecular forces are attractive.

### *Range* I

In this region, the perturbation theory (PT) for calculating the intermolecular interactions cannot be applied. To some extent, the interacting atoms (molecules) lose their individuality because of a large overlap of their electronic shells. The same variational methods, which are used for molecular calculations, can be applied to the calculation of the total energy of interacting system, which can be considered as a 'supermolecule'. The interaction energy is found as a difference:

$$E_{int} = E_{tot} - \sum_{a=1}^{n} E_a \tag{1.23}$$

where $E_a$ is the energy of isolated subsystems (molecules or atoms) that have to be calculated at the same approximation as a whole system. In this region we can separate only two types of interaction energies: the *Coulomb energy* and the *exchange energy*. If we put to zero all integrals containing the exchange or overlap of electron densities, we obtain $E_{Coul}$. Then, the exchange energy is defined as the difference:

$$E_{exch} = E_{int} - E_{Coul} \tag{1.24}$$

### *Range* II

Both repulsive and attractive forces exist in this region. This causes the minimum of intermolecular potential energy and provides a stability of the system. The magnitude of the interaction energy is much smaller than the self-energy of the interacting molecules and the PT can be applied; although the electron exchange, appearing as a consequence of the antisymmetry of the total wave function, is still large. The standard perturbation Rayleigh–Schrödinger and Brillouin–Wigner theories (see Section A3.3) are developed for the zero-order wave function taken as a simple product of the wave functions of interacting atoms (molecules). In the intermediate distance region, the antisymmetric zero-order wave function has to be dealt with. This leads to essential modifications of the standard perturbation schemes. The approaches developed were named *Symmetry Adapted Perturbation Theories* (SAPT): they are discussed in Section 3.2.1 (for details see References [73–75]).

The exchange energy is separated only in the first order of SAPT. The decomposition of $E_{int}$ into a perturbation series can be written as:

$$E_{int} = \varepsilon_{el}^{(1)} + \varepsilon_{exch}^{(1)} + \sum_{n=2}^{\infty} \varepsilon_{pol.exch}^{(n)} \tag{1.25}$$

$\varepsilon_{el}^{(1)}$ is the classical electrostatic interaction energy between two (or more) systems of charges. The only difference to the classical expression is in the charge distribution: instead of point charges, $\varepsilon_{el}^{(1)}$ contains distributed electron densities. $\varepsilon_{exch}^{(1)}$ is the exchange energy in the first order of PT. Its origin is based on the Pauli principle demanding the antisymmetrization of many-electron wave functions. Because of antisymmetrization, electrons have the probability of being located on each interacting atoms (molecules); so, they are delocalized. This is a specific quantum-mechanical effect. Thus, the exchange interaction exists only in the framework of the quantum-mechanical description. In the second and higher orders of SAPT, the exchange effects cannot be separated from the polarization energy. In various formulations of SAPT, the different expressions for the exchange–polarization energy $\varepsilon_{pol.exch}^{(2)}$ are elaborated [73].

### *Range* III

In this region, exchange effects can be neglected. Usually, it is valid for R $\gtrsim 15a_0$. The different types of intermolecular interactions are classified in the frame of the standard Rayleigh–Schrödinger perturbation theory. In the second order of PT, the polarization energy splits on the induction $\varepsilon_{ind}^{(2)}$ and dispersion $\varepsilon_{disp}^{(2)}$ energies. In higher orders, some mixed term $\varepsilon_{ind.disp}$, which results from coupling of the induction and dispersion energies, appears. In the third order of PT, $\varepsilon_{ind,disp}^{(3)}$ was analyzed by Jeziorski *et al.* [74]. Thus, at distances where the exchange effects are negligible, the PT expansion can be written:

$$E_{int} = \varepsilon_{el}^{(1)} + \sum_{n=2}^{\infty} \left[\varepsilon_{ind}^{(n)} + \varepsilon_{disp}^{(n)}\right] + \sum_{n=3}^{\infty} \varepsilon_{ind.disp}^{(n)} \tag{1.26}$$

What is important is that at these distances the multipole expansion of the electrostatic potential is valid. The interaction operator can be expanded in the multipole series. As a result, $\varepsilon_{el}^{(1)}$ represents a direct multipole–multipole electrostatic energy. At sufficiently large distances, only the first term in the multipole expansion is enough to describe the interaction. For polar molecules, it is the dipole term. In this approximation, the electrostatic energy $\varepsilon_{el}^{(1)}$ is given by the classical expression (Equation (1.6)), but the dipole moments have to be calculated with the molecular (atomic) electron densities.

The physical sense of the induction energy $\varepsilon_{ind}^{(2)}$ is the same as in classical physics (see discussion in Section 1.2). The dispersion energy is a pure quantum-mechanical phenomenon. It originates in the quantum-mechanical fluctuations of electronic density. The instant redistribution of electron density leads to the nonzero mean dipole moment even in the cases when the permanent dipole moment is equal to zero (nonpolar molecules, noble gas atoms). This instant dipole moment induces a dipole and more higher moments in the other molecule. In the second order of PT, the multipole expansion of the dispersion energy $\varepsilon_{disp}^{(2)}$ is usually written in the following form:

$$\varepsilon_{disp}^{(2)} = -\sum_{n=6}^{\infty} \frac{C_n}{R^n} \tag{1.27}$$

For atoms, the sum in Equation (1.27) contains only even powers, whereas in the case of the interaction between molecules, it may also contain odd powers of $n$ (see Sections 3.1.2 and 3.1.3). The coefficients $C_n$ are named as *dispersion coefficients*. For their calculation, quite sophisticated quantum-mechanical methods are elaborated [73]. In the higher orders of PT, the interpretation of different terms in Equation (1.26) becomes more complex. The dispersion energy in the third order of PT, $\varepsilon_{disp}^{(3)}$, will be discussed in Chapter 4. This is the well-known Axilrod–Teller–Mutto dispersion energy [76, 77].

Magnetic interactions exist at all distances, but in ranges I and II they are often negligible in comparison with larger electrostatic interactions. On the other hand, magnetic interactions are precisely detected by measurements of the energy level splitting in a magnetic field in the well-elaborated methodics of electron and nuclear paramagnetic resonances. In some special cases, the magnetic interactions become the largest at sufficiently large distances. This is the case for oriented nonpolar molecules possessing the quadrupole moment as the first nonvanishing multipole moment and having, in the ground state, the total electronic spin $S \neq 0$. The first term in the multipole expansion of different electrostatic intermolecular interactions for these molecules has the following distance dependence:

$\sim\frac{1}{R^5}$ in the electrostatic energy $\varepsilon_{el}^{(1)}$ (the quadrupole–quadrupole interaction).

$\sim\frac{1}{R^6}$ in the dispersion energy $\varepsilon_{disp}^{(2)}$ (the dipole–dipole interaction).

$\sim\frac{1}{R^8}$ in the induction energy $\varepsilon_{ind}^{(2)}$ (the quadrupole–induced dipole interaction).

The magnetic spin–spin interaction is relativistic one and has a dipole-dipole distance behavior. It is $\sim\alpha^2/R^3$, where $\alpha = 1/137$ is the fine structure constant. Although $\alpha^2$ is small, it is evident that the magnetic interactions become predominant with increasing intermolecular distance. The described situation is realized for oxygen molecules absorbed on some surface. The oxygen molecule, as all homonuclear diatomic molecules, has no dipole moment and its ground state is triplet, that is, the total electronic spin $S = 1$ [78].

At distances at which the propagation time of the interaction, $R/c$, is of the same order as the average time of the electronic transitions, which is proportional to $\hbar/I_1$ (where $I_1$ is the first ionization potential), that is, $R \sim \hbar c/I$, the *retardation effect* should be taken into account. Usually, this effect becomes appreciable at $R \gg 500$ bohrs.

Consideration of the retardation effect is important in the coagulation theory of colloid solutions, where interactions between macroscopic bodies should be taken into account. In such cases, the London term for the dispersion interactions, $\sim 1/R^6$, must be replaced with the Casimir–Polder term, $1/R^7$ (see Section 2.5).

The classification of the intermolecular interactions presented in the preceding text is arbitrary to a large extent since it is based on expressions of the perturbation theory and the following question may arise: has such classification any physical sense? Nature does not know about our perturbation theory and other approximation methods, and real potentials do contain contributions from all types of interactions.

Fortunately, the answer to such a question is positive. The representation of the interaction energy as a sum of various components permits the terms giving the greatest contribution in a given range to be separated. Each term in the multipole electrostatic interactions or in the dispersion energy series has a clear physical sense. The magnitude of the multipole interactions depends upon the values of multipole moments, the induction and dispersion forces are proportional to the molecular polarizabilities, and so on. These physical quantities can be obtained from experiments. So, each term is related to real physical properties of atoms or molecules. Often it allows the magnitude of intermolecular interactions to be estimated qualitatively without complex quantitative calculations.

In the next chapter the different types of intermolecular interactions are discussed in detail.

## References

1. Cari Titi Lucretii, *De Renum Natura*, Libri Sex, Paris (1563); Lucretius, *On the Nature of the Universe* (translated by R. Latham), Penguin Books, London (1951).
2. I. Newton, *Optics* (3rd ed), William and John Innys, London (1721); corrected.
3. R. Boscovich, *Teoria Philosophica Naturalis Reducta ad Unicam Legem Virium in Natura Existentium*, Vienna (1758).
4. A.C. Clairault, *Théorie de la Figure de la Terre,* Paris (1743); 2nd edition published by Courcier, Paris (1808).
5. M. Laplace, *Traité de Mécanique Céleste,* Courcier, Paris (1805).
6. C.F. Gauss, *Principia Generalia Theoriae Figurue Fluidorumin Statu Aequilibrii,* Mathematical Tracts, Göttingen (1830).
7. R. Clausius, *Ann. der Phys.* **100**, 353 (1857).
8. J.C. Maxwell, *Phil. Trans. Roy. Soc.* **157**, 49 (1867).
9. J.C. Maxwell, *Phil. Mag.* **4**, 129, 185 (1868).
10. L. Boltzmann, *Sitz. Akad. Wiss. Wien* **66**, 275 (1872).
11. H. Margenau and N.R. Kestner, *Theory of Intermolecular Forces*, Pergamon Press, New York (1971).
12. J.H. van der Waals, Doctoral Dissertation, Leiden (1873).
13. W. Sutherland, *Phil. Mag.* **22**, 61 (1886); **24**, 113 (1887).
14. W. Sutherland, *Phil. Mag.* **35**, 211 (1893); **36**, 507 (1893).
15. J.E. Lennard-Jones, *Proc. Royal Soc.* (London) A **106**, 463 (1924).
16. M. Reinganum, *Ann. der Phys.* **38**, 649 (1912).
17. P. Debye, *Phys. Zs.* **13**, 97, 295 (1912).

18. P. Debye, *Phys. Zs.* **21**, 178 (1920).
19. M. Falkenhagen, *Phys. Zs.* **23**, 663 (1922).
20. W.H. Keesom, *Phys. Zs.* **22**, 129 (1921).
21. W. Heitler and F. London, *Zs. f. Phys.* **44**, 445 (1927).
22. S.C. Wang, *Phys. Zs.* **28**, 663 (1927).
23. F. London, *Zs. f. Phys.* **63**, 245 (1930).
24. F. London, *Zs. Phys. Chem.* B **11**, 222 (1930).
25. H.B. Casimir and D. Polder, *Phys. Rev.* **73**, 360 (1948).
26. H.B. Casimir, *Proc. Kon. Ned. Akad. Wetenschap.* **51**, 793 (1948).
27. E.M. Lifshitz, *Dokl. Akad. Nauk USSR* **97**, 643 (1954).
28. E.M. Lifshitz, *Sov. Phys. JETP* **2**, 73 (1956).
29. M. Born and J.R. Oppenheimer, *Ann. der Phys.* (Leipzig) **84**, 457 (1927).
30. M. Born, *Göttingen Nachr. Acad. Wiss. Math. Nat. Kl.* 1 (1951).
31. M. Born and K. Huang, *Dynamical Theory of Crystal Lattice*, Oxford University Press, London (1956).
32. H. Sellars and P. Pulay, *Chem. Phys. Lett.* **103**, 463 (1984).
33. H. Sellars, *Chem. Phys. Lett.* **108**, 339 (1984).
34. N.C. Handy, Y. Yamaguchi and H.F. Schaeffer, *J. Chem. Phys.* **84**, 4481 (1986).
35. N.C. Handy and A.M. Lee, *Chem. Phys. Lett.* **252**, 425 (1996).
36. W. Kutzelnigg, *Mol. Phys.* **90**, 909 (1997).
37. W. Cencek and W. Kutzelnigg, *Chem. Phys. Lett.* **266**, 383 (1997).
38. J. Komasa, W. Cencek and J. Rychlewski, *Chem. Phys. Lett.* **304**, 293 (1999).
39. H.A. Jahn and E. Teller, *Proc. Royal Soc. (London)* A **161**, 220 (1937).
40. E. Teller, *J. Phys. Chem.* **41**, 109 (1937).
41. R. Englman, *The Jahn–Teller Effect in Molecules and Crystals*, John Wiley & Sons, Inc., New York (1972).
42. I.B. Bersuker and V.Z. Polinger, *Vibronic Interactions in Molecules and Crystals*, Springer–Verlag, Berlin (1989).
43. I.B. Bersuker, *Electronic Structure and Properties of Transition Metal Compounds,* John Wiley & Sons, Inc., New York (1996).
44. G. Herzberg and H.C. Longuet-Higgins, *Disc. Farad. Soc.* **35**, 77 (1963).
45. M. Baer, *Chem. Phys.* **259**, 123 (2000).
46. D. Yarkony, *J. Phys. Chem.* A **105**, 6277 (2001).
47. W. Kołos, *Adv. Quant. Chem.* **5**, 99 (1970).
48. J. Rychlewski, *Adv. Quant. Chem.* **31**, 173 (1999).
49. W. Kołos and J. Rychlewski, *J. Chem. Phys.* **98**, 3960 (1993).
50. L. Wolniewicz, *J. Chem. Phys.* **103**, 1792 (1995).
51. A. Balakhrishnan, V. Smith and B.P. Stoicheff, *Phys. Rev. Lett.* **68**, 2149 (1992).
52. Ch. Jungen, I. Dabrowski, G. Herzberg and M. Vervloet, *J. Mol. Spectr.* **153**, 11 (1992).
53. L. Wolniewicz, *J. Chem. Phys.* **78**, 6173 (1983).
54. F. Luo, G. Kim, G.C. Mc Bane, C.F. Giese and W.R. Gentry, *J. Chem. Phys.* **98**, 9687; 10086 (1993).
55. N.F. Mott and H.S.W. Massey, *The Theory of Atomic Collisions*, Clarendon Press, Oxford (1965).
56. J.B. Hasted, *Physics of Atomic Collisions*, Butterworths, London (1964).
57. I. Armdur and R.R. Bertrand, *J. Chem. Phys.* **36**, 10 (1962).
58. P.E. Philipson, *Phys. Rev.* **125**, 1981 (1962).
59. W.R. Thorson, *J. Chem. Phys.* **39**, 1431 (1963); **41**, 3881 (1964).

60. B. Bederson and W.L. Fete (eds), *Methods of Experimental Physics,* Academic Press, New York (1967), Vol. 8, Chapter 3.
61. L.D. Landau, *Physik. Z. Sowjetunion* **2**, 46 (1932).
62. C. Zener, *Proc. Royal Soc.* (London) A **137**, 696 (1932).
63. E.C. G. Stueckelberg, *Helv. Phys. Acta* **5**, 369 (1932).
64. W. Lichten, *Phys. Rev.* **131**, 229 (1963); **139**, A 27 (1965).
65. F.T. Smith, *Phys. Rev.* **179**, 111 (1969).
66. M.Baer, *Chem. Phys. Lett.* **35**, 112 (1975).
67. C.A. Mead and D. G. Truhlar, *J. Chem. Phys.* **77**, 6090 (1982).
68. A. Kuppermann, in *Dynamics of Molecules and Chemical Reactions,* R.E. Wayatt and J.Z.H. Zang (eds), Marcel Dekker, New York (1996), pp. 411–472.
69. R. Baer, D.M. Charutz, R. Kozloff and M. Baer, *J. Chem. Phys.* **105**, 9141 (1996).
70. S. Adhikari and G.D. Billing, *J. Chem. Phys.* **111**, 40 (1999).
71. H. Köppel and W. Domske, in *Encyclopedia of Computational Chemistry*, P.v.R. Schleyer (ed), John Wiley & Sons, Ltd, Chichester (1998), pp. 3166–3182.
72. A.J.C. Varandas, in *Fundamental World of Quantum Chemistry*, E.J. Brändas and E.S. Kryachko (eds), Kluwer Academic Publishers, Dordrecht (2003), Vol. 2, pp. 33–92.
73. I.G. Kaplan, *Theory of Molecular Interactions*, Elsevier, Amsterdam (1986).
74. B. Jeziorski, B. Moszyński and K. Szalewicz, *Chem. Rev.* **94**, 1887 (1994).
75. G. Chałasiński and M.M. Szcześniak, *Chem. Rev.* **94**, 1793 (1994).
76. B.M. Axilrod and E. Teller, *J. Chem. Phys.* **11**, 299 (1943).
77. J. Mutto, *Proc. Phys.–Math. Soc. Japan* **17**, 629 (1943).
78. J. Slater, *Quantum Theory of Molecules and Solids*, Vol. I, McGraw Hill, New York (1963).

# 2 Types of Intermolecular Interactions: Qualitative Picture

## 2.1 Direct Electrostatic Interactions

### 2.1.1 General expressions

The Hamiltonian of a system consisting of two interacting molecules $A$ and $B$, with the center-of-mass motion subtracted (see Section A3.1), can be represented as a sum of the Hamiltonians corresponding to the isolated molecules ($H_0 = H_A + H_B$) and the operator of the electrostatic interaction between these molecules ($V$):

$$H = H_0 + V \tag{2.1}$$

$$V = -\sum_{a=1}^{n_A}\sum_{j=1}^{N_B}\frac{Z_a e}{r_{aj}} - \sum_{b=1}^{n_B}\sum_{i=1}^{N_A}\frac{Z_b e}{r_{bi}} + \sum_{i=1}^{N_A}\sum_{j=1}^{N_B}\frac{e^2}{r_{ij}} + \sum_{a=1}^{n_A}\sum_{b=1}^{n_B}\frac{Z_a Z_b e^2}{R_{ab}} \tag{2.2}$$

where the subscripts $a$ and $b$ label the nuclei, and $i$ and $j$ denote the electrons of molecules $A$ and $B$, respectively. In this and following chapters the adiabatic approximation will be used everywhere, so the subscript $e$ denoting that electrons are moving at fixed nuclei (Equations (A3.8) and (A3.10)) will be omitted.

At large distances, the operator $V$ may be treated as a small perturbation. Neglecting the electron exchange, the zeroth-order wave functions can be approximated as a simple product of the wave functions of isolated molecules:

$$H_0\Psi_n^A\Psi_m^B = \left(E_n^A + E_m^B\right)\Psi_n^A\Psi_m^B = E_{nm}^{(0)}\Psi_n^A\Psi_m^B \tag{2.3}$$

where $n$ and $m$ denote the sets of quantum numbers of the isolated molecules.

The energy of the direct electrostatic interaction is given by the first-order term of perturbation theory (PT) (see Appendix 3, Equation (A3.94)) and is equal to the expectation value of the perturbation operator averaged over the zeroth-order wave functions:

$$E_{el}^{(1)} = \left\langle \Psi_n^A\Psi_m^B \left| V \right| \Psi_n^A\Psi_m^B \right\rangle \tag{2.4}$$

Equation (2.4) may be expressed in terms of the electron density because in quantum mechanics, instead of the point charges of classical mechanics, the charges are distributed in space with the probability density determined by $|\Psi(r, R)|^2$. In

*Intermolecular Interactions* I.G. Kaplan

the adiabatic approximation, the set of nuclear coordinates $R$ is fixed and the nuclei are considered as point charges. The electron charge is distributed in space with the one-electron density defined as:

$$\begin{aligned}\rho_{nn}^{A}(i) &= N_A \int \left|\Psi_n^A(1, \ldots, i, \ldots, N_A)\right|^2 dV^{(i)} \\ \rho_{mm}^{B}(j) &= N_B \int \left|\Psi_m^B(1, \ldots, j, \ldots, N_B)\right|^2 dV^{(j)}\end{aligned} \tag{2.5}$$

for molecules $A$ and $B$, respectively. Numbers in Equation (2.5) denote the electron coordinates of enumerated electrons; $dV^{(i)}$ is an elementary volume of the configuration space of all electrons, belonging to molecule $A$, with the exception of the *ith* electron. In Equation (2.5) the integration includes the summation over the whole spin space. The pre-integral factors are required for the conservation of electric charge, since the integration of $\rho_{nn}^{A}$ over the configuration space of the ith electron must give the total number of electrons belonging to molecule $A$.

Substituting Equation (2.2) into Equation (2.4) and taking into account the definition of the electron density (given by Equation (2.5)), the following expression of the electrostatic energy for the interaction of two molecules is obtained:

$$\begin{aligned}E_{el}^{(1)} = &- \sum_a Z_a e \int \rho_{mm}^{B}(j) \frac{1}{r_{aj}} dV_j - \sum_b Z_b e \int \rho_{nn}^{A}(i) \frac{1}{r_{bi}} dV_i + \\ &\int \rho_{nn}^{A}(i)\, \rho_{mm}^{B}(j) \frac{e^2}{r_{ij}} dV_i dV_j + \sum_{a,b} \frac{Z_a Z_b e^2}{R_{ab}}\end{aligned} \tag{2.6}$$

Expression (2.6) for $E_{el}^{(1)}$ can be interpreted as a classical electrostatic energy for space distributed electron charges and point nuclear charges. The first two terms correspond to the energy of attraction between the nuclei of one molecule and the electron charge distribution of the other. The third and fourth terms represent the repulsion energy of the electron charge distributions and the nuclear charges, respectively.

With large distances between molecules, charge distributions do not overlap. Then, the electrostatic energy may be represented with good accuracy as a sum of a few terms of an expansion series for $E_{el}^{(1)}$ in powers of $1/R$. Such an expansion series is based on the concept of the multipole moments of the charge distribution.

### 2.1.2 Multipole moments

As is well known from electrostatic theory [1–3], the potential created by a system of charges may be represented, at distances greater than the dimensions of the system, as a series expanded in multipole moments. The first five terms of such an expansion, in a cartesian coordinate system, at a point with a radius-vector **R** can be written as follows:

$$\varphi(R) = \sum_i \frac{e_i}{|\mathbf{R} - \mathbf{r}_i|} = \frac{q}{R} + \frac{\mathbf{d}\cdot\mathbf{R}}{R^3} + \sum_{\alpha,\beta} Q_{\alpha\beta}\frac{X_\alpha X_\beta}{R^5} + \sum_{\alpha,\beta,\gamma} \Omega_{\alpha\beta\gamma}\frac{X_\alpha X_\beta X_\gamma}{R^7} + \sum_{\alpha,\beta,\gamma,\delta} \Phi_{\alpha\beta\gamma\delta}\frac{X_\alpha X_\beta X_\gamma X_\delta}{R^9} + \cdots \tag{2.7}$$

Expansion (2.7) is the Taylor series in powers of $r_i/R$, and $X_a$ denotes the cartesian components of $\mathbf{R}$ ($\alpha = 1, 2, 3$). The origin of the coordinate system is located inside the system of charges; $q$ is the total charge of the system; $\mathbf{d}$ is the dipole moment vector; $Q_{\alpha\beta}$, $\Omega_{\alpha\beta\gamma}$, and $\Phi_{\alpha\beta\gamma\delta}$ are the quadrupole,[1] octopole and hexadecapole moment tensors, respectively:

$$q = \sum_i e_i \tag{2.8}$$

$$\mathbf{d} = \sum_i e_i \mathbf{r}_i \tag{2.9}$$

$$Q_{\alpha\beta} = \frac{1}{2}\sum_i e_i\left(3x_{i\alpha}x_{i\beta} - r_i^2\delta_{\alpha\beta}\right) \tag{2.10}$$

$$\Omega_{\alpha\beta\gamma} = \frac{1}{2}\sum_i e_i\left[5x_{i\alpha}x_{i\beta}x_{i\gamma} - r^2\left(x_{i\alpha}\delta_{\beta\gamma} + x_{i\beta}\delta_{\alpha\gamma} + x_{i\gamma}\delta_{\alpha\beta}\right)\right] \tag{2.11}$$

$$\Phi = \frac{1}{8}\sum_i e_i\left[35x_{i\alpha}x_{i\beta}x_{i\gamma}x_{i\delta} - 5r^2\left(x_{i\alpha}x_{i\beta}\delta_{\gamma\delta} + x_{i\alpha}x_{i\gamma}\delta_{\beta\delta} + x_{i\alpha}x_{i\delta}\delta_{\beta\gamma} + x_{i\beta}x_{i\gamma}\delta_{\alpha\delta} + x_{i\beta}x_{i\delta}\delta_{\alpha\gamma} + x_{i\gamma}x_{i\delta}\delta_{\alpha\beta}\right) + r^4\left(\delta_{\alpha\beta}\delta_{\gamma\delta} + \delta_{\alpha\gamma}\delta_{\beta\delta} + \delta_{\alpha\delta}\delta_{\beta\gamma}\right)\right] \tag{2.12}$$

where $x_{i\alpha}$ represent the cartesian components of $\mathbf{r}_i$, the $\delta$'s are the Kronecker symbols. It is evident that the multipole moments depend upon the arrangement of charges and do not depend upon the point $\mathbf{R}$ in which the potential is determined.

The general expression for the $mth$ rank moment cartesian tensor is expressed as:

$$M_{\alpha\beta\ldots\mu} = \frac{(-1)^m}{m!}\sum_i e_i r^{2m+1}\frac{\partial^m}{\partial x_{i\alpha}\partial x_{i\beta}\ldots\partial x_{i\mu}}\left(\frac{1}{r}\right) \tag{2.13}$$

It can be verified that for $m = 1$, Equation (2.13) turns into Equation (2.9) for the dipole moment; for $m = 2$ it turns into Equation (2.10) for the quadrupole moment, etc. The multipole cartesian tensors defined in Equations (2.10) to (2.14) are symmetric in all their indices.

---

[1] In this book the Buckingham [3] definition of the quadrupole moment is chosen. Some authors define the quadrupole moment in different ways. For instance, in books by Landau and Lifshitz [1] and Hirschfelder *et al.* [4], it is defined as twice the quantity given by Equation (2.10).

For a continuous charge distribution, as it takes place for delocalized electron charge in atoms and molecules, the sums over discretes point charges $i$ in Equations (2.8) to (2.13) must be replaced on integrals over the configuration space of atoms or molecules. For instance, the quadrupole moment is expressed as:

$$Q_{\alpha\beta} = \frac{1}{2} \int \rho(\mathbf{r}) \left[3x_\alpha x_\beta - r^2 \delta_{\alpha\beta}\right] dV \tag{2.14}$$

where $\rho(\mathbf{r})$ is the one-electron density (Equation (2.5)).

The first term in Equation (2.7) corresponds to an approximation in which the whole charge is concentrated at the origin, that is, the set of charges is replaced by the total point charge. The second term is determined by the dipole moment and decreases as $1/R^2$. It is expressed via a scalar product of two vectors, so it is proportional to the cosine of the angle between vectors $\mathbf{d}$ and $\mathbf{R}$ (see Equation (A2.12)). The potential of the dipole moment attains its maximum value in the direction of the dipole moment vector. If a given system of charges has a total charge equal to zero (which occurs quite often since atoms and molecules are neutral), then the dipole term becomes the leading one.

The point-charge model of a neutral system with nonvanishing dipole moment is represented by a system of two equal and opposite point charges (Figure 2.1a). The dipole moment is given by:

$$\mathbf{d} = e\mathbf{r}_+ - e\mathbf{r}_- = e\Delta\mathbf{r} \tag{2.15}$$

where $\Delta\mathbf{r}$ is a vector with a modulus equal to the distance between the charges and directed from the negative charge to the positive one. The dipole moment has the same direction and it does not depend on the coordinate origin chosen. Let us verify that this property is valid for any neutral system.

The radius-vectors of the $ith$ charge in two coordinate systems, whose origins are displaced by a vector $\mathbf{a}$, are related to each other as follows:

$$\mathbf{r}'_i = \mathbf{r}_i + \mathbf{a} \tag{2.16}$$

Then, the dipole moment with respect to the primed coordinate system is:

$$\mathbf{d}' = \sum_i e_i \mathbf{r}'_i = \sum_i e_i \mathbf{r}_i + \mathbf{a} \sum_i e_i \tag{2.17}$$

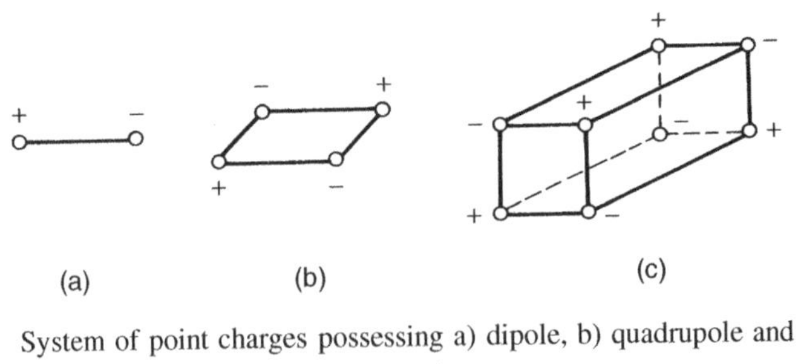


**Figure 2.1** System of point charges possessing a) dipole, b) quadrupole and c) octopole moments

**Table 2.1** Values of the dipole and quadrupole moments for some molecules[a]

| Molecule | Dipole moment ($d \times 10^{18}$ esu cm) | quadrupole moment[b] ($Q \times 10^{26}$ esu cm$^2$) |
|---|---|---|
| KF | 8.585 | −9.29 |
| NaF | 8.152 | −1.97 |
| LiF | 6.330 | 5.82 |
| LiH | 5.882 | −5.0 |
| HCN | 2.986 | 4.4 |
| BrCN | 2.94 | 6.8 |
| ClCN | 2.80 | 6.6 |
| $NH_3$ | 1.47 [5] | $-2.42 \pm 0.04$ [8] |
| HF | 1.736 [5] | $1.75 \pm 0.02$ [9] |
| HCl | 1.07 | 3.8 |
| HBr | 0.788 | 4.0 |
| HI | 0.382 | 6.0 |
| NO | 0.158 | −1.8 |
| CO | 0.1125 [5] | $-1.44 \pm 0.03$ [10] |
| $CO_2$ | 0 | −4.3 |
| $C_2H_2$ [11] | 0 | $4.71 \pm 0.14$ |
| $C_2H_4$ | 0 | 1.5 |
| $C_6H_6$ | 0 | 3.6 |
| $Cl_2$ [12] | 0 | $2.4 \pm 0.12$ |
| $N_2$[12] | 0 | $-1.09 \pm 0.07$ |
| $F_2$ [13] | 0 | 0.56 |
| $H_2$[14] | 0 | $0.46 \pm 0.02$ |
| $O_2$ [15] | 0 | $-0.3 \pm 0.1$ |

[a]The data for molecules without references are taken from Reference [7].
[b]The origin of coordinate system is located at the center of mass.

Therefore, in the case of a neutral system, $\mathbf{d}' = \mathbf{d}$. If a given system has a nonzero net charge, then a coordinate system origin can be chosen in such a manner that the dipole moment becomes zero. In this case one should take:

$$\mathbf{a} = -\sum_i e_i \mathbf{r}_i / \sum_i e_i \tag{2.18}$$

The vector $\mathbf{a}$ determines the position of the *center of charges* in the same way as the inertia center does for a system of masses.

Thus, for neutral systems, the dipole moment is a physical characteristic of the charge distribution. The molecules with a nonvanishing dipole moment are named as *polar*. In Table 2.1, the values of dipole moments for some molecules are presented (for more detailed data, see References [5, 6]).

If the dipole moment of a neutral system vanishes, then the potential (Equation (2.7)) is determined by the quadrupole term. The simplest system of point charges with a nonvanishing quadrupole moment is formed by a set of two positive and two negative charges, with equal absolute values, placed at the vertices of a parallelogram (Figure 2.1(b)). According to the definition (2.10), the quadrupole moment is a symmetric tensor. Therefore, only six of its nine components are independent. From the definition of $Q_{\alpha\beta}$ it also follows that the sum of its diagonal components must be equal to zero:

$$\sum_{\alpha=1}^{3} Q_{\alpha\alpha} = 0 \tag{2.19}$$

The proof of Equation (2.19) follows directly from the following relations:

$$\sum_{\alpha} \delta_{\alpha\alpha} = 3 \qquad \sum_{\alpha} x_{i\alpha}^2 = r_i^2$$

Therefore, the quadrupole moment tensor has only five independent components.

By an appropriate choice of the coordinate system, the tensor $Q_{\alpha\beta}$ can be transformed so that only the diagonal components ($Q_{\alpha\alpha}$) do not vanish. These components are named as the *principal values* of $Q_{\alpha\beta}$. From Equation (2.19), it follows that only two principal values are independent. In the particular case of axial symmetry along the z-axis, the arrangement of the two principal axes in the xy-plane is arbitrary. As a result of the axial symmetry, $Q_{xx} = Q_{yy}$. Then from Equation (2.19) it follows:

$$Q_{zz} = -2Q_{xx} = -2Q_{yy} \tag{2.20}$$

that is, only one component of $Q_{\alpha\beta}$ is independent. Usually, $Q_{zz}$ is chosen as the independent component and is represented by $Q$, being denoted as the quadrupole moment. For example, all the linear nonpolar molecules have axial symmetry and are characterized by the quadrupole moment. The values of the quadrupole moment for some nonpolar molecules are presented in Table 2.1.

The quadrupole moment does not depend on the choice of the origin of the coordinate system if the total charge and the dipole moment of the system vanish, as in the case of nonpolar molecules. If the dipole moment does not vanish, then the origin of coordinates can be chosen in such a manner that the quadrupole moment becomes equal to zero. Such an origin is denoted as the *dipole center* (in an analogy with the charge center). Therefore, for polar molecules the value of the quadrupole moment cannot serve as a characteristic of a molecule. The values of the quadrupole moment presented in Table 2.1 for polar molecules are given relative to the coordinate origin at the center of mass.

The potential of a quadrupole moment, with components given by Equation (2.20), is easily obtained by a direct summation in the appropriate term of Equation (2.7) and is given by:

$$\varphi_Q = \frac{Q}{2R^3}\left(3\cos^2\theta - 1\right) = \frac{Q}{R^3} P_2(\cos\theta) \tag{2.21}$$

where $\theta$ is the angle between the direction of the axial symmetry of the quadrupole charge distribution and a radius-vector $\mathbf{R}$; $P_2(\cos\theta)$ is a Legendre polynomial.

The value of the quadrupole moment ($Q$) represents the deviation of the real charge distribution from the spherical one. In fact, according to the definition (Equation (2.10)), one obtains:

$$Q \equiv Q_{zz} = 1/2 \sum_i e_i \left(3z_i^2 - r_i^2\right) \tag{2.22}$$

In the particular case of the spherical charge distribution:

$$\sum_i e_i z_i^2 = \sum_i e_i x_i^2 = \sum_i e_i y_i^2 = 1/3 \sum_i e_i r_i^2 \tag{2.23}$$

and from Equation (2.22) it follows that $Q = Q_{zz} = Q_{xx} = Q_{yy} = 0$. If $Q$ is positive, it denotes that $\sum_i e_i z_i^2 > 1/3 \sum_i e_i r_i^2$, that is, the charge distribution has a prolate symmetry along the z axis. If $Q$ is negative, it denotes an oblate charge distribution (Figure 2.2).

The relation, $Q_{zz} = Q_{xx} = Q_{yy} = 0$, is not, however, a sufficient condition for the charge distribution to be the spherically symmetric, since the system may possess higher-order moments such as, for example, an octopole moment (or $2^3th$-pole moment). The charge distribution with an octopole moment is depicted in Figure 2.1c. The potential of the field, due to such a distribution, is proportional to $1/R^4$ (see Equation (2.7)).

As the order of moment increases, the use of the multipole moments in the form of the symmetrical tensors in the cartesian coordinate becomes awkward as a result of the difficulty in separating the independent components. Fortunately, it was revealed that $2^l th$ pole moments behave as the irreducible spherical tensors (see Section A2.2.6) and may be expressed in terms of spherical functions:

$$Q_l^m = -\sum_i \left[\frac{4\pi}{2l+1}\right]^{1/2} e_i r_i^l Y_l^m(\theta_i, \varphi_i) \tag{2.24}$$

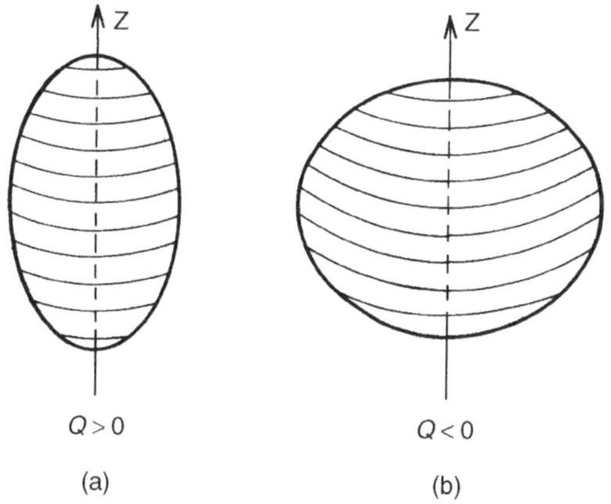


**Figure 2.2** Connection of charge distribution with the sign of quadrupole moment: a) prolate distribution, $Q > 0$; b) oblate distribution, $Q < 0$

Each irreducible spherical tensor is characterized by the value of $l$ and has $(2l+1)$ independent components:

$l = 1$ 2-*pole, dipole*
$l = 2$ $2^2$-*pole, quadrupole*
$l = 3$ $2^3$-*pole, octopole*
$l = 4$ $2^4$-*pole, hexadecapole*

The relations between the spherical and cartesian components for the dipole and quadrupole moments are the following:

$$Q_1 = d_z \qquad Q_1^{\pm} = \pm\frac{1}{\sqrt{2}}\left(d_x \pm i d_y\right)$$
$$Q_2^2 = Q_{zz} \qquad Q_2^{\pm 1} = -\sqrt{\frac{2}{3}}\left(Q_{xz} \pm i\, Q_{yz}\right) \tag{2.25}$$
$$Q_2^{\pm 2} = \frac{1}{\sqrt{6}}\left(Q_{xx} - Q_{yy} \pm 2i\, Q_{xy}\right)$$

The molecular symmetry reduces the number of independent components of the $2^l$-pole moment. The following theorem is valid [7]:

> if a molecule has an axis of n-fold symmetry, only one independent scalar quantity is required to determine any electric multipole tensor of rank $p < n$.

For the quadrupole moment, this result was obtained earlier (Equation (2.20)). For the octopole moment from it follows

$$\Omega \equiv \Omega_{zzz} = -2\Omega_{xxx} = -2\Omega_{yyy} \tag{2.26}$$

For a neutral system, we proved above that the value of the dipole moment does not depend on the origin of the coordinate frame. This is the particular case of the general theorem:

> the first nonvanishing $2^l$-pole moment and only it, does not depend upon the choice of the coordinate system.

Thus, each system of charges is characterized by the value of the first nonvanishing multipole moment. It does not mean that the charge distribution over the multipole series cannot be expanded, but each term in this expansion, except the first, will depend on the choice of the origin of the coordinate frame. Usually it is chosen in the center of the mass.

On the basis of group theory, it is possible to determine multipole moments that do not vanish for a molecule with a given point symmetry. The $2^l$-pole moments transform according to the irreducible representations $D^{(l)}$ of the $\mathbf{R}_3$ group. The expression for the character corresponding to a rotation through an angle $\alpha$ is given in Appendix 2, (Equation (A2.190)). To find characters corresponding to other types of operations of point groups, it is necessary to take into account that the operation of inversion $I$ changes the sign of spherical harmonic if its rank is odd. In general:

$$\chi^{(l)}(I) = (-1)^l (2l+1) \tag{2.27}$$

The characters of reflection ($\sigma$) and rotation – reflection operations ($S_n$) are found if they are expressed as:

$$\sigma = IC_2 \qquad S_n = \sigma C_n \tag{2.28}$$

This results in:

$$\chi^{(l)}(\sigma) = (-1)^l \chi^{(l)}(\pi); \qquad \chi^{(l)}(S_n) = (-1)^l \chi^{(l)}\left(\frac{2\pi}{n} + \pi\right) \tag{2.29}$$

where $\chi^{(l)}(\alpha)$ is given by Equation (A2.190).

Usually the ground state is characterized by the totally symmetric irreducible representation of the molecular point group. It is denoted as $A_1$ or $A_g$. The condition of existence of the $2^l$-pole moment in the ground state of some molecule possessing a point symmetry group **G** is the presence of the totally symmetric irreducible representation of the group **G** in the decomposition of the reducible representation $\Gamma^{(l)}$ formed by spherical harmonics $Y_{lm}$.

As an illustration of how to use the group-theoretical methods to find nonvanishing multipole moments in a field of a given point symmetry, let us consider the point group $\mathbf{C}_{2h}$. This group can be represented as a direct product $\mathbf{C}_{2h} = \mathbf{C}_2 \times \mathbf{C}_i$, so it possesses a center of symmetry. The irreducible representations of $\mathbf{C}_{2h}$ and the characters of reducible representations $\Gamma^{(l)}$ for $l = 1$ to 4, calculated by means of Equations (A2.190), (2.27) and (2.29), are presented in Table 2.2. Using Equation (A2.137) for the decomposition of a reducible representation, the following decompositions are obtained:

$$\begin{aligned}
\Gamma^{(1)} &= A_u + 2B_u \\
\Gamma^{(2)} &= 3A_g + 2B_g \\
\Gamma^{(3)} &= 3A_u + 4B_u \\
\Gamma^{(4)} &= 5A_g + 4B_g
\end{aligned}$$

From these decompositions it follows that the first nonvanishing multipole moment in the field of $\mathbf{C}_{2h}$ symmetry is the quadrupole and the second moment in the multipole expansion is the hexadecapole. This result is the consequence of the general statement:

> for molecules possessing the center of symmetry, all odd-rank tensors vanish.

The first nonvanishing multipole moments in the ground state of molecules possessing different point symmetry are presented in Table 2.3. As follows from Table 2.3, for $d = 0$, molecules must possess more than two symmetry axes or have the center of symmetry, or have the improper rotation axis (with reflection). The quadrupole moment characterizes the charge distribution in all two-atom homoatomic molecules (hydrogen ($H_2$), nitrogen ($N_2$), oxygen ($O_2$), chlorine ($Cl_2$), etc.)

**Table 2.2** Characters of the irreducible representations of the group $\mathbf{C}_{2h}$ and the reducible representations $\Gamma^{(\ell)}$

| $\mathbf{C}_{2h}$ | $E$ | $C_2$ | $\sigma_h$ | $I$ |
|---|---|---|---|---|
| $A_g$ | 1 | 1 | 1 | 1 |
| $A_u$ | 1 | 1 | −1 | −1 |
| $B_g$ | 1 | −1 | −1 | 1 |
| $B_u$ | 1 | −1 | 1 | −1 |
| $\chi^{(1)}$ | 3 | −1 | 1 | −3 |
| $\chi^{(2)}$ | 5 | 1 | 1 | 5 |
| $\chi^{(3)}$ | 7 | −1 | 1 | −7 |
| $\chi^{(4)}$ | 9 | 1 | 1 | 9 |

in benzene ($C_6H_6$), ethylene ($C_2H_4$), and so on. The octopole moment becomes the physical characteristic of charge distribution for molecules with the $\mathbf{T}_d$ symmetry ($CH_4, CF_4$, etc.), which possess fourfold symmetry axis and can be, according to the theorem above, characterized by one scalar value $\Omega$. This value, together with the hexadecapole moment (the second term in the multipole series of $T_d$ symmetry molecules), is presented in Table 2.4 for some molecules with the $T_d$ symmetry.

**Table 2.3** Values of $\ell$ for the first nonvanishing $2^\ell$-pole moments in the ground states of molecules possessing various point symmetries

| Point Group | $\mathbf{C}_n$ | $\mathbf{C}_{nv}$ | $\mathbf{C}_s$ | $\mathbf{C}_{nh}$, $n>1$ | $\mathbf{S}_n$ | $\mathbf{D}_n$ | $\mathbf{D}_{nd}$ | $\mathbf{D}_{nh}$ | $\mathbf{T}_d$ | $\mathbf{O}_h$ | $\mathbf{I}_h$ | $\mathbf{K}_h$ |
|---|---|---|---|---|---|---|---|---|---|---|---|---|
| $\ell$ | 1 | 1 | 1 | 2 | 2 | 2 | 2 | 2 | 3 | 4 | 6 | absent |

**Table 2.4** Molecular octopole and hexadecapole moments [7]

| | $CH_4$ | $CF_4$ | $SiH_4$ | $NH_4^+$ |
|---|---|---|---|---|
| Octopole moment $\Omega \times 10^{34}$ esu cm$^3$ | 4.5 | 4.4 | 6.62 | 6.31 |
| Hexadecapole moment[a] $\Phi \times 10^{42}$ esu cm$^4$ | −6.0 | – | −16.6 | −6.25 |

[a]The origin of coordinate system is at the center of mass. The directions of the axes are parallel to the edges of a cube containing atoms at alternate vertices:
$\Phi = \Phi_{zzzz} = \Phi_{xxxx} = \Phi_{yyyy} = -16\Phi_{zzyy} = -16\Phi_{xxyy}$

As discussed in the preceding text, for molecules with the center of symmetry, all odd-rank tensors vanish. This is the reason that the molecules possessing the $\mathbf{O}_h$ symmetry ($SF_6$, $OsF_8$ etc.) are characterized by the hexadecapole moment. In this case $\Omega = 0$ (the rank $l = 3$) and the first nonvanishing moment is the hexadecapole. For the same reason in the multipole series of such molecules as $H_2$, $N_2$, $O_2$, etc., $C_2H_4$, $C_2H_6$, $C_6H_6$, the next moment after quadrupole term is the hexadecapole. So, the hexadecapole moment is not as exotic as might be expected.

The molecules that belong to the icosahedral symmetry $I_h$ (fullerene and some of its compounds) are characterized by very high multipole moment with $l = 6$. It is $2^6$-pole moment.

## 2.1.3 Multipole–multipole interactions

The energy of the electrostatic interaction of two charge systems $A$ and $B$ may be considered as the potential energy of one system of charges in an external field created by the second one:

$$V_{AB} = \sum_{i \in A} e_i \varphi_B(r_i) \tag{2.30}$$

where $\varphi_B(r_i)$ is the potential due to the charges of system $B$ at those points in which the charges of system $A$ are located. If the systems considered are at a large distance from each other, then the potential $\varphi_B(r_i)$ changes insignificantly over the region of space occupied by the system $A$. In that case, it can be expanded in a Taylor series in power of $r_i$ with respect to the origin, $0_A$, chosen inside the system $A$:

$$\varphi_B(r_i) = \varphi_B(0_A) + \sum_{\alpha} \left(\frac{\partial \varphi_B}{\partial x_\alpha}\right)_{0_A} x_{i\alpha} + \frac{1}{2} \sum_{\alpha,\beta} \left(\frac{\partial^2 \varphi_B}{\partial x_\alpha \partial x_\beta}\right)_{0_A} x_{i\alpha} x_{i\beta} + \cdots . \tag{2.31}$$

Subtracting from $\varphi_B$ the following quantity:

$$\frac{r_i^2}{6} \sum_{\alpha,\beta} \frac{\partial^2 \varphi_B}{\partial x_\alpha \partial x_\beta} \delta_{\alpha\beta} \equiv \frac{r_i^2}{6} \sum_{\alpha} \frac{\partial^2 \varphi_B}{\partial x_\alpha^2} = 0,$$

being equal to zero, since the potential satisfies the Laplace equation $\Delta\varphi = 0$ (the definition of the operator $\Delta$ is given in Appendix 2, Equation (A2.54)) and substituting it into Equations (2.30) and (2.31), $V_{AB}$ can be obtained in terms of the multipole moments of system $A$ as follows:

$$\begin{aligned} V_{AB} &= \varphi_B(0_A) \sum_{i \in A} e_i + \sum_{\alpha} \left(\frac{\partial \varphi_B}{\partial x_\alpha}\right)_{0_A} \sum_{i \in A} e_i x_{i\alpha} + \\ &\quad \sum_{\alpha,\beta} \left(\frac{\partial^2 \varphi_B}{\partial x_\alpha \partial x_\beta}\right)_{0_A} \frac{1}{2} \sum_{i \in A} e_i \left(x_{i\alpha} x_{i\beta} - \frac{1}{3} \delta_{\alpha\beta} r_i^2\right) + \cdots \\ &= q_A \varphi_B(0_A) + \mathbf{d}_A (\nabla \varphi_B)_{0_A} + \frac{1}{3} \sum_{\alpha,\beta} Q_{\alpha\beta}^A \left(\frac{\partial^2 \varphi_B}{\partial x_\alpha \partial x_\beta}\right)_{0_A} + \cdots \end{aligned} \tag{2.32}$$

where the definition of the vector operator $\nabla$ was used.

We assume that the distance between two systems of charges is large, therefore the potential $\varphi_B$ may be expanded into the multipole series (Equation (2.7)) at a point $0_A$. The vector $\mathbf{R}$ in Equation (2.7) corresponds to the vector connecting points $0_A$ and $0_B$ chosen inside systems of charges $A$ and $B$, respectively. Substituting this expansion into Equation (2.32), we find that the first group of terms corresponds to the monopole–multipole interactions, the second one corresponds to the dipole–multipole interactions, etc. The terms corresponding to the monopole–monopole, monopole–dipole and dipole–dipole interactions are:

$$V_{AB} = \frac{q_A q_B}{R} + \frac{q_A(\mathbf{d}_B\mathbf{R}) - q_B(\mathbf{d}_A\mathbf{R})}{R^3} + \frac{\mathbf{d}_A\mathbf{d}_B}{R^3} - \frac{3(\mathbf{d}_A\mathbf{R})(\mathbf{d}_B\mathbf{R})}{R^5} \tag{2.33}$$

For those not familiar with vector calculus it is useful to consider in more detail how the expression for the dipole–dipole interaction energy (the last two terms in Equation (2.33)) was obtained. This is done by substituting into the dipole term in Equation (2.32) the dipole term in the multipole expansion (2.7) of the potential $\varphi_B$ and applying some formulae of vector calculus (Section A2.1). The expression that has to be calculated is:

$$E_{dd} = \mathbf{d}_A \bullet \nabla\left(\frac{\mathbf{d}_B \bullet \mathbf{R}}{R^3}\right) = \frac{\mathbf{d}_A}{R^3} \bullet \nabla(\mathbf{d}_B \bullet \mathbf{R}) + (\mathbf{d}_B \bullet \mathbf{R})\,\mathbf{d}_A \bullet \nabla\left(\frac{1}{R^3}\right)$$

Using Equations (A2.42) and (A2.38), the right-hand part of this expression is transformed into:

$$\frac{\mathbf{d}_A \bullet \mathbf{d}_B}{R^3} - (\mathbf{d}_B \bullet \mathbf{R})\,\mathbf{d}_A \bullet \frac{3\nabla R}{R^4}$$

According to Equation (A2.39), $\nabla R = \mathbf{R}/R$ and we arrive at the well-known result:

$$E_{dd} = \frac{\mathbf{d}_A \bullet \mathbf{d}_B}{R^3} - \frac{3(\mathbf{d}_A \bullet \mathbf{R})(\mathbf{d}_B \bullet \mathbf{R})}{R^5}. \tag{2.34}$$

From the definition of the scalar (dot) product (Equation (A2.12)), it is easy to see that the largest magnitude of the dipole–dipole attraction $\left(E_{dd} = -2d_A d_B/R^3\right)$ corresponds to the collinear parallel configuration, and the larger repulsion $\left(E_{dd} = 2d_A d_B/R^3\right)$ corresponds to the collinear antiparallel configuration. If one of interacting dipoles is perpendicular to the vector $\mathbf{R}$ and the orientation of two dipoles is mutually perpendicular, $E_{dd} = 0$.

The expressions (2.34) and (2.33), as well, correspond to the classical electrostatic interaction. In quantum mechanics, the electrostatic interaction has to be substituted into the matrix element (Equation (2.4)) as an operator. For polar molecules, the dipole–dipole interaction term is the first term in the multipole expansion. If both interacting molecules are in the ground electronic state:

$$E_{el}^{(1)}(dd) = \frac{\mathbf{d}_{00}^A\mathbf{d}_{00}^B}{R^3} - \frac{3\left(\mathbf{d}_{00}^A\mathbf{R}\right)\left(\mathbf{d}_{00}^B\mathbf{R}\right)}{R^5} \tag{2.35}$$

where $\mathbf{d}_{00}^{A}$ is the expectation value of the dipole moment of molecule $A$ in the ground electronic state:

$$\mathbf{d}_{00}^{A} = \left\langle \Psi_{0}^{A} \left| \mathbf{d} \right| \Psi_{0}^{A} \right\rangle \tag{2.36}$$

According to Table 2.3, except for molecules belonging to the point groups $\mathbf{C}_s$, $\mathbf{C}_n$ and $\mathbf{C}_{nv}$, the dipole moment in the ground state of symmetric molecules is zero. However, for excited states of these molecules, the dipole moment may not vanish.

The general formula for the operator of the electrostatic interaction via the irreducible spherical tensors is derived in Section 3.1 and has the following form ($Z \parallel \mathbf{R}$ and $\mathbf{R}$ is directed from $A$ to $B$):

$$V_{AB} = \frac{q_A q_B}{R} + q_B \sum_{l=1}^{\infty} \frac{Q_l^0 (A)}{R^{l+1}} + q_A \sum_{l=1}^{\infty} (-1)^l \frac{Q_l^0 (B)}{R^{l+1}} + \sum_{l_1, l_2 = 1}^{\infty} \sum_{m=-l_<}^{l_<} \frac{F(l_1, l_2, m)}{R^{l_1 + l_2 + 1}} Q_{l_1}^{m} (A)\, Q_{l_2}^{-m} (B) \tag{2.37}$$

$$F(l_1, l_2, m) = (-1)^{l_2} \frac{(l_1 + l_2)!}{\left[(l_1 + m)!\,(l_1 - m)!\,(l_2 + m)!\,(l_2 - m)!\right]^{1/2}} \tag{2.38}$$

where $l_<$ in the last sum in Equation (2.37) is the min $(l_1, l_2)$.

In molecules, the sum over charges in the expression for multipole moments (Equation (2.24)) decomposes on the nuclear and electronic term:

$$Q_l^m (A) = \sum_{a=1}^{n_A} Z_a e \left( \frac{4\pi}{2l+1} \right)^{1/2} R_a^l Y_l^m (\theta_a, \varphi_a) - \sum_{i=1}^{N_A} e \left( \frac{4\pi}{2l+1} \right)^{1/2} r_i^l Y_l^m (\theta_i, \varphi_i) \tag{2.39}$$

for example, the dipole moment vector has the form:

$$\mathbf{d}_A = \sum_{a=1}^{n_A} Z_a e \mathbf{R}_a - \sum_{i=1}^{N_A} e \mathbf{r}_i \tag{2.40}$$

The dependence on the distance for each multipole–multipole term can be easily found from the Table 2.5. It corresponds to the distance dependence $R^{-(l_1+l_2+1)}$ following from the general multipole–multipole Expansion (2.37). The electrostatic interaction between homonuclear diatomic molecules is determined by the quadrupole–quadrupole term proportional to $1/R^5$. In the particular case of tetrahedral molecules, such as methane, the leading term corresponds to the octopole–octopole interaction $\left(1/R^7\right)$. As noted in Section 2.1.2, the hexadecapole moment is not such an exotic one. For instance, the ethylene molecule (with $\mathbf{D}_{2h}$ symmetry) has vanishing dipole and octopole moments. Therefore, the second term of the multipole expansion for the interaction energy in the ethylene dimer corresponds to the quadrupole–hexadecapole interaction $\left(1/R^7\right)$ (see Table 2.5). The

**Table 2.5** Dependence of the multipole–multipole interactions on the distance

| | Monopole | Dipole | Quadrupole | Octopole | Hexadecapole |
|---|---|---|---|---|---|
| Monopole | $1/R$ | $1/R^2$ | $1/R^3$ | $1/R^4$ | $1/R^5$ |
| Dipole | $1/R^2$ | $1/R^3$ | $1/R^4$ | $1/R^5$ | $1/R^6$ |
| Quadrupole | $1/R^3$ | $1/R^4$ | $1/R^5$ | $1/R^6$ | $1/R^7$ |
| Octopole | $1/R^4$ | $1/R^5$ | $1/R^6$ | $1/R^7$ | $1/R^8$ |
| Hexadecapole | $1/R^5$ | $1/R^6$ | $1/R^7$ | $1/R^8$ | $1/R^9$ |

multipole expansion and multipole-multipole interactions are discussed in more detail in Section 3.1.

As follows from Table 2.3, the first nonvanishing multipole moment for $\mathbf{I}_h$ symmetry has $l = 6$. Thus, in the ground state, the fullerene molecule is characterized by $2^6$-pole moment; all multipole moments with $l < 6$ vanish because of a very high symmetry of the molecule. The consequence of this is the short-range character of electrostatic interaction of molecules possessing $\mathbf{I}_h$ point symmetry. At large distances, the direct electrostatic interaction of two fixed fullerene molecules has the distance dependence $\sim 1/R^{13}$, the latter follows from Table 2.5 extended to $2^6$-pole moments, or the expansion (2.37).

The dependence of the multipole expansion for the interaction energy on the choice of the origin of coordinate has been studied by Amos and Crispin [16]. They considered the interaction of proton with the lithium hydride (LiH) molecule. In this case, only the dipole moment does not depend on the choice of the coordinate origin; the other moments do. The best convergence of the monopole–multipole series has been observed when the coordinate origin was chosen to be the charge center or the dipole center (the two expansions are rather close, see Table 2.6), although more often the origin is located at the mass center since its position is found very easily.

As mentioned in the preceding text, the multipole expansion is correct for large separations between the interacting systems. The necessary condition of its validity is the lack of the overlap between the corresponding charge distributions. However, as a result of the quantum mechanical delocalization of the charges, the overlapping takes places in all cases. Since it decreases exponentially with the distance, the evaluation of the interaction energy by means of a multipole series means the neglect of exponentially decreasing terms. This results in the asymptotical convergence of the multipole expansion, that is, it converges exactly only at $R \to \infty$. At any finite distance $R$, the multipole expansion is convergent only up to a certain term; addition of higher terms makes the series divergent (see Section 3.1.4).

The magnitude of the multipole–multipole interactions depends on the mutual orientation of the interacting molecules. If the molecules are not fixed by an external field, then the probability of different orientations is determined by the Boltzmann law, $\exp\left[-V(R, \Omega)/kT\right]$, where $\Omega$ is a set of Euler's angles determining

**Table 2.6** Convergence of the multipole expansion for the LiH-$H^+$ interaction energy [16] (Moments in a.u., energy in $10^{-3}$ hartree[a])

| Origin of coordinate frames is located within LiH molecule | Moments | | | $E^{(1)}_{el}$ | | | | | |
|---|---|---|---|---|---|---|---|---|---|
| | | | | $R' = 8a_0$ | | | $R' = 10a_0$ | | |
| | $d_0$ | $Q_0$ | $\Omega_0$ | 1 | 2 | 3 | 1 | 2 | 3 |
| On atom Li | 2.47 | −5.90 | −13.13 | 0 | 5.76 | 5.76 | 0 | 2.95 | 2.95 |
| On atom H | 2.47 | 9.03 | −27.51 | 11.94 | 7.42 | 5.26 | 6.55 | 3.58 | 2.72 |
| On dipole center | 2.47 | 0 | −2.77 | 5.57 | 5.57 | 5.43 | 2.88 | 2.88 | 2.84 |
| On center of charges | 2.47 | −0.64 | −2.90 | 5.00 | 5.58 | 5.45 | 2.59 | 2.89 | 2.85 |
| On center of masses | 2.47 | −4.03 | −7.71 | 1.82 | 5.72 | 5.58 | 0.93 | 2.93 | 2.89 |

[a] $R$ determines the distance of the proton from the Li atom. Symbols 1, 2, and 3 denote the number of terms which are taken into account in the multipole expansion.

the mutual orientation of the molecules. In the particular case of free rotating molecules, it is necessary to carry out the quantum-mechanical average over the whole set of rotational states for each molecule. It can be proven exactly that this summation over the quantum rotational states of molecules is equivalent to the classical integration over all possible molecular orientation in space. Since at large distances the interaction energy becomes small, at not too low temperatures, such that $kT \gg V$, the Boltzmann exponent may be approximated using the first terms. Hence, the matrix element has to contain the operator, $V(1 - V/kT)$. Expanding $V$ in a multipole series and averaging over orientations gives:

$$\overline{\langle V_{dd}\rangle} = \overline{\langle V_{dq}\rangle} = \overline{\langle V_{qq}\rangle} = \cdots = 0$$

where a line denotes the average over orientations, and brackets mean the quantum-mechanical average. A temperature-dependent term, $-V^2/kT$, provides a nonvanishing contribution. In the particular case of the dipole–dipole interaction, the average over orientations leads to the classical expression for the orientational energy (see Equation (1.10)):

$$-\frac{1}{kT}\overline{\langle V_{dd}^2\rangle} = -\frac{2}{3}\frac{\left(d_{nn}^A\right)^2\left(d_{mm}^B\right)^2}{kTR^6} \tag{2.41}$$

but the values of dipole moments depend upon quantum states of interacting molecules. The $R$-dependence for multipole–multipole terms of the orientational energy can be found from Table 2.5 taking the square quantities in each intersection.

## 2.2 Resonance Interaction

Resonance arises between the ground state of one molecule and the excited state of another if the energy of the transition to an excited state in both molecules is equal, in this case the molecules are said to be *in resonance*. This situation always occurs when identical molecules interact with each other.

Consider a system of two molecules (atoms): a molecule $A$, initially in an excited state, and a molecule $B$, initially in its ground state. In the absence of any interaction, the state of such a system is described by the wave function $\Psi_n^A \Psi_0^B$. By the definition of resonance, the state described by the wave function $\Psi_0^A \Psi_m^B$ corresponds to the same energy. Consequently, there is a degeneracy. The corresponding second-order energy matrix obtained in a first order of degenerate perturbation theory is diagonal if the symmetric ($g$) and antisymmetric ($u$) combinations of the initial zeroth-order functions are used:

$$\Psi_{g,u} = \frac{1}{\sqrt{2}} \left( \Psi_n^A \Psi_0^B \pm \Psi_0^A \Psi_m^B \right) \tag{2.42}$$

The first-order correction to the interaction energy is given by:

$$E_{g,u}^{(1)} = \left\langle \Psi_{g,u} \left| V \right| \Psi_{g,u} \right\rangle = \frac{1}{2} \left[ \left\langle \Psi_n^A \Psi_0^B \left| V \right| \Psi_n^A \Psi_0^B \right\rangle + \left\langle \Psi_0^A \Psi_m^B \left| V \right| \Psi_0^A \Psi_m^B \right\rangle \right.$$
$$\left. \pm 2 \left\langle \Psi_n^A \Psi_0^B \left| V \right| \Psi_0^A \Psi_m^B \right\rangle \right] \tag{2.43}$$

The first two terms in Equation (2.43) are the direct electrostatic energy (see Equation (2.4)) between molecule $A$ in its $nth$ excited state and molecule $B$ in its ground state and molecule $A$ in its ground state and molecule $B$ in its $mth$ excited state. The last term corresponds to the interaction of the transition electron densities of molecules $A$ and $B$:

$$V_{n0,0m} = \left\langle \Psi_n^A \Psi_0^B \left| V \right| \Psi_0^A \Psi_m^B \right\rangle = \int \rho_{n0}^A(i)\, \rho_{0m}^B(j) \frac{e^2}{r_{ij}} dV_i dV_j \tag{2.44}$$

with one-electron transition densities:

$$\rho_{n0}^A(i) = N_A \int \Psi_n^A(1, \ldots, i, \ldots, N_A)^* \, \Psi_0^A(1, \ldots, i, \ldots, N_A)\, dV^{(i)}$$
$$\rho_{0m}^B(j) = N_B \int \Psi_0^B(1, \ldots, j, \ldots, N_B)^* \, \Psi_m^B(1, \ldots, j, \ldots, N_B)\; dV^{(j)} \tag{2.45}$$

due to the transition of the excitation from $A$ to $B$:

$$A^* + B \rightarrow A + B^* \tag{2.46}$$

In the energy-transfer studies, the Expression (2.44) is often known as the *resonance integral*.

At sufficiently large distances between interacting molecules, the interaction energy may be represented by the multipole expansion series. In the case of neutral molecules, the leading term of that series is the dipole–dipole interaction. As a result, it is found that the resonance dipole–dipole interaction exists even between nonpolar molecules. Since the resonance interaction arises in the first-order approximation of the perturbation theory, it decreases with the distance as $1/R^3$ and contributes at longer ranges than the dipole–dipole polarization interaction (see Section 2.3), which decreases as $1/R^6$. It should be kept in mind that dipole–dipole approximation may be invalid when the interacting molecules are close together. Depending on the parity of the stationary state, the energy (Equation (2.44)) may be positive or negative.

The interaction of a molecule in an excited state with a molecule in its ground state leads to the formation of so-called *excimer complexes*, which are manifested via a frequency shift in the absorption and luminescence spectra. One of the very important applications of excimer complexes was the discovery of the excimer laser [17]. The xenon excimer laser is a good example. In this laser, the xenon (Xe) atoms are excited by a pulsed electron beam and form metastable Xe ... $Xe^*$ excimers. The radiative transition to the ground state of the dimer results in a repulsive $Xe_2$ state. This allows a high inverse population to be obtained, which is necessary for effective laser action.

From the form of the functions describing the states of the interacting system (Equation (2.42)), it follows that the excitation corresponds with the same probability to molecule $A$ as it does to molecule $B$. If it is assumed that at a given moment only one molecule has been excited, then this state is not the stationary one and, as a result of the resonance interaction, the excitation will be transferred from one molecule to the other with a frequency proportional to the resonance integral (Equation (2.44)). The resonance interaction is responsible for the existence of the delocalized exciton states in molecular crystals [18, 19]. The width of the exciton band determined by the value of the resonance integral may reach rather large values. For instance, in anthracene crystals, the value of the resonance splitting for the lower excited state is 220 $cm^{-1}$; it is even more, 575 $cm^{-1}$, in naphthacene crystals [19].

When the excitation, transferred from a donor molecule to an acceptor molecule, dissipates so rapidly that an inverse transfer is absent, a one-sided energy transfer takes place because of the disruption of the resonance. It lies in the basis of such phenomena as sensitized luminescence, resonance extinction and others. In this case, the probability of transfer is proportional to the square of the resonance integral, and for the dipole–dipole interactions it falls off as $1/R^6$. The theory of such radiationless energy transfer was developed by Förster [20] and Dexter [21]. A similar excitation transfer, due to the resonance interaction, is responsible for the migration of energy in crystals [18] and along a polymer chain under an initial excitation of one of its fragments [22].

The theory of the charge transfer in a polymer chain or the hole migration in solids:

$$A_n^+ + A_{n'} \rightarrow A_n + A_{n'}^+ \tag{2.47}$$

was developed in Reference [23]. The efficiency of this process is also determined by the resonance interaction. In the absence of dynamical interaction between holes, the Hubbard Hamiltonian [24], which is now widely used in the high $T_c$ superconductivity studies, can be written as follows:

$$H = \epsilon_0 \sum_n b_n^+ b_n + \sum_{n,n'} M_{nn'} b_n^+ b_{n'} \tag{2.48}$$

where $b_n^+$ and $b_n$ are the hole creation and annihilation operators, respectively; $\epsilon_0$ is the energy of hole creation and $M_{nn'}$ is the so-called hopping integral that characterizes the efficiency of charge transfer (hopping) from site $n$ to site $n'$. As was pointed out in Reference [23], the hopping integral in a general many-electron approach has the sense of the resonance integral (Equation (2.44)) and can be expressed as:

$$M_{nn'} = \left\langle \Psi_0\left(A_n^+\right) \Psi_0\left(A_{n'}\right) |V| \Psi_0\left(A_n\right) \Psi_0\left(A_{n'}^+\right)\right\rangle \tag{2.49}$$

## 2.3 Polarization Interactions

The forces due to the polarization of one molecule by the other are denoted as *polarization forces*. They are described by the second- and higher-order terms of the Rayleigh–Schrödinger perturbation theory (see Section A3.3.1). In the second order, the expression for the energy of interaction between two molecules in their ground states is given by Equation (A3.102):

$$E_{pol}^{(2)} = -\sum_{n,m}{}' \frac{\left|\left\langle \Psi_n^A \Psi_m^B |V| \Psi_0^A \Psi_0^B \right\rangle\right|^2}{\left(E_n^A - E_0^A\right) + \left(E_m^B - E_0^B\right)} = E_{ind}^{(2)} + E_{disp}^{(2)} \tag{2.50}$$

where prime in the summation means that the quantum numbers, $n$ and $m$, do not simultaneously take the values that correspond to the ground states of the isolated molecules. The summation over $n$ and $m$ may be divided into two parts, $E_{ind}^{(2)}$ and $E_{disp}^{(2)}$, having a different physical meaning. Let us consider these two parts separately.

### 2.3.1 Induction interactions

The expression for the induction energy in the second order of perturbation theory is a part of the general Formula (2.50):

$$\begin{aligned} E_{ind}^{(2)} &= -\sum_{m\neq 0} \frac{\left|\left\langle \Psi_0^A \Psi_m^B |V| \Psi_0^A \Psi_0^B \right\rangle\right|^2}{E_m^B - E_0^B} - \sum_{n\neq 0} \frac{\left|\left\langle \Psi_n^A \Psi_0^B |V| \Psi_0^A \Psi_0^B \right\rangle\right|^2}{E_n^A - E_0^A} \\ &= -\sum_{m\neq 0} \frac{\left|V_{0m,00}\right|^2}{E_m^B - E_0^B} - \sum_{n\neq 0} \frac{\left|V_{n0,00}\right|^2}{E_n^A - E_0^A} \end{aligned} \tag{2.51}$$

The first term describes the electrostatic interaction of molecule $A$, characterized by the ground state electron density, $\rho_{00}^A(i)$ (see Equation (2.5)), with molecule $B$, characterized by an induced transition electron density:

$$\rho_{m0}^B(j) = N_B \int \Psi_m^B(1, \ldots, j, \ldots, N_B)^* \, \Psi_0^B(1, \ldots, j, \ldots, N_B)\, dV^{(j)} \tag{2.52}$$

Similarly, the second term in Equation (2.51) corresponds to the interaction of molecule $B$ in its ground state with the induced electron density distribution of molecule $A$.

The matrix elements of operator $V$ (see Equation (2.2)), appearing in Equation (2.51), may be represented in a form similar to that of the classical interaction of two space charge distribution. Taking into account the orthogonality condition for the ground and excited state wave functions, we obtain:

$$V_{0m,00} = -\sum_a Z_a e \int \rho_{m0}^B(j) \frac{1}{r_{aj}} dV_j + \int \rho_{00}^A(i) \rho_{m0}^B(j) \frac{e^2}{r_{ij}} dV_i dV_j \qquad (2.53)$$

$$V_{n0,00} = -\sum_b Z_b e \int \rho_{n0}^A(i) \frac{1}{r_{bi}} dV_i + \int \rho_{n0}^A(i) \rho_{00}^B(j) \frac{e^2}{r_{ij}} dV_i dV_j \qquad (2.54)$$

The first term in Equation (2.53) corresponds to the interaction between the electrons of molecule $B$ characterized by an induced electron density, $\rho_{m0}^B$, and the nuclei of molecule $A$. The second term represents the interaction between the same induced electron distribution of molecule $B$ and the electrons of the molecule $A$ characterized by the ground-state electron density, $\rho_{00}^A$. The terms in Equation (2.54) have a similar meaning.

The induction energy for the interaction of atoms and molecules in their ground state is always negative, since $E_0^A$ and $E_0^B$ are the negative ground state energies and the difference in the dominators of the first and second terms in Equation (2.51) is positive. Thus, for molecules in the ground state, the induction energy has an attractive nature. In the case of the interaction between atoms (molecules) in excited states, the induction energy may correspond either to attraction or to repulsion.

At large distances between molecules, the induction energy may be represented via the multipole series by expanding $V$ in powers of $1/R$. The first term of this expansion corresponds to the interaction of an induced dipole with the field of the inducing molecule. The distance dependence is given by the square of the corresponding dipole–multipole interaction and may be found easily from Table 2.5. For example, for the interaction between an ion and a neutral molecule, the leading term becomes $1/R^4$ while for the interaction between a polar molecule with a neutral one, the leading term has the $1/R^6$ dependence. The interaction of the quadrupole moment of one molecule with the induced dipole of the other is characterized by a $1/R^8$ leading term.

The induction interaction energy of two polar molecules in the dipole–dipole approximation may be derived by substitution of Equation (2.34) into Equation (2.51) and averaging over the mutual orientations of the dipoles:

$$E_{ind}^{(2)}(dd) = -\frac{2}{3R^6} \left|d_{00}^A\right|^2 \sum_{m\neq 0} \frac{\left|d_{m0}^B\right|^2}{E_m^B - E_0^B} - \frac{2}{3R^6} \left|d_{00}^B\right|^2 \sum_{n\neq 0} \frac{\left|d_{n0}^A\right|^2}{E_n^A - E_0^A} \qquad (2.55)$$

The sum, appearing in Equation (2.55), may be easily expressed in terms of an experimentally measured quantity – the average statical molecular polarizability:

$$\overline{\alpha}_1(0) = \frac{2}{3} \sum_{n\neq 0} \frac{|d_{n0}|^2}{E_n - E_0} \qquad (2.56)$$

The subindex 1 denotes that it is the dipole polarizability. As a result, an expression identical to the classical expression for the induction interaction of two dipoles is obtained:

$$E_{ind}(dd) = -\frac{1}{R^6}\left[\left(d_{00}^A\right)^2 \overline{\alpha_1^B}(0) + \left(d_{00}^B\right)^2 \overline{\alpha_1^A}(0)\right] \tag{2.57}$$

where $d_{00}^A$ (or $d_{00}^B$) is the matrix element of the dipole moment of molecule $A$ (or $B$) in its ground state.

For neutral atoms in the spherical-symmetry ground states (noble gas, alkaline earths, etc.), that is, atoms not possessing multipole moments, the induction forces are equal to zero if the overlap of the charge distribution of interacting atoms can be neglected. But in the presence of the overlap, the induction forces defined by Equation (2.51) do not equal zero. It should be stressed that the induction interaction between atoms not possessing multipole moments depends only upon the overlap of the charge distributions and exponentially decreases with the distance. Strictly speaking, the term induction loses its physical sense in this case.

For ions, the induction forces can be large enough because of the monopole–multipole interactions. The expression for the interaction of the monopole with the induced dipole and induced quadrupole has the following form:

$$E_{ind}^{(2)}(qd + qQ) = -\frac{1}{2R^4}\left[q_A^2\overline{\alpha}_1^B(0) + q_B^2\overline{\alpha}_1^A(0)\right] - \frac{1}{2R^6}\left[q_A^2\overline{\alpha}_2^B(0) + q_B^2\overline{\alpha}_2^A(0)\right] \tag{2.58}$$

The $qQ$ term is expressed through the static *quadrupole polarizability*:

$$\alpha_2(0) = 2\sum_{n\neq 0}\frac{|\langle n\,|Q_{zz}|\,0\rangle|^2}{E_n - E_0} \tag{2.59}$$

## 2.3.2 Dispersion interactions

*Dispersion interactions* were defined by London [25] in 1930 and can be presented by the rest part of $E_{pol}^{(2)}$ after subtracting $E_{ind}^{(2)}$:

$$E_{disp}^{(2)} = -\sum_{m,n\neq 0}\frac{\left|\langle\Psi_n^A\Psi_m^A\,|V|\,\Psi_0^A\Psi_0^B\rangle\right|^2}{\left(E_n^A - E_0^A\right) + \left(E_m^B - E_0^B\right)} = -\sum_{m,n\neq 0}\frac{\left|V_{nm,00}\right|^2}{\left(E_n^A - E_0^A\right) + \left(E_m^B - E_0^B\right)} \tag{2.60}$$

The matrix element, appearing in Equation (2.60), corresponds to the electrostatic interaction of two mutually induced electron distributions, $\rho_{n0}^A$ and $\rho_{m0}^B$, and may be expressed as follows:

$$V_{nm,00} = \int \rho_{n0}^A(i)\,\rho_{m0}^B(j)\,\frac{e^2}{r_{ij}}dV_i dV_j \tag{2.61}$$

The dispersion energy is determined by the quantum-mechanical fluctuations of the electron density. Fluctuations cause the redistribution of the electron density

**Table 2.7** Values of the dispersion energy (in $cm^{-1}$) for two hydrogen atoms in the ground state [27]

| $R$, *bohrs* | $C_6/R^6$ | $C_8/R^8$ | $C_{10}/R^{10}$ |
|---|---|---|---|
| 8 | 5.44 | 1.63 | 0.67 |
| 9 | 2.68 | 0.63 | 0.21 |
| 10 | 1.43 | 0.27 | 0.07 |

$C_6 = 6.499$ a.u.; $C_8 = 124$ a.u. $C_{10} = (2150.6_{\,dipole-octopole} + 1135.2_{\,quadrupole-quadrupole})$ a.u.

in atoms and molecules. An instantaneous charge distribution, corresponding to an instantaneous dipole (and higher-order multipoles) moment of one molecule, induces a multipole moment on the other. The interaction of these moments defines the dispersion energy. For molecules in their ground states, it is always negative, that is, it corresponds to an attraction. The attractive dispersion forces are often called *the van der Waals forces* in honor of van der Waals who first introduced attractive forces between molecules in his equation of state (Equation (1.3)).

As was shown by London [25, 26] in his fundamental studies of the dispersion forces, the leading dipole–dipole term in the dispersion energy (when retardation is neglected) can be presented as a change, due to the dipole–dipole interaction, in the zero-point vibration energy of electric field created by zero-point vibrations of fluctuating dipole moments of interacting atoms (molecules). The zero-point vibrations are the quantum phenomenon. Thus, the London mechanism reflects the quantum nature of the dispersion interaction.

The multipole expansion for dispersion energy is usually written as a series with coefficients $C_n$ denoted as *dispersion coefficients*:[2]

$$E_{disp}^{(2)} = -\sum_{n=6}^{\infty} \frac{C_n}{R^n} \tag{2.62}$$

For interaction between atoms, the series (2.62) contains only even powers of $n$. The first term, proportional to $1/R^6$, corresponds to the dipole–dipole interaction; the second term $\left(\sim 1/R^8\right)$ corresponds to the dipole–quadrupole interaction; and the third term is due to the dipole–octopole and quadrupole–quadrupole interactions (see Table 2.5). The last two contributions must be taken into account in the calculation of the term $\sim 1/R^{10}$. This fact was confirmed by precise calculations, carried out by Kołos [27], for two hydrogen atoms (see Table 2.7). For interacting molecules, the series (2.62) may also contain odd powers of $n$ (see Section 3.1.3).

It is important to stress that irrelevant to the value of the first nonvanishing permanent multipole moment, the first term in Equation (2.62) is $\sim 1/R^6$ and corresponds to

[2] In this subsection, unless otherwise stated, the atomic unit system is used: $e = m = \hbar = 1$. In atomic units, the transition frequency is given by $\omega_{n0} = E_n - E_0$.

the $d$-$d$ dispersion interaction. The reason for this is in the nonsymmetrical disturbance of the ground state electron distribution by quantum-mechanical fluctuations. As a result, even for such symmetrical molecule as methane ($CH_4$), the leading term in the dispersion multipole expansion is $\sim 1/R^6$, whereas according to Table 2.5 the leading term in the direct electrostatic interaction is octopole–octopole, $\sim 1/R^7$ (for fixed molecules) and $\sim 1/R^{14}$ for rotating molecules (the orientation forces); for induction interaction, the leading term is octopole–dipole, $\sim 1/R^{10}$.

As pointed out in Section 2.1.3, the multipole expansion converges only asymptotically to an exact value of the energy. For finite $R$, beginning from some value of $n$, the expansion terms $C_n/R^n$ increase in absolute value. However, this does not mean that the multipole expansion is invalid. At sufficiently large $R$, the sum of the first terms of the expansion approximates the energy with good accuracy. In many applications, for qualitative estimations, it is enough to use only the first term in the multipole expansion. As follows from Table 2.7, at $R \geq 10$ bohr the energy of the dispersion interaction may be evaluated by the first term, $C_6/R^6$, within an error less than $\sim$20%; for larger $R$, the error rapidly diminishes. This fact explains the existence of a large number of methods devoted to the precise evaluation of $C_6$ (see Reference [28], Chapter 2, Section 1.4).

For spherically symmetric systems, or for arbitrary systems averaged over all orientations, the following expression for the dispersion coefficient of the dipole–dipole interaction (in atomic units) may be obtained from Equation (2.60):

$$C_6^{AB} = \frac{2}{3} \sum_{n,m \neq 0} \frac{\left|d_{n0}^A\right|^2 \left|d_{m0}^B\right|^2}{\omega_{n0}^A + \omega_{m0}^B} \tag{2.63}$$

Equation (2.63) may be expressed as:

$$C_6^{AB} = \frac{3}{2} \sum_{n,m \neq 0} \frac{f_{n0}^A f_{m0}^B}{\omega_{n0}^A \omega_{m0}^B \left(\omega_{n0}^A + \omega_{m0}^B\right)} \tag{2.63a}$$

where:

$$f_{n0} = \frac{2}{3} \omega_{n0} \left|d_{n0}\right|^2 \tag{2.64}$$

are the *dipole oscillator strengths* for the quantum transitions of type $0 \rightarrow n$ for the isolated molecules.

The calculation of $C_6$ (based on Equation (2.63)) requires that the oscillator strengths for transitions to discrete, as well as to continuum, states be known. Therefore, the direct application of Equation (2.63) presents some problems. In practice, the Casimir–Polder formula expressing $C_6$ in terms of the dynamic polarizabilities of molecules is more widely used. This formula is derived from Equation (2.63) with the help of the integral identity:

$$\frac{1}{a+b} = \frac{2}{\pi} \int_0^\infty \frac{ab}{\left(a^2 + z^2\right)\left(b^2 + z^2\right)} dz, \quad a > 0, \ b > 0 \tag{2.65}$$

Equation (2.65) is easily proved by an elementary integration. The integrand should be represented as:

$$\frac{ab}{\left(a^2+z^2\right)\left(b^2+z^2\right)}=\frac{ab}{a^2-b^2}\left(\frac{1}{b^2+z^2}-\frac{1}{a^2+z^2}\right)$$

and take into account the result of integration:

$$\int_0^\infty \frac{dz}{a^2+z^2}=\frac{1}{a}\arctan\frac{z}{a}\Big|_0^\infty=\frac{\pi}{2a}$$

If one denotes in Equation (2.65) $a=\omega_{n0}^A$, $b=\omega_{m0}^B$ and $z=\omega$ and substitutes it into Equation (2.63), the coefficient $C_6$ is presented as:

$$C_6=\frac{3}{\pi}\int\sum_{n,m\neq 0}\frac{f_{n0}^A f_{m0}^B d\omega}{\left(\left(\omega_{n0}^A\right)^2+\omega^2\right)\left(\left(\omega_{m0}^B\right)^2+\omega^2\right)} \tag{2.66}$$

The average dynamic polarizability of atoms (molecules) is defined via the oscillator strength (Equation (2.64)) as:

$$\overline{\alpha_1(\omega)}=\sum_{n\neq 0}\frac{f_{n0}}{\omega_{n0}^2-\omega^2} \tag{2.67}$$

where subindex 1 denotes that $\alpha_1(\omega)$ corresponds to the dipole transitions. The multipole polarizabilities and the generalization of the Casimir–Polder formula on the dispersions coefficients of higher order are discussed in Section 3.1.2. In a complex plane, Equation (2.67) for a dynamic polarizability transforms into:

$$\overline{\alpha_1(i\omega)}=\sum_{n\neq 0}\frac{f_{n0}}{\omega_{n0}^2+\omega^2} \tag{2.68}$$

The definition (2.68) allows Equation (2.66) to be represented via the dynamic polarizabilities $\alpha_1(i\omega)$; the latter can be measured in experiments. As a result, we arrive at the famous Casimir–Polder formula:

$$C_6=\frac{3}{\pi}\int\overline{\alpha^A(i\omega)}\ \overline{\alpha^B(i\omega)}d\omega \tag{2.69}$$

where the integrand contains the average dipole dynamic polarizabilities as functions of an imaginary variable. This formula is employed as a basis for developing various methods for determining $C_6$ via both *ab initio* calculations and semiempirical techniques (Reference [28], Chapter 2).

Two simple approximate formulae, which permit one to estimate qualitatively the coefficient $C_6$, can be derived from Equation (2.63).

If the oscillator strength for one of the transitions in a molecule is significantly larger than all others, then the summations over the excited states in Equation (2.63) may be replaced by a single term. Such a replacement may also be done in a general way, if some effective quantities, $\overline{\omega}_{n0}$ and $\overline{f}_{n0}$, are introduced.

Then:

$$C_6=\frac{3}{2}\frac{\overline{f_{k0}^A}\ \overline{f_{10}^B}}{\overline{\omega_{k0}^A}\ \overline{\omega_{10}^B}\left(\overline{\omega_{k0}^A}+\overline{\omega_{10}^B}\right)} \tag{2.70}$$

and the static polarizability, obtained from Equation (2.67) at $\omega = 0$, also contains only one term:

$$\overline{\alpha_1(0)} = \overline{f_{k0}}/\overline{\omega}_{k0}^2 \tag{2.71}$$

Equation (2.71) permits an effective oscillator strength to be expressed by means of the static polarizability and we obtain the well-known London formula:

$$C_6 = \frac{3}{2}\overline{\alpha_1^A(0)}\ \overline{\alpha_1^B(0)}\frac{\overline{\omega_{k0}^A}\ \overline{\omega_{10}^B}}{\overline{\omega_{k0}^A} + \overline{\omega_{10}^B}} \tag{2.72}$$

Here $\overline{\alpha}_1^A(0)$ is the average dipole static polarizability. The averaged transition frequencies, appearing in Equation (2.72), may be considered as empirical parameters. Usually, they are replaced by the first ionization potentials. Then, the London formula is rewritten as follows:

$$C_6 = \frac{3}{2}\overline{\alpha_1^A(0)}\ \overline{\alpha_1^B(0)}\frac{I_l^A I_l^B}{I_1^A + I_1^B} \tag{2.73}$$

Equation (2.73) allows a qualitative estimate of $C_6$ to be obtained without any difficulty, since all the terms appearing in Equation (2.73) are well known for a majority of molecules. The estimates, obtained in such a manner, represent the lower limit for $C_6$ (see Table 2.8).

Another well-known approximation was proposed by Slater and Kirkwood [29]. It may be obtained from the exact expression (Equation (2.63)) by extracting averaged transition frequencies from the summation and applying the theorem regarding the sum of oscillator strengths, according to which:

$$\sum_n f_{n0}^A = N_A$$

**Table 2.8** Dispersion constant $C_6$ (in a.u.) for some rare gas atoms evaluated via the London (2.73) and Slater–Kirkwood (2.74) formulae, the exact calculation [30] is also presented

| | | | | $C_6$ | | |
|---|---|---|---|---|---|---|
| Systems | $I_1, eV$ | $\alpha_1(0)$ ,a.u. | $N_A$ | Equation (2.73) | Equation (2.74) | Reference [30] |
| He–He | 24.58 | 1.38 | 2 | 1.297 | 1.726 | 1.461 |
| Ne–Ne | 21.55 | 2.67 | 6 | 4.23 | 8.01 | 6.88 |
| Ar–Ar | 15.75 | 11.09 | 6 | 53.28 | 67.85 | 66.90 |
| Kr–Kr | 14.00 | 16.72 | 6 | 107.7 | 125.60 | 135.10 |
| Xe–Xe | 12.13 | 27.34 | 6 | 249.90 | 262.62 | 281.15 |

In this approximation, one obtains the following expression for the static polarizability:

$$\overline{\alpha_1^A(0)} \simeq N_A/\overline{\omega}_{k0}^2 \quad \text{or} \quad \omega_{k0} = \left(N_A/\overline{\alpha_1^A(0)}\right)^{\frac{1}{2}}$$

Then Equation (2.63) is transformed as follows:

$$C_6 = \frac{3}{2}\frac{\overline{\alpha_1^A(0)}\;\overline{\alpha_1^B(0)}}{\left(\overline{\alpha_1^A(0)}/N_A\right)^{\frac{1}{2}} + \left(\overline{\alpha_1^B(0)}/N_B\right)^{\frac{1}{2}}} \tag{2.74}$$

When applying Equation (2.74), for $N_A$ it is recommended that the number of valence electrons and not the total number of electrons is used.

Values of the dispersion coefficient $C_6$ obtained via Equations (2.73) and (2.74) for some noble gases are presented in Table 2.8. The exact values of $C_6$, calculated with the help of Equation (2.69), are also given [30]. The London formula (Equation (2.73)), underestimates the values of $C_6$ for all noble gas atoms. The Slater–Kirkwood formula (Equation (2.74)) overestimates $C_6$ for light atoms and underestimates it for heavy atoms, although its values are in a better agreement with the exact values than those obtained via the London formula.

For polar molecules, the polarization forces contain both dispersion and induction dipole–dipole components. Since both components depend on the distance in the same way, the order of their ratio is estimated easily with the help of Equations. (2.57) and (2.73). For example, in the case of the interaction of identical molecules, the following estimation is obtained:

$$\frac{E_{disp}^{(2)}}{E_{ind}^{(2)}} \simeq \frac{3}{8} I_1 \frac{\overline{\alpha_1(0)}}{d_{00}^2} \tag{2.75}$$

For a long time, it was accepted that the dispersion interaction is a pure quantum phenomenon that does not exist in the framework of classical physics. In the 1970s, more than forty years after London's seminal paper [25], Boyer [31–34] demonstrated that the London expression for the dispersion energy and the Casimir–Polder expression [35] for the retarded dispersion interaction (see Section 2.5) can be derived in the framework of classical electrodynamics but with an additional assumption about the existence of fluctuating classical Lorentz invariant electromagnetic radiation at the absolute zero of temperature (classical electromagnetic zero-point radiation). The latter contained Plank's constant introduced for fixing the scale of classical radiation spectrum to correspond to $\frac{1}{2}\hbar\omega$ per normal mode. Boyer considered polarizable particles and solved the classical Maxwell equations but instead of traditional homogeneous boundary conditions he used the boundary conditions corresponding to the presence of fluctuating classical electromagnetic radiation with the Lorentz invariant spectrum. This allowed him to obtain also the temperature dependence of dispersion interactions in the nonretarded and retarded regions [36]. The influence of temperature effects on the dispersion interaction are discussed in Section 2.5.

Thus, the dispersion forces can be introduced in classical physics although not in a 'pure' (traditional) classical physics. The latter has to be 'contaminated' by introducing, as in quantum mechanics, the zero-point fluctuating radiation scaled to $\frac{1}{2}\hbar\omega$ per normal mode.

## 2.4 Exchange Interaction

The origin of the *exchange interaction* is in the Pauli principle. According to the latter, the wave function of a many-electron system must be antisymmetric with respect to permutations of electrons. This leads to some new terms in the expression of the electrostatic energy that do not appear in Equation (2.4) for the direct electrostatic interaction. Thus, it is a specific quantum-mechanical effect. In the classical limit, the exchange interaction is turned to zero.

As an illustration, let us consider the most simple example – the hydrogen molecule ($H_2$). The electron state of atom $H_a$ is characterized by the coordinate wave function $\varphi_a(\mathbf{r})$ and the spin wave function $\chi_\sigma$ ($\sigma$ is the projection of electron spin $s_z = \pm\frac{1}{2}$ denoted as $\alpha$ and $\beta$). For a two-electron system, the total wave function can be presented as a simple product of the coordinate wave function, $\Phi(1, 2)$, and the spin wave function, $\Omega(1, 2)$. For brevity the sets of electron coordinates are denoted by numbers 1 and 2. The total spin of two-electron system can possess only two values: $S = 0$ (the singlet state) and $S = 1$ (the triplet state). In the theory of angular momentum in quantum mechanics, it is proved that the two-electron spin functions for the singlet and triplet state are:

$$\begin{aligned}\Omega^{(S=0)}(1,2) &= \frac{1}{\sqrt{2}}\left[\chi_\alpha(1)\,\chi_\beta(2) - \chi_\alpha(2)\,\chi_\beta(1)\right]\\ \Omega^{(S=1)}(1,2) &= \frac{1}{\sqrt{2}}\left[\chi_\alpha(1)\,\chi_\beta(2) + \chi_\alpha(2)\,\chi_\beta(2)\right]\end{aligned} \tag{2.76}$$

Thus, the singlet spin function is antisymmetric and the triplet wave function is symmetric with respect to permutation of its arguments.

To fulfill the requirement of antisymmetrization of the total wave function, the coordinate wave function $\Phi(1, 2)$ must possess the permutation symmetry, which is reciprocal to the symmetry of the spin wave function $\Omega(1, 2)$. So, the antisymmetric wave functions for the singlet and triplet state may be presented in the form:

$$\begin{aligned}\Psi^{S=0}(1,2) &= N_{ab}(S=0)\left[\varphi_a(1)\,\varphi_b(2) + \varphi_a(2)\,\varphi_b(1)\right]\Omega^{S=0}(1,2)\\ \Psi^{S=1}(1,2) &= N_{ab}(S=1)\left[\varphi_a(1)\,\varphi_b(2) - \varphi_a(2)\,\varphi_b(1)\right]\Omega^{S=1}(1,2)\end{aligned} \tag{2.77}$$

where $N_{ab}$ is the normalization factor:

$$N_{ab}(S=0) = 1/\sqrt{2\left(1+s_{ab}^2\right)},\; N_{ab}(S=1) = 1/\sqrt{2\left(1-s_{ab}^2\right)} \tag{2.78}$$

and $s_{ab}$ is the overlap integral:

$$s_{ab} = \langle \varphi_a \mid \varphi_b \rangle$$

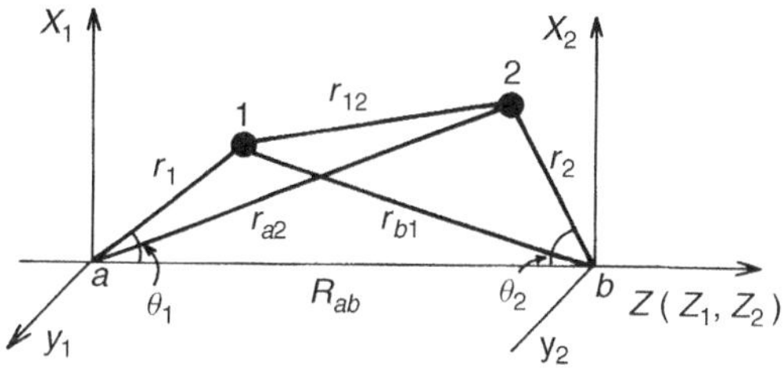


**Figure 2.3** Notations in the system of the two hydrogen atoms

appearing because of the nonorthogonality of the atomic coordinate wave functions located on different atoms.

The Hamiltonian of the hydrogen molecule contains the Hamiltonians of isolated atoms, $H_a(1)$ and $H_b(2)$, and the interaction operator:

$$V(1,2) = -\frac{e^2}{r_{a2}} - \frac{e^2}{r_{b1}} + \frac{e^2}{r_{12}} + \frac{e^2}{R_{ab}} \tag{2.79}$$

(See notations in Figure 2.3). In the first order of perturbation theory (that corresponds to the Heitler–London approximation), the interaction energy for the singlet and triplet state is presented as a sum of two terms:

$$E_{int}^{(1)} = \left\langle \Psi^{S=0,1}(1,2) \left| V(1,2) \right| \Psi^{S=0,1}(1,2) \right\rangle =$$

$$= \begin{cases} \left[1/\left(1+s_{ab}^2\right)\right]\left[E_{el}^{(1)} + E_{exch}^{(1)}\right] & \text{for } S=0 \\ \left[1/\left(1-s_{ab}^2\right)\right]\left[E_{el}^{(1)} - E_{exch}^{(1)}\right] & \text{for } S=1 \end{cases} \tag{2.80}$$

where the electrostatic energy, $E_{el}^{(1)}$, is given by Equation (2.4) and the exchange energy is defined as:

$$E_{exch}^{(1)} = \langle \varphi_a(1)\varphi_b(2) \left| V(1,2) \right| \varphi_b(1)\varphi_a(2) \rangle \tag{2.81}$$

In contrast to the electrostatic energy, the exchange energy has no classical analogue. It is the interaction of two delocalized electron densities: $\rho_{ab}(1) = \varphi_a^*(1)\varphi_b(1)$ and $\rho_{ab}(2) = \varphi_a(2)\varphi_b^*(2)$. According to the wave functions (Equation (2.77)), each electron can, with equal probability, be located in the state $\varphi_a$ or $\varphi_b$. This exchange of electrons between atoms $a$ and $b$ is the reason for appearance of a new quantum-mechanical kind of interaction energy – the exchange energy. In reality, it is a part of the electrostatic energy due to the redistribution of electron density in the state described by the antisymmetrized wave function.

For many-electron systems, the expression for the exchange energy can be also obtained by straightforward calculations. Consider two many-electron molecules (atoms), $A$ and $B$, in the ground state with the wave functions $\Psi_0^A$ and $\Psi_0^B$ (for excited states, the consideration will be the same). We will not separate the spin parts and will consider the total wave functions. The wave functions $\Psi_0^A$ and

$\Psi_0^B$ are already antisymmetric, but the total wave function of the system $AB$ must be antisymmetric with respect to permutations of electrons, not only in each of the molecules, but also between molecules. This can be fulfilled if the product $\Psi_0^A \Psi_0^B$ is acted on by the antisymmetrization operator $\widehat{\mathcal{A}}$:

$$\Psi_0^{AB} = \widehat{\mathcal{A}} \Psi_0^A \Psi_0^B = N_{AB} \sum_Q (-1)^q \, Q \Psi_0^A \Psi_0^B \tag{2.82}$$

where $Q$ are the permutations of electrons between molecules (there are $(N_A + N_B)!/N_A!N_B!$ such permutations), $q$ is the parity of the permutation $Q$, $N_{AB}$ is the normalization factor, which includes the overlap integrals. In the first order of perturbation theory:

$$\begin{aligned} E_{int}^{(1)} &= \left\langle \Psi_0^{AB} \left| V_{AB} \right| \Psi_0^{AB} \right\rangle \\ &= N_{AB}^2 \left\langle \sum_Q (-1)^q \, Q \Psi_0^A \Psi_0^B \left| V_{AB} \right| \sum_Q (-1)^q \, Q \Psi_0^A \Psi_0^B \right\rangle \end{aligned} \tag{2.83}$$

To calculate the matrix elements over antisymmetric functions, it is sufficient to put the antisymmetric function only in one side of the matrix element [37]:

$$\begin{aligned} E_{int}^{(1)} &= N_{AB}^2 \left\langle \Psi_0^A \Psi_0^B \left| V_{AB} \right| \sum_Q (-1)^q \, Q \Psi_0^A \Psi_0^B \right\rangle \\ &= N_{AB}^A \left\{ \left\langle \Psi_0^A \Psi_0^B \left| V_{AB} \right| \Psi_0^A \Psi_0^B \right\rangle \right. \\ &\quad \left. + \left\langle \Psi_0^A \Psi_0^B \left| V_{AB} \right| \sum_{Q \neq I} (-1)^q \, Q \Psi_0^A \Psi_0^B \right\rangle \right\} \end{aligned} \tag{2.84}$$

The first term in braces represents the direct electrostatic energy, $E_{el}^{(1)}$, while the second term corresponds to the exchange energy:

$$E_{exch}^{(1)} = \left\langle \Psi_0^A \Psi_0^B \left| V_{AB} \right| \sum_{Q \neq I} (-1)^q \, Q \Psi_0^A \Psi_0^B \right\rangle \tag{2.85}$$

Direct calculation of the exchange energy for two hydrogen atoms gives the exponential distance dependence. As was shown by Böhm and Alrichs [38], the exchange energy for two many-electron atoms with closed shells can also be approximated by a simple exponential function: $A \exp(-\alpha R)$. Thus, the exchange energy approaches exponentially to zero with the increasing distance between the atoms. The ratio $E_{ex}^{(1)}/E_{el}^{(1)}$ also falls down exponentially with the distance, but at short distances exchange forces dominate. Taking them into account is also important at intermediate distances. The ratio $E_{ex}^{(1)}/E_{el}^{(1)}$ obtained via the first-order perturbation theory for water [39] and ethylene [40] dimers, and via the variational procedure for the hydrogen molecule [41] is presented in Table 2.9. For the ethylene dimer, the exchange contribution is larger than the electrostatic contribution up to $R = 7$ bohrs. For $(H_2O)_2$, the same situation is observed up to $R \leq 5$ bohrs. For

**Table 2.9** Ratio $E_{exch}^{(1)}/E_{el}^{(1)}$ for the water dimer [39], the ethylene dimer [40] and the hydrogen molecule [41]

| *R*, *bohr* | 4 | 4.8 | 5 | 5.2 | 6 | 7 | 8 | 9 | 10 | 16 |
|---|---|---|---|---|---|---|---|---|---|---|
| $(H_2O_2)^a$ | 2.45 | 1.47 | – | 1.08 | – | 0.107 | – | – | – | – |
| $(C_2H_4)_2^b$ | 2.05 | – | 3.87 | – | 2.69 | 1.03 | 0.27 | – | 0.042 | 0.005 |
| $H_2^c$ | – | – | – | – | – | – | 0.45 | 0.18 | 0.15 | – |

[a]Distance refers to the separation between the oxygen atoms.
[b]Distance refers to the separation between centers of mass.
[c]The ratio $E_{exch}^{(1)}/E_{el}^{(1)}$ is evaluated by means of the variational procedure with all the perturbation corrections taken into account.

the hydrogen molecule, the total exchange contribution constitutes approximately 50% of the electrostatic interaction even at $R = 8$ bohrs.

The exchange interaction is positive (repulsive) only for interacting systems with closed electronic shells (subshells), such as noble gas atoms, alkaline earths atoms (beryllium, magnesium, calcium, etc.), molecules in the ground state with coupled spins ($S = 0$). These systems do not have valence electrons. If an atom has an unpaired spin (hydrogen, lithium, etc.), the exchange energy is usually negative and contributes to the formation of molecules. The well-known *covalent bond* is to a great extent created by the exchange interaction. Thus, the approximation of the exchange part of interaction in some model potentials (the Buckingham potential and its modifications, see Chapter 5) by the positive exponential function is not appropriate for systems with free valence electrons forming the covalent compounds.

In the second order of perturbation theory, the exchange and polarization energy cannot be separated, only the exchange corrections to the polarization energy can be calculated. The exchange – polarization energy separates into a sum of the exchange – induction and exchange – dispersion contributions:

$$E_{exch}^{(2)} = E_{exch-ind}^{(2)} + E_{exch-disp}^{(2)}$$

It is important to note that, when the exchange of electrons is taken into account, the standard perturbation theory in the Rayleigh–Schrödinger form becomes incorrect, since the zeroth-order functions (Equation (2.82)) are not eigenfunctions of the zeroth-order Hamiltonian ($H_0 = H_A + H_B$), only the first-order correction (Equation (2.84)), is valid. Beginning from the first paper by Eisenschitz and London [42], numerous different versions of the exchange perturbation theory (its abbreviated name is SAPT – the Symmetry-Adapted Perturbation Theory) has been developed (see Section 3.2.1.2).

The importance of second-order corrections arising from the electron exchange was demonstrated by Murrell and Shaw [43] for the interaction of systems with uncoupled spins by calculating the dispersion energy for the system of two hydrogen atoms in the lowest triplet state, both with and without exchange. In the last case

**Table 2.10** Contribution of the exchange energy to the dispersion energy for the hydrogen–hydrogen system [43]

| $R$, bohr | $C_6/R^{6a}$ | $E_{exch-.disp} = \widetilde{C}_6/R^6$ | $\widetilde{C}_6$ |
|---|---|---|---|
| 4 | $0.0^2 1586$ | $0.0^3 571$ | 2.340 |
| 5 | $0.0^3 416$ | $0.0^3 262$ | 4.089 |
| 6 | $0.0^3 139$ | $0.0^3 114$ | 5.349 |
| 7 | $0.0^4 55$ | $0.0^4 51$ | 6.015 |
| 8 | $0.0^4 25$ | $0.0^4 24$ | 6.296 |
| 9 | $0.0^4 12$ | $0.0^4 12$ | 6.398 |

[a]The exact value of the dispersion constant $C_6$ is 6.499 a.u.

they obtained a significant overestimate in the dispersion energy for $R < 6$ bohrs (Table 2.10). In particular, the error is approximately of the order of 60% at $R = 5$ bohrs. In the region of the van der Waals minimum, $R \simeq 8$ bohrs, the error becomes considerably smaller, being of the order of ~4%, but this distance is unusually large for the van der Waals minimum. The values for the coefficient $\widetilde{C}_6$, defined formally by the relation $E_{exch.disp} = \widetilde{C}_6/R^6$, are also presented in Table 2.10. Such a definition neglects the exchange interactions, the contribution of higher multipoles and the errors of the multipole expansion in the overlap region of atomic wave functions. So, it is not surprising that instead of a constant value of $C_6$, a strong dependence on $R$ is obtained.

For closed-shell systems, the exchange contributions to the dispersion energy in the second-order of the perturbation theory is quite small. In Table 2.11 the different energy interaction components are presented for $(He)_2$ at the van der Waals minimum calculated by SAPT [44]. The exchange energy, $E^{(1)}_{exch}$, is much larger than the electrostatic energy, $E^{(1)}_{el}$. The latter is natural because helium atoms do not possess any multipole electric moments. The exchange contribution to the dispersion energy, $E^{(2)}_{exch-disp}$, is of the order of 3% of the pure dispersion energy, $E^{(2)}_{disp}$. However, the exchange contribution to the induction energy, $E^{(2)}_{exch-ind}$, is very large and constitutes about 82% of $E^{(2)}_{ind}$. It is evident that it cannot be neglected even in qualitative calculations [45]. Note also that the exchange contributions to

**Table 2.11** Contributions of the different energy interaction components to the interaction energy of the helium dimer evaluated at the van der Waals minimum $R_0 = 5.6$ bohrs [44] (energy in $K$)

| $E^{(1)}_{el}$ | $E^{(1)}_{exch}$ | $E^{(2)}_{ind}$ | $E^{(2)}_{exch-ind}$ | $E^{(2)}_{disp}$ | $E^{(2)}_{exch-disp}$ |
|---|---|---|---|---|---|
| −1.560 | 11.251 | −0.252 | 0.207 | −17.098 | 0.514 |

the negative induction and dispersion energies are positive; so, the account of the exchange diminishes the absolute value of the polarization energy.

The total exchange contribution to interaction energy grows rapidly as the molecules approach. At short distances, the interacting molecules can no longer be considered as separated objects. They form a single quasi-molecule (or supermolecule) that may be studied using a variational approach, which gives the total energy, $E_{tot}$, of the system. For a system of two molecules $A$ and $B$, the interaction energy is defined as a difference:

$$E_{int} = E_{tot} - (E_A + E_B) \tag{2.86}$$

In these calculations is very important to properly take into account the so-called Basis Set Superposition Error (BSSE) (Section 3.2.2.2).

Neglecting all terms which contain the exchange electron density gives the energy of the Coulomb interaction, $E_{Coul}$, which includes all the multipole interactions. Then the exchange interaction energy is given by the difference (see Equation (1.24)).

For a diatomic molecule with two valence electrons, the Coulomb and the exchange energies are usually defined by means of the energies of the lowest singlet and triplet terms [46]:

$$E_{Coul} = \frac{1}{2}\left[E\left({}^1\Sigma_g^+\right) + E\left({}^3\Sigma_u^+\right)\right] \tag{2.87}$$

$$E_{exch} \equiv \Delta E^{ST} = \frac{1}{2}\left[E\left({}^1\Sigma_g^+\right) - E\left({}^3\Sigma_u^+\right)\right] \tag{2.88}$$

For an atom, the energies determined by Equations (2.87) and (2.88) coincide exactly with the atomic Coulomb and exchange energies. However, this is not true for molecules due to the presence of the overlap integrals in the normalization constant (compare with Equation (2.80)), although the main contributions to Equations (2.87) and (2.88) are given by the Coulomb and the exchange energies, respectively.

In the first-order perturbation treatment, the ground state of the hydrogen molecule is described by the Heitler–London wave function. At large $R$, the Heitler–London function provides qualitatively the correct asymptotic expression for the ground-state energy. Hence, it can be expected that it would approximate $\Delta E^{ST}$ (equal according to Equation (2.88) to $E_{exch}$) more exactly when the distance $R$ is larger. However, as was shown by Herring [47], the Heitler–London function leads to an incorrect asymptotic behavior of $\Delta E^{ST}$. The correct asymptotic expression has the following form [48]:

$$\Delta E_{as}^{ST} = -0.82R^{5/2}e^{-2R} + 0\left(R^2e^{-2R}\right) \tag{2.89}$$

whereas in the Heitler–London approach one obtains:

$$\left(\Delta E_{as}^{ST}\right)_{HL} = \left\{-\frac{28}{45} + \frac{2}{15}(C + \ln R)\right\} R^3e^{-2R} + 0\left(R^2e^{-2R}\right) \tag{2.90}$$

where $C = 0.5772$ is the Euler constant.

**Table 2.12** Singlet–triplet splitting $\Delta E^{ST}$ of two hydrogen atoms in different approximations (energy in hartrees)

| $R, a_0$ | $E^{ST}_{var}$[50] | $(\Delta E^{ST}_{as})_{HL}$ | $\Delta E^{ST}_{as}$ | $6.5\ R^{-6}$ |
|---|---|---|---|---|
| 7 | $0.0^4968$ | $0.0^482$ | $0.0^488$ | $0.0^455$ |
| 8 | $0.0^4176$ | $0.0^415$ | $0.0^417$ | $0.0^425$ |
| 9 | $0.0^5312$ | $0.0^528$ | $0.0^530$ | $0.0^412$ |
| 10 | $0.0^6537$ | $0.0^649$ | $0.0^654$ | $0.0^565$ |
| 11 | $0.0^791$ | $0.0^784$ | $0.0^792$ | $0.0^537$ |
| 12 | $0.0^716$ | $0.0^714$ | $0.0^715$ | $0.0^522$ |
| 15 | – | $0.0^{10}58$ | $0.0^{10}67$ | $0.0^657$ |
| 20 | – | $0.0^{14}50$ | $0.0^{14}62$ | $0.0^610$ |
| 50 | – | $0.0^{39}11$ | $0.0^{39}54$ | $0.0^942$ |
| 60 | – | $-0.0^{49}11$ | $0.0^{47}18$ | $0.0^914$ |

$0.0^n968 = \underbrace{0.0\cdots0}_{n}968$

At $R > 60$ bohrs, the Heitler–London energy for the ${}^3\sum_u^+$ state becomes even smaller than the energy of the ground singlet state ${}^1\sum_g^+$. Such behavior is due to the electron correlation, which is not properly accounted for in the Heitler–London function. Even the simplest function, constructed from 1s-orbitals and taking into account ionic terms (the so called Weinbaum function), gives the nonintersecting singlet and triplet terms. The calculation of the singlet-triplet splitting in the hydrogen molecule, taking into account the superposition of excited and ionic configurations and carried out at $R = 8$ bohrs [49], significantly improves the agreement with the exact variational calculation by Kołos and Wołniewicz [50].

The values of $\Delta E^{ST}$, obtained by the asymptotically correct formula (Equation (2.89)), and by the Heitler–London expression (Equation (2.90)), are presented in Table 2.12. For comparison, the value of the dipole–dipole term of the dispersion energy and the accurate values of $\Delta E^{ST}_{var}$ [50] are also given. As observed in Table 2.12, the deviations in the Heitler–London approximation are not so large at $R = 7$ to 12 bohrs, where $\Delta E^{ST}$ is not too small. However, considerable deviations appear at $R \gtrsim 20$ bohrs, where $\Delta E^{ST}$ becomes substantially smaller than the dispersion energy. At distances $R > 30$ bohrs, the contribution due to the magnetic spin–spin interactions becomes appreciable also. It should be noted that at distances $R \geq 12$ bohrs the hyperfine interactions within the hydrogen atom become greater than the exchange interactions, and the molecular terms of the hydrogen molecule may not be classified according to the total electronic spin, since the nature of the coupling is changed. Thus, at large distances $\Delta E^{ST}$ does not determine the actual splitting of terms. It should be concluded that study of the asymptotic behavior of the singlet–triplet splitting is only of a methodical interest.

In ion–atom or ion–molecule collisions, the charge transfer has a large cross section. In this process, a valence electron is transferred from a neutral atom to an ion. The charge–exchange cross section is determined by the exchange splitting of the symmetric (gerade) and antisymmetric (ungerade) term. For $H_2^+$, this exchange splitting equals:

$$\Delta E_{exch} = \frac{1}{2}\left[E\left({}^2\Sigma_g^+\right) - E\left({}^2\Sigma_u^+\right)\right] \tag{2.91}$$

Holstein [51] and Herring [47] obtained a general formula for the exchange energy, $\Delta E_{exch}\left(H_2^+\right)$. The Holstein–Herring formula was studied by many authors; an interesting discussion on this topic can be found in Reference [52].

Electron transfer may also occur in an interaction between neutral molecules [53, 54]. This transfer occurs when the two potential curves, which at $R = \infty$ correspond to neutral molecules, $A$ and $B$, and to ions, $A^+$ and $B^-$, approach each other at intermediate separations (the potential curves may intersect in the absence of interaction). The effects of the electron transfer were detected in studies on collisions of alkali metal atoms with halogen molecules and some other molecules with a large electron affinity. The energy necessary for electron transfer comes not only from the initial kinetic energy of the reactants, but also from the energy of the ionizing dissociation process [55].

Consideration of the electron transfer may lead to the stabilization of a system. In the zeroth-order interaction treatment, a degeneracy occurs in the region, where the terms, $AB$ and $A^+B^-$, intersect. Hence, the wave function of the system may be written as the superposition of two antisymmetric functions corresponding to neutral and ionic states, respectively:

$$\Psi_0 = \Psi_0(AB) + \alpha\Psi_0\left(A^+B^-\right)$$

where the coefficient $\alpha$ determines the contribution of the ionic states.

When molecule $A$ has a low ionization potential and molecule $B$ has a large electron affinity, the contribution of the charge–transfer states may lower appreciably the total energy of the system. This promotes the formation of stable donor–acceptor complexes [55]. However, the largest contribution to the stabilization energy of such complexes comes from the electrostatic interactions [56]. Quinone–hydroquinone, $I_2$–benzene and complexes between conjugate double-bond molecules are examples of such compounds.

## 2.5 Retardation Effects in Long-Range Interactions and the Influence of Temperature

All interactions discussed in the preceding sections are considered as instant. Retardation due to the finite velocity of light has not been taken into account. However, at large distances between molecules, retardation effects may be of a great importance. In the case of dispersion forces, retardation effects qualitatively change their dependence on the distance. The retardation effects become significant when the distance between the molecules becomes comparable to the wave lengths ($\lambda$) of the molecular transition from the ground to excited states.

As mentioned in the historical survey (Chapter 1), the effect of retardation on dispersion forces was first considered in the coagulation theory of colloids. Colloidal particles usually carry a charge that attracts ions of the opposite sign in solution. As a result, each particle is surrounded by a double electric layer. These double layers repel each other when the particles get closer. The van der Waals attraction forces[3] compete with this Coulomb repulsion: a decrease in the thickness of the double layer leads to the dominance of the attractive forces. In this case, the colloid coalesces and precipitates (gel formation). The experiments [57] had demonstrated that to explain the regularities observed, it was necessary to assume that there is some weakening of the London dispersion forces at $R \gtrsim 400$ bohrs. Overbeck [57] supposed that the weakening is due to the retardation effect, related to the finite time of propagation of the interaction. This hypothesis stimulated the theoretical study by Casimir and Polder [35], who calculated the dipole–dipole interaction taking into account the retardation effect. Their calculation up to the fourth order of PT dealt with the interaction of a molecule with an electromagnetic field.[4] The expression obtained has the following form (in atomic units):

$$E_{ret}(dd) = -\frac{4}{\pi} \sum_{m,n} \left|d_{n0}^{A}\right|^2 \left|d_{m0}^{B}\right|^2 \omega_{n0}^{A}\omega_{m0}^{B}$$
$$\times \int \frac{u^4 e^{-2uR}}{R^2\left(\omega_{n0}^{A} + u^2\right)\left(\omega_{m0}^{B} + u^2\right)} \left\{1 + \frac{2}{uR} + \frac{5}{u^2R^2} + \frac{6}{u^3R^3} + \frac{3}{u^4R^4}\right\} du \tag{2.92}$$

It may be expressed in a closed form by means of integral sine or cosine [60, 61]. If $R$ is much larger than the reduced averaged wavelength of excitation, $\bar{\lambda} = \bar{\lambda}/2\pi$, then from Equation (2.92) the asymptotic Casimir–Polder formula follows:

$$E_{ret.as} = -\frac{23}{4\pi} \frac{\alpha_1^A(0)\,\alpha_1^B(0)}{\alpha R^7}, \qquad R \gg \bar{\lambda} \tag{2.93}$$

where $\alpha = e^2/\hbar c \simeq 1/137$ is the fine structure constant and $\alpha_1^A(0)$ is the static dipole polarizability of molecule $A$.

At distances $R \gtrsim \bar{\lambda}$, the fluctuations of the vacuum electromagnetic field become an important factor. At these distances, in addition to the fluctuating dipole moments of interacting systems that are responsible for the nonretarded dispersion interaction (see Section 2.3), it is necessary to take into account the dipole moments induced by the fluctuations of the vacuum electromagnetic field at the points of location of interacting systems. The space correlations of the fluctuating moments induced by

---

[3] In spectroscopy and condensed matter sciences, the van der Waals forces usually mean the attractive dispersion forces. In general case, the van der Waals forces can include, in addition to the dispersion forces, the induction attractive forces and the electrostatic attraction.

[4] More direct derivation of the Casimir–Polder formula was given later by McLachlan [58]. Dzyaloshinskii [59] obtained it with the help of quantum field theory methods (see also References [60, 61]).

the vacuum field are the main mechanism of the retardation dispersion interaction. In the limit $R \gg \bar{\lambda}$, the dispersion forces are completely conditional upon the vacuum field fluctuations. In this region, the term proportional to $R^{-6}$ vanishes due to the contribution of transverse photons, and only the term proportional to $R^{-7}$ remains. Thus, the Casimir–Polder asymptotic forces are not corrections to the London dispersion forces $\sim R^{-6}$; in the region $R \gg \bar{\lambda}$, the latter is replaced by Equation (2.93).

The simplified picture for the weakening of dispersion interaction due to the retardation can be described in the following manner. The field of the instantaneous dipole moment, $\mathbf{d}^A$, of molecule $A$ reaches molecule $B$ at time $R/c$ and induces a dipole moment, $\mathbf{d}^B$, which interacts with $\mathbf{d}^A$ after a time $2R/c$. During this time, $\mathbf{d}^A$ can change its direction; in particular, it can rotate in such direction that would lead to a vanishing interaction. Thus, it is natural that the value of the retarded interaction is less than the value of the instantaneous one.

Let us estimate the magnitude of $\bar{\lambda}$. It can be expressed via the mean energy of excitation of interacting molecules $\overline{\Delta E_{ex}}$ as $\bar{\lambda} = \hbar c/\overline{\Delta E_{ex}}$, or in atomic units, $\bar{\lambda} = 137/\overline{\Delta E_{ex,hartree}}$. For helium, $\overline{\Delta E_{ex}} = 1.14$ hartree and $\bar{\lambda} = 120$ bohr. For the Lyman transition in hydrogen, $\bar{\lambda} = 245$ bohr, while for organic molecules, $\bar{\lambda}$ is larger, of the order of 500 bohr. Thus, $R \gg \bar{\lambda}$ corresponds to $R \geq 2.10^3$ Å or 0.2 μm. At these distances, the Casimir–Polder law, $\sim 1/R^7$, is fulfilled very well. It should be recalled that in condensed media, the speed of light is slower and retardation effects come in at smaller distances.

Expression (2.93) is only the first term of an asymptotic expansion in the powers $\bar{\lambda}/R$. The next terms are proportional to $\bar{\lambda}^2 R^{-9}$, $\bar{\lambda}^4 R^{-11}$, and so on. For the helium–helium interaction, the first terms of the asymptotic expansion were calculated [62] and are:

$$E_{ret.as} = -3.48\,\alpha^{-1}R^{-7} + 21.51\,\alpha^{-3}R^{-9} - 395.3\,\alpha^{-5}R^{-11} + \cdots. \tag{2.94}$$

At $R = 7.5\,\bar{\lambda}$, inclusion of only the Casimir–Polder term leads to an error of ~10% by comparison with the exact formula (Equation (2.92)). Inclusion of two terms gives an error of 2%, and for three terms an error less than 1% is obtained.

* *
*

The definitions of different kinds of intermolecular forces follow from quantum-mechanical perturbation theory and do not contain temperature. As discussed in Section 2.1.3, direct electrostatic interactions between *non-fixed* molecules depend upon temperature via the Boltzmann factor and vanishes at $kT \gg V_{int}$. It is important to know the influence of temperature on the dispersion interactions [36, 63, 64].

The temperature dependent potential can be obtained from the expression for the Helmholtz free energy $F(R, T)$. Boyer [36] from the classical viewpoint and recently Ninham and Daicic [63] in terms of quantum electrodynamics studied both the nonretarded and retarded regimes. In spite of the different approaches, they came to the similar main conclusions. In the nonretarded regime:

$$R \ll \bar{\lambda} = \frac{\hbar c}{\overline{\Delta E}_{ex}} \tag{2.95}$$

the London expression is valid for all $T$ with only small temperature corrections. In contrast, in the retarded regime:

$$R \gg \frac{\hbar c}{\overline{\Delta E}_{ex}} \tag{2.96}$$

the Casimir–Polder expression (2.93) is a correct asymptotic result only at $T = 0$.

At finite $T$ there is an interplay of retardation and temperature effects and the condition for separation includes also temperature. Namely, for:

$$\frac{kTR}{\hbar c} \gg 1 \quad \text{or} \quad R \gg \frac{\hbar c}{kT} \tag{2.97}$$

the leading term in the dispersion interaction has the form [36, 64]:

$$E_{disp}(R, T) = -3\,kT\frac{\alpha_1^A(0)\,\alpha_1^B(0)}{R^6} \tag{2.98}$$

For any nonzero finite temperature, the separation $R$ can be chosen sufficiently large such that the condition (2.97) is satisfied. Hence, at finite temperatures, the correct asymptotic expression has the $R^{-6}$ dependence and the Casimir–Polder formula is not valid.

The change of the distance dependence of the retarded dispersion forces on a more slowly decreasing one indicates that the simplified explanation of the more rapid decrease of the dispersion interaction at large distances presented above is not valid when the condition (2.97) is satisfied. The thermal (blackbody) radiation with the average energy per normal mode equal to $kT$ dominates the zero-point radiation and changes the distance dependence [36]. At sufficiently large separation, the system should show a classical behavior even at low temperatures. From the thermodynamic viewpoint, the temperature dependence of the dispersion interaction has a pure entropic origin [64, 65].

Nevertheless, for each system with given $\overline{\Delta E}_{ex}$, some interval of separation exists for which the retardation condition (2.96) is fulfilled but the condition (2.97) is changed to the opposite, that is:

$$\frac{\hbar c}{\overline{\Delta E}_{ex}} \ll R \ll \frac{\hbar c}{kT} \tag{2.99}$$

For these separations, the Casimir–Polder $R^{-7}$ dependence can be expected. In this case, a curious situation occurs: with increasing distance, the nonretarded $R^{-6}$ potential goes over to the retarded $R^{-7}$ potential at the distances in the interval (2.99) and with a further increase in separation, it comes, at last, back to the $R^{-6}$ distance dependence. The similar situation was first revealed by Lifshitz [66] in his theory of interaction between macroscopic bodies (see Section 2.7).

Because of the large magnitude of $\overline{\Delta E}_{ex}$ in comparison with $kT$, the condition (2.99) is not exotic. For instance, for organic molecules, $\overline{\Delta E}_{ex} \simeq 7eV = 7.91 \times$

$10^4 K$ and the right-hand side of the condition (2.99) can be easily fulfilled at room temperature:

$$10^{-2}\frac{\hbar c}{800K} \ll R \ll \frac{\hbar c}{300K} = 7.6 \times 10^{-4} \text{ cm} \tag{2.100}$$

For the same reason, the high-temperature limit:

$$kT \gg \overline{\Delta E}_{ex} \tag{2.101}$$

discussed in References [36, 64, 67] cannot be practically realized. For organic molecules having the least $\overline{\Delta E}_{ex}$, it corresponds to temperatures about 500 000 K. At these temperatures molecules do not exist and atoms are stripped to a great extent and converted to multicharged ions.

* *
*

As discussed in the beginning of this section, the idea of taking into account the retardation in intermolecular interaction was suggested by Overbeck [57] in his studies of colloid suspensions. The classical Derjaguin–Landau [68] and Verwey–Overbeck [57] theory, usually denoted as the DLVO theory, is based on the pairwise approach and linear screening approximation of the Poisson–Boltzmann equation. The DLVO theory predicts that an isolated pair of charged colloidal microspheres is characterized by a pure repulsive screen Coulomb interaction at large separations. Although this prediction was confirmed in experiments [69, 70], the DLVO theory has been frequently questioned both on theoretical [71] and experimental [72–74] grounds. Besides the restriction to linear screening, the theory neglects double-layer fluctuations and many-body interactions. These factors are important at high volume fractions [72]. In spite of the long history of the problem, the interaction of colloidal particles still presents a challenge for theorists and experimenters.

The last statement is supported by recent publications about measurements in several laboratories, in which a strong long-range attraction between colloidal microspheres confined to a plane by charged glass walls have been revealed [75–78]; as well as a strong attraction between like-charged latex spheres leading to formation of colloidal crystals [79]. The measured magnitude of attraction was about 0.1–1.5 $kT$ for different particle diameters and distances between planes, and the interparticle separation corresponded to the micron range. This attraction does not follow from the DLVO theory that predicts only repulsion. As was rigorously proved [80, 81], the Poisson–Boltzmann equation for the potential between like-charged spheres in a salt solution does not admit attractive interaction even under very general conditions. On the other hand, the conventional dispersion interaction cannot provide the observed attraction, because its contribution to the interaction potential of colloidal microspheres at the separations in the micron range is less than 0.01 $kT$ [77].

One possible explanation of the attraction between like-charged colloidal spheres confined with charged walls has been suggested in Reference [82]. It was demonstrated analytically, and by Brownian dynamic simulation, that beyond some critical separation hydrodynamic coupling due to the electrostatic repulsion from the charged wall overcomes the mutual repulsion of microspheres, so that the separation between microspheres decreases. But this mechanical theory does not explain the attraction far from walls. Nor does it apply to previous experiments [75–78], since the suspensions in these experiments were in equilibrium and the spheres fluctuated about the equilibrium position between the two walls.

The original theory of attraction between particles in fluids based on hydrodynamic thermal fluctuations was proposed by Ivlev [83]. He studied two possible mechanisms that could lead to an attraction between two microbodies placed in a hydrodynamic medium: an attraction stemming from the hydrodynamic fluctuation of sound waves and an attraction based on the dependence of the effective masses of the microbodies on their separation. The derivation was performed for uncharged particles like hard spheres. The suggested new mechanism of attraction has to be taken into account in studies of neutral-particle interactions in fluids and charge-particle interactions in colloidal suspensions, although the latter at distance ranges smaller than the micron range. The point is that the hydrodynamic attraction [83] is efficient at relatively small distances, $(R - \delta)/\delta \lesssim 0.1$, where $\delta$ is the diameter of microspheres and $R$ is the center-to-center distance. At larger distances, hydrodynamic attraction is suppressed by Coulomb repulsion. The latter may be the reason that in experiments without confinement only a pure repulsion was observed [69, 70].

At present, it is not clear how confinement can influence the electromagnetic or hydrodynamic fluctuation mechanisms of attraction and increase its magnitude. On the other hand, the theory of attraction caused by some hydrodynamic nonequilibrium effects [82, 84] cannot be applied for an explanation of experiments [75–78], since the suspensions in these experiments were in equilibrium. The latter was confirmed by a thermodynamic self-consistent check in a recent experiment [85], where the confinement-induced attraction between like-charged colloidal spheres was again observed. Thus, the confinement-induced attraction experimentally observed [75–78, 85] still cries out for an explanation.

For isolated atoms and molecules, the magnitude of the van der Waals forces is very small. It is very difficult to measure their distance dependence and is even more difficult to measure the effect of retardation in intermolecular interactions. Nevertheless, the effect of retardation was recently detected in very precise spectroscopic measurements of the high Rydberg states in the helium atom [86], although in respect to the induction, but not to the dispersion forces. This remarkable experiment is described briefly below.

The idea of the experiment, in which retardation effects in microscopic atomic interactions can be revealed, was suggested in the theoretical paper by Kelsey and Spruch [87]. The authors considered the interaction between the remote Rydberg electron and the rest of the atom. In this case in the expansion of the electrostatic energy, there is the main Coulomb term $\sim 1/R$, but the second is the monopole–dipole induction term caused by the polarization of the atomic core by the Rydberg electron. According to Equation (2.58):

$$E_{ind}^{(2)}(qd) = -\frac{1}{2R^4} e^2 \overline{\alpha}_1^{A^+}(0) \tag{2.102}$$

If the charged particle is at rest, there is no retardation of the induction interaction. The authors [87] calculated the correction to the monopole–dipole induction interaction at large distances taking into account the motion of the Rydberg electron. In contrast to dispersion interactions, where at sufficiently large distances the nonretarded dispersion interactions vanish due to the contribution of the transverse photon, for induction interactions the nonretarded monopole–dipole term persists as the relativistic theory is introduced but has to be supplemented by some additive correction. The correction due to the exchange of two photons (at least one of which is transverse, travelling with a speed $c$) between the Rydberg electron and the atomic core has the following leading term [87]:

$$E_{ret}\,(qd) = \frac{11}{4\pi}\frac{\hbar e^2}{mc}\frac{\overline{\alpha}_1^{A^+}(0)}{R^5} \tag{2.103}$$

In the simplest case of the helium atom, Equation (2.103) is valid at distances corresponding to the Rydberg states with $n \gtrsim (1/\alpha)^{\frac{1}{2}}$ where $n$ is the principal quantum number and $\alpha$ is the fine structure constant.

Although the contribution of $E_{ret}$ is much smaller than from the $R^{-4}$ term (Equation (2.102)), it was observed in very precise spectroscopic experiments with the high Rydberg states in the helium atom by Hessel *et al.* [86]. The measurement of energy intervals between the Rydberg states in helium with $n = 10$ and $L = 4$–$8$ with precision $10^{-13}$ a.u. (!) confirmed the contributions from 'retardation forces' to better than 10%.

Measurements of the retardation effects in the interaction between macroscopic bodies is discussed in Section 2.7.

## 2.6 Relativistic (Magnetic) Interactions

In the preceding section, the modifications in the dispersion interaction due to the retardation effect at large distances between molecules were discussed. These modifications are related to the finite velocity of light and have a relativistic nature. However, the relativistic effects are revealed not only at large distances but at shorter distances, $R < \lambda\!\!\!^{-}$, as well. They are related, first of all, to magnetic interactions, since the magnetic moment possesses a relativistic nature.

A system of moving charges is characterized by *magnetic* multipole moments. To obtain analytical expressions for these moments, the vector potential of the magnetic field created by the system of moving charges in the Taylor series should be decomposed, as was done in Section 2.1 for electrostatic potentials created by a system of charges. The first term in this expansion is the dipole magnetic moment, which is usually called the *magnetic moment*. It is given by the following expression [1]:

$$\mathbf{M} = \frac{e}{2mc}\sum_i \mathbf{r}_i \times \mathbf{p}_i = \frac{e}{2mc}\mathbf{L} \tag{2.104}$$

where $\mathbf{p}_i$ is the linear momentum of the *ith* electron and **L** is the angular momentum of the system. At large distances the magnetic moment quite precisely characterizes the magnetic properties of the system.

In classical physics, according to Equation (2.104), a moving electron is characterized by the magnetic moment proportional to its angular momentum. In quantum mechanics, each electron possesses two kinds of magnetic moments–an *orbital magnetic moment*:

$$\mathbf{m}_{orb} = \frac{e}{2mc}(\mathbf{r} \times \mathbf{p}) \tag{2.105}$$

and a *spin magnetic moment*:

$$\mathbf{m}_{sp} = \frac{e}{mc}\hbar\mathbf{s} \tag{2.106}$$

where $\mathbf{p} = -i\hbar\nabla$ is the linear momentum operator, $\mathbf{r} \times \mathbf{p}$ is the orbital angular momentum, and $\hbar\mathbf{s}$ is the spin angular momentum of electron.

The energy of interaction of two magnetic moments is expressed in the same way as the energy of the dipole–dipole interaction (Equation (2.34)). For instance, the energy of spin–spin magnetic interaction is equal to:

$$E_{ss} = \frac{e^2\hbar^2}{m^2c^2}\left[\frac{\mathbf{s}_1 \bullet \mathbf{s}_2}{R^3} - \frac{3(\mathbf{s}_1 \bullet \mathbf{R})(\mathbf{s}_2 \bullet \mathbf{R})}{R^5}\right] \tag{2.107}$$

where **R** is a vector connecting the points of spin localization.

In 1957 in order to explain the parity in conserving interactions in weak decay, Zel'dovich [88] pointed out that particle with spin $\frac{1}{2}$ must possess, in addition to a magnetic moment, another dipole moment–the so-called *anapole moment* that corresponds to an odd parity magnetic distribution with the rank one. He proposed a toroidal solenoid as a macroscopic model for an anapole. Parity violating electron–nuclear and electron–electron interactions also give rise to anapole moments in atoms and molecules [89–92]. The anapole moments caused by parity nonconservation are proportional to the Fermi weak interaction constant and are very small in the comparison with the atomic scale.

Meanwhile, it was realized that the magnetic toroidal moments determined by the usual electromagnetic forces with particular symmetry conditions (e.g., the chiral symmetry) exist in crystals [93–97] and chiral molecules (molecules not possessing a center of symmetry) [98]. These toroidal moments have the normal atomic order of magnitude. As revealed by Dubovik *et al.* [93–95], the toroidal moments appear in the multipole expansion of the current density as do the well-known magnetic moments. It is a third family of multipole moments. The first member of this family, a toroidal dipole, can be represented as a solenoid folded into a torus and has the space–time symmetry similar to the anapole moment. The concept of toroidal moments proved to be useful in the study of the antiferromagnetic ordering in the second-order phase transitions in crystals without space inversion center [94, 95]. The detection of toroidal multipole moments in X-ray synchrotron experiments is discussed elsewhere [96, 97].

Relativistic corrections to the electrostatic energy can be obtained with the precision $O\left(\alpha^2\right)$, where $\alpha$ is the fine structure constant, by using the Breit–Pauli Hamiltonian [99]:

$$H = H_0 + \alpha^2 H_{rel} \tag{2.108}$$

where $H_0$ is the nonrelativistic Hamiltonian and $\alpha^2 H_{rel}$ is the relativistic correction to it, which for light atoms can be considered as a small perturbation. If the solution of the nonrelativistic problem:

$$H_0 \Psi_0 = E_0 \Psi_0 \tag{2.109}$$

is known, then in the first order of the perturbation theory, the energy of the system is given by:

$$E = E_0 + \alpha^2 \langle \Psi_0 \left| H_{rel} \right| \Psi_0 \rangle \tag{2.110}$$

where $E_0$ is the nonrelativistic energy of the system.

The Breit–Pauli Hamiltonian was derived for a two-electron atom by Bethe and Salpeter [99] and generalized for a molecular system in References [100–102]. It may be represented as a sum of physically relevant terms, which are usually denoted:

$$H_{rel} = H_{LL} + H_{SS} + H_{SL} + H_P + H_D \tag{2.111}$$

where (in atomic units):

$$H_{LL} = -\frac{1}{2} \sum_{k>j} \left(1/r_{jk}^3\right) \left[r_{jk}^2 \mathbf{p}_j \cdot \mathbf{p}_k + \mathbf{r}_{jk} \cdot \left(\mathbf{r}_{jk} \cdot \mathbf{p}_j\right) \mathbf{p}_k\right] \tag{2.112}$$

$$\begin{aligned} H_{SS} = \sum_{k>j} \Big\{ &- (8\pi/3) \left(\mathbf{s}_j \cdot \mathbf{s}_k\right) \delta^{(3)} \left(\mathbf{r}_{jk}\right) \\ &+ \left(1/r_{jk}^5\right) \left[\mathbf{r}_{jk}^2 \mathbf{s}_j \cdot \mathbf{s}_k - 3 \left(\mathbf{s}_j \cdot \mathbf{r}_{jk}\right) \left(\mathbf{s}_k \cdot \mathbf{r}_{jk}\right)\right] \Big\} \end{aligned} \tag{2.113}$$

$$\begin{aligned} H_{SL} = &\frac{1}{2} \sum_{\beta, j} \left(Z_\beta / r_{j\beta}^3\right) \left(\mathbf{r}_{j\beta} \times \mathbf{p}_j\right) \cdot \mathbf{s}_j \\ &- \frac{1}{2} \sum_{k>j} \left(1/r_{jk}^3\right) \left[\left(\mathbf{r}_{jk} \times \mathbf{p}_j\right) \cdot \mathbf{s}_j - 2 \left(\mathbf{r}_{jk} \times \mathbf{p}_k\right) \cdot \mathbf{s}_j\right] \end{aligned} \tag{2.114}$$

$$H_p = -\frac{1}{8} \sum_j p_j^4 \tag{2.115}$$

$$H_D = \frac{1}{2} \pi \left\{ \sum_{\beta, j} Z_\beta \delta^{(3)} \left(\mathbf{r}_{j\beta}\right) - 2 \sum_{k>j} \delta^{(3)} \left(\mathbf{r}_{jk}\right) \right\} \tag{2.116}$$

The various terms in $H_{rel}$ have the following significance:

$H_{LL}$ corresponds to the classical electromagnetic coupling of the electrons through the interaction of the magnetic fields created by their motion. It contains terms which couple the orbital magnetic moments of the electrons.

$H_{SS}$ gives the interaction between the spin magnetic moments of the electrons. The electron–electron Fermi-type contact term, involving the three-dimensional delta function, gives the behavior when $r_{jk} = 0$. The second terms, corresponding to the usual magnetic dipole–magnetic dipole interaction, is applicable when $r_{jk} \neq 0$.

$H_{SL}$ represents the spin–orbit magnetic coupling between electrons. The first term gives the interaction between the spin of an electron and the magnetic moment associated with its motion. The remainder of $H_{SL}$ gives the spin–other orbit coupling between the spin of one electron and the orbital magnetic moment of another.

$H_p$ is the relativistic correction due to the variation of mass with velocity.

$H_D$ is a term characteristic of the Dirac theory [62] which appears to have no obvious interpretation.

In Equation (2.111) the term related to the interaction with the nuclear spins and responsible for the so-called *hyperfine splitting* of levels has been omitted. Its contribution to intermolecular interactions is discussed at the end of this section. Nuclear coordinates in Equations (2.114) and (2.116) are assumed to be fixed in accordance with the adiabatic approximation, whose accuracy can be estimated as $\alpha^2/M$, where $M$ is the mass of the nuclei (see also Section 1.3).

The Breit–Pauli approximation is valid for systems with $\alpha Z \ll 1$, or $Z \ll 137$. In this case, the term $\alpha^2 H_{rel}$ may be considered as the perturbation and its contribution is calculated using the solution of the nonrelativistic problem. For large $Z$, the relativistic terms will not be small and the perturbation theory may not be applied. The relativistic Schrödinger equation should be solved. This can be done by variational methods. In the last two decades there has been essential progress in this direction. The Sucher formulation of the correct relativistic Hamiltonian [103, 104] was used for the calculation of energy levels and transition probabilities in heavy and even super-heavy atoms [105, 106]. The programs created were applied to some molecules with heavy atoms: AuH [107] and $SnH_4$ [108]; relativistic potentials for AuH, $AuH^-$ and $Au_2$ were also calculated [106]. For more recent achievements see References [109–112].

For light atoms, the Breit–Pauli approach (Equation (2.110)) is quite good. The precise calculation of the contributions of all five components in $H_{rel}$ (Equation (2.111)) to the ground state energy of the hydrogen molecule at small distances $0.6\ a_0 \leq R \leq 3.6a_0$, was performed by Kołos and Wołniewicz [113] (Table 2.13). At the equilibrium distance, $R_0 = 1.4$ bohrs, the total relativistic correction to the ground state energy is equal to $-2.398\ \text{cm}^{-1}$. This gives a correction to the dissociation energy of $-0.524\ \text{cm}^{-1}$. For the hydrogen molecule, these corrections are negligible, but for molecules consisting of heavier atoms they become larger [114].

Meath and Hirschfelder [62] obtained the multipole expansion for all the terms present in $H_{rel}$. The first terms of this expansion appear to decrease more slowly with $R$ than the first terms of the nonrelativistic expansion. For example, for the interaction of two atoms in nondegenerate states, for which the first-order electrostatic corrections to the energy vanish and only the dispersion energy remains, the

**Table 2.13** Values of the relativistic corrections to the ground-state energy of the hydrogen molecule [113] (energy in $cm^{-1}$)

| $R, a_0$ | 0.6 | 1.40 | 2.40 | 3.60 | $\infty$ |
|---|---|---|---|---|---|
| $E_{LL}$ | -1.074 | -0.559 | -0.270 | -0.092 | -0.000 |
| $E_{SS}$ | 3.765 | 1.244 | 0.383 | 0.083 | 0.000 |
| $E_{SL}$ | 0.000 | 0.000 | 0.000 | 0.000 | 0.000 |
| $E_P$ | -45.983 | -19.355 | -13.243 | -14.160 | -14.609 |
| $E_D$ | 37.379 | 16.273 | 11.082 | 10.668 | 11.687 |
| $E_{rel}$ | -5.913 | -2.398 | -2.048 | -2.502 | -2.922 |

expression for $R < \lambda\!\!\!^{-}$ and with accuracy of the order of $\alpha^2$ has the form:

$$E_{disp} = -\frac{C_6}{R^6} - \frac{C_8}{R^8} - \cdots + \alpha^2 \left[ \frac{W_4}{R^4} + \frac{W_6}{R^6} + \cdots \right] \tag{2.117}$$

The main relativistic correction to the London energy is given by the term $\alpha^2 W_4/R^4$, where the coefficient $W_4$ is given by $H_{LL}$ and, similar to the case of the coefficient $C_6$ [see Equation (2.63)], expressed via oscillator strengths and transition frequencies:

$$W_4 = \frac{1}{4} \sum_{n,m \neq 0} \frac{f_{n0}^A f_{m0}^B}{\omega_{n0}^A + \omega_{m0}^B} \tag{2.118}$$

But while the London dispersion energy is negative, the relativistic corrections are positive, which tends to reduce the nonrelativistic dispersion attraction. The numerical estimates of the interaction of noble gas atoms [62, 115] demonstrate that $C_6$ and $W_4$ are of the same order of magnitude and so the ratio of the leading terms in the relativistic and nonrelativistic parts of the interaction energy (Equation (2.117)) is $\alpha^2 R^2 W_4/C_6$. Hence, at large distances, the relativistic contribution to the interaction energy becomes important.

Let us estimate that ratio for the helium–helium interaction. According to Johnson *et al.* [115], $W_4/C_6 \simeq 0.45$ is obtained. Hence, at $R \simeq 100$ bohrs, the relativistic repulsion compensates a quarter of the dispersion attraction. It should be emphasized that Equation (2.117) is valid in the absence of retardation, i.e., at $R < \lambda\!\!\!^{-}$ (for helium–helium, $\lambda\!\!\!^{-} \sim 120$ bohrs). Higher corrections of $\alpha$ may be obtained from the exact electrodynamic expression. They are proportional to $\alpha^3/R^3$ and $\alpha^4/R^2$. According to the estimates obtained by Meath and Hirschfelder [62], the Breit–Pauli approximation (Equation (2.117)) differs from the exact formula by no more than 5% at $R < 0.6\,\lambda\!\!\!^{-}$.

For systems in degenerate states, the first order electrostatic corrections do not vanish. In these cases the interaction may be considered at all $R$ without taking into account the retardation [116]. At a qualitative level, it follows from the fact that

$\lambda\!\!\!^{-} = \hbar c/\Delta E = \infty$. For molecules with dipole electrostatic and magnetic moments:

$$E_{int} = \frac{C_3}{R^3} + \frac{\alpha^2 W_3}{R^3} + \cdots. \tag{2.119}$$

The terms, such as $H_{LL}$, $H_{SS}$, and $H_{SL}$ of the Hamiltonian (Equation (2.111)), give the contributions to the second term of Equation (2.119).

Meath [117] calculated the interaction energy of two hydrogen atoms in their ground states ($L_a = L_b = 0$) for different spin states of the whole system:

$$\begin{aligned} E_{int}\left({}^1\textstyle\sum_0\right) &= -\frac{6.50}{R^6} + \frac{0.46\alpha^2}{R^4} + \cdots \\ E_{int}\left({}^3\textstyle\sum_0\right) &= -\frac{6.50}{R^6} + \frac{0.46\alpha^2}{R^4} + \frac{\alpha^2}{R^3} + \cdots \\ E_{int}\left({}^3\textstyle\sum_1\right) = E_{int}\left({}^3\textstyle\sum_{-1}\right) &= -\frac{6.50}{R^6} + \frac{0.46\alpha^2}{R^4} - \frac{\alpha^2}{2R^3} \end{aligned} \tag{2.120}$$

The magnetic spin–spin interactions exist only for the triplet states. For the term ${}^3\sum_0$, the total relativistic energy becomes about 10% of the London energy already at $R \sim 22$ bohrs, and for the terms ${}^3\sum_{\pm 1}$ at $R \sim 30$ bohrs.

The spin–spin interactions dominate at $R > 100$ bohrs. Since these interactions are static, their form is preserved at $R > \lambda\!\!\!^{-}$. Hence, at large distances the term $\alpha^2 W_3/R^3$ becomes the leading one in the interaction energy of systems with vanishing dipole moments and nonvanishing spins. The hydrogen atom and alkali atoms, and the oxygen molecule as well, are examples of such systems.

According to the estimates of Chang [118], the contribution of the relativistic interactions to the energy of the term $O_g^+$ of the oxygen molecule for different levels is of the order of 10 to 20% at $R = 30$ bohrs and 35 to 60% at $R = 50$ bohrs for the quadrupole–quadrupole interaction. An account of the relativistic interactions in the study of interactions of electronic-excited states of molecules and atoms and, in particular, of the resonance interactions, is of great importance [102].

The term corresponding to the hyperfine interactions, $H_{SI}$, which describes the interaction between electron and nuclear magnetic moments in an atom (or molecule), is not included in the Hamiltonian (2.111). It might be thought that this term is not directly relevant to intermolecular interactions. However, it appears to be of importance at distances where the intermolecular interaction energy becomes of the same order of magnitude as the hyperfine splitting, since it leads to a change in the nature of the spin coupling. At such distances (which are of the order of $R \gtrsim 10$ bohrs for the interaction of two hydrogen atoms, and are larger for alkali metals), the electron spin ceases to be a good quantum number. The effect of the spin interactions on molecular terms is studied in detail for the interaction of two hydrogen atoms in Reference [119]. The effective Hamiltonian for spin states of two electrons with spins $\mathbf{s}_1$ and $\mathbf{s}_2$, and two protons with spins $\mathbf{I}_a$ and $\mathbf{I}_b$, may be represented as follows:

$$H = E_{Coul} - \left[\frac{1}{2} + 2\mathbf{s}_1 \cdot \mathbf{s}_2\right] E_{ex} + \left[\mathbf{s}_1 \cdot \mathbf{s}_2 - 3s_{1z}s_{2z}\right] \frac{\alpha^2}{R^3} + A\left[\mathbf{s}_1 \cdot \mathbf{I}_a + \mathbf{s}_2 \cdot \mathbf{I}_b\right] \tag{2.121}$$

where $E_{Coul}$ is the Coulomb interaction energy, $E_{ex}$ is the exchange interaction energy and $A$ is the hyperfine interaction constant (=0.047 cm$^{-1}$). At $R < 9$ bohr, the contribution of the hyperfine interaction is negligible and it is sufficient to take into account only the electron spins. At 9 bohr $< R <$ 12 bohr, all terms in Equation (2.121) are of importance. At $R = 12$ bohr, $E_{ex} = (1/12)A$, and the singlet–triplet splitting becomes smaller than the hyperfine splitting. Therefore, at $R \geq 12$ bohr, the molecular terms cannot be classified by means of the total electron spin as singlet and triplet ones. At such distances it is necessary to start from atomic states, described by an angular momentum, which is the sum of the electron and the nuclear spins. Similar results for HD and $D_2$ are obtained in Reference [120]. Consideration of the hyperfine interactions is important in the study of scattering processes with spin exchange, since the hyperfine interaction influences the intensity of the well-known in radiospectroscopy 21 cm line, the polarization in the EPR, the optical pumping in hydrogen masers and some other processes related to spin interactions.

## 2.7 Interaction Between Macroscopic Bodies

In this section we briefly discuss the interaction between macroscopic bodies including some recent experimental and theoretical developments; more details can be found in books [121–124].

Attractive forces similar to the van der Waals forces between atoms and molecules appear as macroscopic bodies approach each other. First, the interaction between macroscopic bodies was studied by Casimir [125]. He considered the zero-point energy of the vacuum field between two metallic plates and predicted that they should attract each other. At that time, the Casimir paper [125] did not draw great attention, in spite of his two subsequent publications [126, 127]. Much more attention was paid to the paper published in the same year by Casimir and Polder [35] where the effect of retardation of the dispersion interactions was introduced. The importance of the Casimir effect and his concepts of the vacuum field was fully realized later on in molecular physics and quantum electrodynamics [128, 129]. It became clear that the Casimir effect is extremely important in hadron physics, the supersymmetry physics, and even in cosmology [130].

Casimir [125] showed that two uncharged metallic plane conductors, placed parallel to each other in a vacuum and separated by the distance $L$ attract one another by a force per unit area:

$$F(L) = -\frac{\partial U(L)}{\partial L} = -\frac{\pi^2 \hbar c}{240 L^4} \tag{2.122}$$

This force arises from the interaction energy or potential:

$$U(L) = -\frac{\pi^2 \hbar c}{720 L^3} \tag{2.123}$$

Twenty-one years later, Brown and Maclay [131] developed a modern theoretical interpretation of the Casimir effect by means of the electromagnetic energy-stress tensor in the region between the plates; a further development in this direction see in paper by González [132].

A different approach to the interaction between macroscopic bodies was developed by Lifshitz [66]. The Lifshitz theory is based on the concept that the interaction is due to fluctuations of the electromagnetic field inside a body and beyond its boundaries. Such fluctuations always exist and have both a thermal and a quantum-mechanical nature. It is assumed that the distance between body surfaces is small, but significantly larger than the interatomic distances in the bodies. So, the latter can be treated as a continuum. In this case, the interaction force is determined by the unique macroscopic characteristic–the dielectric permittivity $\epsilon(\omega) = \epsilon'(\omega) + i\epsilon''(\omega)$; information about this can be obtained from independent spectroscopic measurements. The imaginary part of the dielectric permittivity is connected with spectral density of the oscillator strength $f(\omega)$ by the relationship that is well-known in spectroscopy:

$$\omega\epsilon''(\omega) = \frac{2\pi e^2}{m} n_0 f(\omega) \tag{2.124}$$

where $n_0$ is the number of molecules (atoms) per unit volume. The theory developed is applicable to arbitrary bodies, independent of their molecular structure. It automatically takes into account retardation effects since it is based on the exact equations of the electromagnetic field theory, but the nonretarded regime is also included.

The comparatively simple derivation of the Lifshitz equation for the interaction energy of two plates was proposed by van Kampen *et al.* [133]. The Lifshitz theory was generalized for a system of two macroscopic bodies divided by an arbitrary dielectric layer with the help of quantum-field methods [134]. In the particular case of two metallic plates in vacuum, the general Lifshitz expression turns to the Casimir formula (2.122) by approaching the dielectric constant to infinity. Parsegian and Ninham [135, 136] have applied the Lifshitz theory to the study of dispersion forces between biological membranes.

It is instructive to note that the expressions for the interaction of separate atoms (molecules) can be obtained from the general equations for the interaction of macroscopic bodies at the low density limit [66]. At this limit, the London formula with the coefficient $C_6$ (Equation (2.72)) is obtained at $R \ll \hbar c/\overline{\Delta E}_{ex}$ and the Casimir–Polder asymptotic formula (Equation (2.93)) is obtained at $R \gg \hbar c/\overline{\Delta E}_{ex}$. This fact indicates that the macroscopic interaction is determined by the microscopic London (Casimir–Polder) forces.

It must be pointed out that although the Casimir formula (Equation (2.122)) is obtained as a particular case of the Lifshitz theory for two metallic plates in vacuum, the Casimir force and the van der Waals attraction are different physical

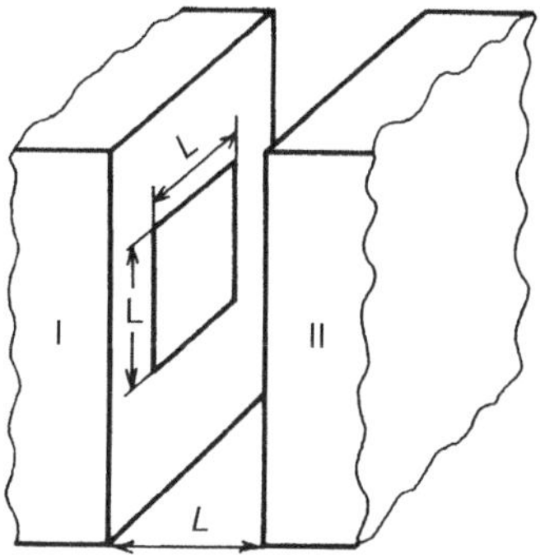


**Figure 2.4** To the derivation of the Equation (2.125) for the distance dependence of the interaction of the macroscopic bodies

phenomena. The van der Waals forces are always attractive, while the Casimir forces can be either attractive or repulsive. If instead of two plates one may consider the system of two halves of a metallic sphere that are being brought together to form a full sphere, they will experience a *repulsive* pressure. The sign of the Casimir force is positive or negative depending on the particular geometry of the boundaries [129, 130].

The dispersion interaction between the molecules, which constitute a macroscopic body, decreases with the distances as $R^{-6}(R^{-7})$. But the interaction between bodies usually falls more slowly with the distance. This can be shown easily with the help of the following qualitative treatment [137]. Consider the interaction between two plane macroscopic plates separated by a distance $L$ from each other (Figure 2.4) and assume that the distance $L$ is considerably smaller than the linear size of the plates. Then, their interaction can be considered as the interaction of two semispaces. Hence, it is sufficient to determine the interaction per unit of surface. Let us single out a square with an area $L^2$ on the surface of each plate. Since the interaction energy of two molecules rapidly decreases as $R^{-6}(R^{-7})$, it is enough to take into account the interaction of two volumes $L^3$, under the assumption that the interaction of molecules far away is negligible. The average distance, $\overline{R}$, between molecules of different volumes is $\overline{R} = \beta L$, where $\beta$ is a small numerical factor with a magnitude between 3 and 1 (the maximum distance is 3.36 $L$, the minimum distance is $L$). Denoting the number of molecules in the unit volume of plates I and II by $n_1$ and $n_2$, one obtains the average interaction energy for a unit surface:

$$\overline{E}_{int} \simeq -\frac{C_6}{\overline{R}^6}\frac{n_1 L^3 n_2 L^3}{L^2} = -\frac{C_6}{\beta^6}\frac{n_1 n_2}{L^2} \tag{2.125}$$

where it was assumed that molecules interact via the London law. The exact expression of the pairwise summation [122] is:

$$\overline{E}_{int} = -\frac{\pi C_6}{12}\frac{n_1 n_2}{L^2} \tag{2.126}$$

This result corresponds to the interaction across vacuum in nonretarded regime at distances large enough that the dispersion interaction is described only by the dipole–dipole term in the Expansion (2.62). Comparing Equations (2.125) and (2.126), it is found that they coincide if $\beta^6 = 12/\pi$ or $\beta \simeq 1.25$. At larger distances, molecules interact via the Casimir–Polder law, the attraction energy between plates decreases with distance as $L^{-3}$, in accordance with the Casimir result (Equation (2.123)).

The pairwise additive method for calculating the nonretarded dispersion forces acting between colloidal particles was developed by de Boer [138] and Hamaker [139] before the theoretical studies by Casimir and Lifshitz. The simple de Boer–Hamaker additive approach makes it comparatively easy to obtain the distance dependence for interaction between bodies with different geometry [121, 122, 124]. For instance, the interaction between a spherical particle (with radius $R$) and a plane plate at the distance $L(L \gg R)$ for the London interaction law is given by the formula [122]:

$$E_{int}(L) = -\frac{2}{9}A_{12}\left(\frac{R}{L}\right)^3 \tag{2.127}$$

where $A_{12}$ is the so-called *Hamaker constant* [139], which depends on the atomic composition of the spherical particle and plate. It is defined as:

$$A_{12} = \pi^2 C_6 n_1 n_2 \tag{2.128}$$

In terms of the Hamaker constant, the interaction between two plane plates, (Equation (2.126)), is expressed as:

$$E_{int} = -\frac{A_{12}}{12\pi}\frac{1}{L^2} \tag{2.126a}$$

The distance dependence of the interaction found in the framework of the additive approach remains valid in the Lifshitz theory, but the Expression (2.128) for the Hamaker constant is changed. It will depend upon the dielectric permittivity of interacting bodies as a function of frequency [122, 124]. The dependence of the interaction energy on the distance between bodies of different shapes is represented in Table 2.14.

**Table 2.14** Dependence of the interaction energy on the separation between bodies with different shape[a]

| | Half-space | Sphere | Cylinder[b] |
|---|---|---|---|
| Half-space | $L^{-2}(L^{-3})$ | | |
| Sphere | $L^{-3}(L^{-4})$ | $L^{-6}(L^{-7})$ | |
| Cylinder | $L^{-3}(L^{-4})$ | $L^{-5}(L^{-6})$ | $L^{-5}(^{-6})$ |

[a]The dependence of the interaction energy on the separation, taking into account the retardation effects, is given in parentheses.
[b]The cylinder is parallel to the plane of the half-space or to the axis of the other cylinder.

To evaluate the adsorption energies and the kinetic energies of gases in pores and capillaries, it is important to know the interaction potentials of atoms with surfaces of different geometrical forms. Several examples of such potentials are presented below [137, 140].

The potential energy of the interaction of an atom $A$ with a macroscopic sphere of radius $R$ at distance $L$ from the center of sphere for $L \gg R$ and $L \gg \lambda\!\!\!^{-}$ is presented as:

$$E_{int}(L) = -\frac{23\hbar c}{16\pi}\frac{\epsilon_0 - 1}{\epsilon_0 + 2}\frac{\alpha^A(0)}{R^4}\left(\frac{R}{L}\right)^7 \tag{2.129}$$

where $\alpha^A(0)$ is the static polarizability of atom $A$, and $\epsilon_0 \equiv \epsilon(0)$ is the static dielectric permittivity of the macroscopic sphere. In metals, $\epsilon_0 \gg 1$ and Equation (2.129) is transformed into:

$$E_{int}(L) = -\frac{23\hbar c}{16\pi}\frac{\alpha^A(0)}{R^4}\left(\frac{R}{L}\right)^7 \tag{2.130}$$

For the interaction of an atom $A$ with a cylindrical macroscopic body with radius $R$ at the conditions $L \gg R$ and $L \gg \lambda\!\!\!^{-}$, the interaction energy is equal to:

$$E_{int}(L) = -\frac{14\hbar c}{15\pi}\frac{(\epsilon_0 - 1)(\epsilon_0 + 4)}{\epsilon_0 + 1}\frac{\alpha^A(0)}{R^4}\left(\frac{R}{L}\right)^6 \tag{2.131}$$

According to Equations (129) to (131), at large distances the dependence on $L$ for the interaction of bodies with atoms is the same as for the interaction with a macroscopic sphere (see Table 2.14).

The interaction energy of an atom inside a cylindrical channel of radius $R$ in the neighborhood of channel axis ($L \ll R$) has the following form for a metal case ($\epsilon_0 \gg 1$) and $R \gg \lambda\!\!\!^{-}$

$$E_{int}(L) = -\frac{3\hbar c}{4}\frac{\alpha^A(0)}{R^4}\left[1 + \frac{9}{2}\left(\frac{L}{R}\right)^2\right] \tag{2.132}$$

* *
*

Lifshitz [66] was the first to study the influence of temperature on the interaction between macroscopic bodies. Later, Mehra [141] derived the temperature dependent expressions of interaction energy that was different from the Lifshitz expressions. The Mehra results confirmed by Brown and Maclay [131] raised doubts concerning the validity of the Lifshitz theory. As was shown by Schwinger *et al.* [142], the general Lifshitz expression, including the temperature dependence, is correct and the errors arise only in the limit taken to recover the conductor case. All qualitative conclusions in Reference [66] about the temperature behavior of the interaction between macroscopic bodies were confirmed in the subsequent theoretical studies [63, 64, 141].

At present, it is well established that the temperature corrections in the nonretarded regime are very small. In the retarded regime their magnitude depends upon the value of the parameter $kTL/\hbar c$. If the condition (2.99) is fulfilled, the contribution of the temperature-dependent terms can be also neglected. The situation is drastically changed for such $L$ that the opposite condition:

$$L \gg \frac{\hbar c}{kT} \tag{2.133}$$

is fulfilled. In this case, the leading order contribution in the interaction energy of two metallic plates per unit area is:

$$U(L, T) = -\zeta(3)\frac{kT}{8\pi L^2} \tag{2.134}$$

where $\zeta(z)$ is the zeta function of Riemann [143], $\zeta(3) = 1.202$. For sufficiently large separations, the interaction again follows the law $L^{-2}$, as in the nonretarded distance range. This physical situation is similar to the situation for the dispersion interaction of two atoms (molecules) described in Section 2.5. It corresponds to a classical limit and the attraction has a pure entropic nature [64, 65].

In the case of two parallel plates, the transition from nonretarded to retarded regime changes the $L^{-2}$ dependent potential on the $L^{-3}$ dependent potential. But with a further increase of separation, the potential comes back to the $L^{-2}$ dependence. This behavior was predicted by Lifshitz [66] about 50 year ago. To the best of our knowledge, it has still not been checked experimentally. The reason is in the weakness of interaction forces at such large distances. At room temperatures, $\hbar c/kT = 7.6.10^{-4}$cm $= 7.6$ μm. The condition (2.133) corresponds to relatively large distances, more than 50 μm. At smaller distances, experimental measurements of the magnitude of interbody interaction and its dependence on separation have been performed in many laboratories.

The first direct measurements of the van der Waals forces between two bodies were performed by Abrikosova and Derjaguin [144, 145]. They used two quartz plates; the separations between them in the range from 100 nm to 400 nm were measured by optical interference, and the attractive force with the help of an elaborated feedback mechanism. After their publications, several groups [146–148] published similar experiments with glass, aluminium and quartz plates. The smallest separation achieved was above 100 nm; so it corresponded to retarded forces. In all these experiments, it was observed that the measured forces were in a reasonable agreement with the retarded limit of the Lifshitz theory.

Later on, Tabor and Winterton [149] performed measurements at separations considerably less than 100 nm. They used very smooth mica surfaces and multiple-beam interferometry that allowed them to determine the separation between two crossed cylindrical sheets of mica to an accuracy of $\pm 0.4$ nm and carry out the measurements at separations 30 nm to 5 nm. These were the first measurements of the nonretarded van der Waals forces. These measurements and measurements by Israelachvili and Tabor [150] performed in a larger distance range (1.4 nm to 130 nm) made it possible to study the transition from nonretarded to retarded regime. In the range 2 to 12 nm the forces are completely nonretarded and have the distance dependence $L^{-n}$ with $n = 2.0 \pm 0.1$ [150]. For separations greater than 12 nm, the power law increases above 2.0 and by $L = 50$ nm has reached $n = 2.9$. Thus, the transition between nonretarded and retarded regime occurs between 12 nm and 50 nm. Above 50 nm, forces are retarded with $n = 3.0 \pm 0.1$. Experiments [149, 150] have demonstrated that there is a gradual transition between nonretarded and retarded regime.

Further experimental verification of the Lifshitz theory was performed by Sabisky and Anderson [151]. They carried out accurate measurements of the thickness of helium films on cleaved surfaces of alkaline-earth fluoride crystals at $T = 1.38K$. The thickness of the films in the range between 1.0 nm to 25 nm was measured by an acoustic interferometry technique. As was noted by Schiff [152], the relatively thick liquid helium films are formed by the van der Waals forces. This problem was studied in detailed by Dzyaloshinskii *et al.* [134]. On the basis of the general Lifshitz theory of the van der Waals forces, they derived the expression relating the dispersion attraction to the film thickness. The experimental data [151], as a function of the thickness of the helium film, showed an excellent agreement with theoretical predictions [134].

The existence of the Casimir force between two conducting flat surfaces in the form of a plate and a sphere in the 0.6 μm to 6 μm range was conclusively demonstrated by Lamoreaux [153]. The exact agreement with theory at the level of 5% was obtained. Several years earlier, the closely related effect, the attraction of a neutral atom to a conducting plate, had been directly measured by Sukenik *et al.* [154]. They studied the deflection of sodium atoms passing through a micron-sized gold cavity with adjustable width from 0.5–8 μm. Precise measurements of the intensity of a sodium atomic beam as a function of cavity plate separation ($L$) revealed the $L^{-4}$ retarded dependence and excluded the $L^{-3}$ nonretarded potential.

# References


1. L.D. Landau and E.M. Lifshitz, *The Classical Theory of Fields* (Fourth Revised English Edition), Pergamon Press, Oxford (1994).
2. A.D. Buckingham, *Quart. Rev. Chem. Soc. London* **13**, 183 (1959).
3. A.D. Buckingham, *Adv. Chem. Phys.* **12**, 107 (1967).
4. J.O. Hirschfelder, C.F. Curtiss and R.B. Bird, *Molecular Theory of Gases and Liquids*, John Wiley & Sons, Inc., New York (1954).
5. A.L. McClellan, *Tables of Experimental Dipole Moments*, Freeman and Co., San Francisco (1963).
6. R.D. Nelson, D.R. Lide and A.A. Mayolt, *Selected Values of Dipole Moments*, NBS Date Reference Series **10** (1967).
7. D.E. Strogryn and A.P. Strogryn, *Mol. Phys.* **11**, 371 (1966).
8. J.G. van Duijneveldt-van de Rijdt and F.B. van Duijneveldt, *J. Mol. Str. (Theochem)* **89**, 185 (1982).
9. J.S. Muenter, *J. Chem. Phys.* **56**, 5409 (1972); F.H. de Leeuw and A. Dymanus, *J. Mol. Spectr.* **48**, 427 (1973).
10. W.L. Meerts, F.H. de Leeuw and A. Dymanus, *Chem. Phys.* **22**, 319 (1977).
11. A.J. Russell and M.A. Spackman, *Mol. Phys.* **88**, 1109 (1996).
12. A.D. Buckingham, C. Graham and J.M. Williams, *Mol. Phys.* **49**, 703 (1983).

13. J.F. Ely, H.J.M. Hanley and J.C. Straty, *J. Chem. Phys.* **59**, 842 (1973).
14. A.D. Buckingham and J.E. Cordle, *Mol. Phys.* **4**, 1037 (1974).
15. A.D. Buckingham, R.L. Disch and D.A. Dummur, *J. Amer. Chem. Soc.* **90**, 3104 (1968).
16. A.T. Amos and R.J. Crispin, in *Theoretical Chemistry, Advances and Perspectives* (eds), H. Eyring and D. Henderson, Academic Press, New York (1976), Vol. 2, pp. 1–66.
17. P.W. Milonni and J.H. Eberly, *Lasers*, John Wiley & Sons, Inc., New York (1988).
18. A.S. Davydov, *Théorie du Solide*, Mir, Moscou (1980).
19. A.S. Davydov, *Theory of Molecular Excitons*, Plenum Press, New York (1971).
20. T. Förster, *Fluoreszenz organischer Verbindungen*, Göttingen (1951).
21. D.L. Dexter, *J. Chem. Phys.* **21**, 836 (1953).
22. I.G. Kaplan and V.G. Plotnikov, *Khimia Vysokikh Energii*, **1**, 507 (1967).
23. I.G. Kaplan and O. Navarro, *J. Phys: Condens. Matter* **11**, 6187 (1999).
24. J. Hubbard, *Proc. Roy. Soc. (London)* A **276**, 238 (1963); A **277**, 237 (1964).
25. F. London, *Zs. Phys. Chem.* B **11**, 222 (1930).
26. F. London, *Trans. Faraday Soc.* **33**, 8 (1937).
27. W. Kołos, *Int. J. Quant. Chem.* **1**, 169 (1967).
28. I.G. Kaplan, *Theory of Molecular Interactions,* Elsevier, Amsterdam (1986).
29. J.C. Slater and J.G. Kirkwood, *Phys. Rev.* **37**, 682 (1931).
30. P.T. Pack, *J. Chem. Phys.* **46,** 1959 (1967).
31. T.H. Boyer, *Phys. Rev.* A **5**, 1799 (1972).
32. T.H. Boyer, *Phys. Rev.* A **6**, 314 (1972).
33. T.H. Boyer, *Phys. Rev.* A **7**, 1832 (1973).
34. T.H. Boyer, *Phys. Rev.* A **9**, 2078 (1974).
35. H.B.G. Casimir and D. Polder, *Phys. Rev.* **73**, 360 (1948).
36. T.H. Boyer, *Phys. Rev.* A **11**, 1650 (1975).
37. I.G. Kaplan, *Symmetry of Many-Electron Systems*, Academic Press, New York (1975).
38. H. Böhm and R. Ahlrichs, *J. Chem. Phys.* **77**, 2028 (1982).
39. B. Jeziorski and M. van Hemert, *Mol. Phys.* **31,** 713 (1976).
40. P.E.S. Wormer and A. van der Avoird, *J. Chem. Phys.* **62**, 3326 (1975).
41. W. Kołos and L. Wołniewicz, *J. Chem. Phys.* **43**, 2429 (1965).
42. R. Eisenschitz and F. London, *Zs. f Phys.* **60**, 491 (1930).
43. J.N. Murrell and G. Shaw, *J. Chem. Phys.* **49**, 4731 (1968).
44. T. Korona *et al.*, *J. Chem. Phys.* **106**, 5109 (1997).
45. V.F. Lotrich *et al.*, *J. Chem. Phys.* **103**, 6076 (1995).
46. P.O. Löwdin, *Rev. Mod. Phys.* **34**, 631 (1962).
47. C. Herring, *Rev. Mod. Phys.* **34**, 80 (1962).
48. C. Herring and H. Flicker, *Phys. Rev.* **134A**, 362 (1964).
49. M.H. Alexander and L. Salam, *J. Chem. Phys.* **46**, 430 (1967).
50. W. Kołos and L. Wołniewicz, *Chem. Phys. Lett.* **24**, 457 (1974).
51. T. Holstein, *J. Phys. Chem.* **56**, 832 (1952).
52. T.C. Scott, J.F. Babb, A. Dalgarno and J.D. Morgan III, *Chem. Phys. Lett.* **203**, 175 (1993); *J. Chem. Phys.* **99**, 2841 (1993).
53. D.R. Herschbach, *Adv. Chem. Phys.* **10**, 319 (1966).
54. V.B. Leonas and A.P. Kalinin, *Sov. Phys. Usp.* **20**, 279 (1977).
55. R.S. Mulliken and W.B. Person, *Molecular Complexes*, John Wiley & Sons, Inc., New York (1969).

56. H. Umeyama, K. Morokuma and S. Yamabe, *J. Am. Chem. Soc.* **99**, 330 (1977).
57. E.J. Verwey and J. Overbeck, *Theory of The Stability of Lyophobic Colloids*, Elsevier, Amsterdam (1948).
58. A.D. McLachlan, *Proc. Roy. Soc. (London)* A **271**, 387 (1963).
59. I.E. Dzyaloshinskii, *Sov. Phys. JETP* **3**, 977 (1957).
60. E.A. Power, *Adv. Chem. Phys.* **12**, 167 (1967).
61. E.A. Power, *Phys. Rev.* A **10**, 756 (1974).
62. W.J. Meath and J.O. Hirschfelder, *J. Chem. Phys.* **44**, 3197, 3210 (1966).
63. B.W. Ninham and J. Daicic, *Phys. Rev.* A **57**, 1870 (1998).
64. H. Wennerström, J. Daicic and B.W. Ninham, *Phys. Rev.* A **60**, 2581 (1999).
65. M. Revzen, R. Opher, M. Opher and A. Mann, *J. Phys. A: Math. Gen.* **30**, 7783 (1997).
66. E.M. Lifshitz, *Sov. Phys. JETP* **2**, 73 (1956).
67. P.W. Milonni and A. Smith, *Phys. Rev.* A **53**, 3484 (1996).
68. B.V. Derjaguin and L.D. Landau, *Acta Physicochimica* (USSR) **14**, 633 (1941).
69. J.C. Crocker and D.G. Grier, *Phys. Rev. Lett.* **73**, 352 (1994).
70. K. Vondermassen, J. Bongers, A. Mueller, and H. Versmold, *Langmuir* **10**, 1351 (1994).
71. H. Löwen, P.A. Madden and J.P. Hansen, *Phys. Rev. Lett.* **68**, 1081 (1992).
72. E.B. Sirota, H.D. Ou-Yang, S.K. Sinha and P.M. Chaikin, *Phys. Rev. Lett.* **62**, 1524 (1989).
73. N. Ise and H. Matsuoka, *Macromolecules* **27**, 5218 (1994).
74. A.E. Larsen and D.G. Grier, *Phys. Rev. Lett.* **76**, 3862 (1996).
75. G.M. Kepler and S. Fraden, *Phys. Rev. Lett.* **73**, 356 (1994).
76. M.D. Carbajal-Tinoco, F. Castro-Román and J.L. Arauz-Lara, *Phys. Rev.* E **53**, 3745 (1996).
77. J.C. Crocker and D.G. Grier, *Phys. Rev. Lett.* **77**, 1897 (1996).
78. D.G. Grier, *Nature* **393**, 621 (1998).
79. A.E. Larsen and D.G. Grier, *Phys. Rev. Lett.* **76**, 3862 (1996).
80. J.C. Neu, *Phys. Rev. Lett.* **82**, 1072 (1999).
81. E. Trizac and J.-L. Raimbault, *Phys. Rev.* E **60**, 653 (1999).
82. T.M. Squires and M.P. Brenner, *Phys. Rev. Lett.* **85**, 4976 (2000).
83. B.I. Ivlev, *J. Phys.: Condens. Matter* **14**, 4829 (2002).
84. Y.O. Popov, *J. Colloid Interface Sci.* **252**, 320 (2002).
85. Y. Han and D.G. Grier, *Phys. Rev. Lett.* **92**, 148301 (2004).
86. E.A. Hessels, P.W. Arcuni, F.J. Deck and S.R. Lundeen, *Phys. Rev.* A **46**, 2622 (1992).
87. E. Kelsey and L. Spruch, *Phys. Rev.* A **18**, 15 (1978).
88. Ya. B. Zel'dovich, *Sov. Phys. JETP* **6**, 1184 (1958).
89. V.A. Alekseev, Ya.B. Zel'dovich and I.I. Sobel'man, *Sov. Phys. Usp.* **19**, 207 (1976).
90. S.M. Apenko and Y.E. Losovik, *J. Phys. B* **15**, L57 (1982).
91. E.R. Boston and P.G.H. Sandars, *J. Phys. B* **23**, 2663 (1990).
92. I.B. Khriplovich, *Parity Nonconservation in Atomic Phenomena,* Gordon and Breach, London (1991).
93. V.M. Dubovik and A.A. Cheshkov, *Sov. J. Part. Nucl.* **5**, 318 (1974).
94. V.M. Dubovik, L.A. Tosunyan and V.V. Tugushev, *Sov. Phys. JETP* **63**, 344 (1986).
95. V.M. Dubovik and V.V. Tugushev, *Phys. Reports* **187**, 145 (1990).
96. J. Coulon *et al.*, *Phys. Rev. Lett.* **88**, 237401 (2002).
97. S. Di Matteo, *Phys. Rev. B* **70**, 165115 (2004).
98. I.B. Khriplovich and M.E. Pospelov, *Z. Phys. D* **17**, 81 (1990).

99. H.A. Bethe and E.E. Salpeter, *Quantum Mechanics of One- and Two-Electron Atoms*, W.A. Benjamin, New York (1957).
100. J.O. Hirschfelder, Ch.F. Curtiss and R.R. Bird, *Molecular Theory of Gases and Liquids*, John Wiley & Sons, Inc., New York (1954).
101. T. Itoh, *Rev. Mod. Phys.* **37**, 159 (1965).
102. J.O. Hirschfelder and W.J. Meath, *Adv. Chem. Phys.* **12**, 3 (1967).
103. J. Sucher, *Phys. Rev.* A **22**, 348 (1980).
104. J. Sucher, *Phys. Scr.* **36**, 271 (1987).
105. Y. Ishikawa and U. Kaldor, in *Computational Chemistry: Review of Current Trends*, J. Leszczysnki (ed) World Scientific, Singapore, 1996, Vol. 1, pp. 1–52.
106. U. Kaldor and E. Eliav, *Adv. Quant. Chem.* **31**, 313 (1999).
107. U. Kaldor and B.A. Hess, *Chem. Phys. Lett.* **230**, 1 (1994).
108. E. Eliav and U. Kaldor, *Chem. Phys. Lett.* **248**, 405 (1996).
109. N.S. Mosyagin, A.V. Titov, E. Eliav and U. Kaldor, *J. Chem. Phys.* **115**, 2007 (2001).
110. A.N. Petrov *et al.*, *Phys. Rev. Lett.* **88**, 073001 (2002).
111. P. Schwerdtfeger (ed), *Relativistic Electronic Structure Theory, Part 1, Fundamentals*, Elsevier, Amsterdam (2002).
112. P. Schwerdtfeger (ed), *Relativistic Electronic Structure Theory, Part 2, Applications*, Elsevier, Amsterdam (2004).
113. W. Kołos and L. Wołniewicz, *J. Chem. Phys.* **41**, 3663 (1964).
114. N.G. Van Kampen, B.R.A. Nijboer and K. Schram, *Phys. Lett.* A **26**, 307 (1968).
115. E.E. Johnston, S.T. Epstein and W.J. Meath, *J. Chem. Phys.* **47**, 1271 (1967).
116. L. Gomberoff and E.A. Power, *Proc. Roy Soc. (London)* A **295**, 476 (1966).
117. W.J. Meath, *J. Chem. Phys.* **45**, 4519 (1966).
118. T.Y. Chang, *Rev. Mod. Phys.* **39**, 911 (1967).
119. J.E. Harriman, M. Twerdochlib, M.B. Milleur and J.O. Hirschfelder, *Proc. Natl. Acad. Sci. USA* **57**, 1558 (1967).
120. M.B. Milleur, L.A. Curtiss, M. Twerdochlib and J.O. Hirschfelder, *J. Chem. Phys.* **48**, 4261 (1968).
121. D. Langbein, *Theory of Van der Waals Attraction*, Springer–Verlag, Heidelber (1974).
122. J. Mahanty and B.W. Ninham, *Dispersion Forces*, Academic Press, London (1976).
123. B.V. Derjaguin, *Theory of Stability of Colloids and Thin Films*, Consultants Bureau, New York (1989).
124. J. Israelachvili, *Intermolecular and Surface Forces*, 2nd edn, Academic Press, London (1991).
125. H.B.G. Casimir, *Proc. Ned. Acad. Wetenschap* **51**, 793 (1948).
126. H.B.G. Casimir, *J. Chim. Phys.* **46**, 407 (1949).
127. H.B.G. Casimir, *Physica* **19**, 846 (1953).
128. G. Plunien, B. Müller and W. Greiner, *Phys. Rep.* **134**, 87 (1986).
129. E. Elizalde and A. Romero, *Amer. J. Phys.* **59**, 711 (1991).
130. V.M. Mostepanenko and N.N. Trunov, *Sov. Phys. Usp.* **31**, 965 (1988).
131. L.S. Brown and G.J. Maclay, *Phys. Rev.* **184**, 1272 (1969).
132. A.E. González, *Ann. Phys.* (N.Y.) **168**, 79 (1986).
133. N.G. Van Kampen, B.R.A. Nijboer and K. Schram, *Phys. Lett.* A **26**, 307 (1968).
134. I.E. Dzyaloshinskii, E.M. Lifshitz and L.P. Pitaevskii, *Adv. Phys.* **10**, 165 (1961).
135. V.A. Parsegian and B.W. Ninham, *J. Chem. Phys.* **52**, 4578 (1970).
136. V.A. Parsegian and B.W. Ninham, *Biophys. J.* **10**, 646, 664 (1970).
137. R.H.S. Winterton, *Contemp. Phys.* **11**, 559 (1970).

138. J.H. de Boyer, *Trans. Farad. Soc.* **32**, 21 (1936).
139. H.C. Hamaker, *Physica* **4**, 1058 (1937).
140. V.M. Nabutovskii, V.P. Belosludov and A.M. Korotkikh, *Sov. Phys. JETP* **50**, 352 (1979).
141. J. Mehra, *Physica* **37**, 145 (1967).
142. J. Schwinger, L.I. DeRead, Jr. and K.A. Milton, *Ann. Phys.* (N.Y.) **115**, 1 (1978).
143. M. Abramowitz and I.A. Stegun (eds), *Handbook of Mathematical Functions*, National Bureau of Standards, Washington (1964).
144. I.I. Abrikosova and B.V. Derjaguin, *Sov. Phys. Dokl.* **90**, 1055 (1953).
145. B.V. Derjaguin and I.I. Abrikosova, *Discuss. Faraday Soc.* **18**, 33 (1954); *Sov. Phys.* JETP **3**, 819 (1957).
146. J.A. Kitschener and A.P. Drosser, *Proc. Roy. Soc. (London)* A **242**, 403 (1957).
147. M.J. Sparnaay, *Physica* **24**, 751 (1958).
148. W. Black, J.G.V. de Jongh, J. Overbeck and M.J. Sparnaay, *Trans. Faraday Soc.* **56**, 1597 (1960).
149. D. Tabor and R.H.S. Winterton, *Proc. Roy Soc. (London)* A **312**, 435 (1969).
150. J.N. Israelachvili and D. Tabor, *Proc. Roy. Soc. (London)* A **331**, 19 (1972).
151. E.S. Sabisky and C.H. Anderson, *Phys. Rev.* A **7**, 790 (1973).
152. L. Schiff, *Phys. Rev.* **59**, 839 (1941).
153. S.K. Lamoreaux, *Phys. Rev. Lett.* **78**, 5 (1997).
154. C.I. Sukenik *et al.*, *Phys. Rev. Lett.* **70**, 560 (1993).

# 3 Calculation of Intermolecular Interactions

## 3.1 Large Distances

### 3.1.1 Derivation of the general expression for the multipole expansion of the Coulomb interaction energy operator

In Chapter 2, the general formula for the operator of the electrostatic interaction expressed via the irreducible spherical tensors was presented without derivation (Equation (2.37)). In this section, its detailed derivation is given.

For simplicity, the interaction between two atoms, $A$ and $B$, is considered. The general expression for the operator of the electrostatic interaction between two molecules was presented in Chapter 2, (Equation (2.2)). For two atoms, in atomic units ($\hbar = m = e = 1$), it reduces to:

$$V = -\sum_{j=1}^{N_B} \frac{Z_A}{r_{aj}} - \sum_{i=1}^{N_A} \frac{Z_B}{r_{bi}} + \sum_{i=1}^{N_A} \sum_{j=1}^{N_B} \frac{1}{r_{ij}} + \frac{Z_A Z_B}{R} \tag{3.1}$$

where the index $i$ numbers the electron of atom $A$, and $j$ numbers the electrons of atom $B$.

In the expansion of $V_{AB}$ in a series of $R^{-n}$, the main difficulties are related to the expansion of $r_{ij}^{-1}$. The expansion of $r_{aj}^{-1}$ and $r_{bi}^{-1}$ in a series of inverse powers of $R$ is well-known in the mathematical literature. The coefficients of that series are the Legendre polynomials, $P_l(\cos\theta)$ [1], where $\theta$ is the angle between vectors $\mathbf{r}_i$ $(\mathbf{r}_j)$ and $\mathbf{R}$ (see Figure 3.1):

$$\frac{1}{r_{b1}} = \frac{1}{\sqrt{r_1^2 + R^2 - 2r_1 R\cos\theta_1}} = \frac{1}{R}\sum_{l=0}^{\infty}\left(\frac{r_1}{R}\right)^l P_l(\cos\theta_1) \tag{3.2}$$

$$\frac{1}{r_{a2}} = \frac{1}{\sqrt{r_2^2 + R^2 - 2r_2 R\cos\theta_2}} = \frac{1}{R}\sum_{l=0}^{\infty}(-1)^l\left(\frac{r_2}{R}\right)^l P_l(\cos\theta_2) \tag{3.3}$$

In an arbitrary coordinate frame in which the $z$-axis is different from the vector $\mathbf{R}$, the orientations of the vectors $\mathbf{r}_1$, $\mathbf{r}_2$, and $\mathbf{R}$ are given by the spherical angles $(\theta_1, \varphi_1) \equiv \Omega_1$, $(\theta_2, \varphi_2) \equiv \Omega_2$, and $(\theta_R, \varphi_R) \equiv \Omega_R$. The Legendre polynomial

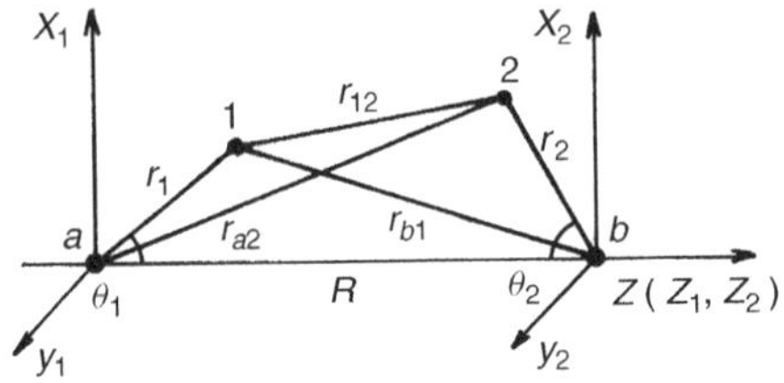


**Figure 3.1** Notations for the two-center coordinate system

is expressed in terms of a product of spherical functions:

$$\frac{1}{r_{b1}} = \frac{4\pi}{R} \sum_{l=0}^{\infty} \sum_{m=-l}^{l} \left(\frac{r_1}{R}\right)^l \frac{Y_l^m(\Omega_1)\, Y_l^{m*}(\Omega_R)}{2l+1} \tag{3.4}$$

$$\frac{1}{r_{a2}} = \frac{4\pi}{R} \sum_{l=0}^{\infty} \sum_{m=-l}^{l} (-1)^l \left(\frac{r_2}{R}\right)^l \frac{Y_l^m(\Omega_2)\, Y_l^{m*}(\Omega_R)}{2l+1} \tag{3.5}$$

The problem of the expansion of $r_{ij}^{-1}$ in an inverse power series in $R$ is called the *bipolar* or *multipole expansion*. It is valid only when the effective diameters of the electronic shells of atoms $A$ and $B$ is much smaller than $R$; in this case, the effect of the overlapping of the electronic shells of the interacting systems is negligible. There are different ways of obtaining the bipolar expansion, of which the most elegant was proposed by Rose [2]. It is based on the irreducible spherical tensor technique [3] (see also Section A2.2.6). A more elementary, although a rather more refined, derivation based on the approaches of References [4, 5] is presented here.

If $r_{12} \neq 0$, the Coulomb potential $r_{12}^{-1}$ satisfies the Laplace equation relative to the coordinates of the first and second electrons:

$$\Delta_1 \left(\frac{1}{r_{12}}\right) = 0 \qquad \Delta_2 \left(\frac{1}{r_{12}}\right) = 0 \tag{3.6}$$

It is well-known that the particular solutions of the Laplace Equations (3.6) in the spherical system of coordinates are functions:

$$r_1^{l_1} Y_{l_1}^{m_1}(\Omega_1) \qquad r_2^{l_2} Y_{l_2}^{m_2}(\Omega_2)$$

Because $r_{12}$ depends on both sets $r_1, \Omega_1$ and $r_2, \Omega_2$, it would be expected that, at large $R$, the function $r_{12}^{-1}$ could be expressed in terms of products of the particular solutions of the Laplace equation mentioned above:

$$\frac{1}{r_{12}} = \sum_{l_1, l_2} \sum_{m_1, m_2} \frac{S(l_1, l_2, m_1, m_2)}{R^{l_1+l_2+1}} r_1^{l_1} Y_{l_1}^{m_1}(\Omega_1)\, r_2^{l_2} Y_{l_2}^{m_2}(\Omega_2) \tag{3.7}$$

where $S(l_1, l_2, m_1, m_2)$ are coefficients to be determined.

To obtain these coefficients, it is necessary to take into account that $r_{12}$ is invariant with respect to rotations and translations of the coordinate frame, due to the scalar properties. Consider the first property. For an infinitely small rotation by an angle $d\varphi$ around the $z$-axis, an arbitrary function $\Phi(\mathbf{r})$ is transformed into the following one [6]:

$$\Phi' = \left(1 + i d\varphi \widehat{L}_z\right) \Phi \tag{3.8}$$

where $\widehat{L}_z = \widehat{L}_{1z} + \widehat{L}_{2z}$ is the operator of the projection of the angular momentum $\mathbf{L}\,(\mathbf{L} = \mathbf{L}_1 + \mathbf{L}_2)$ on the $z$-axis. It should be remembered that the spherical functions are defined by the following relations (in a.u.) [6]:

$$\widehat{\mathbf{L}}^2 Y_l^m(\Omega) = l(l+1) Y_l^m(\Omega) \qquad \widehat{L}_z Y_l^m(\Omega) = m Y_l^m(\Omega) \tag{3.9}$$

It follows from Equation (3.8), that the invariance property of $r_{12}^{-1}$ with respect to rotations is equivalent to the expression:[1]

$$\widehat{L}_z \left(\frac{1}{r_{12}}\right) = \left(\widehat{L}_{1z} + \widehat{L}_{2z}\right)\left(\frac{1}{r_{12}}\right) = 0 \tag{3.10}$$

On the other hand, it follows from Equations (3.9) and (3.10), that operating with $\widehat{L}_z$ on the product of spherical functions in the Expansion (3.7) leads to the following relation for the indexes $m_1$ and $m_2$:

$$m_1 + m_2 = 0$$

Therefore, the double summation over $m_1$ and $m_2$ in Equation (3.7) can be replaced by a single one. Let us write Equation (3.7) in the form:

$$\frac{1}{r_{12}} = \sum_{l_1, l_2 = 0}^{\infty} \sum_{m=-l_<}^{l_<} \frac{F(l_1, l_2, m)\, 4\pi r_1^{l_1} r_2^{l_2} Y_{l_1}^m(\Omega_1) Y_{l_2}^{-m}(\Omega_2)}{R^{l_1+l_2+1}\left[(2l_1+1)(2l_2+1)\right]^{1/2}} \tag{3.11}$$

where $l_< = \min(l_1, l_2)$ and $F(l_1, l_2, m)$ are appropriate coefficients that should be found.

It can be proven that the function:

$$Z_{12}(l_1, l_2) = \sum_{m=-l_<}^{l_<} F(l_1, l_2, m)\, r_1^{l_1} Y_{l_1}^m(\Omega_1)\, r_2^{l_2} Y_{l_2}^{-m}(\Omega_2) \tag{3.12}$$

is the eigenfunction of the operator $\widehat{\mathbf{L}}^2 = \left(\widehat{\mathbf{L}}_1 + \widehat{\mathbf{L}}_2\right)^2$, with the corresponding eigenvalue:

$$\widehat{\mathbf{L}}^2 Z_{12} = (l_1 + l_2)(l_1 + l_2 + 1) Z_{12} \tag{3.13}$$

From Equation (3.10) follows:

$$\widehat{L}_z Z_{12} = 0 \tag{3.14}$$

In order to prove Equation (3.13), consider the following vectorial identity [5]:

$$\begin{aligned}&[\mathbf{r}_1 \times \nabla_1] \cdot [(\mathbf{r}_1 + \mathbf{r}_2) \times (\nabla_1 + \nabla_2)] \\ &\quad = i(\mathbf{r}_1 \times \nabla_1) \cdot \mathbf{L} - (\mathbf{r}_1 \cdot \nabla_1)(\mathbf{r}_2 \cdot \nabla_2) + (\mathbf{r}_1 \cdot \mathbf{r}_2)\nabla_1^2 + (\mathbf{r}_1 \times \nabla_1) \cdot (\mathbf{r}_1 \times \nabla_2)\end{aligned} \tag{3.15}$$

---

[1] The correctness of Equation (3.10) follows from a direct inspection, since $r_{12}$ depends on $\varphi$ only through $\cos(\varphi_1 - \varphi_2)$ and $\widehat{L}_z$ has the form $\widehat{L}_z = -i\left(\frac{\partial}{\partial\varphi_1} + \frac{\partial}{\partial\varphi_2}\right)$.

and use the invariance property of $r_{12}^{-1}$ under displacements. As is known [6, 7], the operator for an infinitely small displacement is expressed in terms of $\nabla$. So, similarly to Equation (3.10):

$$(\nabla_1 + \nabla_2)\left(\frac{1}{r_{12}}\right) = 0 \tag{3.16}$$

The validity of Equation (3.16) is verified directly.

Operation with the left-hand side of Equation (3.15) on $r_{12}^{-1}$ yields a vanishing result, due to Equation (3.16). To operate with the right-hand side of Equation (3.15) on $r_{12}^{-1}$, it is necessary to take into account the definition of the operator $\widehat{\mathbf{L}}_1 = -i\,(\mathbf{r}_1 \times \nabla_1)$, Equation (3.16) and also the equality:

$$\nabla_1\left(\frac{1}{r_{12}}\right) = -\nabla_2\left(\frac{1}{r_{12}}\right) \tag{3.16a}$$

which follows from Equation (3.16). Then one obtains:

$$\widehat{\mathbf{L}}_1 \cdot \widehat{\mathbf{L}}\left(\frac{1}{r_{12}}\right) = \left\{(\mathbf{r}_1 \cdot \nabla_1)(\mathbf{r}_2 \cdot \nabla_2) + \widehat{\mathbf{L}}_1^2\right\}\left(\frac{1}{r_{12}}\right)$$

Similarly it can be written:

$$\widehat{\mathbf{L}}_2 \cdot \widehat{\mathbf{L}}\left(\frac{1}{r_{12}}\right) = \left\{(\mathbf{r}_1 \cdot \nabla_1)(\mathbf{r}_2 \cdot \nabla_2) + \widehat{\mathbf{L}}_2^2\right\}\left(\frac{1}{r_{12}}\right)$$

Combining these formulae and expressing $\mathbf{r} \cdot \nabla = r\dfrac{\partial}{\partial r}$ in spherical coordinates, one obtains:

$$\widehat{\mathbf{L}}^2\left(\frac{1}{r_{12}}\right) = \left\{2\left(r_1\frac{\partial}{\partial r_1}\right)\left(r_2\frac{\partial}{\partial r_2}\right) + L_1^2 + L_2^2\right\}\left(\frac{1}{r_{12}}\right) \tag{3.17}$$

Substituting the Expansion (3.11) into the left- and right-hand sides of Equation (3.17) and taking the equality:

$$2\left(r_1\frac{\partial}{\partial r_1}\right)\left(r_2\frac{\partial}{\partial r_2}\right) r_1^{l_1} r_2^{l_2} = 2 l_1 l_2 r_1^{l_1} r_2^{l_2}$$

and the first Equation (3.9) for the spherical functions into account, Equation (3.13) is obtained.

Thus, according to Equations (3.13) and (3.14), the function $Z_{12}$ is the eigenfunction of the operators of the square of the total angular momentum and its zero projection on the $z$-axis. The spherical functions, $Y_{l_1}^m$ and $Y_{l_2}^{-m}$, which enter into the definition of $Z_{12}$, are the eigenfunctions, according to Equation (3.9), of the angular momentum operators $\widehat{\mathbf{L}}_1^2$, $\widehat{L}_{1z}$ and $\widehat{\mathbf{L}}_2^2$, $\widehat{L}_{2z}$, respectively. It is well known that the construction of the eigenfunctions of $\widehat{\mathbf{L}}^2$ and $\widehat{L}_z$ is carried out in quantum mechanics with the help of the products of the eigenfunctions of $\widehat{\mathbf{L}}_1$, $\widehat{L}_{1z}$ and $\widehat{\mathbf{L}}_2^2$, $\widehat{L}_{2z}$ via the Clebsch–Gordan coefficients (see Section A2.2.4). Therefore, the coefficients $F(l_1, l_2, m)$ have to be proportional to the Clebsch–Gordan coefficients:

$$F(l_1, l_2, m) = C_{l_1 l_2}\langle l_1 m,\ l_2 - m \mid L0\rangle \tag{3.18}$$

through a factor, which is independent of $m$. The Clebsch–Gordan coefficients in Equation (3.18) are equal to [8–10]:

$$\langle l_1 m, l_2 - m \mid L\ 0\rangle = \left[\frac{(2l_1)!\,(2l_2!)}{(2l_1+2l_2)!\,(l_1+m)!\,(l_1-m)!\,(l_2+m)!\,(l_2-m)!}\right]^{1/2} \tag{3.19}$$

To obtain $C_{l_1 l_2}$, it is sufficient to consider any particular case. Assuming that $\theta_1 = \theta_2 = 0$. Then $r_{12} = R + r_2 - r_1$, and the expansion in powers of $R^{-1}$ is reduced to a geometrical progression, which allows, by comparison with the progression with the series Equation (3.11) at $\theta_1 = \theta_2 = 0$ and the coefficients (Equation (3.19)), $C_{l_1 l_2}$ to be found:

$$C_{l_1 l_2} = (-1)^{l_2}\left[\frac{(2l_1+2l_2)}{(2l_1)!\,(2l_2)!}\right]^{1/2}(l_1+l_2)! \tag{3.20}$$

Then, from Equations (3.18)–(3.20) it follows that:

$$F(l_1, l_2, m) = (-1)^{l_2}\frac{(l_1+l_2)!}{[(l_1+m)!\,(l_1-m)!\,(l_2+m)!\,(l_2-m)!]^{1/2}} \tag{3.21}$$

The Expansions (3.4) and (3.5) follow from Equations (3.11) and (3.21) as particular cases. For instance, the Expansion (3.4) corresponds to $l_2 = 0$ and $m = 0$, and the expansion (3.5) to $l_1 = 0$ and $m = 0\left(Y_{00} = 1/\sqrt{4\pi}\right)$.

Let us derive the general expression for the term in the Hamiltonian, which corresponds to the electron–electron interaction. Substituting the coefficients (Equation (3.21)) into Equation (3.11) and separating the terms with $l_1 = l_2 = 0$ and $l_1 = 0$, $l_2 \neq 0$, or $l_1 \neq 0$, $l_2 = 0$, we obtain

$$\sum_{i=1}^{N_A}\sum_{j=1}^{N_B}\frac{1}{r_{ij}} = \frac{N_A N_B}{R} - N_A\sum_{l=1}^{\infty}\frac{(-1)^l}{R^{l+1}}Q_l^0(B) - N_B\sum_{l=1}^{\infty}\frac{Q_l^0(A)}{R^{l+1}}$$
$$+\sum_{l_1,l_2=1}^{\infty}\sum_{m=-l_<}^{l_<}\frac{F(l_1,l_2,m)}{R^{l_1+l_2+1}}Q_{l_1}^m(A)\,Q_{l_2}^{-m}(B) \tag{3.22}$$

where $2l+1$ of the quantities $Q_l^m$ account for the $2^l$th-pole moment of a system of charges:[2]

$$Q_l^m(A) = -\sum_{i=1}^{N_A}\left[\frac{4\pi}{2l+1}\right]^{1/2} r_i^l Y_l^m(\Omega_i) \tag{3.23a}$$

[2] It should be remembered that all the expressions are given in a system of atomic units, where an electron charge is equal to $-1$, and the nuclear charge to $Z_A$. Under a transformation to the dimensional units in the summations over $i$ and $j$ in Equations (3.23a) and (3.23b), the charges $e_i$ and $e_j$, respectively, must be added. This fact should be taken into account also in Equation (3.25) in the definition of the ionic charges, $q_A$ and $q_B$.

$$Q_l^{-m}(B) = -\sum_{j=1}^{N_B} \left[\frac{4\pi}{2l+1}\right]^{1/2} r_j^l Y_l^{-m}(\Omega_j) \tag{3.23b}$$

where $Q_0^0(A) = -N_A$. The dipole moment is obtained for $l = 1$, the quadrupole moment for $l = 2$, the octopole moment for $l = 3$, etc. The expressions for the dipole moment and up to the hexadecapole moment in terms of cartesian coordinates are given in Section 2.1.2; the general expression for the mth rank cartesian tensor is also presented there (Equation (2.13)). The relationships between the spherical and cartesian components of the dipole and quadrupole moment are given in Equation (2.25).

The expansions of $r_{bi}^{-1}$ and $r_{aj}^{-1}$ (Equations (3.4) and (3.5)) after summation over $i$ and $j$ can be written via the multipole moments. Separating the term with $l = 0$, as in Equation (3.22), gives:

$$\begin{aligned} -\sum_{i=1}^{N_A} \frac{Z_B}{r_{bi}} &= -\frac{N_A Z_B}{R} + Z_B \sum_{l=1}^{\infty} \frac{Q_l^0(A)}{R^{l+1}} \\ -\sum_{j=1}^{N_B} \frac{Z_A}{r_{aj}} &= -\frac{N_B Z_A}{R} + Z_A \sum_{l=1}^{\infty} (-1)^l \frac{Q_l^0(B)}{R^{l+1}} \end{aligned} \tag{3.24}$$

All the expressions needed to write the interaction energy operator (Equation (3.1)) in terms of the multipole moments are now available. Substituting Equations (3.22) and (3.24) into Equation (3.1) and taking into account that the atomic charges (in atomic units) are equal to:

$$q_A = Z_A - N_A \qquad q_B = Z_B - N_B \tag{3.25}$$

we arrive at the final result:

$$\begin{aligned} V = {} & \frac{q_A q_B}{R} + q_B \sum_{l=1}^{\infty} \frac{Q_l^0(A)}{R^{l+1}} + q_A \sum_{l=1}^{\infty} (-1)^l \frac{Q_l^0(B)}{R^{l+1}} \\ & + \sum_{l_1, l_2=1}^{\infty} \sum_{m=-l_<}^{l_<} \frac{F(l_1, l_2, m)}{R^{l_1+l_2+1}} Q_{l_1}^m(A) Q_{l_2}^{-m}(B) \end{aligned} \tag{3.26}$$

The first terms of the Expansion (3.26) up to the term $R^{-3}$, inclusively, can now be written out explicitly. Expressing the spherical components in terms of the cartesian ones (see Equation (2.25)) gives:

$$\begin{aligned} V = {} & \frac{q_A q_B}{R} + \frac{q_B d_z^A - q_A d_z^B}{R^2} + \frac{q_B Q_{zz}^A + q_A Q_{zz}^B}{R^3} \\ & + \frac{1}{R^3}\left(-2 d_z^A d_z^B + d_y^A d_y^B + d_x^A d_x^B\right) + 0\left(R^{-4}\right) \end{aligned} \tag{3.27}$$

The first three terms represent the monopole–monopole, monopole–dipole, and monopole–quadrupole interactions, respectively. The fourth term is the dipole–dipole interaction. Its form differs from the Expression (1.6) in Chapter 1, since the Expansion (3.26) has been derived relative to a coordinate system with the axis $z$ directed along $\mathbf{R}$ from nucleus $A$ to $B$. The asymmetry of the indexes of nucleus $A$ and $B$ arises also from this fact. If local coordinate systems on nucleus $A$ and $B$ are introduced in such a way that the $z_A$- and $z_B$-axes are directed towards each other, the Expression (3.26), and following from it Equation (3.27), become symmetric with respect to $A$ and $B$.

Generalization of the Expansion (3.26) to molecules presents no problems. The positions of nuclei in molecules must be assigned by means of the vectors $\mathbf{R}_a$ and $\mathbf{R}_b$ ($a$ and $b$ ranging in value from 1 to $n_A$ and $n_B$, respectively). Each of these vectors is characterized by a set of spherical angles, $\Omega_a$, $\Omega_b$. And, finally, a summation over the nuclei in the definition of the multipole moments (Equation (3.23)) must be included. Then, for a molecule $A$:

$$Q_l^m(A) = \sum_{a=1}^{n_A} Z_a \left(\frac{4\pi}{2l+1}\right)^{1/2} R_a^l Y_l^m(\Omega_a) - \sum_{i=1}^{N_A} \left(\frac{4\pi}{2l+1}\right)^{1/2} r_i^l Y_l^m(\Omega_i) \tag{3.28}$$

The multipole expansion for interacting molecules, expressed in terms of the multipole moments (Equation (3.28)), takes the same form as for atoms. For instance, in the particular case of the interaction of neutral molecules, the first three terms in Equation (3.26) vanish, and the multipole expansion starts from the term, proportional to $R^{-3}$, as follows:

$$V = \sum_{l_1,l_2=1}^{\infty} \sum_{m=-l_<}^{l_<} \frac{F(l_1, l_2, m)}{R^{l_1+l_2+1}} Q_{l_1}^m(A) Q_{l_2}^{-m}(B) \tag{3.29}$$

### 3.1.2 Interaction energy of two atoms in S-states

Consider in detail the interaction of two atoms in S-states, i.e. assume that $L = M = 0$. The eigenfunction of the Hamiltonian $H_0 = H_A + H_B$ of the noninteracting atoms becomes the product of the atomic wave functions:

$$\Psi_0^{(0)} = \Psi_0^A \Psi_0^B \tag{3.30}$$

The function (Equation (3.30)) serves as the zeroth-order wave function in the perturbation theory treatment. The first-order perturbation correction, evaluated by an integration of the Expansion (3.29) over the electronic density distribution with the wave function (Equation (3.30)), vanishes, since atoms do not possess static multipole moments in the spherically-symmetric states. The induction energy vanishes by the same reason, if the overlap between the electron densities of interacting atoms can be neglected (see Section 2.3.1).

To construct the multipole expansion for the dispersion energy in the second-order perturbation treatment, Expansion (3.29) is substituted into the formula for the dispersion energy (Equation (2.60)). The matrix elements in the sum are then divided into the products of the following atomic matrix elements:

$$\left\langle 0_a \left| Q_{l_1}^{m} \right| s_a \right\rangle \left\langle s_a \left| Q_{l_1'}^{m'} \right| 0_a \right\rangle \left\langle 0_b \left| Q_{l_2}^{-m} \right| t_b \right\rangle \left\langle t_b \left| Q_{l_2'}^{-m'} \right| 0_b \right\rangle \tag{3.31}$$

where the notation $|s_a\rangle$ is introduced for simplicity. In atoms, the set of quantum numbers $s_a$ corresponds to $\gamma_a L_a M_a$, where $L_a$ and $M_a$ are an angular momentum in the atomic state and its projection; $\gamma_a$ is a set of additional quantum numbers that enumerate the states with the same $L_a$ and $M_a$. Thus:

$$|s_a\rangle \equiv \left| \Psi_{\gamma_a L_a M_a}^{A} \right\rangle$$

For an S-state, $L = M = 0$. According to the Wigner–Eckart theorem (see Section A2.2.6), the matrix element takes the following form:

$$\left\langle 0_a \left| Q_{l_1}^{m} \right| \gamma_a L_a M_a \right\rangle = \langle 00 \mid l_1 m,\ L_a M_a \rangle \left\langle 0_a \left\| Q_{l_1} \right\| \gamma_a L_a \right\rangle \tag{3.32}$$

where the double bar in the right-hand matrix element indicates its independence of the angular momentum projections. The Clebsch–Gordan coefficient, entering into Equation (3.32), does not vanish only if $L_a = l_1$ and $M_a + m = 0$ and similar $L_a = l_1'$ and $M_a + m' = 0$. Then the following important relations exist: $l_1 = l_1'$, $m = m'$ and $l_2 = l_2'$. As a result, the product (Equation (3.31)) is equivalent to:

$$\left| \left\langle s_a \left| Q_{l_1}^{m} \right| 0_a \right\rangle \right|^2 \left| \left\langle t_b \left| Q_{l_2}^{-m} \right| 0_b \right\rangle \right|^2 \tag{3.33}$$

The Clebsch–Gordan coefficient in Equation (3.32) is given in Appendix 2, Equation (A2.211). Namely:

$$\langle 00 \mid l_1 m,\ l_1 - m \rangle = (-1)^{l_1 - m} \frac{1}{\sqrt{2l_1 + 1}} \tag{3.34}$$

Therefore, the dependence on $m$ in Equation (3.33) is removed. This fact permits to take the multipole moments $Q_l^m$ at $m = 0$. Consequently, the following expression for the dispersion energy of the interaction of two S-state atoms is obtained:

$$E_{disp}^{(2)} = -\sum_{l_1=1}^{\infty} \sum_{l_2=1}^{\infty} \frac{D^{AB}(l_1, l_2)}{R^{2(l_1+l_2+1)}} \tag{3.35}$$

$$D^{AB}(l_1, l_2) = \sum_{m=-l_<}^{l_<} F(l_1, l_2, m)^2 {\sum_{s_a, t_b}}' \frac{\left| \left\langle s_a \left| Q_{l_1}^{0}(A) \right| 0_a \right\rangle \right|^2 \left| \left\langle t_b \left| Q_{l_2}^{0}(B) \right| 0_b \right\rangle \right|^2}{\omega_{s0}^{A} + \omega_{t0}^{B}} \tag{3.36}$$

where the prime in the summation over $s_a$ and $t_b$ means, that the states $0_a$ and $0_b$ are omitted. The transition frequencies included in the denominator of Equation (3.36) are equal in the system of atomic units to the difference of the energies of the excited and ground states of atom:

$$\omega_{s0}^{A} = E_s^A - E_0^A \qquad \omega_{t0}^{B} = E_t^B - E_0^B \tag{3.37}$$

It may be proved that:

$$\sum_{m=-l_<}^{l_<} F(l_1, l_2, m)^2 = \frac{(2l_1 + 2l_2)!}{(2l_1)!\,(2l_2)!} \tag{3.38}$$

The coefficients $D(l_1, l_2)$ can be expressed via the integral of the polarizabilities of isolated atoms. Such a representation was first obtained in a famous paper by Casimir and Polder [11]. The authors replaced the sum of the frequencies in the denominator of Equation (3.36) by their product, based on the following integral identity:

$$\frac{1}{a+b} = \frac{2}{\pi}\int_0^\infty \frac{ab}{\left(a^2+z^2\right)\left(b^2+z^2\right)}dz; \; a > 0, \; b > 0 \tag{3.39}$$

This identity has been proved in Section 2.3.2. Taking $\omega_{a0}^A$ and $\omega_{t0}^B$ as $a$ and $b$ in Equation (3.39), then for the coefficient $D(l_1, l_2)$ the following expression is valid (cf. the derivation for the dipole case in Section 2.3.2):

$$D^{AB}(l_1, l_2) = \frac{(2l_1 + 2l_2)!}{(2l_1)!\,(2l_2)!}\frac{2}{\pi}\int_0^\infty \alpha_{l_1}^A(i\omega)\,\alpha_{l_2}^B(i\omega)\,d\omega \tag{3.40}$$

where $\alpha_l(i\omega)$ is the *multipole dynamical polarizability*, which depends on an imaginary argument and is determined according to:

$$\alpha_l(i\omega) = \sum_n{}' \frac{f_{n0}^l}{\omega_{n0}^2 - (i\omega)^2} = \sum_n{}' \frac{f_{n0}^l}{\omega_{n0}^2 + \omega^2} \tag{3.41}$$

The quantities $f_{n0}^l$ represent the oscillator strengths for the $2^l$th-pole transitions:

$$f_{n0}^l = 2\omega_{n0}\left|\langle n\left|Q_l^0\right|0\rangle\right|^2 \tag{3.42}$$

In the particular case of the dipole transitions, i.e. $l = 1$, the formula (3.42) is transformed into:

$$f_{n0}^1 = 2\omega_{n0}\,|\langle n\,|z|\,0\rangle|^2 = \frac{2}{3}\omega_{n0}\,|\langle n\,|\mathbf{r}|\,0\rangle|^2 \tag{3.43}$$

Usually, Expansion (3.35) is written as a sum in powers $R^{-n}$, see Equation (2.62):

$$E_{disp}^{(2)} = -\sum_{n=6}^{\infty}\frac{C_n^{AB}}{R^n} \tag{3.44}$$

where for atom–atom interactions only even powers $n$ enter into the sum. The constants $C_n^{AB}$ are denoted as *dispersion coefficients*. To transform Equation (3.35) into Equation (3.44), the notation $2(l_1 + l_2 + 1) = n$ is used and the summation over $l_1$ and $l_2$ is transformed into the summation over $l_1$ and $n$, replacing $l_2$ by $n/2 - l_1 - 1$. At fixed $n$, the maximum value of $l_1$ is attained for minimum $l_2 = 1$. Therefore, the summation over $l_1$ is carried out until $n/2 - 2$. Then the

Expression (3.35) transforms into:

$$E_{disp}^{(2)} = -\sum_{n=6}^{\infty}\sum_{l_1=1}^{n/2-2} \frac{D^{AB}(l_1, n/2 - l_1 - 1)}{R^n} \tag{3.45}$$

From the last expression it follows that:

$$C_n^{AB} = \sum_{l_1=1}^{n/2-2} D^{AB}(l_1, n/2 - l_1 - 1) \tag{3.46}$$

In particular, for the first three dispersion coefficients one has:

$$\begin{aligned} C_6^{AB} &= D^{AB}(1,1) \\ C_8^{AB} &= D^{AB}(1,2) + D^{AB}(2,1) \\ C_{10}^{AB} &= D^{AB}(1,3) + D^{AB}(2,2) + D^{AB}(3,1) \end{aligned} \tag{3.47}$$

The Equalities (3.40) and (3.47) provide the following integral formulae for these coefficients:

$$C_6^{AB} = \frac{3}{\pi}\int_0^{\infty} \alpha_1^A(i\omega)\,\alpha_1^B(i\omega)\,d\omega \tag{3.48}$$

$$C_8^{AB} = \frac{15}{2\pi}\int_0^{\infty} \left[\alpha_1^A(i\omega)\,\alpha_2^B(i\omega) + \alpha_2^A(i\omega)\,\alpha_1^B(i\omega)\right] d\omega \tag{3.49}$$

$$\begin{aligned} C_{10}^{AB} = {} & \frac{35}{\pi}\int_0^{\infty} \alpha_2^A(i\omega)\,\alpha_2^B(i\omega)\,d\omega \\ & + \frac{14}{\pi}\int_0^{\infty} \left[\alpha_1^A(i\omega)\,\alpha_3^B(i\omega) + \alpha_3^A(i\omega)\,\alpha_1^B(i\omega)\right] d\omega \end{aligned} \tag{3.50}$$

where $\alpha_1(i\omega)$, $\alpha_2(i\omega)$, and $\alpha_3(i\omega)$ are the dipole, quadrupole and octopole polarizabilities, respectively, which are defined by the Formulae (3.41) and (3.42).

It should be emphasized that the general Expression (3.40) and the Formulae (3.48)–(3.50), derived from it, are obtained for ground states for atoms, since the two quantities $\omega_{s0}^A$ and $\omega_{t0}^B$ are only positive in this case, as assumed in Equation (3.39).

For the interaction of charged atoms (ions), in addition to the dispersion energy, the induction energy and the direct electrostatic interaction energy, obtained as a first-order correction, make also nonvanishing contributions. For closed-shell (or S-state) ions, the electrostatic energy is determined only by the Coulomb term in Equation (3.27). The induction energy is expressed in terms of the polarizabilities, similar to the interaction of neutral systems. But for charged systems, the first term in the induction energy becomes proportional to $R^{-4}$ (see Equation (2.58)). As a result, the following expression is obtained for the interaction energy, retaining only terms up to $R^{-6}$:

$$E_{int}^{AB} = \frac{q_A q_B}{R} - \frac{1}{2R^4}\left[q_A^2 \alpha_1^B(0) + q_B^2 \alpha_2^A(0)\right]$$
$$- \frac{1}{2R^6}\left[q_A^2 \alpha_2^B(0) + q_B^2 \alpha_2^A(0)\right] - \frac{C_6^{AB}}{R^6} + 0\left(R^{-8}\right) \quad (3.51)$$

The expression for the static dipole polarizability is given by Equation (2.56). The quadrupole static polarizability is derived from the general formula (Equation (3.41)), taking into account the relations between the spherical and cartesian components (Equation (2.25)); it has the following form:

$$\alpha_2(0) = 2\sum_n{}' \frac{|\langle n\,|Q_{zz}|\,0\rangle|^2}{E_n - E_0} \quad (3.52)$$

### 3.1.3 Dispersion and induction interactions of molecular systems

The interaction of atoms in spherically-symmetric states has been considered in the preceding section. If the interaction of molecules, or atoms in degenerate electronic states is examined, these is, in addition to dispersion energy, induction energy and the energy of direct electrostatic interaction of multipole moments. The molecular case was first analyzed in References [12–14].

Consider the dispersion energy in the approximation of the second-order perturbation theory. In the case of molecules, the product of the matrix elements (Equation (3.31)) does not reduce to the product of the squares of the absolute values, as in Equation (3.33), since the relations $l_1 = l_1'$ and $l_2 = l_2'$ are not satisfied in general. Instead of Equations (3.35) and (3.36), the more general formulae hold:

$$E_{disp}^{(2)} = -\sum_{l_1=1}^{\infty}\sum_{l_2=1}^{\infty}\sum_{l_1'=1}^{\infty}\sum_{l_2'=1}^{\infty} \frac{D^{AB}\left(l_1, l_2; l_1', l_2'\right)}{R^{l_1+l_2+l_1'+l_2'+2}} \quad (3.53)$$

$$D^{AB}\left(l_1, l_2; l_1', l_2'\right) = \sum_{m=-l_<}^{l_<}\sum_{m'=-l_<'}^{l_<'} F\left(l_1, l_2, m\right) F\left(l_1', l_2', m\right)$$
$$\times \sum_{s,t}{}' \frac{\left\langle 0_A \left| Q_{l_1}^m(A) \right| s_A \right\rangle \left\langle s_A \left| Q_{l_1'}^{m'}(A) \right| 0_A \right\rangle \left\langle 0_B \left| Q_{l_2}^{-m}(B) \right| t_B \right\rangle \left\langle t_B \left| Q_{l_2'}^{-m'}(B) \right| 0_B \right\rangle}{\omega_{s0}^A + \omega_{t0}^B} \quad (3.54)$$

The first term of the Expansion (3.53) is proportional to $R^{-6}$ and corresponds to $l_1 = l_1' = l_2 = l_2' = 1$. Contrary to the case of the interaction of S-state atoms, the following term of this expansion can be proportional to $R^{-7}$, whenever $l_1 = l_1' =$

$l_2 = 1$ and $l_2' = 2$ or $l_2 = l_2' = l_1 = 1$ and $l_1' = 2$. It occurs if the product:

$$\left\langle 0_B \left| Q_1^{-m}(B) \right| t_B \right\rangle \left\langle t_B \left| Q_2^{-m'}(B) \right| 0_B \right\rangle \tag{3.55}$$

does not vanish. The multipoles $Q_1$ and $Q_2$ have different parities. Therefore, one of the factors in Equation (3.55) vanishes for molecules with an inversion center. Thus, the term $\sim R^{-7}$ exists only for molecules that have not inversion center.

It is convenient to rewrite the Expansion (3.53) in the form of Equation (3.44) by a summation of the coefficients. This operation is carried out successively, using the notations $l_1 + l_2 + 1 = j$ and $l_1' + l_2' + 1 = j'$ and choosing $l_1$, $j$ and $l_1'$, $j_1'$ as summation indexes. For a fixed $j\,(j')$, the index $l_1\,(l_1')$ ranges from 1 to $j-2$ $(j'-2)$. Then, the Expansion (3.53) is transformed as follows:

$$E_{disp}^{(2)} = -\sum_{j=3}^{\infty}\sum_{j'=3}^{\infty}\frac{\Delta(j, j')}{R^{j+j'}} = -\sum_{n=6}^{\infty}\frac{C_n}{R^n} \tag{3.56}$$

$$\Delta(j, j') = \sum_{l_1=1}^{j-2}\sum_{l_1'=1}^{j'-2} D\left(l_1, j-l_1-1;\ l_1', j'-l_1'-1\right) \tag{3.57}$$

where the coefficient $C_n$ is expressed through $\Delta(j, j')$ by replacing $j$ and $j'$ with $j$ and $n = j + j'$:

$$C_n = \sum_{j=3}^{n-3}\Delta(j, n-j) \tag{3.58}$$

According to Equation (3.58), one obtains:

$$\begin{aligned} C_6 &= \Delta(3, 3) \\ C_7 &= \Delta(3, 4) + \Delta(4, 3) \\ C_8 &= \Delta(3, 5) + \Delta(4, 4) + \Delta(5, 3) \end{aligned} \tag{3.59}$$

The expressions for the induction energy have similar structure and are obtained from Equation (3.54), assuming $s_A$ or $t_B$ equal to zero.

The case of the interaction of an S-state atom with a linear molecule has been studied extensively. The coefficients of the Expansion (3.56) for the dispersion and induction energies are anisotropic and can be represented as a sum of isotropic and anisotropic components, the latter being expanded in terms of the Legendre polynomials $P_l(\cos\theta)$. The detailed form of these coefficients is given by Pack [14]. In particular, the first dispersion and induction coefficients have the following form:

$$C_6^{AM} = C_6^{AM}(0) + C_6^{AM}(2)\, P_2(\cos\theta) \tag{3.60}$$

where the atom is denoted by the symbol $A$ and the linear molecule by $M$, and $\theta$ is the angle between the molecular axis and a line connecting the atom with the molecular center of mass. The expressions for the isotropic and anisotropic dispersion coefficients have the following form:

$$C_6^{AM}(0) = \frac{3}{\pi} \int_0^\infty \alpha_1^A(i\omega)\,\overline{\alpha}_1^M(i\omega)\,d\omega \tag{3.61}$$

$$C_6^{AM}(2) = \frac{1}{\pi} \int_0^\infty \alpha_1^A(i\omega)\left[\alpha_{1,\parallel}^M(i\omega) - \alpha_{1,\perp}^M(i\omega)\right]d\omega \tag{3.62}$$

where $\overline{\alpha}_1$ is the average dipole polarizability, $\alpha_{1,\parallel}$ and $\alpha_{1,\perp}$ are the longitudinal and transverse dipole polarizabilities, and:

$$\overline{\alpha}_1 = \frac{1}{3}\left(\alpha_{1,\parallel} + 2\alpha_{1,\perp}\right) \tag{3.63}$$

The coefficient $C_7$ contains only the anisotropic terms:

$$C_7^{AM} = C_7^{AM}(1)\,P_1(\cos\theta) + C_7^{AM}(3)\,P_3(\cos\theta) \tag{3.64}$$

$$C_7^{AM}(1) = \frac{18}{5\pi} \int_0^\infty \alpha_1^A(i\omega)\left[\alpha_{12,0}^M(i\omega) + \sqrt{3}\alpha_{12,1}^M(i\omega)\right]d\omega \tag{3.65a}$$

$$C_7^{AM}(3) = \frac{12}{5\pi} \int_0^\infty \alpha_1^A(i\omega)\left[\alpha_{12,0}^M(i\omega) - \frac{2\sqrt{3}}{3}\alpha_{12,1}^M(i\omega)\right]d\omega \tag{3.65b}$$

where $\alpha_{12,m}(\omega)$ is the mixed dipole–quadrupole polarizability of the molecule:

$$\alpha_{12,0}(\omega) = \sum_t{}' \frac{2\omega_{t0}\langle 0|Q_1^0|t\rangle\langle t|Q_2^0|0\rangle}{\omega_{t0}^2 - \omega^2} \tag{3.66a}$$

$$\alpha_{12,1}(\omega) = \sum_t{}' \frac{2\omega_{t0}\langle 0|Q_1^1|t\rangle\langle t|Q_2^{-1}|0\rangle}{\omega_{t0}^2 - \omega^2} \tag{3.66b}$$

The dispersion coefficient $C_8$ is represented as the sum of the isotropic and two anisotropic terms:

$$C_8^{AM} = C_8^{AM}(0) + C_8^{AM}(2)\,P_2(\cos\theta) + C_8^{AM}(4)\,P_4(\cos\theta) \tag{3.67}$$

The isotropic component, $C_8(0)$, has the same form as for atoms (see Equation (3.49)), but with the average polarizability of molecule $M$. The average quadrupole polarizability is given by:

$$\overline{\alpha}_2 = \frac{1}{5}\left(\alpha_{20} + 2\alpha_{21} + 2\alpha_{22}\right) \tag{3.68}$$

In the notation used for the polarizability, $\alpha_{lm}$, $l$ defines the multiplicity and $m$ a magnitude of the projection of the angular momentum $l$ on the molecular axis.

For anisotropic components, the following are obtained:

$$C_8^{AM}(2) = \frac{2}{\pi}\int_0^\infty \alpha_2^A(i\omega)\left[\alpha_{1,\parallel}^M(i\omega) - \alpha_{1,\perp}^M(i\omega)\right]d\omega$$
$$+ \frac{12}{7\pi}\int_0^\infty \alpha_1^A(i\omega)\left[\alpha_{20}^M(i\omega) + \alpha_{21}^M(i\omega) - 2\alpha_{22}^M(i\omega)\right]d\omega$$
$$+ \frac{36}{7\pi}\int_0^\infty \alpha_1^A(i\omega)\left[\alpha_{13,0}^M(i\omega) + \sqrt{\frac{8}{3}}\alpha_{13,1}^M(i\omega)\right]d\omega \tag{3.69}$$

and

$$C_8^{AM}(4) = \frac{20}{7\pi}\int_0^\infty \alpha_1^A(i\omega)\left[\alpha_{13,0}^M(i\omega) - \sqrt{\frac{3}{2}}\alpha_{13,1}^M(i\omega)\right]d\omega$$
$$+ \frac{9}{7\pi}\int_0^\infty \alpha_1^A(i\omega)\left[\alpha_{20}^M(i\omega) - \frac{4}{3}\alpha_{21}^M(i\omega) + \frac{1}{3}\alpha_{22}^M(i\omega)\right]d\omega \tag{3.70}$$

where $\alpha_{13,m}$ is the mixed dipole–octopole polarizability, which is given by formulae, analogous to Equation (3.66), with $Q_2^m$ replaced with $Q_3^m$.

The induction coefficients take the following form (they are marked with a tilde to distinguish them from the dispersion coefficients):

$$\tilde{C}_6^{AM} = \left(d_0^M\right)^2 \alpha_1^A(0)\left[1 + P_2(\cos\theta)\right] \tag{3.71}$$

$$\tilde{C}_7^{AB} = \frac{6}{5} d_0^M Q_0^M \alpha_1^A(0)\left[3P_1(\cos\theta) + 2P_3(\cos\theta)\right] \tag{3.72}$$

$$\tilde{C}_8^{AB} = \frac{5}{2}\left(d_0^M\right)^2 \alpha_2^A(0) + \frac{3}{2}\left(Q_0^M\right)^2 \alpha_1^A(0)$$
$$+ \left[\frac{36}{7} d_0^M \Omega_0^M \alpha_1^A(0) + \frac{12}{7}\left(Q_0^M\right)^2 \alpha_1^A(0) + 2\left(d_0^M\right)^2 \alpha_2^A(0)\right] P_2(\cos\theta)$$
$$+ \left[\frac{20}{7} d_0^M \Omega_0^M \alpha_1^A(0) + \frac{9}{7}\left(Q_0^M\right)^2 \alpha_1^A(0)\right] P_4(\cos\theta) \tag{3.73}$$

where $\alpha_l(0)$ is the static multipole polarizability, which is given by Equation (3.41) at $\omega = 0$, and:

$$d_0 = \langle 0 | Q_1^0 | 0 \rangle, \quad Q_0 = \langle 0 | Q_2^0 | 0 \rangle, \quad \Omega_0 = \langle 0 | Q_3^0 | 0 \rangle \tag{3.74}$$

are the expectation values of the dipole, quadrupole and multipole moments.

In conclusion of this section, the formula for the dispersion energy are presented for the case of two tetrahedral molecules, such as methane ($CH_4$). The angles between the coordinate axes of molecule $M_1$ with a line, connecting the carbon atoms, can be designated by $\theta_x(M_1)$, $\theta_y(M_1)$, and $\theta_z(M_1)$. The first terms of the multipole expansion for the dispersion energy are [15]:

$$E_{disp}(\mathrm{CH_4-CH_4}) = -\frac{C_6}{R^6} - \frac{C_7}{R^7}\left[\cos\theta_x(M_1)\cos\theta_y(M_1)\cos\theta_z(M_1) + \cos\theta_x(M_2)\cos\theta_y(M_2)\cos\theta_z(M_2)\right] + \cdots \quad (3.75)$$

The variational procedure gives the following values for the dispersion coefficients [15]:

$$C_6(\mathrm{CH_4-CH_4}) = 160\ a.u. \qquad C_7(\mathrm{CH_4-CH_4}) = 568\ a.u.$$

The expressions for the dispersion energy of the interaction of different types of molecules are given elsewhere [16, 17]. The methods of evaluating dispersion coefficients have been described previously (Reference [18], Chapter 2).

## 3.1.4 Convergence of the multipole expansion

### 3.1.4.1 Perturbation series and the multipole expansion

The detailed derivation of the multipole expansion of the intermolecular interaction potential $V(R)$ was given in Section 3.1.1. Then this expansion was substituted into the expression for a second-order correction to the perturbation energy that gave the possibility of obtaining the Formulae (3.35) and (3.36) for S-state atoms and Formulae (3.53) and (3.54) for a general case. The conditions of validity of the multipole expansion, as well as of the perturbation theory as a whole, will be considered in detail in this section.

At distances $R \gtrsim 10$–15 bohrs, the intermolecular interaction potential $V(R)$ can, with a good accuracy, be considered as a perturbation[3] to the sum of the Hamiltonians of the isolated molecules $H_A + H_B = H_0$:

$$H = H_0 + V(R) \quad (3.76)$$

At such distances, the exchange effects become negligible. Therefore, the Rayleigh–Schrödinger perturbation theory may be applied to the evaluation of the interaction energy. This interaction energy is represented as a series over various orders:

$$E_{Coul}(R) = E_{el}^{(1)}(R) + \sum_{k=2}^{\infty} E_{pol}^{(k)}(R) \quad (3.77)$$

where the terms are given by the standard quantum-mechanical formulae (see Section A3.3.1). It should be stressed that in Equation (3.77), the exchange of electrons is not taken into account. Therefore, the energy (Equation (3.77)) has a pure Coulombic nature, reflected by the subscript 'Coul'. The meaning of the other subscripts was explained in Chapter 2.

[3] The limit of validity of the perturbation theory for longer molecules must be increased. The numerical methods for the evaluation of the interaction energies of such molecules are given in Reference [18], Chapter 2; see also the discussion in Section 3.1.5.

The convergence of the perturbation theory series (Equation (3.77)) can be rigorously studied only for simple systems such as $H_2^+$ [19–21] or $H_2$ [22, 23]. Tables 3.1 and 3.2 have been prepared on the basis of the calculations by Kołos for the ground ${}^1\sum_g^+$ and excited $B\,{}^1\sum_u^+$ states of the system H–H [22, 23] where $\Delta(n)$ is defined as follows:

$$\Delta(n) = \frac{1}{E_{Coul}}\left\{E_{Coul} - \left[E_{el}^{(1)} + \sum_{k=2}^{n} E_{pol}^{(k)}\right]\right\} 100\% \qquad (3.78)$$

The Coulomb energy, $E_{Coul}$, was calculated via the variational method. As follows from Table 3.1, at $R \geq 8$ bohrs the first two terms of the perturbation series approximate well the Coulomb energy. At $R = 10$ bohrs, taking into account the second-order perturbation correction only permits almost 99% of the variational Coulomb energy to be obtained.

The results of the numerical study of the excited $B\,{}^1\sum_u^+$ – state are presented in Table 3.2 for three distances [23]. The large magnitude of the exchange energy, even at 15 bohrs, should be mentioned here. This is explained by the large radius of the electronic shells of the excited states. The contribution of the higher-order

**Table 3.1** Estimates of the first-order correction terms in the perturbation expansion of the interaction energy for the ground state of H–H (energy in $cm^{-1}$) [22]

| $R$, $a_0$ | 4 | 6 | 8 | 10 | 12 |
|---|---|---|---|---|---|
| $E_{pol}^{(2)}$ | 560.0 | −53.667 | −8.028 | −1.809 | −0.557 |
| $E_{pol}^{(3)}$ | −109.854 | 1.396 | 0.298 | 0.025 | 0.0024 |
| $\Delta(2)$ | – | 6.5% | 1.8% | 1.1% | – |
| $\Delta(3)$ | – | 8.5% | 1.8% | 0.27% | – |

The definition of $\Delta(n)$ is given by Equation (3.78).

**Table 3.2** Energy components of the total interaction energy of two hydrogen atoms in $B\,{}^1\sum_u^+$ state (energy in $cm^{-1}$) [23]

| $R$, $a_0$ | 12 | | 15 | | 18 | |
|---|---|---|---|---|---|---|
| $E_{int}^{tot}$ | −927.2 | 100% | −117.3 | 100% | −45.2 | 100% |
| $E_{Coul}$ | −210.5 | 22.7% | −81.2 | 69.2% | −44.0 | 97.3% |
| $E_{ex}$ | −716.7 | 77.3% | −36.1 | 30.8% | −1.2 | 2.7% |
| $E_{el}^{(1)}$ | −155.4 | 16.8% | −73.4 | 62.5% | −41.9 | 92.7% |
| $E_{pol}^{(2)}$ | −44.8 | 4.8% | −7.4 | 6.3% | −1.9 | 4.2% |
| $\sum_{i=3}^{\infty} E_{pol}^{(i)}$ | −10.3 | 1.1% | −0.4 | 0.3% | −0.2 | 0.4% |
| $\Delta(2)^a$ | 4.8% | | 0.58% | | 0.41% | |

$^a\Delta(2)$ is defined by Equation (3.78).

corrections, by comparison to the second-order perturbation correction, is rather small and decreases rapidly with the distance.

The multipole approximation is valid at large distances. Each term of the perturbation series (Equation (3.77)) is decomposed in a series of inverse powers of the intermolecular distance, yielding the series:

$$E_{Coul.as}(R) = -\sum_{n=k}^{\infty} \frac{C_n}{R^n} \tag{3.79}$$

where $k$ is determined by the first nonvanishing multipole moment of the interacting molecules (see Table 2.3).

The expression for the multipole expansion of the operator $1/r_{12}$ (Equation (3.11)) was derived under the assumption that $R > (r_1 + r_2)$ (see Figure 3.1). In the general case, the bipolar expansion for the operator of the Coulomb interaction of two electrons is written in terms of the associated Legendre polynomials [4, 24]:

$$\frac{1}{r_{12}} = \sum_{l_1=0}^{\infty} \sum_{l_2=0}^{\infty} \sum_{m=-l_<}^{l_<} B_{l_1 l_2}^{[m]}(r_{1a}, r_{2b}, R)\, P_{l_1}^{m}(\cos\theta_{1a})\, P_{l_2}^{m}(\cos\theta_{2b})\; e^{-m(\varphi_{1a}-\varphi_{2b})} \tag{3.80}$$

where $l_<$ is the smallest value of $l_1$ and $l_2$. The expressions for the coefficients $B_{l_1 l_2}^{[m]}$ are different in the four possible ranges of variables, which correspond to different cases of overlap of the electronic shells (Figure 3.2):

$$B_{l_1 l_2}^{[m]I} = \frac{(-1)^{l_2+|m|}\,(l_1+l_2)!\; r_{1a}^{l_1}\, r_{2b}^{l_2}}{(l_1+|m|)!\;(l_2+|m|)!\; R^{l_1+l_2+1}} \tag{3.81a}$$

$$B_{l_1 l_2}^{[m]II} = \frac{1}{D_{l_1 l_2}^{[m]}} \sum_{s,t=0}^{2(l_1+l_2+1)} A_{l_1 l_2, st}^{[m]}\, r_{2a}^{s-l_1-1}\, r_{2b}^{t-l_2-1}\, R^{l_1+l_2-s-t+1} \tag{3.81b}$$

$$B_{l_1 l_2}^{[m]III} = \begin{cases} \dfrac{(-1)^{l_1+l_2}\,(l_2-|m|)!}{(l_1+|m|)!\,(l_2-l_1)!} \dfrac{r_{1a}^{l_1}}{r_{2b}^{l_2+1}} R^{l_2-l_1}, & l_2 \geq l_1, \\ 0, & l_2 < l_1, \end{cases} \tag{3.81c}$$

$$B_{l_1 l_2}^{[m]IV} = \begin{cases} \dfrac{(l_1-|m|)!}{(l_2+|m|)!\,(l_1-l_2)!} \dfrac{r_{2b}^{l_2}}{r_{1a}^{l_1+1}} R^{l_1-l_2}, & l_1 \geq l_2, \\ 0, & l_1 < l_2. \end{cases} \tag{3.81d}$$

The quantities $D_{l_1 l_2}^{[m]}$ and $A_{l_1 l_2, st}^{[m]}$ are tabulated for $l_1, l_2 = 0, 1, 2, 3$ and a closed formula [25] exists for $B_{l_1 l_2}^{[m]II}$. In the range I, the expansion Equation (3.11) coincides with Equation (3.80).

The multipole series (Equation (3.79)) corresponds to the substitution of the expansion (3.80) with the coefficients $B_{l_1 l_2}^{[m]I}$ in the form of Equation (3.81a) into

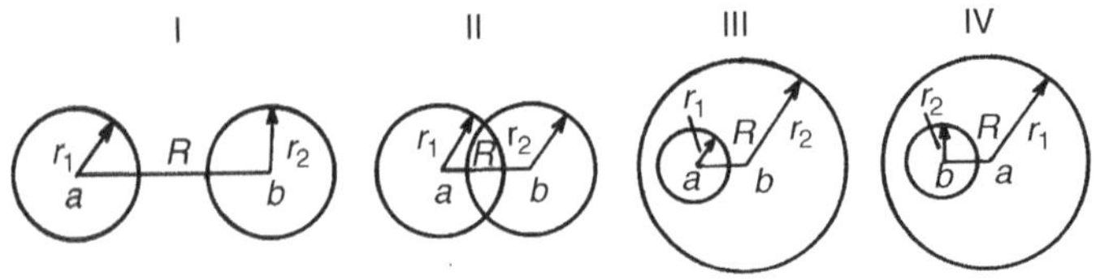


**Figure 3.2** Four possible cases of overlapping of electronic shells in the system of two interacting spherical distributions

the matrix elements of the interaction energy and, therefore, corresponds to a neglect of the overlap of the electronic shells of the interacting molecules. It means, in fact, that the exponentially decreasing terms are disregarded.

The summation of the expansions in the powers of $R^{-1}$ for each order of the perturbation theory leads to the expansion (3.79). If only the second order of the perturbation theory is considered, the danger exists that the remaining terms will be of the same order as the terms in the subsequent orders of the perturbation theory, which are not taken into account. For instance, the multipole series expansion for $H_2^+$ has the form [26]:

$$\begin{aligned} E^{(2)}(R) &= -\frac{9}{2R^4} - \frac{15}{R^6} - \frac{525}{4R^8} - \frac{2835}{4R^{10}} - 0\left(R^{-12}\right) \\ E^{(3)}(R) &= -\frac{213}{2R^7} - \frac{1773}{R^9} - 0\left(R^{-11}\right) \\ E^{(4)}(R) &= -\frac{3555}{32R^8} - \frac{80379}{8R^{10}} - 0\left(R^{-12}\right) \end{aligned} \tag{3.82}$$

It follows from Equation (3.82) that it makes sense to take into account terms up to $R^{-10}$ within the framework of the second-order perturbation theory, only when the third- and fourth-order corrections to the interaction energy are evaluated. Such a situation exists in the case of H–H$^+$. For neutral systems, a larger number of terms may be taken into account within the framework of the second-order perturbation treatment. For instance, for the H–H system [27, 28]:

$$\begin{aligned} E^{(2)} &= -\frac{6.49}{R^6} - \frac{1.24 \cdot 10^2}{R^8} - \frac{3.28 \cdot 10^3}{R^{10}} - \frac{1.21 \cdot 10^2}{R^{12}} - 0\left(R^{-14}\right) \\ E^{(3)} &= \frac{3.47 \cdot 10^3}{R^{11}} + \frac{2.91 \cdot 10^5}{R^{13}} + 0\left(R^{-15}\right) \\ E^{(4)} &= -\frac{1.24 \cdot 10^2}{R^{12}} - 0\left(R^{-14}\right) \end{aligned} \tag{3.83}$$

The terms up to $R^{-10}$ can be maintained if the only second-order perturbation treatment is performed.

It should be mentioned, however, that preserving a larger number of terms in the expansion in the powers of $R^{-1}$ can lead to worsening of the results, because the multipole series corresponds to a class of asymptotically divergent series.

A rigorous proof of that statement was given by Ahlrichs [29]. The divergence of the multipole expansions for the simplest systems has been proved in References [30–32]. Let us consider this problem in detail.

### 3.1.4.2 Study of the convergence of the multipole expansion

The problem of convergence of the multipole expansion was investigated first by Brooks [30] for the model system consisting of two three-dimensional harmonic oscillators. In this case, the integrals, which arise from substitution of the multipole expansion for the interaction operator within the framework of the second-order perturbation treatment, can be evaluated explicitly. Brooks obtained the expansion:

$$E_{disp.as}^{(2)} = \frac{1}{\omega R^2} \sum_{L=2}^{\infty} \frac{(2L)!\left(2^L - 2\right)}{2^{2L} L! L} \frac{1}{(\omega\alpha)^L} \frac{1}{R^{2L}} \tag{3.84}$$

where $\omega$ is the oscillator frequency and $\alpha$ is the polarizability. For finite $R$, the general term of the series (Equation (3.84)) does not approach zero, as $L$ increases. Applying the d'Alambert convergence criterion, one obtains the limit of the ratio of the $(n+1)$th to the $n$th term:

$$\frac{c_{n+1}}{c_n} = \frac{(2n+1)(2n+2)}{2(n+1)R^2} \underset{n \gg 1}{\simeq} \frac{2n}{R^2}$$

At finite $R$ a term for which $2n > R^2$ can always be found. Hence, the series diverges for finite $R$. The wave function of the harmonic oscillator decreases with the distance as $\exp\left(-\beta r^2\right)$, i.e. it undergoes a more rapid damping than molecular wave functions, which decrease as $\exp(-\beta r)$. Thus, a divergence of the multipole expansion for real molecules should also be expected.

It follows from Series (3.84) that at large $R$ the first terms of the series decrease. Moreover, the decrease becomes more rapid as $R$ increases, and the approximation provided by the first terms is better for larger $R$. Nevertheless, at any given $R$ there always exists such $n$ that the subsequent terms will increase. Such a series belongs to the class of *asymptotic* (Poincaré) or *semi-convergent* (Stieltjes) *series* [33]. The definition of asymptotic series is given in Section A3.3.5 (see Equations (A3.160) and (A3.161)).

The features of the general proof about the asymptotic property of the multipole series were outlined by Brooks [30]. However, his proof is incorrect, since he neglects the difference between the eigenfunctions of the exact Hamiltonian and those of a Hamiltonian in which the interaction operator is replaced by the finite multipole expansion. In particular, Brooks does not take into account the antisymmetry of the exact wave function. In subsequent studies, the proof of the divergence of the multipole series for particular systems has been identified with the proof of its asymptotic property. The grounds for such a suggestion is based on the fact that the interaction energy at large $R$ is approximated, with good accuracy, by the first series terms for all the systems studied and, therefore, the divergence of the series means that it belongs to the class of asymptotic, or semi-convergent, series. As mentioned above, the correct proof of the asymptotic property of the multipole series for any nonrelativistic system was given by Ahlrichs [29]. We will not present his proof here but refer the reader to the original paper and consider instead the two well-studied simple systems, $H_2^+$ and $H_2$.

The problem of convergence of the multipole expansion for the system hydrogen atom–proton was investigated in detail by Dalgarno *et al.* [26, 31, 34]. In this case, the explicit analytical expressions for the coefficients of the multipole series can be obtained [31]:

$$E^{(2)}_{pol.as}\left(H_2^+,{}^2\sum\nolimits_g\right) = -2\sum_{l=1}^{\infty}\frac{(2l+2)!\,(l+2)}{l\,(l+1)}\frac{1}{(2R)^{2l+2}} \tag{3.85}$$

It can be proved easily, for example, with the help of the d'Alambert criterion that the Series (3.85) diverges for finite $R$.

Although the asymptotic series diverges for any given $R$, there exists an optimal $n$ which provides the best approximation of the function via a series. In practice, the multipole series is usually cut off at the term after which an increase occurs. Further, the sum of all the terms up to the smallest one is taken and one-half of the smallest term is added to this sum [26]. Expressing the energy as a series in powers of $R^{-1}$ means, as mentioned above, that the exponentially decreasing terms are neglected.

For molecular systems, which are more complex than $H_2^+$, it is impossible to obtain the explicit expression for the multipole expansion. Young [32] proved the divergence of the multipole expansion within a framework of the second-order perturbation treatment for the hydrogen molecule that based on the construction of the dominating series. Young's proof is of interest from the methodological point of view as an example of the application of the Hylleraas variational principle (see Section A3.3.4). Below its basic arguments are reproduced.

Consider the system consisting of two hydrogen atoms in their ground state. The zeroth-order function takes the following form:

$$\Psi^{(0)} = \Psi_a^{1s}(r_{1a})\,\Psi_b^{1s}(r_{2b}) \tag{3.86}$$

The general expression for the dispersion energy (Equation (3.35)) can be rewritten as:

$$E^{(2)}_{disp} = -\sum_{l_1=1}^{\infty}\sum_{l_2=1}^{\infty}\frac{(2l_1+2l_2)!}{(2l_1)!\,(2l_2)!}E^{(2)}_{l_1l_2}R^{-2(l_1+l_2+1)} \tag{3.87}$$

where $E^{(2)}_{l_1l_2}$ denotes the double sum in Equation (3.36), which is represented in the form:

$$E^{(2)}_{l_1l_2} = \left\langle\Psi^{(0)}\left|Q^m_{l_1}(A)\,Q^{-m}_{l_2}(B)\right|\Psi^{(1)}_{l_1l_20}\right\rangle \tag{3.88}$$

if Equation (A3.103) is used. For further transformations, it is convenient to introduce the following notation:

$$\theta^m_{l_1l_2} = Q^m_{l_1}(A)\,Q^{-m}_{l_2}(B) \tag{3.89}$$

The upper bound of the $E_{l_1 l_2}^{(2)}$ can be obtained with the help of the Hylleraas variational principle (see Equations (A3.146) and (A3.147)):

$$E_{l_1 l_2}^{(2)} \leq \widetilde{E}_{l_1 l_2}^{(2)} \tag{3.90}$$

where:

$$\widetilde{E}_{l_1 l_2}^{(2)} = \left\langle \widetilde{\Psi}_{l_1 l_2}^{(1)} \left| H_0 - E^{(0)} \right| \widetilde{\Psi}_{l_1 l_2}^{(1)} \right\rangle + 2 \left\langle \Psi^{(0)} \left| \theta_{l_1 l_2}^{0} \right| \widetilde{\Psi}_{l_1 l_2}^{(1)} \right\rangle \tag{3.91}$$

According to Kirkwood [35], the function $\widetilde{\Psi}_{l_1 l_2}^{(1)}$ can be chosen in the following form (see Equation (A3.153)):

$$\widetilde{\Psi}_{l_1 l_2}^{(1)} = c\chi \tag{3.92}$$

$$\chi = \frac{\theta_{l_1 l_2}^{0} \Psi^{(0)}}{\left\langle \theta_{l_1 l_2}^{0} \Psi^{(0)} \mid \theta_{l_1 l_2}^{0} \Psi^{(0)} \right\rangle^{1/2}} \tag{3.93}$$

where the function $\chi$ is normalized. The coefficient $c$ is determined from the condition of minimum $E_{l_1 l_2}^{(2)}$ [36]. One obtains:

$$c = \frac{\left\langle \chi \left| \theta_{l_1 l_2}^{0} \right| \Psi^{(0)} \right\rangle}{\left\langle \chi \left| H^{(0)} \right| \chi \right\rangle - E^{(0)}} \tag{3.94}$$

The substitution of Equations (3.92)–(3.94) into Equation (3.91) leads to:

$$\widetilde{E}_{l_1 l_2}^{(2)} = -\frac{\left\langle \chi \left| \theta_{l_1 l_2}^{0} \right| \Psi^{(0)} \right\rangle^2}{\left\langle \chi \left| H^{(0)} \right| \chi \right\rangle - E^{(0)}} \tag{3.95}$$

The evaluation of the integrals of Equation (3.95) allows the general term that dominates Series (3.87) to be obtained explicitly:

$$\begin{aligned} \frac{(2l_1 + 2l_2)!}{(2l_1)!\,(2l_2)!} E_{l_1 l_2}^{(2)} R^{-2(l_1+l_2+1)} &\leq \frac{(2l_1 + 2l_2)!\,(2l_1 + 2)\,(2l_2 + 2)}{\left[\dfrac{l_1}{l_1 + 1} + \dfrac{l_2}{l_2 + 1}\right] R^{2(l_1+l_2+1)}} \\ &< -2\frac{2l_1 + 2l_2}{2l_1} \frac{(l_1 + 1)\,(2l_1)!}{(2R)^{2l_1+1}} \frac{(l_2 + 1)\,(2l_2)!}{(2R)^{2l_2+1}} \end{aligned}$$

The upper bound, which is negative, can be very large (in absolute value) at rather large $R$. Therefore, the nth expansion term does not approach zero as $n \to \infty$. The Series (3.87) diverges for all finite $R$.

Nevertheless, at rather large $R$, the multipole expansion describes satisfactorily the behavior of the interaction energy. The first expansion terms decrease sufficiently fast and provide a good approximation for the polarization energy, see Table 3.3 where the error of the multipole expansion of the polarization energy[4] in

[4] The magnitude of the error given by the multipole expansion is denoted sometimes as *penetration energy*: $E_{pen}^{(2)} = E_{pol}^{(2)} - E_{pol.as}^{(2)}$.

**Table 3.3** Estimates of the terms in the multipole expansion of the polarization energy for the ground state of H–H (energy in $cm^{-1}$) [22]

| $R$ | $E^{(2)}_{pol}$ | $E^{(2)}_{pol.as}$ | $\alpha^{(2)}$ |
|---|---|---|---|
| 6 | −53.67 | −52.8 | 1.62% |
| 8 | −8.028 | −8.28 | 3.13% |
| 10 | −1.809 | −1.83 | 1.16% |
| 12 | −0.557 | −0.557 | 0 |

$\alpha^{(2)}$ is determined by Equation (3.96a).

the second order of PT is designated as:

$$\alpha^{(2)} = \frac{E^{(2)}_{pol} - E^{(2)}_{pol.as}}{E^{(2)}_{pol}} 100\% \tag{3.96a}$$

Convergence of the multipole sum to the exact energy value becomes worse for excited states, due to the large size of the electronic density distribution in the former. Kołos [23] has shown that the multipole expansion poorly approximates $E_{pol}$ for the state $B^1\sum_u^+$ of the system H–H, dissociating onto $1s$- and $2p$-state hydrogen atoms, up to $R \sim 20$ bohrs. Although, the approximation given by the first-order electrostatic energy, $E^{(1)}_{el}$, designated as:

$$\alpha^{(1)} = \frac{E^{(1)}_{el} - E^{(1)}_{el.as}}{E^{(1)}_{el}} 100\%, \tag{3.96b}$$

is satisfactory (see Table 3.4). It should be noticed that the hydrogen molecule is a special case, since its polarizability changes drastically as a result of the transition from the ground to an excited state. In the case of other many-electron molecules, such transitions do not affect the polarizability so strikingly and the multipole approximation should be expected to be more satisfactory.

**Table 3.4** Estimates of the terms in the multipole expansion of the interaction energy for the $B^1\sum_u^+$-state of H–H (energy in $cm^{-1}$) [23]

| $R, a_0$ | $E^{(1)}_{el}$ | $E^{(1)}_{el.as}$ | $\alpha^{(1)}$ | $E^{(2)}_{pol}$ | $E^{(2)}_{pol.as}$ | $\alpha^{(2)}$ |
|---|---|---|---|---|---|---|
| 12 | −155.41 | −140.96 | 9.36% | −44.8 | – | – |
| 15 | −73.37 | −72.17 | 1.6% | −7.4 | −5.6 | 24% |
| 18 | −41.86 | −41.77 | 0.2% | −1.9 | −1.6 | 15.8% |
| 20 | −30.46 | −30.45 | 0.03% | −0.88 | −0.71 | 19.3% |

$\alpha^{(1)}$ and $\alpha^{(2)}$ are determined by Equations (3.96).

### 3.1.5 Elimination of divergence in the multipole expansion

As discussed in Section 3.1.4, the expressions for multipole expansion (Equations (3.4), (3.5) and (3.11)) and a general formula (Equation (3.26)) are valid only if the effective sizes of the interacting systems are smaller than the intersystem distance $R$. It means that the overlap of the electronic shells of the interacting systems is negligible. On the other hand, for large nonspherical systems having some dimensions much larger than others (for instance, polymers), the convergence condition $R > \left(r_i + r_j\right)$ can be violated even in the absence of overlap, because of large values of $r_i$ or $r_j$. This case will be discussed at the end of this section.

For atoms and small molecules, all possible cases of overlapping of their electronic shells are depicted in Figure 3.2. The divergence of the multipole Expansion (3.11) arises from using this expansion, which is valid only in the region I (Figure 3.2), over the whole space. It can be expected that the application of the proper bipolar expansion in each region depicted in Figure 3.2 must lead to a series which converges over $l_1$ and $l_2$. Cusachs [37] was the first who showed it. He carried out the calculation of $E_{pol}^{(2)}$ for $H_2^+$ (in this case $E_{pol}^{(2)}$ reduces to the induction energy). The system considered has only one electron. Hence, only the operator $r_{b1}^{-1}$ is expanded. Expansion (3.2) is valid only in the region $r_1 < R$. Cusachs chose in each region the corresponding expansions:

$$\frac{1}{r_{1b}} = \begin{cases} \sum_{1=0}^{\infty} \frac{r^l}{R^{l+1}} P_l\left(\cos\theta\right), & r < R, \\ \sum_{1=0}^{\infty} \frac{R^l}{r^{l+1}} P_l\left(\cos\theta\right), & r > R. \end{cases} \tag{3.97}$$

For $E_{pol}^{(2)}$, its expression via the first-order correction of wave function $\Psi^{(1)}$ is presented in Appendix 3 (Equation (A3.103)). The function $\Psi^{(1)}$ was found from the Equation (A3.149). Substitution of Expansion (3.97) into the perturbation operator $V$ provided the possibility of representing $E_{ind}^{(2)}$ in the form of a convergent expansion. Cusachs obtained the explicit expressions for the dipole and quadrupole components of the induction energy, which at $R \to \infty$ transform to those of the multipole components obtained by Dalgarno *et al.* [31, 34].

The general approach was developed by Kreek and Meath [38] (see also Reference [39]). The authors also started from the expression for $E^{(2)}$, Equation (A3.103) where $\Psi^{(1)}$ satisfies Equation (A3.149). The operators $r_{ia}^{-1}$, and $r_{jb}^{-1}$, appearing in the perturbation operator, were expanded as Series (3.97) and the operator $r_{ij}^{-1}$ was expanded in the series Equation (3.80) where the coefficients are determined in each region in accordance with Equation (3.81). In this case, the perturbation operator can be represented as a sum of components with definite $l_1$ and $l_2$ as follows:

$$V = \sum_{l_1=0}^{\infty} \sum_{l_2=0}^{\infty} V_{l_1 l_2} \tag{3.98}$$

where $V_{l_1 l_2}$, is the sum of the corresponding terms of the Expansions (3.80), (3.81) and (3.97) over $m$ and all the interacting electrons, labelled by $i$ and $j$.

$V$ may be expanded in a series of products of the corresponding factors. Each factor is characterized by a definite symmetry with respect to rotations and is labelled by the definite value of an angular momentum. Then the function $\Psi^{(1)}$, which is determined from Equation (A3.149), can be represented also in the form of the corresponding expansion:

$$\Psi^{(1)} = \sum_{l_1=0}^{\infty} \sum_{l_2=0}^{\infty} \Psi_{l_1 l_2}^{(1)} \tag{3.99}$$

Thus, Equation (A3.149) is decomposed into the equations for each $\Psi_{l_1 l_2}^{(1)}$:

$$\left(H_0 - E^{(0)}\right) \Psi_{l_1 l_2}^{(1)} + \left(V_{l_1 l_2} - E^{(1)} \delta_{l_1 0} \delta_{l_2 0}\right) \Psi^{(0)} = 0 \tag{3.100}$$

If the states of the interacting molecules are spherically symmetric, the selection rules yield:

$$E^{(1)} = \left\langle \Psi^{(0)} \left| V \right| \Psi^{(0)} \right\rangle = \left\langle \Psi^{(0)} \left| V_{00} \right| \Psi^{(0)} \right\rangle \tag{3.101}$$

Substituting Expansions (3.98) and (3.99) into the expression for the second-order energy correction (Equation (A3.103)) and taking the selection rules for angular momentum into account, one obtains:

$$E^{(2)} = \sum_{l_1=0}^{\infty} \sum_{l_2=0}^{\infty} E_{l_1 l_2}^{(2)} \tag{3.102}$$

$$E_{l_1 l_2}^{(2)} = \left\langle \Psi^{(0)} \left| V \right| \Psi_{l_1 l_2}^{(1)} \right\rangle = \left\langle \Psi^{(0)} \left| V_{l_1 l_2} \right| \Psi_{l_1 l_2}^{(1)} \right\rangle \tag{3.103}$$

It is not a simple problem to solve Equation (3.100) for the majority of interactions. Usually, the different variational principles, in particular the Hylleraas variation principle, are applied (see Equations (3.90) and (3.91)). The variational function $\widetilde{\Psi}_{l_1 l_2}^{(1)}$ is assumed to be orthogonal to $\Psi^{(0)}$, and it is determined by the condition of energy minimum. A similar procedure, but with more complex expressions, is also used for higher-order perturbation corrections.

Kreek and Meath [38] called their approach as the method of 'non-expanded' energies, and the terms in Equation (3.102) as the individual 'non-expanded' energies. It means that they are not expanded in powers of $R^{-n}$, although, in fact, the energy is expanded but in a series of partial components, $E_{l_1 l_2}^{(2)}$. The partial energies $E_{l_1 l_2}^{(2)}$ involve exponentially decreasing terms and coincide with the multipole components only in the limit $R \to \infty$. Hence, the convergence problem for the Series (3.102) does not arise.

In the case of spherically-symmetric states of neutral atoms, the partial energies approach the corresponding dispersion energies at large $R$. For example, for hydrogen–hydrogen [38]:

$$\lim_{R\to\infty} E^{(2)}_{1,1} = C^{(2)}_6 R^{-6}$$
$$\lim_{R\to\infty} 2E^{(2)}_{1,2} = C^{(2)}_8 R^{-8} \qquad (3.104)$$
$$\lim_{R\to\infty} \left(2E^{(2)}_{1,3} + E^{(2)}_{2,2}\right) = C^{(2)}_{10} R^{-10}$$

A comparison of the behavior of the partial energies $E^{(2)}_{l_1 l_2}$ with the multipole components of the dispersion energy yields the values of $R$ where the multipole expansion becomes incorrect (Figure 3.3). Further development of the non-expanded energy method was performed in References. [40–44], where it was also spread on the molecule-molecule interaction that depend not only upon distance $R$ but also upon the mutual orientations.

Consider the partial wave expansion (Equation (3.102)) for the second-order dispersion energy of two interacting neutral systems:

$$E^{(2)}_{disp}(R) = \sum_{l_1=1}\sum_{l_2=1} E^{(2)}_{l_1 l_2}(R) \qquad (3.105)$$

The large-distance limit of individual non-expanded energies in Equation (3.105) can be expressed via the non-expanded dispersion coefficients $C_{l_1 l_2}$:

$$\lim_{R\to\infty} E^{(2)}_{l_1 l_2}(R) = -\frac{C_{l_1 l_2}}{R^{2(l_1+l_2+1)}} \qquad (3.106)$$

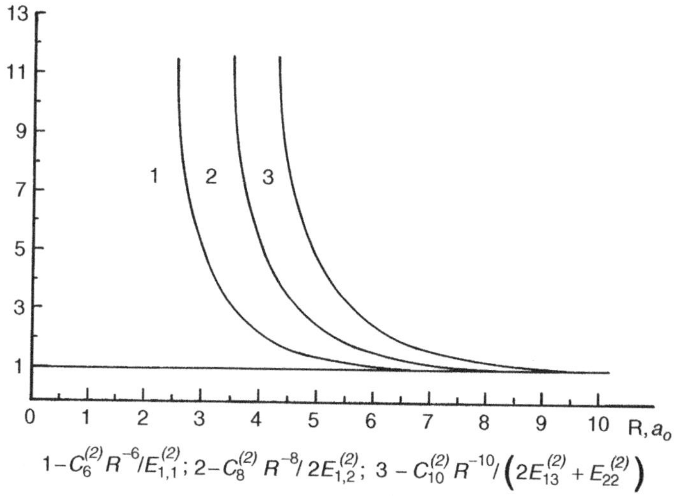


**Figure 3.3** Dependence of the ratios of the multipole series components of the dispersion energy to the corresponding partial energies on the distance for H–H [38]

They are connected with the wide-used dispersion coefficients $C_n$ by a simple summation:

$$C_n = \sum_{l_1, l_2}{}' C_{l_1 l_2} \tag{3.107}$$

where the prime in the sum means that $l_1$ and $l_2$ satisfy an obvious condition:

$$2 (l_1 + l_2 + 1) = n \tag{3.108}$$

For instance, for $n = 10$ there two terms in Sum (3.107), cf. Equation (3.104).

For finite $R$, the deviation of the individual non-expanded dispersion energies from their asymptotic limit (Equation (3.106)) can be described by so-called *damping function*. It is defined as a factor in the expression for $E^{(2)}_{l_1 l_2}(R)$. Namely:

$$E^{(2)}_{l_1 l_2}(R) = \frac{C_{l_1 l_2} D_{l_1 l_2}(R)}{R^{2(l_1+l_2+1)}} \tag{3.109}$$

At finite distances, the dispersion energy (Equation (3.105)), with $E^{(2)}_{l_1 l_2}$ expressed by Equation (3.109), can be represented in an equivalent but a more commonly encountered form:

$$E^{(2)}_{disp}(R) = \sum_{n=6} \frac{C_n D_n(R)}{R^n} \tag{3.110}$$

This form with the damping function $D_n(R)$ is widely used in semiempirical [45, 46] and *ab initio* [47, 48] model potentials (see Chapter 5). For sufficiently large values of distance $R$, the damping function approaches to unity and the dispersion energy may be rigorously described by the multiple series.

The expression for $D_n(R)$ via the noncomposed quantities may be obtained by comparing Equations (3.105) and (3.109) with the Representation (3.110). It follows:

$$D_n(R) = \sum_{l_1, l_2}{}' \frac{C_{l_1 l_2} D_{l_1 l_2}(R)}{C_n} \tag{3.111}$$

where $l_1$ and $l_2$ in sum run only over the values satisfying the Condition (3.108).

In the approach of the non-expanded energies, the problem is faced of solving the equations for the first order wave function $\Psi^{(1)}$ that can be solved only for simple systems. Calculations with highly correlated wave functions were carried out by Wheatley and Meath [44] for six pair interactions between hydrogen, helium and lithium. They calculated the values of the damping function (Equation (3.111)) for a broad range of distances. It is instructive to estimate the distance where the multiple expansion can be used without the damping factor. In Table 3.5 the values of distance $R_c$, at which the damping functions differs from unity no more than in 5%, are represented for three pair interactions.

The values for the helium–helium interaction can be used as a tentative limit in the case of closed-shell system interactions, the values of the hydrogen–hydrogen and lithium–lithium interactions model the open-shell system interactions, although the fact that the value of $R_c$ increases with the atom size has to be taken into

**Table 3.5** The values of distance $R_c$ that provide the deviation of the first terms in the multipole expansion of the dispersion energy from their asymptotic values less than 5% (based on the results of Reference [44], distance in bohrs)

| | He–He | H–H | Li–Li |
|---|---|---|---|
| $D_6$ | 5.0 | 7.0 | 14.0 |
| $D_8$ | 6.0 | 8.5 | 16.0 |
| $D_{10}$ | 7.0 | 10.0 | 18.0 |

account. For one-valence-electron atoms, according to Table 3.5, it is almost two times larger for the lithium–lithium interaction than for the hydrogen–hydrogen interaction.

As mentioned in the beginning of this section, in the case of interaction between large systems (polymers, biological structures, molecular crystals), the convergence condition $R > r_1 + r_2$ for the multipole expansion can be violated even if the overlap of the charge densities of the interacting systems is negligible. The typical distances of interaction are often comparable to the molecular sizes or smaller, although they are large enough that the exchange of electrons and overlapping effects may be neglected. This kind of divergence is termed *shape divergence* as opposed to the *extension* or *penetration divergence* that always exists in the multipole expansion.

At present, different approximate methods for calculating the interactions between large molecules have been developed: some of them are discussed in detail in Reference [18], Chapter 2. All these methods are based on a common idea: the molecule is divided into separate fragments (atoms, a small group of atoms like a methyl group, or bonds); and each fragment is described by its own multipole moments (it can be only monopoles, that is, point charges). The multisite multipole expansion converges much better than a monosite molecular multipole expansion. The calculation methods created for large molecule interaction can be classified as the point-charge methods [49–54], the bond–bond interaction methods [5, 55–57], the distributed multipole expansions [58–61] based on the similar ideas and its improved modification called by authors [62–64] *topologically partitioned electric properties* and based on the Bader topological theory [65, 66].

Although the convergence in the multisite multipole expansion is considerably improved, it has to be remembered that these methods still contain the asymptotic multipole expansions for each fragment (or domain) and also suffer from the approximations depending on the division model. The latter is inevitable in the case of large systems; although, the asymptotic divergence of the multipole expansion can be avoided. For small systems, it was done by Meath and coworkers, as discussed in this section. For large systems the promising approach was developed by Wheatley [67]. He also started from the partition of a large molecule into separate parts but then instead of the multipole expansion he introduced so-called *Gaussian multipoles*. The latter are defined by differentiation of spherical Gaussian functions. A Gaussian multipole expansion is readily obtained if the molecular wave function is calculated using a Gaussian basis set that is now widely exploited.

The electrostatic field due to a Gaussian multipole includes the charge overlap effects at small distances; for sufficiently large distances it is equal to the point multipole electrostatic field. Hence, Gaussian multipoles should be regarded as a generalization of point multipoles that includes short-range penetration effects. An implementation of the Gaussian multipole technique was performed in Reference [68]. It was shown that rather accurate results can be obtained with this technique, although the Gaussian multipole expansion is more time-consuming than standard multipole-expansion schemes.

# 3.2 Intermediate and Short Distances

## 3.2.1 Perturbation theory with exchange

### 3.2.1.1 Ambiguity of the exchange-perturbation theory series

In the proceeding section, the case of large separations where the exchange effects are negligible was considered. At intermediate distances, that is, in the range 4 bohr $\leq R \leq$ 15 bohr, where the exchange effects become appreciable, the interaction between molecules (or molecular fragments) that are not large can still be considered as a small perturbation and the Hamiltonian of two interacting molecules is presented as:

$$H = H_0 + V \tag{3.112}$$

$$H_0 = H_A + H_B \tag{3.113}$$

where $H_0$ is the zeroth-order Hamiltonian with eigenfunctions as a simple product of the eigenfunctions of the isolated molecules:

$$\Phi_{nm}^{(0)} = \Psi_n^A \Psi_m^B \tag{3.114}$$

$$H_0 \Phi_{nm}^{(0)} = \left(E_n^A + E_m^B\right) \Phi_{nm}^{(0)} \tag{3.115}$$

However, presentation of the zeroth-order wave functions as a product (Equation (3.114)) does not provide the exchange energy contribution. The latter is negligible at large distances but becomes important at intermediate distances. As was demonstrated by Claverie [69], the presentation (Equation (3.114)) leads to nonphysical solutions. The reason for this nonphysical behavior is based on the *Pauli exclusion principle* [70, 71]. The many-electron wave must satisfy the Pauli exclusion principle at all orders of perturbation theory. It means that the zeroth-order wave functions (Equation (3.114)) must be antisymmetrized. The function (Equation (3.114)) is antisymmetric only with the respect to permutations of the electrons of molecule $A$ among themselves and of the electrons of molecule $B$

among themselves. The total antisymmetrization is performed by the operator permutating electrons between molecules:

$$\widehat{\mathcal{A}} = N_{AB} \sum_{Q} (-1)^q\, Q \tag{3.116}$$

where $N_{AB}$ is the normalization factor and $q$ is the parity of the exchange permutation $Q$, the total number of such permutations is equal to:

$$N_Q = \frac{(N_A + N_B)!}{N_A! N_B!} \tag{3.117}$$

The Pauli-permitted zeroth-order wave function is constructed by means of the operator (3.116):

$$\Psi_{nm}^{(0)}\,(antisymm.) = \widehat{\mathcal{A}}\Phi_{nm}^{(0)} = N_{AB} \sum_{Q} (-1)^q\, Q \Psi_n^{(0)} \Psi_m^{(0)} \tag{3.118}$$

Nevertheless, this seemingly natural physical choice of the zeroth-order functions leads to serious difficulties in the development of the perturbation theory. The antisymmetric zeroth-order functions (Equation (3.118)) are not eigenfunctions of the zeroth-order Hamiltonian $H_0$ due to its non-invariance with respect to permutations of electrons between molecules. For instance, for the case of the simplest system of two hydrogen atoms:

$$H_0 = H_A + H_B = -\frac{1}{2}\nabla\frac{2}{2} - \frac{1}{r_{a1}} - \frac{1}{2}\nabla\frac{2}{2} - \frac{1}{r_{b2}} \tag{3.119}$$

It is evident that $H_0$ is nonsymmetric relative to permutations of the electrons. While, the total Hamiltonian obtained by addition of the interaction operator:

$$V = -\frac{1}{r_{a2}} - \frac{1}{r_{b1}} + \frac{1}{R_{ab}} + \frac{1}{r_{12}} \tag{3.120}$$

to $H_0$ is symmetric.

The symmetry group of the total Hamiltonian contains that of the zeroth-order Hamiltonian as a subgroup. In other words, the total Hamiltonian $H$ commutes with the antisymmetrizer $\widehat{\mathcal{A}}$:

$$\left[H, \widehat{\mathcal{A}}\right] = 0 \tag{3.121}$$

while the zeroth-order Hamiltonian and the perturbation operator $V$ do not commute with it:

$$\left[H_0, \widehat{\mathcal{A}}\right] \neq 0 \qquad \left[V, \widehat{\mathcal{A}}\right] \neq 0 \tag{3.122}$$

Therefore, the use of the antisymmetrized functions in the form of Equation (3.118) as zeroth-order functions, does not allow the standard techniques of the Rayleigh–Schrödinger or Brillouin–Wigner perturbation theories (Section A3.3) to be applied.

The set of antisymmetrized functions $\{\widehat{\mathcal{A}}\Phi_{nm}\}$, where $\Phi_{nm}$ is the product of the eigenfunctions of the isolated molecules, has another inconvenient shortcoming. The functions $\widehat{\mathcal{A}}\Phi_{nm}$ are nonorthogonal to each other. In addition to numerical inconveniences, it means that the functions $\widehat{\mathcal{A}}\Phi_{nm}$ cannot be eigenfunctions of any Hermitian operator.

These shortcomings lead to the possibility of different versions of the perturbation theory based on the Functions (3.118). As shown by Lekkerkerker and Laidlaw [72], the coefficients $c_{nm}$ of the expansion of any function in the antisymmetric basis (Equation (3.118)) are defined with the precision up to an arbitrary function of the electronic coordinates. In the first paper, devoted to the exchange perturbation theory, Eisenschitz and London [73] chose the coefficients $c_{nm}$ under the condition that the sum of the squares of the absolute values of $c_{nm}$ should be minimum. Generally speaking, that choice is an arbitrary one and offers no advantages by comparison with the choices made in other formalisms [72].

In connection with the ambiguity of the expansion over antisymmetric functions, a large number of different *exchange perturbation theories* (EPT) were developed. They can be divided into two main groups, depending on the zeroth-order Hamiltonian used. The first group combines the methods employing the nonsymmetric zeroth-order Hamiltonian (Equation (3.113)) and named *Symmetry Adapted Perturbation Theories* (SAPT), since at each order of PT the antisymmetric zeroth-order wave function must be used. The second group includes the methods that allow to apply the standard Rayleigh–Schrödinger PT, either because of the special choice of symmetric zeroth-order Hamiltonian for which the antisymmetric functions are the eigenfunctions, or for any other reasons.

#### 3.2.1.2 Symmetry adapted perturbation theories

The first solution of the problem of developing the perturbation theory with exchange was given in 1930 by Eisenschitz and London [73]. After their seminal paper, many different approaches in this direction were published [74–86]. They are discussed in detail in the author's previous book [18, Chapter 3] and in recent reviews by Jeziorski and Szalewicz [87, 88]. Here only a short outline is presented.

The often used SAPT approaches are usually denoted by abbreviations. Some of them are listed below:

| | |
|---|---|
| EL–HAV | are the Eisenschitz–London method [73], the equivalent treatment due to Hirschfelder [75], and the reformulation of this approach in terms of wave operators by van der Avoird [76]. |
| MS–MA | are the Murrell–Shaw [77] and Musher–Amos [78] methods; as was shown in Reference [89] they are equivalent. |
| AM | is the Amos–Musher method [79]. |
| HS | is the Hirschfelder–Silbey method [74]. |
| CBH | is the Chipman–Bowman–Hirschfelder approach [84]. |
| JK | is the Jeziorski–Kołos approach [86]. |

A comparative analysis of different SAPT approaches was performed in References [84, 89–98]. It was shown that the correct wave function, which is the solution of the Schrödinger equation with the total Hamiltonian $H$, can be represented as:

$$^{\nu}\Psi = {}^{\nu}\widehat{A}\Phi_{\nu} \tag{3.123}$$

where $^{\nu}\widehat{A}$ is the symmetrizer operator, $\Phi_{\nu}$ is the nonsymmetric function that can be different for each irreducible representation $^{\nu}\Gamma$ of the symmetry group $\mathbf{G}$ of the Hamiltonian. The group $\mathbf{G}$ is the direct product of the permutation group with the point symmetry group of the system. Thus, the wave function (Equation (3.123)) fulfills the requirements of the Pauli principle and the point symmetry of the molecule as well.

The Schrödinger equation for the function (Equation (3.123)) can be written in the form:

$$^{\nu}\widehat{A}\left(H - {}^{\nu}E\right)\Phi_{\nu} = 0 \tag{3.124}$$

that is equivalent [84] to:

$$\left(H - {}^{\nu}E\right)\Phi_{\nu} = \sum_{\mu \neq \nu} {}^{\mu}\widehat{A}F_{\nu} \tag{3.125}$$

where $F_{\nu}$ is an arbitrary function determining the components of $\Phi_{\nu}$ in the subspaces with symmetrics $\mu \neq \nu$. The different modifications of EPT correspond to different choices of $F_{\gamma}$, namely [94]:

$$\begin{aligned}
&\text{EL–HAV:} && {}^{\mu}\widehat{A}F_{\nu} = {}^{\mu}\widehat{A}\left(V - {}^{\nu}E + E_0\right)\Phi_{\nu} \\
&\text{AM:} && {}^{\mu}\widehat{A}F_{\nu} = {}^{\mu}\widehat{A}V\Phi_{\nu} \\
&\text{MS–MA:} && {}^{\mu}\widehat{A}F_{\nu} = {}^{\mu}\widehat{A}\left(H_0 + V - {}^{\nu}E\right)\Phi_{\nu} \\
&\text{HS:} && {}^{\mu}\widehat{A}F_{\nu} = \left({}^{\mu}E - {}^{\nu}E\right){}^{\mu}\widehat{A}\Phi_{\nu}, \qquad \mu \neq \nu.
\end{aligned} \tag{3.126}$$

In the CBH approach [84], the more general form of $F_{\gamma}$ is chosen:

$$F_{\gamma} \sim \sum_{\mu} {}^{\mu}\widehat{A}\left[{}^{\mu}\alpha_{\nu}V - {}^{\mu}\beta_{\nu}\right]\Phi_{\nu} \tag{3.127}$$

with the variational parameters $^{\mu}\alpha_{\nu}$ and $^{\mu}\beta_{\nu}$.

Which approach is the best one? This problem can be solved only under an independent criterion for the selection of nonsymmetric function. Many different criteria for the optimal choice of a primitive function were suggested [90, 91, 94–98]. Some of them are based on the condition of the smallest deviation of the function $\Phi_{\nu}$ from the eigenfunction (Equation (3.114)) of the unperturbed Hamiltonian (Equation (3.113)). It should be mentioned that the Adams criterion [91] does not allow the best approach to be selected. The analysis performed in References [95, 96] did not give a preference to any of the SAPT approaches studied. On the other hand, other criteria lead to contradictions. Chipman's criterion [94] justifies the MS–MA approach up to the second order, while according to Kutzelnigg [98] the MS–MA approach and the EL–HAV as well are not justified. This means that an ambiguity is peculiar not only to the different SAPT approaches, but to the criteria of their selection as well. The latter is related to the nonphysical meaning of the primitive function concept.

Jeziorski and Kołos [86] proposed an original approach based on an iterative procedure for the solution of the Schrödinger equation. On this way, they obtained the general expressions connecting the energy and wave functions at each iterative step. These expressions can be written in the following form (the detailed derivation is presented in Reference [18]):

$$\mathcal{E}_n = E_n - E_0 = \left\langle \Phi_0 \left| V \right| \widehat{G}\Psi_{n-1} \right\rangle \tag{3.128}$$

$$\Psi_n = \Phi_0 + \widehat{R}_0 \left(\mathcal{E}_n - V\right) \widehat{F}\Psi_{n-1} \tag{3.129}$$

where $\Phi_0$ is the eigenfunction and $\widehat{R}_0$ is the reduced resolvent of the zeroth-order Hamiltonian $H_0$ (Equation (A3.140)):

$$\widehat{R}_0 = \sum_{k\neq 0} \frac{|\Phi_k\rangle \langle \Phi_k|}{E_0^{(0)} - E_k^{(0)}} \tag{3.130}$$

$E_k^{(0)}$ is the energy of an excited state of $H_0$, and $\widehat{F}$ and $\widehat{G}$ are symmetry forcing operators. Different choice of these symmetry operators lead to different SAPT approaches.

In the JK scheme [86], the symmetry forcing procedure was chosen to be intermediate to those that generate the EL–HAV and MS–MA approaches. The authors [86] applied the iteration procedure (Equation (3.129)) with operators $\widehat{G} = 1$ and $F = \widehat{A}$, choosing the initial function $\Psi_0 = \left\langle \Phi_0 \mid \widehat{A}\Phi_0 \right\rangle^{-1} \widehat{A}\Phi_0$. This choice leads to the JK perturbation series, the particular terms of which have the following form:

$$E^{(n)} = \left\langle \Phi_0 \left| V \right| \Psi^{(n-1)} \right\rangle \tag{3.131}$$

$$\Psi^{(1)} = \left\langle \Phi_0 \mid \widehat{A}\Phi_0 \right\rangle^{-1} \widehat{R}_0\widehat{A}\left(E^{(1)} - V\right)\Phi_0 \tag{3.132}$$

$$\Psi^{(n)} = -\widehat{R}_0 V\widehat{A}\Psi^{(n-1)} + \sum_{r=1}^{n-1} E^{(r)}\widehat{R}_0\widehat{A}\Psi^{(n-r)}, \quad n \geq 2. \tag{3.133}$$

In the next study by Jeziorski *et al.* [99], the authors suggested that the symmetry is forced in the energy corrections but not in Equation (3.129) for the wave function. They put $\widehat{G} = \left\langle \Phi_0 \mid \widehat{A}\Phi_0 \right\rangle^{-1} \widehat{A}$, $F = 1$, and $\Psi_0 = \Phi_0$. Note that the new choice of symmetry forcing operators was in some sense opposite to the previous JK choice. If the result of this modernized iteration procedure is expanded in powers of $V$, a perturbation expansion for the energy is obtained with the *nth* order term given by the recurrence formula:

$$E^{(n)} = \frac{1}{\left\langle \Phi_0 \mid \widehat{A}\Phi_0 \right\rangle} \left[ \left\langle \Phi_0 \left| V \right| \widehat{A}\Psi^{(n-1)} \right\rangle - \sum_{r=1}^{n-1} E^{(r)} \left\langle \Phi_0 \mid \widehat{A}\Psi^{(n-r)} \right\rangle \right] \tag{3.134}$$

The corrections to the wave function coincide with those in the standard Rayleigh–Schrödinger theory:

$$\Psi^{(n)} = -R_0 V \Psi^{(n-1)} + \sum_{r=1}^{n} \langle \Phi_0 | V | \Psi^{(r-1)} \rangle \widehat{R}_0 \Psi^{(n-r)} \quad (3.135)$$

This approach was named by authors [99] *symmetrized Rayleigh–Schrödinger* (SRS) theory.

As discussed above, there are no theoretical criteria for selection of the best SAPT method. Hence, only numerical calculations can provide a criterion to identify which approach is better. The comparative calculations given by different SAPT schemes are analyzed in References [18, 87, 88]. The results demonstrate that each of them has its advantages and shortcomings. It is important to note that, because of computational problems, the discussed SAPT schemes, except the SRS approach, were applied only to small molecular systems with number of electrons $N \leq 4$. As claimed by the authors [87, 88], all practical applications of the SAPT method performed thus far for the many-electron molecules have been based on the SRS theory approach.

In fact, it is not the initial SRS method [99], but a modified method named a *hybrid* approach [100]. It includes the *supermolecular* approach for the Hartree–Fock part of the interaction energy and the corrections in the correlation part on the *intramonomer correlations* within the framework of the double perturbation theory [101]. The interaction energy is represented as a sum of two main contributions:

$$E_{int} = E_{int}^{HF} + \Delta E_{SAPT}^{corr} \quad (3.136)$$

The separate calculation of $E_{int}^{HF}$ considerably improved the convergence of the SAPT series, since the Hartree–Fock interaction energy contains most of the exchange, induction and higher-order induction-exchange contributions. The Hartree–Fock contribution is calculated within the framework of the variational method, hence it is defined as a difference of the total Hartree–Fock energies:

$$E_{int}^{HF}(AB) = E^{HF}(AB) - \left[E^{HF}(A) + E^{HF}(B)\right] \quad (3.137)$$

and the problem of the Basis Set Superposition Error (BSSE) becomes significant (see Section 3.2.2.2).

The correlation part of the interaction energy is decomposed in the SAPT series with account of the intramonomer correlation [100]:

$$\Delta E_{SAPT}^{corr} = E_{intra}^{(1)} + E_{ind-intra}^{(2)} + E_{disp}^{(2)} + E_{disp-exch}^{(2)} + E_{disp-intra}^{(2)} + \cdots \quad (3.138)$$

where $E_{intra}^{(1)}$ collects the intramonomer correlation contributions to $E^{(1)}$, $E_{ind-intra}^{(2)}$ represents analogous contributions to $E_{ind}^{(2)}$, $E_{disp-exch}^{(2)}$ is the exchange part of the dispersion energy in the second order that usually reduce the magnitude of the dispersion interaction (see Table 3.7 in Section 3.2.2).

This SAPT scheme with some modifications, e.g. with inclusion of the coupled HF response of perturbed system (orbital relaxation effects [102]), has been applied for calculations of intermolecular interactions in many-electron complexes and clusters, for instance, $(H_2O)_2$ [100, 103], He–$C_2H_2$ [104], $H_2$–CO and $D_2$–CO [105],

Ne–HCN [106]; see also a recent review [107]. At the present status of computational technology, the SAPT codes allow rigid-monomer intermolecular potentials to be calculated with an accuracy of a few percent (this requires the fourth order of PT) for dimers containing up to 50 electrons [88]. At the second order of PT, which provides a qualitative level of obtained potentials, calculations can be made for even larger systems [108].

The comparison of the SAPT calculations with the *ab initio* calculations by the Møller–Plesset perturbation theory (Section A3.3.2) or a more advance coupled cluster, single, double and noniterative triple (CCSD(T)) method ( Section A3.2.2.3), reveals a good agreement, to within a few percent in the region of the van der Waals minimum and for larger intermolecular separations. For small separations the difference increases, since in the repulsive region, the SAPT results are less accurate than the supermolecular highly correlated calculations. The reason for this can be in the cancellation of the basis-set effect and exchange effect [109].

It should be stressed that the SAPT schemes do not have a rigorous theoretical base. As shown by Adams [110, 111], the SAPT series is divergent. Nevertheless, it does not prevent the SAPT method demonstrating accurate results in practical calculations. Certainly, it can be explained by some peculiarities of the SAPT scheme. But it is worthwhile to comment that, as often happens in molecular physics and quantum chemistry, approximations based on some assumptions that do not look well substantiated become more precise than expected. It was so with the concept of the linear combinations of atomic orbitals (LCAO). Perhaps the most striking example is the Born–Oppenheimer adiabatic approximation that was originally suggested as an expansion over a small parameter $(m/M)^{1/4}$ [112]. In the most unfavorable case of the hydrogen molecule, this parameter is equal to 0.15. As follows from the precise *ab initio* calculations, the error of adiabatic approximation in the hydrogen molecule case has the order $10^{-4}$–$10^{-5}$ (see Section 1.3).

The supermolecular MPPT approach, competing with SAPT, belongs to the second group of EPT methods, which allow the standard RS perturbation theory to be applied. This approach and its modification by Chałasiński and Szcześniak [113, 114], connecting MPPT with SAPT, will be discussed in next section.

#### 3.2.1.3 Methods allowing the standard Rayleigh–Schrödinger perturbation theory to be applied

As discussed in Section 3.2.1.1, the main problem arising in the EPT is related to the fact that the Pauli-permitted antisymmetric wave functions (Equation (3.118)) are not the eigenfunctions of the zeroth-order Hamiltonian $H_0$. One way to remove this difficulty is to construct the symmetrized Hamiltonian $\widetilde{H}_0$. Such an approach was first proposed in References [115, 116]. The Hamiltonian $\widetilde{H}_0$ has the antisymmetric wave functions (Equation (3.118)) as the eigenfunctions, but the latter form an nonorthogonal set. Hence, $\widetilde{H}_0$ cannot be Hermitian. The employment of the non-Hermitian operators is not ordinary in quantum mechanics, nevertheless, they can be applied in practical calculations.

The procedure suggested in References [115, 116] is simple in principle. The total Hamiltonian is presented as a sum of kinetic $T$ and potential $U$ operators (the latter divided on the unperturbed and perturbed parts):

$$H = T + U = T + U_0 + V = H_0 + V \tag{3.139}$$

is identically transformed to:

$$H = \widetilde{H}_0 + V \tag{3.140}$$

in such a way that:

$$\widetilde{H}_0 \widehat{A} \Phi_0 \equiv \left(T + \widetilde{U}_0\right) \widehat{A} \Phi_0 = E_0 \widehat{A} \Phi_0 \tag{3.141}$$

It follows formally from Equation (3.141) that:

$$\widetilde{U}_0 = E_0 - T \widehat{A} \Phi_0 / \widehat{A} \Phi_0 \tag{3.142}$$

The potential (3.142) is denoted usually as the *Sternheimer potential.* It can be transformed into:

$$\widetilde{U}_0 = \widehat{A} U_0 \Phi_0 / \widehat{A} \Phi_0 \tag{3.143}$$

where the commutation of $T$ with $\widehat{A}$ and the identity $T\Phi_0 = E_0\Phi_0 - U_0\Phi_0$ are applied. The perturbation potential becomes:

$$\widetilde{V} = H - \widetilde{H}_0 = U - \frac{\widehat{A} U_0 \Phi_0}{\widehat{A} \Phi_0} = U - \frac{\widehat{A} U \Phi_0}{\widehat{A} \Phi_0} + \frac{\widehat{A} V \Phi_0}{\widehat{A} \Phi_0}$$

Since $U$ and $\widehat{A}$ commute with each other and the action of the potential energy operator on an arbitrary function reduces to a multiplication, for $\widetilde{V}$ an expression that is similar to Equation (3.143) is obtained:

$$\widetilde{V} = \widehat{A} V \Phi_0 / \widehat{A} \Phi_0 \tag{3.144}$$

The perturbation operator (Equation (3.144)) was applied to the evaluation of the two-electron system hydrogen–hydrogen via the RS formalism [117]. However, the Sternheimer approach is not applicable to many-electron systems [118]. Different modifications of the Sternheimer Hamiltonian have been proposed [119–122]. In spite of some restrictions in applications to many-electron system [123], this non-Hermitian formalism was successfully exploited in several studies [124, 125] of some many-electron systems.

A matrix formalism allowing to apply the standard RS perturbation theory was developed in References [126–129]. In these approaches, the molecular orbitals belonging to different molecules are subjected to a preliminary orthogonalization. The terms due to the nonorthogonality as well as the terms due to the intramolecular correlation are involved in the perturbation. Hence, for this approach to be valid, the nonorthogonal terms have to be small. But if large basis sets are employed with diffuse basis functions included, the overlap effects are large enough; they cannot be considered as a small perturbation. Another problem is that the orthogonalization process leads to contamination of the orbitals of one molecule by those of other. The similar orbital contamination problems should exist in the biorthogonal formalism proposed by Sarjan *et al.* [130, 131].

However, the rigorous perturbation method based on the RS perturbation theory and taking into account the exchange was developed more than 70 years ago; it is the *Møller-Plesset perturbation theory* (MPPT) [132] (Section A3.3.2). Various modifications of the MPPT approach have been developed and grouped as *many-body perturbation theory* (MBPT) [133–136]. The application of the MPPT to molecular systems was initiated by several groups in the 1970s [137–141] and became widespread after the Møller–Plesset code was embedded in the suite of programs 'Gaussian'.

The main idea of the Møller–Plesset approach [132] was to use the Hartree–Fock Hamiltonian as the zeroth-order approximation. For the intermolecular interaction problem, it means that instead of being the sum of the Hamiltonians of noninteracting molecules (Equation (3.113)), $H_0$ is chosen as a sum over the one-electron $h_i^{HF}$ Hamiltonians of the system of interacting molecules (*'supermolecular' approach*):

$$H_0 = \sum_{i=1}^{N} h_i^{HF} \tag{3.145}$$

The zeroth-order wave function is the antisymmetric Hartree–Fock function that is naturally an eigenfunction of the Hamiltonian (Equation (3.145)). Usually, the unrestricted Hartree–Fock approach (UHF, Section A3.2.1) is employed:

$$\Psi^{(0)} = \frac{1}{\sqrt{N!}} det \left| \psi_{1\alpha}\psi_{2\beta}\psi_{3\alpha} \cdots \psi_{N\beta} \right| \tag{3.146}$$

The perturbation is the difference between the total Hamiltonian $H$ (in the adiabatic and nonrelativistic approximations) and the HF Hamiltonian (Equation (3.145)):

$$V = H - \sum_{i=1}^{N} h_i^{HF} \tag{3.147}$$

The intermolecular interaction energy is defined as a difference of the total energy of the 'supermolecule' and the energies of constituent molecules. For a two-molecule (two-atom) system:

$$E_{int} = E\,(AB) - [E\,(A) + E\,(B)] \tag{3.148}$$

Each term in Equation (3.148) is calculated in the same order of MPPT using the standard Rayleigh–Schrödinger formulae. The sum of the zeroth and first orders corresponds to the Hartree–Fock level (see Appendix 3, Equation (A3.115)). The second order of the MPPT (MP2) gives, in many cases, quite satisfactory qualitative results. For more precise calculations it is necessary to use the fourth order of the MPPT with single, double, triple and quadruple excitations denoted as MP4(SDTQ).

The application of the MPPT at intermediate and not very short distances does not lead to any problems. Certainly, it may be applied at large distances as well, but as long as the magnitude of $E_{int}$ remains larger than the relative errors in computation of the total energies in Equation (3.148). The potential energy surfaces for atomic and molecular clusters can be calculated for a broad range of distances. Such types of calculation were performed by Tao and Klemperer [142]. Using the full MP4 approach with a large basis set containing the bond functions, they obtained the potential energy surface for the dimer $(HCl)_2$ with a high level of precision (see also the MP4 calculations of the potential energy surfaces for the complexes of Ar–HF [143] and He–Co [144] performed by this group).

The potential energy surfaces for alkaline-earth trimers $Be_3$, $Mg_3$ and $Ca_3$ were calculated in a broad range of distances at the MP4(SDTQ) level by Kaplan *et al.* [145]. In this study, the two-body and three-body potential energy surfaces were calculated separately and these data were used to construct many-body model potentials [146] (see also Section 5.1.13).

The many-body decomposition, separation of interaction energy on the Hartree–Fock and electron correlation contributions, and the orbital population analysis allow the elucidation of the nature of binding in the compounds composed of atoms with closed electron shells, that is, with atoms not possessing the valence electrons in their ground state [145, 147, 148] (see Section 4.4.3).

The modification of the MPPT approach was developed by Pulay *et al.* [149–154]. They demonstrated that employing the localized occupied orbitals allows the computation time to be decreased considerably and, to a great extent, to reduce the basis set superposition error (for its definition and discussion see Section 3.2.2). The localized MPPT, mainly in its second order version (LMP2), was successfully applied to study the intermolecular interactions [155–158].

In Reference [156], LMP2 was applied to study the aurophilic interactions in the $[X\text{–}Au\text{–}PH_3]_2$ dimers for $X = H$ and Cl. The decomposition of the correlation energy elaborated in Reference [155] reveals that the dispersion attraction is accompanied by the important ionic component at short distances. This correlation energy decomposition is based on the localization of excitation on monomers. For instance, simultaneous single excitations on two different monomers corresponds to the dispersion interaction between these monomers [155].

Fomine *et al.* [157, 158] studied the intermolecular interactions between aromatic molecules employing the LMP2 technique. In Reference [158], the comparative calculations by MP2, LMP2 and MP4(SDTQ) were performed. The comparison with MP4(SDTQ) shows that for the geometry optimization of aromatic dimers, the LMP2 method appears to be preferable to the MP2 method.

Chałasiński and Szcześniak [113, 114] combined the MPPT approach with the intermolecular SAPT scheme that allowed the interaction energy calculated at different orders of MPPT to be decomposed into physically meaningful contributions: exchange, polarization (dispersion and induction), polarization-exchange and deformation. This decomposition was based on the symmetry-adapted *double* perturbation theory [101], in which the intramolecular correlations and intermolecular interactions were treated together as perturbations.

The zeroth-order approximation in the MPPT method is the Hartree–Fock approximation. Let us denote it as the self-consistent field (SCF) approximation. The SCF interaction energy can be decomposed as [159]:

$$E_{int}^{SCF} = \varepsilon^{HL} + \Delta E_{def}^{SCF} = \varepsilon_{el}^{(1)} + \varepsilon_{exch}^{(1)} + \Delta E_{def}^{SCF} \tag{3.149}$$

where the Heitler–London energy, $\varepsilon^{HL}$, is the sum of electrostatic, $\varepsilon_{el}^{(1)}$, and exchange, $\varepsilon_{exch}^{(1)}$, energies calculated in the first order of the standard PT. The deformation energy, $\Delta E_{def}^{SCF}$, is the correction due to the relaxation of orbitals in the self-consistent field. It originates from the induction–exchange interactions. This effect is not allowed in the first order of the standard PT. Let us denote this correction to the first order of the standard PT as $\Delta E_{def}^{(1)} \equiv \Delta E_{def}^{SCF}$.

In the second order of the MPPT:

$$E_{int}^{(2)} = \varepsilon_{el,r}^{(12)} + \varepsilon_{disp}^{(20)} + \Delta E_{def}^{(2)} + \Delta E_{exch}^{(2)} \tag{3.150}$$

where $\varepsilon_{el,r}^{(12)}$ is the electrostatic-correlation energy caused by the intramonomer correlation in the second order, $\varepsilon_{disp}^{(20)}$ is the second order dispersion energy calculated over the UHF functions (as is common in the MPPT approach), $\Delta E_{def}^{(2)}$ includes the second order intramonomer correlation corrections to the SCF deformation term and $\Delta E_{exch}^{(2)}$ gathers the second order exchange-correlation effects. These terms have the same meaning in the higher orders of the MPPT.

The general expression for the *nth* order correction is presented in the following form [114]:

$$E_{int}^{(n)} = \varepsilon_{el}^{(1,n)} + \sum_{i=0}^{n-2} \varepsilon_{disp}^{(n-i,i)} + \Delta E_{def}^{(n)} + \Delta E_{exch}^{(n)} \tag{3.151}$$

where $\varepsilon_{el}^{(1,n)}$ is the electrostatic intramolecular correlation correction in the *nth* order, $\varepsilon_{disp}^{(n-i,i)}$ represents the terms that couple the intermonomer correlation terms in the $(n-i)\,th$ order and intramonomer terms in the *ith* order. $\Delta E_{def}^{(n)}$ represents the inter- and intramonomer components to the deformation term and $\Delta E_{exch}^{(n)}$ represents the exchange–dispersion energy and intramonomer correlation contributions in the *nth* order to the exchange energy.

The described supermolecular approach combined with SAPT was applied to analyze the nonadditive effects in trimers of different nature [114, 160]. In Reference [161] the potential surfaces for Ar-CO and He-CO were calculated at the fourth order of MPPT using extended basis sets with the bond functions. The minimum for both complexes was obtained at approximately T-shaped geometry. The geometrical parameters obtained agree well with experimental data. According to analysis by the energy decomposition Equations (3.149)–(3.151), both complexes are bound by dispersion forces, whereas the anisotropy of interaction is determined by the exchange repulsion component $\varepsilon_{exch}^{HL}$ of $E_{int}^{SCF}$. Other systems of interest are discussed in a recent review [109].

### 3.2.2 Variational methods

In a molecular complex, the interacting molecules loose their identity at short distances. The system should be considered as a 'supermolecule' and studied with the help of the methods used for isolated molecules. Variational methods are very appropriate in these studies. In these methods the interaction energy, $E_{int}$, is defined as the difference between the total energy $E$ of the system under study and the sum of the energies of the isolated molecules:

$$E_{int} = E\,(ABC\ldots Q) - \sum_{a=A}^{Q} E_a \tag{3.152}$$

The supermolecular approach is valid not only at short distances but also for any separation and this is its advantage in comparison with the perturbation approaches. On the other hand, $E_{int}$ becomes small at large distances. The determination of the interaction energy by means of the Formula (3.152), as a small difference of two large quantities, imposes strict requirements on the accuracy of the energies involved in Equation (3.152) and restricts the intermolecular distance R within a range of short and intermediate distances $\left(R \lesssim 20 \text{ bohrs}\right)$.

To obtain results of calculations having the experimental precision, the variational function must be multiparametric and explicitly include the interelectronic distance $r_{12}$. This kind of calculation was performed for the simplest two-electron molecular system, the hydrogen molecule, by Kołos and Wolniewicz [162] and by Kołos and Rychlewski [163] (see also References [164, 165] and the discussion in Section 1.3). These calculations demonstrated an exact agreement with the experiment data within the limits of the experimental error (Table 1.1).

Unfortunately, the calculations, similar to mentioned above, are possible only for few-electron systems, see Section A3.2.2.1. In the case of many-electron systems, the practically applicable methods are based on the self-consistent Hartree–Fock method with allowance made for the electron correlation effects.

#### 3.2.2.1 The Hartree–Fock approximation and accounting for the electron correlation

A brief outline of the Hartree–Fock method is presented in Section A3.2.1. Here we discuss the application of the algebraic approach of this method, often denoted as the LCAO MO SCF approach, to the numerical studies of intermolecular interactions and potential energy surfaces.

A large number of calculations of potential energy surfaces by the LCAO MO SCF approach, initiated by the pioneering papers of Clementi [166] and Morokuma and Pedersen [167], were published in the 1970s. Later on, the rapid development of computational technology made the employment of beyond HF *ab initio* methods a routine procedure. And, at present, the SCF approach is used by nobody as a final step, even in cases when it is quite satisfactory.

But the first SCF studies that took into account all electrons and relatively large basis sets were rather impressive. The calculations with the basis sets close to the Hartree–Fock limit predicted satisfactorily the geometry of the complex under study and its electron density distribution, the accuracy of the electron density distribution being estimated at 2–5% [167]. The analysis of numerical studies of the potential energy surfaces at that time are presented in References [168–170].

The accuracy of the solution depends strongly on the basis set used. The calculations of potential surfaces in the neighborhood of the equilibrium geometry showed that the correct results require a rather large basis set, in which case the results will approach the Hartree–Fock limit. The basis set must include polarization functions with a large angular momentum, as minimum, d-functions. For instance, the calculation of the reaction $F + CH_3F$ without d-functions has predicted an intermediate complex with an energy of 13 kJ/mol higher than the energy of the reactants; a calculation with d-functions led to an activation energy of 29 kJ/mol [171]. The height of the inversion barrier in ammonia was raised from 5.0 to 21.3 kJ/mol with the inclusion of d-functions [172], agreeing very closely with the experimental value.

On the other hand, it is well known that the Hartree–Fock energy does not include the electron correlation (see discussion below). Hence, the good prediction of the barrier heights and minima means that in these regions the SCF energy surface parallels the exact one. This takes place for chemical reactions that occur without a change in the number of paired valence electrons. For instance [173]:

$$NH_3 + HCl \rightarrow NH_4Cl$$

or

$$CN^- + CH_3F \rightarrow [NC - CH_3 - F]^- \rightarrow CH_3CN + F^- \tag{3.153}$$

However, as the distance between the molecules, or atoms, increases, the relative error due to the correlation energy usually increases. As a result, in some cases incorrect dissociation products are predicted. In particular, the HF approximation predicts dissociation into products with unpaired electrons (radicals) for the molecules lithium hydride and nitrogen [174] and hydrogen [175]. In the case of $H_2$, the one-determinant description of the Hartree–Fock method leads to an energy, for an infinite separation of the nuclei, corresponding to a mixture of the states $H^+H^-$ and $2H(^2S)$ that exceeds the dissociation energy by 3 eV, i.e. with an error of ~60%.

The interactions of closed-shell molecules (atoms) with atoms possessing a half-occupied shell correspond to that class of reaction for which the Hartree–Fock approximation predicts correctly the dissociation products. Nevertheless, the Hartree–Fock potential curves will be parallel to the exact ones only for the reactions that do not involve breaking or formation of chemical bonds, such as:

$$\begin{gathered} Na\left(^2S\right) + He\left(^1S\right) \rightarrow NaHe\left(^2\sum^+\right) \\ F\left(^2P\right) + He\left(^1S\right) \rightarrow FHe\left(^2\Pi\right) \end{gathered} \tag{3.154}$$

For instance, the SCF approach to reaction [176]:

$$H_2 + Cl \rightarrow HCl + H \tag{3.155}$$

or to reaction [177]:

$$H + F_2 \rightarrow HF + F \tag{3.156}$$

leads to large errors. In particular, the SCF numerical study of Reaction (3.156) gives the height of the reaction barrier at the saddle point of 51 kJ/mol, while inclusion of the electron correlations leads to a value of 4.2 kJ/mol (the experimental value of the activation energy is 5.0 kJ/mol).

The molecules and clusters that are predominately stabilized by the electron correlation effects are unstable at the SCF level. The well-known examples are the $F_2$ molecule [178] and the $Be_n$, $Mg_n$ and $Ca_n$ ($n = 2, 3$) dimers and trimers [145].

At large distances, the correlation energy is completely reduced to the dispersion energy. The numerical proof of this was presented in Reference [145]; see Table 3.7 and the following discussion on page 124. Hence, the SCF potential energy surfaces do not contain the dispersion contribution, whereas the latter is a very important component in the interaction of neutral systems at large distances.

From the discussion above it follows that, except some particular cases, the intermolecular interactions must be studied with the electron correlation effects being taken into account. The formal definition of the electron correlation energy was given in 1959 by Löwdin [175]:

> The correlation energy of a state of a particular Hamiltonian is defined as the difference between the exact eigenvalue of the Hamiltonian for that state and the value, which is obtained in the Hartree–Fock approximation.

Namely:

$$\Delta E^{corr} = \langle \Psi | H | \Psi \rangle - \left\langle \Psi^{HF} | H | \Psi^{HF} \right\rangle \tag{3.157}$$

Since the exact eigenvalue for systems with more than two electrons cannot be obtained, this definition is quite conventional. The value of $\Delta E^{corr}$ depends upon the approximation employed in the electron-correlation calculation. For example, if the calculation is performed at the full configuration interaction (FCI) level, the latter should be noted in Equation (3.157):

$$\Delta E^{corr}(FCI) = E^{FCI} - E^{HF} \tag{3.158}$$

The value of $\Delta E^{corr}$ depends not only on the calculation method used but also to a great extent on the basis set. The use of large basis sets is often more important than the precision of the calculation method used.

The main methods taking the electron correlation into account are briefly discussed in Section A3.2.2. Among these methods used to study the intermolecular interactions are different versions of *configuration interaction* (CI) method, including the *multiconfiguration self-consistent field* (MC SCF) method and *coupled cluster* (CC) method with accounting of single, double and triple (noniterative) excitations (CCSD(T)). On the other hand, the application of the density functional method to molecular complexes bound by the dispersion interactions meets with serious problems (see Section 3.2.2.3). Although, this section is devoted to the variational methods of intermolecular interaction studies, it is worthwhile mentioning that the *Møller–Plesset Perturbation Theory* (MPPT) discussed in preceding section is also very appropriate for studying the electron correlation effects, since its first order corresponds to the Hartree–Fock approximation. Thus, the electron-correlation correction is defined as a sum over all orders of MPPT starting from the second one:

$$\Delta E^{corr}(MPn) = \sum_{k=2}^{n} \varepsilon_{MP}^{(k)} \tag{3.159}$$

In the CI approach that has been widely used, the electron correlation can be easily divided into intermolecular and intramolecular components. Consideration of the intermolecular component leads to the dispersion forces, while inclusion of the intramolecular one leads to a decrease of the dipole moment of the molecule and an increase in its polarizability [179]. It should be borne in mind that the excited configurations, which provide the dominant contribution to the dispersion energy, and the excited configurations, which determine the intramolecular correlation, are, as a rule, different. This is even more evident in the asymptotic limit for the products. For instance, to describe adequately the dissociation of $H_2O(^1A_1)$ into $O(^1D)$ and $H_2\left(^1\Sigma_g^+\right)$, only two excited configurations must be added to the Hartree–Fock determinant. The contribution of these configurations to the ground state is negligibly small in the region of equilibrium but becomes dominant at large distances.

An essential improvement of the convergence of the CI expansion is achieved in the MC SCF method (see Section A3.2.2.2.). Although, the simultaneous variation of orbital and configuration coefficients in the MC SCF procedure introduces complications in comparison with the direct CI method, the good results, obtained with a small number of configurations, constitute quite a reward. In early 1970s, Wahl and co-workers [180–183] obtained the potential curves for the complexes $(He)_2$, $(Ne)_2$ and ArH in the region of the van der Waals minimum. They applied the procedure of localization of molecular orbitals, that gives the possibility of selecting those excited configurations, which correspond to the dispersion excitations, that is, to transitions to unoccupied orbitals of electrons belonging to each atom. Five configurations: the Hartree–Fock ground state and four dispersion configurations were sufficient for a good agreement with the experimental data for $(He)_2$ [180]. In the case of ArH [182], they chose two charge-transfer and nine dispersion configurations.

In further developments of the MC SCF method it was realized that it is much more effective to divide the orbitals into *active* and *inactive* ones (the latter are usually the double-occupied inner-shell orbitals that are frozen in respect to the CI procedure). Then the full configuration calculations are performed at the MC SCF level only in the active space. This approach was designated as *complete active space* SCF method (CAS SCF) [184] (Section A3.2.2.2.). Effective procedures improving the convergence of the MC SCF method were developed. This has allowed a very large complete space, more than $10^5$ configurations, to be used [185, 186].

In the last few years, the coupled cluster method (see Section A3.2.2.3) in its CCSD(T) version has become popular in the intermolecular potential energy surface calculations. The potential energy surfaces were calculated at the CCSD(T) level for different kind of complexes and clusters; among them the precise calculations of Ne–CO [187], Ar–$C_6H_6$ [188], Ar–$N_2$ [189], He–ICl and Ne–ICl [190], $C_6H_6$–$N_2$ [191], $C_6H_6$–He [192] and Ne–$N_2$ [193] should be mentioned. The use of the Dunning-type [194, 195] augmented correlation-consistent polarized valence double and triple zeta (aug-cc-pVDZ, aug-cc-pVTZ) basis sets with an additional set of bond functions provides a good agreement with experimental infrared spectra and microwave rotational spectra. Further refinement of the basis sets in Reference [193], where QZ and 5Z, which is close to convergence, were exploited, led to a decrease in the frequency errors, down to 0.6%. Thus, the calculations at CCSD(T) level with sufficiently large and refined basis sets can predict the spectroscopic properties of van der Waals complexes.

It should be pointed out that the single reference CC (SR CC) approach, as CCSD(T), is quite satisfactory for closed-shell systems in the entire range of geometry. In the case of chemical-bond breaking, the SR CC methods often yield wrong results. For instance, SR CCSD leads to a completely wrong dissociation limit for the bond breaking in the molecules HF, $F_2$ and $N_2$ [196, 197]. In these cases, the multireference CC (MR CC) approaches are recommended (see Section A3.2.2.3).

As pointed out in the preceding text, the alkaline-earth dimers and trimers are not stable at the SCF level. It is the electron correlation that stabilizes these clusters. This is not surprising, since the alkaline-earth atoms (beryllium, magnesium, calcium, etc.) in the ground state have the closed outer ns subshell, that is, they do not have the valence electrons. But for clusters containing metallic atoms possessing the valence electrons in the ground states, the electron correlation still plays the crucial role in their stability. A good illustration is the $Li_n$ clusters. In Table 3.6, the calculation of the correlation energy (3.159) at MP4 level for $Li_n$ $(n = 2-4)$ clusters is presented [198]. The correlation contribution to the binding energy is a dominant factor for all clusters considered.

At large distances, the electron correlation energy can be interpreted as the dispersion energy; while at intermediate distances, where the overlap of the atomic valence shells becomes essential, the dispersion forces cannot be defined without allowing for exchange effects. At these distances, the multipole expansion is not valid (see Section 3.1.4). It is instructive to compare the magnitudes of the pure dispersion energy and the electron correlation energy at different distances. Let us consider it for the alkaline-earth dimers.

**Table 3.6** The electron correlation contribution to the binding energy of $Li_n$ clusters (energy in kcal/mol)

| | $E_b^{MP4}$ | $E_b^{SCF}$ | $\Delta E^{corr}$ | $\Delta E^{corr}/E_b$ |
|---|---|---|---|---|
| $Li_2$ | 22.0 | 3.9 | 18.1 | 0.82 |
| $Li_3$ | 30.4 | 11.9 | 18.4 | 0.60 |
| $Li_4$ | 63.8 | 19.5 | 44.3 | 0.69 |

**Table 3.7** Comparison of $\epsilon_{disp}^{(2)}$ (Equation (3.160)), with $\Delta E^{corr}$ for the alkaline-earth dimers at equilibrium and large distances (in kcal/mol)

| | $Be_2$ | | $Mg_2$ | | $Ca_2$ | |
|---|---|---|---|---|---|---|
| $R$,Å | $\epsilon_{disp}^{(2)}$ | $\Delta E^{corr}$ | $\epsilon_{disp}^{(2)}$ | $\Delta E^{corr}$ | $\epsilon_{disp}^{(2)}$ | $\Delta E^{corr}$ |
| 2.56 | −79.62 | −7.94 | | | | |
| 3.92 | | | −8.95 | −2.27 | | |
| 4.56 | | | | | −14.40 | −3.19 |
| 5.00 | −0.35 | −0.36 | −1.29 | −0.87 | −6.85 | −2.23 |
| 6.00 | −0.10 | −0.11 | −0.35 | −0.33 | −1.68 | −0.92 |
| 7.00 | −0.03 | −0.04 | −0.11 | −0.12 | −0.55 | −0.39 |
| 8.00 | | | −0.04 | −0.04 | −0.22 | −0.17 |
| 9.00 | | | | | −0.10 | −0.08 |
| 10.00 | | | | | −0.05 | −0.04 |

At large distances, the energy of the dispersion interaction between two atoms can be presented with a fine precision as a sum of three terms:

$$\epsilon_{disp}^{(2)} = -\left(\frac{C_6}{R^6} + \frac{C_8}{R^8} + \frac{C_{10}}{R^{10}}\right) \tag{3.160}$$

where the dipole–dipole $(R^{-6})$, dipole–quadrupole $(R^{-8})$ and dipole–octopole plus quadrupole–quadrupole $(R^{-10})$ dispersion interactions are taken into account. The dispersion coefficients $C_n$ for the beryllium, magnesium and calcium atoms were estimated in Reference [199] by the Padé approximant method (see Section A3.3.5). Using the values of $C_n$ converted to $\left[\frac{kcal}{mol}\text{Å}^n\right]$ units, the sum (Equation (3.160)) has been found at equilibrium and at large distances. These are presented in Table 3.7 together with $\Delta E^{corr}$ [145]. It follows from Table 3.7, that at large distances the electron correlation energy coincides with the pure dispersion energy with a very good precision: for $Be_2$ at $r \geq 5$ Å, for $Mg_2$ at $r \geq 6$ Å, and for $Ca_2$ at $r \geq 9$ Å. Note: this coincidence takes place for two different physical quantities calculated by different methods. From this follows that it is based on the physical ground: the dispersion forces have the electron correlation origin.

At equilibrium distances, the absolute value of the pure dispersion energy is much larger than $\Delta E_{corr}$ (in the case of $Be_2$, 10 times!). The exchange and overlap contributions to the electron correlation energy are repulsive and cause a decrease in dispersion attraction.

#### 3.2.2.2 Basis set superposition error

The calculation of the interaction energy in the supermolecular approach (Equations (3.148) and (3.152)) possesses one serious defect, which is connected with a basis set inconsistency that leads to an artificial enhancement of the intermolecular interaction energy. This was revealed in the supermolecular SCF approach in the late 1960s [200]. Consider this problem in the case of the dimer AB. In the standard approach, it is natural to define:

$$E_{int}^{St}(R) = E^{AB}(R) - \left[E^{A}(\{A\},\infty) + E^{B}(\{B\},\infty)\right] \tag{3.161}$$

where $R$ is the separation between monomers in the dimer $AB$ and $E^{A}(\{A\},\infty)$ denotes that the monomer energy is calculated at infinite separation (for isolated monomer) using the monomer basis set $\{A\}$. Note that the latter is never complete. The dimer basis $\{AB\}$ is evidently larger than those of monomers. This causes the artificial stabilization deepening the dimer potential energy curve.

The remedy for this defect was proposed by Jansen and Ros [201] for a particular reaction and, as a general method, by Boys and Bernardi [202]. They suggested that the supermolecular basis is used for all terms in Equation (3.161). Their approach was called *function counterpoise* [202]; it is usually designated by an abbreviation CP:

$$E_{int}^{CP}(R) = E^{AB}(R) - \left[E^{A}(\{AB\},R) + E^{B}(\{AB\},R)\right] \tag{3.162}$$

where $E^{A}(\{AB\},R)$ is the energy of the monomer $A$ calculated using the dimer basis set $\{AB\}$. It means that A is calculated using not only its basis set $\{A\}$ but also a set of ghost orbitals $\{B\}$ located in the space of $B$ at distance $R$. In practice, this calculation is identical to the dimer calculation except that the nuclear charges are put at zero for all nuclei belonging to $B$ and the appropriate number of electrons are subtracted from the calculation.

Thus, the CP procedure adds an 'extra' stability to monomers that makes their calculation consistent with the dimer calculation. The error in Equation (3.161) caused by the basis inconsistency is named the *basis set superposition error* (BSSE) [203]. It is defined as a difference:

$$\begin{aligned} BSSE = E_{int}^{CP} - E_{int}^{St} &= \left[E^{A}(\{A\},\infty) + E^{B}(\{B\},\infty)\right] \\ &\quad - \left[E^{A}(\{AB\},R) + E^{B}(\{AB\},R)\right] \end{aligned} \tag{3.163}$$

Soon after the publication of the CP scheme, it was employed by several groups [204–207]. It was shown that the magnitude of the BSSE decreases as the size of the basis set increases. Table 3.8, shows the binding energy of the $(H_2O)_2$ dimer calculated with different basis sets using the standard approach (Equation

**Table 3.8** Effect of enlarging the basis set on the magnitude of BSSE in the binding energy calculations of $(H_2O)_2$, $R_{OO} = 3.00$ Å, energy in kcal/mol.

| basis | $E_{int}^{st}$ | $E_{int}^{CP}$ | $BSSE$ |
|---|---|---|---|
| minimum [206] | −5.14 | −4.33 | 0.81 |
| intermediate [208] | −4.02 | −3.87 | 0.15 |
| extended [209] | −3.90 | −3.9 | <0.05 |

(3.161)), and Equation (3.162), which takes into account the superposition error. The magnitude of the BSSE is maximum for the minimum basis set and approaches to zero for the calculation performed with an extended, close to the HF limit, basis set [209].

The dramatic illustration of the basis set inconsistency for small basis sets was performed by Ostlund and Merrifield [207]. The calculation for the $(He)_2$ dimer with small basis set using non-corrected BSSE (Equation (3.161)) results in a potential curve with a deep minimum (Figure 3.4). The use of the CP procedure (Equation (3.162)) led to the disappearance of this minimum, yielding a potential curve very close to that calculated with an extended basis set [203].

Accounting for BSSE is important not only to evaluate $E_{int}$, but also to evaluate other physical properties connected with intermolecular interactions. The latter was rather obviously demonstrated in the evaluation of the incremental polarizability as the two helium atoms approach [207, 210]. Even gradient optimization should be performed to take proper account of BSSE [211, 212].

The CP procedure was first applied to the intermolecular interaction studies at the SCF level by Johansson *et al.* [204]. But in this paper the first criticism that the CP procedure overestimates the magnitude of BSSE appeared. The concept about the overcorrection of BSSE in the CP scheme was then accepted as evident in many subsequent papers [129, 213–215]. The main objection to the CP procedure was made on the grounds that in the dimer AB, the Pauli exclusion principle prevents the monomer $A$ from using the double occupied orbitals of $B$ and *vice versa,* while in the CP procedure the monomers occupy all ghost orbitals of the other monomer. This leads to the overestimation of BSSE. In Reference [129], it was proposed using

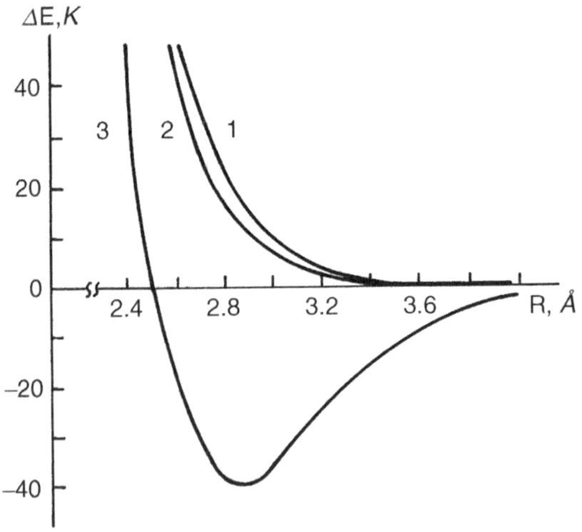


**Figure 3.4** The potential curve of $He_2$ calculated with a small basis set [207]: 3 – by Equation (3.161) and 2 – by Equation (3.162); 1 – calculation with an extended basis set [203].

only virtual ghost orbitals (*virtual counterpoise*, VCP) instead of using all ghost orbitals as in the Boys–Bernardi procedure (called *full counterpoise*, FCP). This concept was supported by numerical estimates in which the SCF interaction energy corrected by FCP procedure deviated significally from the HF value calculated with a relatively large basis set [216].

A good theoretical argument against the objection above was presented by Gutowski *et al.* [217]. They pointed out that in the ideal calculation, one would use a *complete* basis set. The latter can be located at an arbitrary point of space; it can be also constructed from two incomplete basis set located at different points. In any case, a complete basis set may be used to calculate the dimer AB and the monomers A and B as well. The Pauli principle must be applied for the dimer and composed monomers independently; it may not forbid the electrons of isolated monomers from occupying the ghost orbitals of other monomer that are occupied in the dimer.

The original CP (FCP) method and the alternative to it, the VCP scheme, were numerically studied and analyzed in References [217–219]; a detailed discussion of these studies and the BSSE problem as a whole was presented by Duijneveldt *et al.* in two comprehensive reviews [220, 221]. The comparison of the BSSE-free interaction energies found by SAPT with the CP-corrected supermolecular interaction energies for some van der Waals complexes demonstrated a very good agreement for different basis sets used. These studies have led to the conclusion that the overcorrection conception is wrong and the CP method is well-founded.

An additional confirmation of the validity of the CP procedure was obtained by Mayer and Valiron [222] by the BSSE-free *chemical Hamiltonian approach* (CHA) at the MP2 level. The CHA concept was suggested in 1983 by Mayer [223] and then elaborated in different calculation schemes. [224–226, 222]. The main idea of CHA is to separate the BSSE effects before calculating the interaction energy. By means of the second quantization formalism in the nonorthogonal basis, the interaction Hamiltonian is divided into the counterpoise correction and the BSSE-free term. This allows the BSSE-free wave functions to be obtained, which is then used to calculate the BSSE-free interaction energy as an expectation value of the standard adiabatic Hamiltonian. As was shown in Reference [222], the BSSE-free interaction energies obtained by the CHA MP2 method are in very good agreement with the CP-corrected supermolecular MP2 energies for the van der Waals dimers and some hydrogen-bonded complexes as well.

Nevertheless, papers criticizing the CP method have been still published see, for instance, References [227–231]. The reason for this criticism is connected with an incompleteness of the exploited basis sets, which besides BSSE contain other drawbacks based on their incompleteness. Therefore, removing the BSSE does not always make the incomplete basis set appropriate to describe the physics of interaction. For instance, the employment of a basis set that is not sufficiently good can lead to significant errors in the CP-corrected direct electrostatic interaction; in particular, the dipole moments of monomers are changed as a result of CP procedure [114, 221]. This is why most of the claims about the CP-procedure are related to the calculations of hydrogen-bonded systems in which the electrostatic interaction plays the dominant role, namely, $(HF)_2$ [216], $(HF)_3$ [231], or some other hydrogen-bonded complexes [229]. On the other hand, the most precise CP verifications [217–219] were performed for systems with spherical interacting atoms, such as $He_2$ or $He{-}Li^+$, which do not possess any multipole moments.

When the CP-corrected interaction energy is not satisfactory, it is first necessary to try to improve the basis set, instead of introducing corrections to the CP procedure. For large and well-balanced basis sets, the CP-corrected energies usually approach the basis set limit.

Careful studies of the BSSE dependence for hydrogen-bonded complexes upon the basis extension at the SCF and electron correlated levels were performed in Reference[232–235]. The basis-set-limit energy was estimated by the explicitly correlated linear R12-method [236–238] (see Section A3.2.2.1). In this approach, the interelectronic distance $r_{12}$ is incorporated into the wave function explicitly, as required by the electronic Coulomb cusp relation, and this allows the correlation energy close to the basis set limit where BSSE vanishes to be obtained. The convergence to this basis set limit in the case of calculations employing the aug-cc-pVXZ basis sets [195] follows the form [239]:

$$\Delta E^{corr}(X) = \Delta E_{lim}^{corr} + AX^{-3} \tag{3.164}$$

The main conclusions from the studies [232–235] can be systematized as follows:

a. Both the CP-corrected and uncorrected interaction energies converge toward the basis set limit, although from different sides (see Figure 3.5).

b. For all complexes studied, the BSSE in the SCF part of the interaction energy is significantly smaller than that in the correlation part.

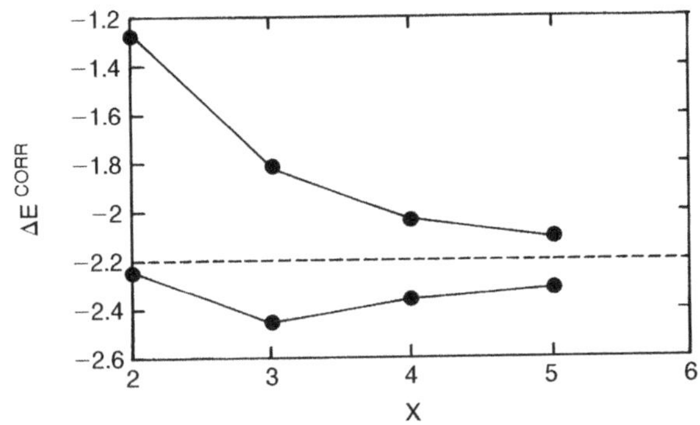


The horizontal dashed line is the basis–set–limit energy obtained by the linear R12– method, (energy in mhartree.)

**Figure 3.5** CP-corrected (upper curve) and uncorrected (lower curve) correlation contributions for $(H_2O)_2$ at the MP2 level [233] as a function of number X in the employed aug-cc-pVXZ basis sets

c. As was stressed in Reference [233], the basis set convergence of the CP-corrected correlation contribution is always monotonic, smooth and systematic, while the uncorrected correlation contribution is always non-monotonic and very unsystematic.

d. In spite of the unsystematic behavior of the uncorrected correlation contribution, it approaches the basis set limit with the same or better level of accuracy than the CP-corrected correlation contributions. The uncorrected correlation contribution calculated at the MP2 level with the aug-cc-pVDT basis set nearly coincides with the basis set limit (Figure 3.5). The latter due to a fortuitous compensation of BSSE and the error caused in the incomplete description of the Coulomb interaction, since they have opposite signs.

Thus, although the CP-procedure eliminates the BSSE, in the systems with large electrostatic interaction, as the hydrogen-bonded complexes, the elimination of the BSSE can make the results worse, since the BSSE and the error due to an inadequate description of the Coulomb interaction have opposite signs. In these cases, the noCP calculations can be employed, especially with aug-cc-pVDZ basis set.

However, the predictive value of noCP calculations will be always doubtful as long as significant BSSE is present. In complexes composed of the closed electronic shell atoms, as van der Waals dimers, the direct electrostatic interactions are negligible and the elimination of BSSE by the CP procedure always improves the interaction energy.

#### 3.2.2.3 Density functional theory

In the last fifteen years, the density functional theory (DFT) approaches (Section A3.2.2.4) have been widely used in atomic, molecular and solid-sate physics. It is connected with an ability of the DFT method to be applied to large-scale many-electron systems without large time-consuming expenses, and based on the one-electron nature of the Kohn–Sham (KS) equation. The latter, as the HF equation, is the one-electron approximation to the many-electron Schrödinger equation but with a special exchange-correlation potential. Because of this, the KS equation, by contrast with the HF equation, takes to some extent the electron correlation into account.

The development, within the framework of the local spin-density approximation (LSDA), of the gradient-corrected exchange and correlated functionals has considerably improved the precision of the DFT method. Among widely used gradient corrected functionals, should be mentioned the Becke exchange functional [240] denoted as B88, or B, and the Lee–Yang–Parr (LYP) [241] and Perdew–Wang (PW91) [242] correlation functionals. Later on, the hybrid functionals (a superposition of the HF and DFT exchange functionals) [243, 244], B3PW91 and B3LYP, became very popular; these and some other modern functionals are discussed in Section A3.2.2.4.

Early applications of DFT methods to intermolecular interaction problems have been presented in book [18], Chapter 3. Here the applications in the last two decades are considered. The numerous results of DFT calculations performed in this period demonstrated that its gradient-corrected versions yield quite reliable results in the prediction of the global minima on the potential energy surfaces and give reasonable binding energies for different classes of atomic and molecular complexes in their ground states [245–253].

Systematic comparative studies of different gradient-corrected density functionals in calculations of structural, energetic and magnetic properties of $(HF)_n$, $n = 1 - 6$, clusters were performed by Maerker *et al.* [252]. It was found that the hybrid approaches (particularly B3LYP) provide rather useful results. However, in general, they are not competitive with the more quantitative MP2 calculations with large basis sets. On the other hand, the cost of carrying out such MP2 calculations increases very rapidly with the cluster size.

Handy and collaborators [253] obtained a full six-dimensional potential surface for the water dimer employing the hybrid B3WP91 potential with one parameter adjusted to reproduce the correct $R_{OO}$ length. They obtained the global minimum and transition states on the surface that were in agreement with *ab initio* calculations, while the binding energy and barrier heights were slightly underestimated.

Another even more impressive example, demonstrating that the semiempirical hybrid functionals can produce results comparable with *ab initio* methods, is the employment of the ten-parameter hybrid functional developed by Becke [244, 254]. The fitting of this functional to accurate thermochemical data provided, see Reference [254], very good agreement with the experimental data for the heat of formation of 148 molecules (so-called extended G2 test set [255] employed for the verification of *ab initio* calculations by the Pople group).

But, as is typical for semiempirical methods, the semiempirical functionals with parameters adjusted to definite classes of molecules are not universal. The hybrid potential adjusted in Reference [253] for $(H_2O)_2$ was tested in Reference [256]. It was revealed that, for the $OH^-{-}H_2O$ complex, this adjusted hybrid potential produces results worse than the standard hybrid functionals.

As was shown in Reference [256], the best available hybrid functionals at that time provide the accurate potential surfaces for many-atomic compounds only in the case of strongly bound ionic hydrogen-bonded complexes like $OH^-{-}H_2O$. For other systems studied, including $(H_2O)_2$, DFT calculations failed to reproduce the correct angular dependence of the potential surfaces obtained at the CCSD(T) level; note that for the LDA approach it has been long known that LDA gives an inadequate description of the anisotropic effects [257]. For the van der Waals complexes not only the angular dependence, but many other characteristics of potential surfaces were reproduced wrong. For a typical van der Waals complex like $He{-}CO_2$, none of the hybrid functionals used reproduces the weak binding minima. This is not surprising, because the van der Waals complexes are bound by the dispersion forces. In a recent paper [258], Kohn *et al.* pointed out that the commonly used gradient DFT methods designed for nonuniform electron gas fail to capture the essence of van der Waals energies.

As demonstrated in Section 3.2.2.1, the dispersion forces have a purely electron correlation origin and at large distances the energy of dispersion interaction coincides with the correlation energy (Table 3.7). In general, it may not be concluded *a priori* that nonlocal gradient-corrected functionals cannot include, if only partly, the dispersion interactions. It can be clarified only in a direct calculation.

Such studies were performed in the middle of the 1990s by Kristyan and Pulay [259] and Pérez-Jordán and Becke [260] on the 'classical' van der Waals dimers $He_2$, $Ne_2$ and $Ar_2$, and some of their heterocompounds. It was revealed that the gradient-corrected BLYP, BPW91 and the hybrid B3PW91 functionals yield purely repulsive potential curves. Only the so-called *half-and-half Becke functional* [261] gives some binding, though it essentially underestimates the dissociation energy.

The lack of the dispersion energy in the DFT formalism at the more crude LDA level was known much earlier [262]. It simulated some authors to add the dispersion terms directly to the DFT functionals [263, 264]. Recently, Gianturco *et al.* applied the DFT method for study thermodynamical properties of the CO–He mixtures [265] and the intermolecular interactions in the rather large $CO(He)_n$ clusters with $n$ up to 12 [266]. They added the dispersion interaction terms to the half-and-half-Becke functional [261]. The long-range dispersion contribution was taken from the earlier work by Perdew and Pack [267], where it was represented as a sum of three terms:

$$V_{disp}(\theta, R) = \frac{C_6(\theta)}{R^6} + \frac{C_7(\theta)}{R^7} + \frac{C_8(\theta)}{R^8} \qquad (3.166)$$

and the values of the coefficients $C_6$, $C_7$ and $C_8$ were calculated for interaction of CO with all rare-gas atoms.

Meanwhile, intensive studies to create new effective functionals were going on. In 1998, Adamo and Barrone [268] constructed a new MPWPW exchange-correlation functional (the short description of this and other functionals mentioned below is given in Section A3.2.2.4). This functional yields rather good binding energies for the van der Waals dimer $Ne_2$ and overestimate the $He_2$ binding energy. Also in 1998, Hamprecht *et al.* [269] reparameterized the hybrid functional of Becke [244] creating two exchange-correlation functionals named in literature B97-1 and B97-2. The application of B97-1 to the van der Waals complexes led to the best result among other DFT approaches [270].

In 2003, Tao, Perdew, Staroverov and Scuseria [271] constructed a nonempirical exchange-correlation functional based on the *meta-generalized gradient approximation* (meta-GGA). This TPSS functional (labelled according to the author names) was tested on molecules, solids and solid surfaces [272, 273] and demonstrated an uniform accuracy for diverse properties and systems.

Several rather precise new hybrid functionals were created in 2004. Xu and Goddard [274, 275] developed the X3LYP functional that performs better than the popular B3LYP functional and, in contract with the latter, yields the stable van der Waals complexes. Truhlar and collaborators [276, 277] created the MPW1B95 and MPWB1K functionals. The parameters in MPW1B95 were adjusted for application in thermochemistry, while MPWB1K was optimized against the kinetic database [276] for calculating the barrier heights, saddle points, *etc.* on the potential energy surfaces.

Recently, Zhao and Truhlar [270] have tested 44 DFT functionals using four benchmark databases for hydrogen bonding, charge transfer, dipole interactions and weak interactions. For weak interactions (where the dominant role is played by the dispersion forces), the best performance was given by B97-1 [269]; in the overall results for all four types of nonbonding interactions, the best was MPWB1K [276].

In spite of providing the stability of van der Waals complexes by new hybrid potentials, the precision of the binding energy obtained is not enough for quantitative predictions. The best B97-1 functional has a mean error in the calculation of nine weakly bound complexes of about 40% [270]. The two nearest best performers, MPWB1K and MPW1B95, have a mean error between 50% and 40%. Thus, the conclusion by Mourik and Gdanitz made in 2002 in their critical survey [278] that state-of-the-arts DFT methods are incapable of accounting of dispersion effects in a quantitative way is still valid.

All applications of the DFT approaches discussed above are related to systems in the nondegenerate ground state. The calculation of the excited states is still a problem in DFT, though studies in this direction are in progress [279–281]. On the other hand, the excited states are often degenerate and this is an insoluble problem for DFT. It is well known that the adiabatic approximation is valid only for nondegenerate electronic states (Appendix 3, Section 1). The adiabatic approximation breaks down for degenerate or near-degenerate electronic states. In this case, the electron and nuclear motions cannot be separated because the electron-nuclear interaction terms in the coupled equations (A3.13) are not negligible; these off-diagonal terms cannot be presented by electron densities. As was shown by Bersuker [282], the version of DFT currently known cannot be applied to degenerated or pseudodegenerate states.

It is worthwhile mentioning that within DFT formalism, the space and spin multiplet structure cannot be calculated directly. The procedures developed for the multiplet structure calculation [283, 284] are all beyond the scope of DFT and cannot be rigorously based. This drawback of DFT descends from the invariance of the electron density with respect to the space and permutation symmetry of the wave functions describing the quantum state. In particular, the expression of the electron density does not depend upon the value of the total spin $S$ [285] (the latter is uniquely connected with the permutation symmetry of the coordinate wave function), see the discussion in the end of Section A3.2.2.4. This is the reason that it is necessary to be very careful when studying magnetic properties by DFT methods.

# References

1. M. Abramowitz and I.A. Stegun, *Handbook of Mathematical Functions*, Dover Publishing, New York (1965).
2. M.E. Rose, *J. Math. Phys.* **37**, 215 (1958).

3. U. Fano and G. Racah, *Irreducible Tensorial Sets*, Academic Press, New York (1959).
4. B.C. Carlson and G.S. Rushbrooke, Proc. *Cambridge Phil. Soc.* **44**, 626 (1950).
5. A.T. Amos and R.J. Crispin, in *Theoretical Chemistry: Advances and Perspectives*, H. Eyring and D. Henderson (eds), Academic Press, New York (1976), Vol. 2, pp. 2–66.
6. L.D. Landau and E.M. Lifshitz, *Quantum Mechanics* (Nonrelativistic Theory), 3rd edn, Pergamon Press, Oxford (1977).
7. L.D. Landau and E.M. Lifshitz, *Mechanics*, Pergamon Press, Oxford (1989).
8. M.E. Rose, *Elementary Theory of Angular Momentum,* John Wiley & Sons, Inc., New York (1957).
9. A.R. Edmonds, *Angular Momentum in Quantum Mechanics,* Princeton University Press, Princetone (1957).
10. D.M. Brink and G.R. Satchler, *Angular Momentum*, 3rd edn, Clarendon Press, Oxford (1993).
11. H.B. Casimir and D. Polder, *Phys. Rev.* **73**, 360 (1948).
12. A.D. Buckingham, *Disc. Faraday Soc.* II, **40**, 232 (1965).
13. A.D. Buckingham, *Adv. Chem. Phys.* **12**, 107 (1967).
14. R.T. Pack, *J. Chem. Phys.* **64**, 1659 (1976).
15. H.N.W. Lekkerkerker, P. Coulon and R. Luyckx, *J. Chem. Soc. Faraday Trans.* II, **9**, 1328 (1977).
16. P. Isnard, D. Robert and L. Calatry, *Mol. Phys.* **31**, 1789 (1976).
17. F. Mulder, G. Van Dijk and C. Huiszoon, *Mol. Phys.* **38** 577 (1979).
18. I.G. Kaplan, *Theory of Molecular Interactions*, Elsevier, Amsterdam (1986).
19. P.D. Robinson, *Proc. Phys. Soc. (London)* **78**, 537 (1961).
20. P.R. Certain and W. Byers Brown, *Int. J. Quant. Chem.* **6**, 131 (1972).
21. R. Ahlrichs and P. Claverie, *Int. J. Quant. Chem.* **6**, 1001 (1972).
22. W. Kołos, *Int. J. Quant. Chem. Symp.* **8**, 241 (1974).
23. W. Kołos, *Int. J. Quant. Chem.* **9**, 133 (1975).
24. R.J. Buehler and J.O. Hirschfelder, *Phys. Rev.* **83**, 628 (1951); *Ibid* **85**, 149 (1952).
25. A. van der Avoird, P.E.S. Wormer, F. Mulder and R.M. Berns, *Van der Waals Systems*, Topics Curr. Chem., Springer-Verlag, Berlin (1980).
26. A. Dalgarno and J.T. Lewis, *Proc. Phys. Soc. (London)* **69**, 57 (1956).
27. G.P. Arrighini, F. Biondi, and C. Guodotti, *Mol. Phys.* **26**, 1137 (1973).
28. J.F. Bukta and W.J. Meath, *Mol. Phys.* **27**, 1235 (1974).
29. R. Ahlrichs, *Theoret. Chim. Acta*, **41**, 7 (1976).
30. F.C. Brooks, *Phys. Rev.* **86**, 92 (1952).
31. A. Dalgarno and A.L. Stewart, *Proc. Roy. Soc. (London)* A **238**, 276 (1956).
32. R.N. Young, *Int. J. Quant. Chem.* **9**, 47 (1975).
33. A. Erdélyi, *Asymptotic Expansions*, Dover Publishing, New York (1956).
34. A. Dalgarno and N. Lynn, *Proc. Phys. Soc. (London)* **70**, 223 (1957).
35. J.G. Kirkwood, *Zs. f. Phys.* **65**, 209 (1930).
36. J.O. Hirschfelder, W. Byers Brown and S.T. Epstein, *Adv. Quant. Chem.* **1**, 255 (1964).
37. L.C. Cusachs, *Phys. Rev.* **125**, 561 (1962).
38. H. Kreek and W.J. Meath, *J. Chem. Phys.* **50**, 2289 (1969).
39. T.R. Singh, H. Kreek and W.J. Meath, *J. Chem. Phys.* **52**, 5565 (1970).
40. K.C. Ng, W.J. Meath and A.R. Allnatt, *Mol. Phys.* **32**, 177 (1976); **38**, 449 (1979).
41. A. Koide, W.J. Meath and A.R. Allnatt, *Chem. Phys.* **58**, 105 (1981).
42. P.J. Knowles and W.J. Meath, *Chem. Phys. Lett.* **124**, 164 (1986).
43. P.J. Knowles and W.J. Meath, *Mol. Phys.* **59**, 965 (1986); **60**, 1143 (1987).

44. R.J. Wheatley and W.J. Meath, *Mol. Phys.* **80**, 25 (1993).
45. R. Ahlrichs, R. Penco and G. Scoles, *Chem. Phys.* **19**, 119 (1977).
46. K.T. Tang and J.P. Toennies, *J. Chem. Phys.* **80**, 3726 (1984).
47. J. Hernández-Cobos, I.G. Kaplan and J.N. Murrell, *Mol. Phys.* **92**, 63 (1997).
48. I.G. Kaplan, J.N. Murrell, S. Roszak and J. Leszczynski, *Mol. Phys.* **100**, 843 (2002).
49. G.G. Hall, *Chem. Phys. Lett.* **20**, 501 (1973).
50. A.D. Tait and G.G. Hall, *Theoret. Chim. Acta* **31**, 311 (1973).
51. L.L. Shipman, *Chem. Phys. Lett.* **31**, 361 (1975).
52. A.T. Amos and J.A. Joffe, *Theoret. Chim. Acta* **40**, 221 (1975).
53. F. Mulder, J.F. Tomas and W.J. Meath, *Mol. Phys.* **41**, 249 (1980).
54. G.G. Hall and D. Martin, *Theoret. Chim. Acta* **59**, 281 (1981).
55. L. Salem, *Mol. Phys.* **3**, 441 (1962).
56. R. Bonaccorsi, R. Cimiraglia, E. Scrocco and J. Tomasi, *Theoret. Chim. Acta* **33**, 57 (1974).
57. A.T. Amos and R.J. Crispin, *Mol. Phys.* **31**, 147, 159 (1976).
58. W.A. Sokalski and R.A. Poirier, *Chem. Phys. Lett.* **98**, 86 (1983).
59. A.J. Stone and M. Alderton, *Mol. Phys.* **56**, 1047 (1985).
60. F. Vigne-Maeder and P. Clavérie, *J. Chem. Phys.* **88**, 4934 (1988).
61. C.R. Le Sauer and A.J. Stone, *Mol. Phys.* **78**, 1267 (1993).
62. J.G. Angyán, G. Jansen, M. Loos, C. Hättig and B.A. Hess, *Chem. Phys. Lett.* **219**, 267 (1994).
63. G. Jansen, C. Hättig, B.A. Hess and J.G. Angyán, *Mol. Phys.* **88**, 69 (1996).
64. C. Hättig, G. Jansen, B.A. Hess and J.G. Angyán, *Mol. Phys.* **91**, 145 (1997).
65. R.W.F. Bader, *Atoms in Molecules–A Quantum Theory*, Oxford University Press, Oxford (1990).
66. R.W.F. Bader, *Chem. Rev.* **91**, 893 (1991).
67. R.J. Wheatley, *Mol. Phys.* **79**, 597 (1993).
68. R.J. Wheatley and J.B.O. Mitchell, *J. Comput. Chem.* **15**, 1187 (1994).
69. P. Claverie, *Int. J. Quant. Chem.* **5**, 273 (1971).
70. I.G. Kaplan in *Handbook of Molecular Physics and Quantum Chemistry*, S. Wilson (ed), John Wiley & Sons, Ltd, Chichester (2003), Vol. 2, Chapter 2.
71. I.G. Kaplan, *Int. J. Quant. Chem.* **89**, 268 (2002).
72. H.N.W. Lekkerkerker and W.G. Laidlaw, *J. Chem. Phys.* **52**, 2953 (1970)
73. R. Eisenschitz and F. London, *Zs. f. Phys.* **60**, 491 (1930).
74. J.O. Hirschfelder and R. Silbey, *J. Chem. Phys.* **45**, 2188 (1966).
75. J.O. Hirschfelder, *Chem. Phys. Lett.* **1**, 326, 363 (1967).
76. A. van der Avoird, *J. Chem. Phys.* **47**, 3649 (1967).
77. J.N. Murrell and G. Shaw, *J. Chem. Phys.* **46**, 1768 (1967).
78. J.I. Musher and A.T. Amos, *Phys. Rev.* **164**, 31 (1967).
79. A.T. Amos and J.I. Musher, *Chem. Phys. Lett.* **3**, 721 (1969).
80. B. Kirtman and R.L. Mowery, *J. Chem. Phys.* **55**, 1447 (1971).
81. T.J. Venanzi and B. Kirtman, *J. Chem. Phys.* **59**, 523 (1973).
82. D.J. Klein, *Int. J. Quantum Chem. Symp.* **4**, 271 (1971).
83. F.A. Matsen and B.R. Junker, *J. Phys. Chem.* **75**, 1878 (1971).
84. D.M. Chipman, J.D. Bowman and J.O. Hirschfelder, *J. Chem. Phys.* **59**, 2830 (1973).
85. R. Peierls, *Proc. Roy. Soc.(London)* A **333**, 157 (1973).
86. B. Jeziorski and W. Kołos, *Int. J. Quantum Chem.* **12**, Suppl. 1, 91 (1977).

87. B. Jeziorski and K. Szalewicz, in *Encyclopedia of Computational Chemistry*, P. von R. Schleyer (ed), John Wiley & Sons, Ltd, Chichester (1998), Vol.2, 1376–1398.
88. B. Jeziorski and K. Szalewicz, in *Handbook of Molecular Physics and Quantum Chemistry*, S. Wilson (ed), John Wiley & Sons, Ltd Chichester (2003), Vol. 3, Chapter 9.
89. S.T. Epstein and R.E. Johnston, *Chem. Phys. Lett.* **1**, 602 (1968).
90. A.T. Amos, *Chem. Phys. Lett.* **5**, 587 (1970).
91. W.H. Adams, *Phys. Rev. Lett.* **32**, 1093 (1974).
92. N. Suzuki and Y.J. I'Haya, *Chem. Phys. Lett.* **36**, 666 (1975).
93. D.M. Chipman, *Chem. Phys. Lett.* **40**, 147 (1976).
94. D.M. Chipman, *J. Chem. Phys.* **66**, 1830 (1977).
95. W.H. Adams and E.E. Polymeropoulos, *Phys. Rev.* A **17**, 11 (1978).
96. E.E. Polymeropoulos and W.H. Adams, *Phys. Rev.* A **17** 18, 23 (1978).
97. W. Kutzelnigg, *Chem. Phys.* **28**, 293 (1978).
98. W. Kutzelnigg, *J. Chem. Phys.* **73**, 343 (1980).
99. B. Jeziorski, K. Szalewicz and G. Chałasiński, *Int. J. Quant. Chem.* **14**, 271 (1978).
100. S. Rybak, B. Jeziorksi and K. Szalewicz, *J. Chem. Phys.* **95**, 6576 (1991).
101. K. Szalewicz and B. Jeziorski, *Mol. Phys.* **38**, 191 (1979).
102. R. Moszynski, S.M. Cybulski and G. Chałasiński, *J. Chem. Phys.* **100**, 4998 (1994).
103. E.M. Mas, K. Szalewicz, R. Bukowski and B. Jeziorksi, *J. Chem. Phys.* **107**, 4207 (1997).
104. R. Moszynski, P.E.S. Wormer and A. van der Avoird, *J. Chem. Phys.* **102**, 8385 (1995).
105. P. Jankowski and K. Szalewicz, *J. Chem. Phys.* **108**, 3554 (1998).
106. G. Murdachaew, A.J. Misquitta, R. Bukowski and K. Szalewicz, *J. Chem. Phys.* **114**, 764 (2001).
107. R. Moszynski, P.E.S. Wormer and A. van der Avoird, in *Computational Molecular Spectroscopy* P. Jensen and P. Bunker (eds), John Wiley & Sons, Inc., New York, 2000, 69–108.
108. R. Bukowski, K. Szalewicz and C. Chabalowski, *J. Phys. Chem.* A **103**, 7322 (1999).
109. G. Chałasiński and M.M. Szcześniak, *Chem. Rev.* **100**, 4227 (2000).
110. W.N. Adams, *Chem. Phys. Lett.* **229**, 472 (1994).
111. W.N. Adams, *Int. J. Quant. Chem.* **60**, 273, 1279 (1996).
112. M. Born and J.R. Oppenheimer, *Ann. der Phys. (Leipzig)* **84**, 457 (1927).
113. G. Chałasiński and M.M. Szcześniak, *Mol. Phys.* **63**, 205 (1988).
114. G. Chałasiński and M.M. Szcześniak, *Chem. Rev.* **94**, 1723 (1994).
115. R.E. Makinson and J.S. Turner, *Proc. Phys. Soc. (London)* **66**, 857 (1953).
116. R.M. Sternheimer, *Phys. Rev.* **96**, 951 (1954).
117. P.R. Certain, J.O. Hirschfelder, W Kołos and L. Wolniewicz, *J. Chem. Phys.* **49**, 24 (1968).
118. S.T. Epstein and J.N. Karl, *J. Chem. Phys.* **44**, 4347 (1966).
119. E. Cordinaldesti, *Nuovo Cim.* **25**, 1190 (1962), **30**, 105 (1963).
120. L. Jansen, *Phys. Rev.* **162**, 63 (1967).
121. L. Jansen and L. Lombardi, *Chem. Phys. Lett.* **1**, 417 (1967).
122. W. Byers Brown, *Chem. Phys. Lett.* **2**, 105 (1968).
123. L. Piela, Int. *J. Quant. Chem.* **5**, 85 (1971).
124. E.E. Polymeropoulos, J. Brickmann, L. Jansen and R. Block, *Phys. Rev.* A **30**, 1593 (1984).

125. J.Juanós i Timoneda and K.L.C. Hunt, *J. Chem. Phys.* **84**, 3954 (1986).
126. M.V. Basilevsky and M.M. Berenfeld, *Int. J. Quantum Chem.* **23**, 555 (1972).
127. J.F. Gouyet, *J. Chem. Phys.* **59**, 4637 (1973); *Ibid.*, **60**, 3690 (1974).
128. V. Kvasnička, V. Laurinc and I. Hubač, *Phys. Rev.* A **10**, 2016 (1974).
129. J.P. Daudey, P. Claverie and J.P. Malrieu, *Int. J. Quantum Chem.* **8**, 1 (1974).
130. P.R. Surjan, I. Mayer and I. Lukovits, *Chem. Phys. Lett.* **119**, 538 (1985).
131. P.R. Surjan and I. Mayer, *Theochem* **72**, 47 (1991).
132. C. Møller and M.S. Plesset, *Phys. Rev.* **46**, 611 (1934).
133. K.A. Bruckner, *Phys. Rev.* **97**, 1353 (1953); **100**, 361 (1955).
134. J. Goldstone, *Proc. Roy. Soc. (London)* A **239**, 267 (1957).
135. M.P. Kelly, *Phys. Rev.* **131**, 684 (1963); **144**, 39 (1966).
136. N.H. March, W.H. Young and S. Sampanthar, *The Many-body Problem in Quantum Mechanics*, Cambridge University Press, London (1967).
137. R.J. Bartlett and D.H. Silver, *J. Chem. Phys.* **62**, 3258 (1974).
138. R.J. Bartlett and D.H. Silver, *Int. J. Quant. Chem. Symp.* **9**, 183 (1975).
139. J.A. Pople, J.S. Binkley and R. Seeger, *Int. J. Quant. Chem.* Symp. **10**, 1 (1976).
140. S. Wilson and D.H. Silver, *Phys. Rev.* A **14**, 1949 (1976).
141. R. Krishnan, M.J. Frish and J.A. Pople, *J. Chem. Phys.* **72**, 4244 (1980).
142. F.-M. Tao and W. Klemperer, *J. Chem. Phys.* **103**, 950 (1995).
143. F.-M. Tao and W. Klemperer, *J. Chem. Phys.* **101**, 1129 (1994).
144. F.-M. Tao, S. Drucker, R.C. Cohen and W. Klemperer, *J. Chem. Phys.* **101**, 8680 (1994).
145. I.G. Kaplan, S. Roszak and J. Leszczynski, *J. Chem. Phys.* **113**, 6245 (2000).
146. I.G. Kaplan, J.N. Murrell, S. Roszak and J. Leszczynski, *Mol. Phys.* **100**, 843 (2002).
147. I.G. Kaplan, S. Roszak and J. Leszczynski, *Adv. Quant. Chem.* **40**, 287 (2001).
148. C.C. Diaz, I.G. Kaplan and S. Roszak, *J. Mol. Modeling* **11**, 330 (2005).
149. P. Pulay, *Chem. Phys. Lett.* **100**, 151 (1983).
150. S. Saebø and P. Pulay, *J. Chem. Phys.* **86**, 914 (1987).
151. S. Saebø, W. Tong, and P. Pulay, *J. Chem. Phys.* **98**, 2170 (1993).
152. A. El Azhary, G. Rauhut, P. Pulay and H.-J. Werner, *J. Chem. Phys.* **108**, 5185 (1998).
153. G. Rauhut, P. Pulay and H.-J. Werner, *J. Comput. Chem.* **18**, 1241 (1998).
154. G. Hetzer, P. Pulay and H.-J. Werner, *Chem. Phys. Lett.* **290**, 143 (1998).
155. M. Schütz, G. Rauhut and H.-J. Werner, *J. Phys. Chem.* A **102**, 5997 (1998).
156. N. Runeberg, M. Schütz and H.-J. Werner, *J. Chem. Phys.* **110**, 7210 (1999).
157. S. Fomine, M.A. Tlenkopatchev, S. Martínez and L. Fomina, *J. Phys. Chem.* A **106**, 3941 (2002).
158. A. Reyes *et al.*, *J. Phys. Chem.* **107**, 7027 (2003).
159. G. Chałasiński, S.M. Cybulski and M.M. Szcześniak, *J. Chem. Phys.* **92**, 2481 (1990).
160. M.M. Szcześniak and G. Chałasiński, *J. Mol. Str. (Theochem)* **261**, 37 (1992).
161. B. Kukawska-Tamawska, G. Chałasiński and K. Olsewski, *J. Chem. Phys.* **101**, 4964 (1994).
162. W. Kołos and L. Wolniewicz, *J. Chem. Phys.* **41**, 3663 (1964); **43**, 2429 (1965); *Ibid.* **49**, 404 (1968).
163. W. Kołos and J. Rychlewski, *J. Chem. Phys.* **98**, 3960 (1993).
164. L. Wolniewicz, *J. Chem. Phys.* **99**, 1851 (1993); **103**, 1792 (1995).
165. J. Rychlewski, *Adv. Quant. Chem.* **31**, 173 (1999).
166. E. Clementi, *J. Chem. Phys.* **46**, 3842, 3851 (1967); **47**, 2323 (1967).
167. K. Morokuma and L. Pedersen, *J. Chem. Phys.* **48**, 3275 (1968).

168. R.F.W. Bader and R.A. Gangi, in *Theoretical Chemistry (A Review of The Recent Literature)*, Burlington House, London (1975), Vol.2.
169. P. Schuster, W. Jakubetz and W. Marius, *Top. Curr. Chem.*, **60**, 1 (1975).
170. P. Schuster, in *The Hydrogen Bond, Recent Developments in Theory and Experiments* P. Schuster, G. Zundel and C. Sandorfy (eds), North Holland, Amsterdam (1976), p. 25.
171. A.J. Duke and R.F.W. Bader, *Chem. Phys. Lett.*, **10**, 631 (1971).
172. A. Rauk, L.C. Allen and E. Clementi, *J. Chem. Phys.* **52**, 4133 (1970).
173. S.F. Boys and F. Bernardi, *Mol. Phys.* **19**, 553 (1970).
174. R.K. Nesbet, *Phys. Rev.* **122**, 1497 (1961).
175. P.-O. Löwdin, *Adv. Chem. Phys.* **2**, 207 (1959).
176. S. Rothenberg and H.F. Schaefer, *Chem. Phys. Lett.*, **10**, 565 (1971).
177. S.V. O'Neil, P.K. Pearson, H.F. Schaefer and C.F. Bender, *J. Chem. Phys.* **58**, 1126 (1973).
178. A.C. Wahl, *J. Chem. Phys.* **41**, 2600 (1964).
179. H. Lischka, *J. Am. Chem. Soc.* **96**, 4761 (1974).
180. P. Bertocini and A.C. Wahl, *Phys. Rev. Lett.* **25**, 991 (1970).
181. P. Bertocini and A.C. Wahl, *J. Chem. Phys.* **58**, 1259 (1973).
182. A.F. Wagner, G. Das and A.C. Wahl, *J. Chem. Phys.* **60**, 1885 (1974).
183. W.J. Stevens, A.C. Wahl, M.A. Gardner and A.M. Karo, *J. Chem. Phys.* **60**, 2195 (1974).
184. B.O. Roos, *Adv. Chem. Phys.* **69**, 399 (1987).
185. H.-J. Werner, *Adv. Chem. Phys.* **69**, 1 (1987).
186. R. Shepard, *Adv. Chem. Phys.* **69**, 63 (1987).
187. G.C. McBane and S.M. Cybulski, *J. Chem. Phys.* **110**, 11734 (1999).
188. M. Koch, B. Fernández and J. Makarewicz, *J. Chem. Phys.* **111**, 198 (1999).
189. B. Fernández, M. Koch and J. Makarewicz, *J. Chem. Phys.* **110**, 8525 (1999).
190. R. Prosmiti, C. Cunha, P. Villarreal and G. Delgado-Barrio, *J. Chem. Phys.* **117**, 7017 (2002).
191. S. Lee *et al.*, *J. Chem. Phys.* **118**, 1230 (2003).
192. S. Lee *et al.*, *J. Chem. Phys.* **119**, 12256 (2003).
193. C.R. Munteance, J.L. Cacheiro and B. Fernández, *J. Chem. Phys.* **120**, 9104 (2004).
194. D.E. Woon and T.H. Dunning, Jr., *J. Chem. Phys.* **103**, 4572 (1995).
195. T.H. Dunning, Jr., K.A. Peterson and D.E. Woon, *Basis Sets: Correlation Consistent Sets*, in *Encyclopedia of Computational Chemistry*, P.v.R. Schleyer (ed), John Wiley & Sons, Ltd, Chichester (1998), pp. 88–115.
196. J. Paldus and X. Li, *Adv. Chem. Phys.* **110**, 1 (1999).
197. X. Li and J. Paldus, *J. Chem. Phys.* **115**, 5759 (2001).
198. I.G. Kaplan, J. Hernández-Cobos, I. Ortega-Blake and O. Novaro, *Phys. Rev.* A **53**, 2493 (1996).
199. J.M. Standart and P.R. Certain, *J. Chem. Phys.* **83**, 3002 (1985).
200. N. R. Kestner, *J. Chem. Phys.* **48**, 252 (1968).
201. H.B. Jansen and P. Ros, *Chem. Phys. Lett.* **3**, 140 (1969).
202. S.F. Boys and F. Bernardi, *Mol. Phys.* **19**, 553 (1970).
203. B. Liu and A.D. McLean, *J. Chem. Phys.* **59**, 4557 (1973).
204. A. Johansson, P. Kollman and S. Rothenberg, *Theoret. Chim. Acta* **29**, 167 (1973).
205. M. Urban and P. Hobza, *Theoret. Chim. Acta,* **36**, 207, 215 (1975).
206. G.F.H. Dierksen, W.P. Kramer and B.O. Roos, *Theoret. Chim. Acta* **36**, 249 (1975).

207. N.S. Ostlund and D.L. Merrifield, *Chem. Phys. Lett.* **39**, 612 (1976).
208. B. Jeziorski and M. van Hemert, *Mol. Phys.* **31**, 713 (1974).
209. H. Popkie, H. Kistenmacher and E. Clementi, *J. Chem. Phys.* **59**, 1325 (1973).
210. P.J. Fortune and R.P. Certain, *J. Chem. Phys.* **61**, 2620 (1974).
211. S. Simon, M. Duran and J.J. Dannenberg, *J. Chem. Phys.* **105**, 11024 (1996).
212. P. Hobza, O. Bludsky and S. Suhai, *Phys. Chem. Chem. Phys.* **1**, 3073 (1999).
213. F. Spiegelmann and J.P. Malrieu, *Mol. Phys.* **40**, 1273 (1980).
214. P.W. Fowler and P.A. Madden, *Mol. Phys.* **49**, 913 (1983).
215. B.H. Wells and S. Wilson, *Mol. Phys.* **50**, 1295 (1983).
216. D.W. Schwenke and D.G. Truhlar, *J. Chem. Phys.* **82**, 2418 (1985).
217. M. Gutowski, F.B. van Duijneveldt, G. Chałasiński and L. Piela, *Mol. Phys.* **61**, 233 (1987).
218. S.M. Cybulski and G. Chałasiński, *Chem. Phys. Lett.* **197**, 591 (1992).
219. M. Gutowski, J.G.C.M. van Duijneveldt-van de Rijdt, J.H. van Lenthe and F.B. van Duijneveldt, J. *Chem. Phys.* **98**, 4728 (1993).
220. F.B. van Duijneveldt, J.G.C.M. van Duijneveldt-van de Rijdt and J.H. van Lenthe, *Chem. Rev.* **94**, 1873 (1994).
221. F. B. van Duijneveldt, in *Molecular Interactions*, S. Scheiner (ed), John Wiley & Sons, Inc., New York (1997), pp. 81–104.
222. I. Mayer and P. Valiron, *J. Chem. Phys.* **109**, 3360 (1998).
223. I. Mayer, *Int. J. Quant. Chem.* **23**, 341 (1983).
224. I. Mayer and P.R. Surjan, *Int. J. Quant. Chem.* **36**, 225 (1989).
225. P. Valiron, A. Vibók and I. Mayer, *J. Comput. Chem.* **11**, 1 (1993).
226. I. Mayer and A. Vibók, *Mol. Phys.* **92**, 503 (1997).
227. J. Yang and N.R. Kestner, *J. Phys. Chem.* **95**, 9214, 9221 (1991).
228. J.E. Del Bene, *Int. J. Quant. Chem. Symp.* **26**, 527 (1992).
229. D. Feller, *J. Chem. Phys.* **96**, 6104 (1992).
230. J.J. Novoa, M. Planas and M.-H. Whangbo, *Chem. Phys. Lett.* **225**, 240 (1994).
231. K.R. Liedi, *J. Chem. Phys.* **108**, 3199 (1998).
232. J.G.C.M. van Duijneveldt-van der Rijdt and F.B. van Duijneveldt, in *Theoretical Treatments of Hydrogen Bond*, D. Hadžy (ed), John Wiley & Sons, Inc., New York (1997), pp. 13–47.
233. A. Halkier *et al.*, *J. Chem. Phys.* **111**, 9157 (1999).
234. S.S. Xantheas, C.J. Burnham and R.J. Harrison, *J. Chem. Phys.* **116**, 1493 (2002).
235. J.G.C.M. van Duijneveldt-van de Rijdt, W. T.M. Mooij and F.B. van Duijneveldt, *Phys. Chem. Chem. Phys.* **5**, 1169 (2003).
236. J. Noga, W. Klopper and W. Kutzelnigg, *Chem. Phys. Lett.* **199**, 497 (1992).
237. H. Müller, W. Kutzelnigg and J. Noga, *Mol. Phys.* **92**, 535 (1997).
238. W. Klopper in *Encyclopedia of Computational Chemistry* P.v.R. Schleyer (ed), John Wiley & Sons, Ltd, Chichester (1998), pp. 2351–2375.
239. A. Halkier *et al.*, *Chem. Phys. Lett.* **286**, 243 (1998).
240. A. D. Becke, *Phys. Rev. A* **38,** 3098 (1988).
241. C. Lee, W. Yang and R.G. Parr, *Phys. Rev. B*, **37**, 785 (1988).
242. J. P. Perdew, in *Electronic Structure of Solids*, P. Ziesche and H. Eschring (eds), Akademik Verlag, Berlin (1991).
243. A. D. Becke, *J. Chem. Phys.* **98**, 5648 (1993).
244. A. D. Becke, *J. Chem. Phys.* **107**, 8554 (1997).

245. T. Ziegler, *Chem. Rev.* **91**, 651 (1991).
246. M. Kieninger and S. Suhai, *Int. J. Quant. Chem.* **52**, 465 (1994).
247. R. Santamaria, I.G. Kaplan and O. Novaro, *Chem. Phys. Lett.* **218**, 395 (1994).
248. C. Lee, C. Sosa and J.J. Novoa, *J. Chem. Phys.* **103**, 4360 (1995).
249. R.G. Parr and W. Yang, *Ann. Rev. Phys. Chem.* **46**, 701 (1995).
250. S.S. Xantheas, J. *Chem. Phys.* **102**, 4505 (1995).
251. H. Guo, S. Sirois, E. I. Proyanov, and D.R.Salahub, in *Theoretical Treatments of Hydrogen Bonding*, D. Hadži (ed), John Wiley & Sons, Ltd, Chichester (1997), pp. 49–74.
252. C. Maerker *et al.*, *J. Comput. Chem.* **16**, 1695 (1997).
253. D.K.W. Mok, N.C. Handy and R.D. Amos, *Mol. Phys.* **92**, 667 (1997).
254. H.L. Schmider and A.D. Becke, *J. Chem. Phys.* **108**, 9624 (1998).
255. L.A. Curtiss, K. Raghavachari, P.C. Redfern and J.A. Pople, *J. Chem. Phys.* **106**, 1063 (1997).
256. A. Milet, T. Korona, R. Moszynski and E. Kochanski, *J. Chem. Phys.* **111**, 7727 (1999).
257. D.J. Singh, K. Schwarz and P. Blaha, *Phys. Rev. B* **46**, 5349 (1992).
258. W. Kohn, Y. Meir and D.E. Makarov, *Phys. Rev. Lett.* **80**, 4152 (1998).
259. S. Kristyan and P. Pulay, *Chem. Phys. Lett.* **229**, 175 (1994).
260. J. M. Pérez-Jordán and A.D. Becke, *Chem. Phys. Lett.* **233**, 134 (1995).
261. A.D. Becke, *J. Chem. Phys.* **98**, 1372 (1993).
262. R.G. Parr and W. Yang, *Density-Functional Theory of Atoms and Molecules*, Oxford University Press, New York (1989).
263. Y. S. Kim and R.G. Gordon, *J. Chem. Phys.* **61,** 1(1974).
264. J.S. Cohen and R.T. Pack, *J. Chem. Phys.* **61**, 2372 (1974).
265. F.A. Gianturco *et al.*, *Mol. Phys.* **92**, 957 (1997); **94**, 605 (1998).
266. F.A. Gianturco, M.Lewerenz, F.Paesani and J.P. Toennies, *J. Chem. Phys.* **112**, 2239 (2000).
267. G.A. Parker and R.T. Pack, *J. Chem. Phys.* **69**, 3268 (1978).
268. C.A. Adamo and V. Barone, *J. Chem. Phys.* **108**, 664 (1998).
269. F.A. Hamprecht, A.J. Cohen, D.J. Tezer and N.C. Handy, *J. Chem. Phys.* **109**, 6264 (1998).
270. Y. Zhao and D.G. Truhlar, *J. Chem. Theory Comput.* **1,** 415 (2005).
271. J. Tao, J.P. Perdew, V.N. Staroverov and G.E. Scuseria, *Phys. Rev. Lett.* **91**, 146401 (2003).
272. V.N. Staroverov, G.E. Scuseria, J. Tao and J.P. Perdew, *J. Chem. Phys.* **119**, 12129 (2003).
273. J.P. Perdew, J. Tao, V.N. Staroverov and G.E. Scuseria, *J. Chem. Phys.* **120**, 6898 (2004).
274. X. Xu and W.A. Goddard, III, *Proc. Nat. Acad. Sci. (USA)*, **101**, 2673 (2004).
275. X. Xu and W.A. Goddard, III, *J. Phys. Chem. A*, **108**, 2305 (2004).
276. Y. Zhao, J. Lynch and D.G. Truhlar, *J. Phys. Chem. A*, **108**, 2715, 4786 (2004).
277. Y. Zhao and D.G. Truhlar, *J. Phys. Chem. A*, **108**, 6908 (2004).
278. T. van Mourik and R.J. Gdanitz, *J. Chem. Phys.* **116**, 9620 (2002).
279. K. Burke and E.K.U. Gross, in *Density Functionals: Theory and Applications*, Springer-Verlag, Berlin (1998), pp. 116–146.
280. M. Levy and A. Nagy, *Phys. Rev. Lett.* **83**, 4361 (1999).

281. A. Görling, *Phys. Rev. Lett.* **85**, 4229 (2000).
282. I.B. Bersuker, *J. Comput. Chem.* **18**, 260 (1997).
283. T. Ziegler, *Chem. Rev.* **91**, 651 (1991).
284. A. Nagy, *Phys. Reports* **298**, 1 (1998).
285. I.G. Kaplan, *Int. J. Quant. Chem.* **89**, 268, (2002).

# 4 Nonadditivity of Intermolecular Interactions

## 4.1 Physical Nature of Nonadditivity and the Definition of Many-Body Forces

Most physical laws established for many-particle systems are additive. The well-known examples are the Coulomb law:

$$V = \sum_{a<b} \frac{q_a q_b}{r_{ab}} \tag{4.1}$$

and the Newton gravitational law:

$$V = \gamma \sum_{a<b} \frac{m_a m_b}{r_{ab}} \tag{4.2}$$

where $r_{ab}$ is the distance between two interacting objects. These laws imply that the charges or bodies can be described as point objects. The interaction of point objects is always characterized by the pair additivity. There is also an additivity for space-extended objects if they are rigid. In general, the potential energy of a rigid-particle system can be always represented as a sum of pair potentials, $g_{ab}$:

$$V = \sum_{a<b} g_{ab} \tag{4.3}$$

irrespective of an interaction law.

However, in quantum mechanics, the charges are not points and they are not rigid. The interacting atoms (molecules) have an internal electronic structure that can be modified in different environments. There are two kinds of interatomic forces that lead to nonadditivity: polarization and exchange forces (a general classification and description of different interatomic interactions are given in Section 1.4 and Chapter 2).

The nonadditivity arising from the polarization forces is the most evident. The interaction energy of two atoms depends upon the location of other atoms because the latter polarize the electronic charge distribution of both interacting atoms. For a three-atom system each pair interaction depends on coordinates of all three atoms (Figure 4.1):

$$V(\mathbf{r}_1, \mathbf{r}_2, \mathbf{r}_3) = V_{12}(r_{12}, r_{13}, r_{23}) + V_{13}(r_{13}, r_{12}, r_{23}) + V_{23}(r_{23}, r_{12}, r_{13}) \tag{4.4}$$

*Intermolecular Interactions* I.G. Kaplan

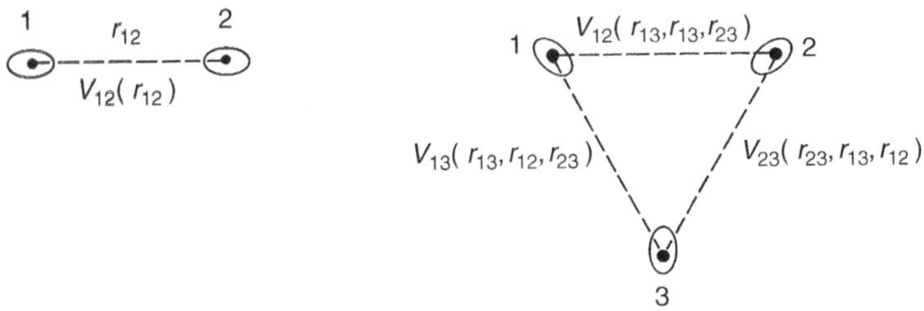


**Figure 4.1** Polarization forces. In a three-particle system, the two-body interaction energies depend upon coordinates of all three particles

So, $V_{ij}$ in Equation (4.4) cannot be considered as a pure two-body interaction. However, we may always represent the energy (Equation (4.4)) as a sum of two-body interactions of isolated pairs and a remainder depending upon coordinates of three atoms:

$$V(\mathbf{r}_1, \mathbf{r}_2, \mathbf{r}_3) = V_{12}(r_{12}) + V_{13}(r_{13}) + V_{23}(r_{23}) + V_{123}(r_{12}, r_{13}, r_{23}) \quad (4.5)$$

This supplementary term to the two-body interaction energies originates from three-body interactions and is called the *three-body interaction energy*.

A second type of interatomic force which leads to nonadditivity comprises exchange forces of two kinds. The first kind has its origin in the Pauli principle, which requires the antisymmetrization of the many-electron wave function. The exchange of electrons belonging to three or more atoms results in nonadditive terms in the interaction energy. The second kind of exchange interaction is connected with the direct exchange through the transverse photons. It results in the nonadditivity of the Casimir–Polder electrodynamical interaction [1] in a many-atom system. The physical picture in the case of three atoms is the following: atom A emits a photon that is scattered by atom B and then absorbed by atom C (Figure 4.2).

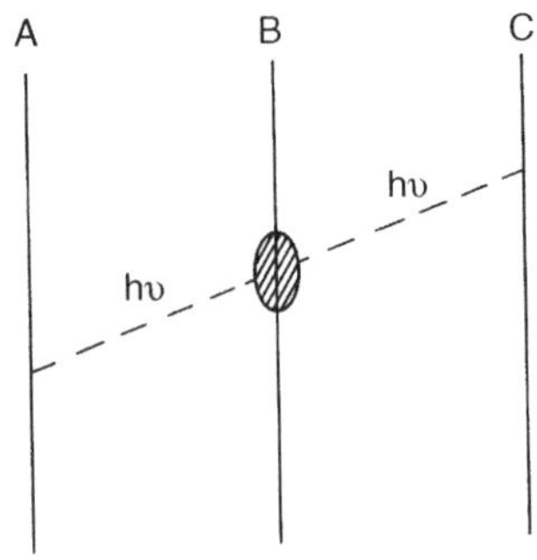


**Figure 4.2** The interaction through the transverse photon exchange gives rise to the three-body forces

One of the necessary conditions for a many-body description is the validity of the decomposition of the system under consideration on separate subsystems. In the case of very large collective effects, we cannot separate the individual parts of the system and only the total energy of the system can be defined. However, in atomic systems, the inner-shell electrons are to a great extent localized. Therefore, even in metals with strong collective valence-electron interactions, atoms (or ions) can be identified as individuals and many-body interactions can be defined. The important role in this separation is played by the validity for molecular systems of the Born–Oppenheimer or adiabatic approximation, which allows the potential energy of an N-atom system to be described as a functional of the positions of atomic nuclei (Section 1.3).

In variational calculations, the interaction energy is computed as a difference of the total energy of the system $E(N)$ and the energy of isolated subsystems (atoms, molecules etc.) $E_1(N)$ :

$$E_{int}(N) = E(N) - E_1(N) \equiv E(N) - \sum_{a=1}^{N} E(a) \tag{4.6}$$

This energy can be decomposed into the energies of pair, triplet and so on, up to N-body interactions. The total energy of an N-particle system can be represented as a finite sum:

$$E(N) = E_1(N) + E_2(N) + E_3(N) + \cdots + E_N(N) \tag{4.7}$$

The equivalent decomposition for the interaction energy is:

$$E_{int} = E_2(N) + E_3(N) + \cdots + E_N(N) \tag{4.8}$$

In the general case, it cannot be expected that the terms in Equation (4.8) have diminishing values, and it is not surprising that the decomposition (Equation (4.8)) does not converge entirely, as was shown by Heine *et al.* [2] in the case of solids. For metal clusters, it was revealed that three-body and even four-body interaction energies can be larger then two-body ones [3, 4] and this was confirmed recently in more precise calculations [5, 6]. From this it follows that in theoretical many-body decompositions of small systems, all terms in Equation (4.8) have to be preserved. Nevertheless, in semiempirical potentials, the truncated (to the three-body) decomposition or only one pairwise term are successfully used. The reason is that the potentials of such kind have fitting parameters. So, it is not real potentials but effective ones.

There are many empirical and semiempirical pair potentials that describe the properties of liquids and solids quite satisfactorily (Chapter 5). The parameters in these potentials are not real parameters of a true two-body interaction; their values depend upon properties of a medium. So, these effective two-body potentials include nonadditive interactions through their parameters. As a result, the latter cannot be directly related to definite physical properties [7]. For instance, the coefficient of the term $R^{-6}$ in the Buckingham or the Lennard-Jones potentials is not equal to the dispersion constant $C_6$. In many cases, to obtain good agreement with experimental data, the effective potentials for clusters and liquids must be constructed with three-body terms [8–10] and even four-body ones [11]. Effective potentials, which take into account many-body interactions and satisfactorily describe properties of metals and superconductors, are discussed in Section 5.1.12.

Note that the decomposition (Equation (4.8)) is exact and can be performed at arbitrary interatomic distances, although the relative weights of $m$-body contributions are method-dependent. To obtain the reliable values of $m$-body interaction energies, the calculation method must be as precise as possible and has to include the electron correlation.

The analytical expressions for many-body interaction energies are defined in a recurrence manner:[1]

$$E_2(N) = \sum_{a<b} \epsilon_{ab} \tag{4.9}$$

where

$$\epsilon_{ab} \equiv E(ab) - [E(a) + E(b)] = E(ab) - E_1(ab) \tag{4.10}$$

The number of pairs in Equation (4.9) is equal to[2] $C_N^2 = N(N-1)/2$:

$$E_3(N) = \sum_{a<b<c} \epsilon_{abc} \tag{4.11}$$

$$\epsilon_{abc} = E(abc) - E_1(abc) - E_2(abc) \tag{4.12}$$

a similar definition holds for the energy of four-body interactions:

$$E_4(N) = \sum_{a<b<c<d} \epsilon_{abcd} \tag{4.13}$$

$$\epsilon_{abcd} = E(abcd) - E_1(abcd) - E_2(abcd) - E_3(abcd) \tag{4.14}$$

and so on up to $E_N(N)$. In Equations (4.9), (4.11) and (4.13), $E_k(N)$ designates the sum of $k$-body interactions for all $C_N^k = N!/(N-k)!k!$ k-atom clusters, which can be separated in the original N-atom cluster; $\epsilon_{ab\ldots k}$ is the $k$-body interaction energy in a concrete k-atom cluster. These k-atom clusters are treated as isolated with the geometry taken as in the original N-atom cluster.

For $N \geq 4$, the calculation according to algorithm (Equations (4.9)–(4.14)) becomes rather cumbersome. A closed formula for the energy of the $m$-body interactions can be obtained [5]. For this, it is necessary to express the energy of the $m$-body interactions through $k$-body ($k \leq m-1$) interactions.

According to the definition:

$$E_m(N) = \sum_{a<b<\cdots<m} \epsilon_{ab\ldots m} \tag{4.15}$$

---

[1] Note that in the formulae below, the total energy and the sum of its many-body contributions is designated by the capital Latin letter $E$. The interaction energy of each separated $m$-particle part of the system is designated by the Greek letter $\epsilon$.

[2] The notation $C_N^n = N!/(N-n)!n!$ designates the number of different combinations of $N$ objects $n$ at time.

This sum contains $C_N^m = N!/(N-m)!m!$ terms. Consider one of them:

$$\epsilon_{ab\ldots m} = E(ab\ldots m) - \sum_{k=1}^{m-1} E_k(ab\ldots m) \tag{4.16}$$

The energy of the $k$-body interactions in an $m$-particle system is equal to:

$$E_k(ab\ldots m) = \sum_{a<b<\cdots<k} \epsilon_{ab\ldots k} \tag{4.17}$$

Substituting Equations (4.16) and (4.17) into Equation (4.15), one obtains:

$$E_m(N) = \sum_{a<b<\cdots<m} E(ab\ldots m) - \sum_{k=1}^{m-1}\left(\sum_{a<b<\cdots<m}\ \sum_{a<b<\cdots<k} \epsilon_{ab\cdots k}\right) \tag{4.18}$$

The double sum in parentheses in Equation (4.18) contains $C_N^m C_m^k$ terms. Collecting terms on $C_N^k$ membered sets, each corresponding to the energy of $k$-body interactions in an N-particle system:

$$E_k(N) = \sum_{a<b<\cdots<k} \epsilon_{ab\ldots k} \tag{4.19}$$

we obtain the expression sought:

$$E_m(N) = \sum_{a<b<\cdots<m} E(ab\ldots m) - \sum_{k=1}^{m-1} a_{mN}^k E_k(N) \tag{4.20}$$

where

$$a_{mN}^k = \frac{C_N^m C_m^k}{C_N^k} = \frac{(N-k)!}{(N-m)!\,(m-k)!} \tag{4.21}$$

For $m = N$, all $a_{NN}^k = 1$ in full accordance with the decomposition (Equation (4.7)).

Thus, the calculation of $m$-body contributions to the interaction energy of N-particle system can be performed by the following recurrent procedure:

$$E_2(N) = \sum_{a<b} E(ab) - a_{2N}^1 E_1(N) \tag{4.22}$$

$$E_3(N) = \sum_{a<b<c} E(abc) - a_{3N}^1 E_1(N) - a_{3N}^2 E_2(N) \tag{4.23}$$

$$E_4(N) = \sum_{a<b<c<d} E(abcd) - a_{4N}^1 E_1(N) - a_{4n}^2 E_2(N) - a_{4N}^3 E_3(N) \tag{4.24}$$

. . . . . . . . . .

$$E_N(N) = E(ab\ldots N) - E_1(N) - E_2(N) - \cdots - E_{N-1}(N) \tag{4.25}$$

The coefficients in Equations (4.22)–(4.24) are given by Formula (4.21).

To estimate the convergence of the many-body expansion, it is convenient to express the $m$-body energy in relative quantities. The rate of decrease of $m$-body energy is given by the ratio:

$$\alpha_m(m-1, N) = \frac{E_m(N)}{E_{m-1}(N)} \tag{4.26}$$

The many-body decomposition (Equation (4.8)) is often expressed as the ratio to the additive two-body energy:

$$E_{int}(N) = E_2(N)\left[1 + \alpha_3(2, N) + \alpha_4(2, N) + \cdots + \alpha_N(2, N)\right] \tag{4.27}$$

The case of a charged system is worthy of special treatment. The main problem is that in a charged system, the location of the charge in its constituent $m$-body parts is not known (the only exceptions are the heteroatomic ionic crystals). The approximate approach for partition of the total interaction energy of a charged system on its additive and nonadditive components is elaborated in Reference [12].

## 4.2 Manifestations of Nonadditive Effects

As discussed in Section 4.1, the additivity exists only if the charge distributions of interacting pairs are not disturbed by the other particles. This can be realized in rarefied gases due to the very small probability of triple collisions. The contribution of many-body forces for a larger density can be estimated from experiments with shock waves, in which the obtained PV-dependence may be interpreted with the help of different empirical potentials. For instance, Ross and Alder [13] analyzed the experiments on the shock compression of argon. The repulsive branch of the potential plays a dominant role in the measured range of pressure. These authors approximated the potential by different model potentials, with parameters chosen by fitting the thermodynamic properties. It was shown that at low pressures, corresponding to a density approximately half that of liquid argon, the potential curve is reproduced by the Buckingham potential exp-6, with an exponent $\alpha = 12 - 12.5$. This value coincides with that obtained from experiments with argon atomic beams. This fact emphasizes the smallness of many-body interactions at that pressure range. However, to reproduce the experimental results obtained at higher density, the exponent value has to be increased up to $\alpha = 13.5$. Hence, the potential obtained in such a way is not a real pair potential but an effective one. An increase in the parameter $\alpha$ indicates a decrease in the repulsion due to the many-body interactions.

There are a number of macroscopic properties (other than energy) that are in principle sensitive to many-body forces. For instance, viscosity, thermal conductivity and diffusion in gases are dependent on many-body forces. But it is very difficult to measure these quantities sufficiently accurately and then compare the results with calculated values from pairwise additive potentials [14]. More often, the experimental information about the existence of many-body forces is obtained from the measurements of the third virial coefficient. The virial coefficients[3] appear in the expansion of the equation of state of a nonideal gas in powers of $V^{-1}$ [15]:

$$\frac{PV}{RT} = 1 + \frac{B(T)}{V} + \frac{C(T)}{V^2} + \cdots . \tag{4.28}$$

The magnitude of the second virial coefficient depends only on pair interactions. In the case of a central potential $V(R)$, it has the following form:

$$B(T) = 2\pi N_A \int_0^{\infty} \left[1 - \exp\left(-\frac{V(R)}{kT}\right)\right] R^2 dR \tag{4.29}$$

where $N_A$ is the Avogadro constant and $B(T)$ is measured in moles. Many-body interactions manifestate themselves in the behavior of the higher virial coefficients. For instance, the expression for the third virial coefficient can be divided into additive and nonadditive terms [16]:

$$C(T) = C^{add} + \Delta C \tag{4.30}$$

$$C^{add} = -\frac{8}{3}\pi^2 N_A^2 \iiint f_{12} f_{13} f_{23} R_{12} R_{13} R_{23} dR_{12} dR_{13}, dR_{23} \tag{4.31}$$

$$\Delta C = -\frac{8}{3}\pi^2 N_A^2 \iiint \exp\left(-\frac{V_2}{kT}\right) \times \left[\exp\left(-\frac{V_3}{kT}\right) - 1\right] R_{12} R_{13} R_{23} dR_{12} dR_{13} dR_{23} \tag{4.32}$$

where $f_{ij} = \exp\left[-V_{ij}/kT\right] - 1$, $R_{ij}$ is the distance between the $i$th and $j$th molecules; $V_2$ and $V_3$ are the two-body and three-body potentials:

$$V_2 = V_{12} + V_{13} + V_{23}$$

$$V_3 = V - V_2 \tag{4.33}$$

If the three-body corrections are negligible, $V_3 = 0$, and the nonadditive term $\Delta C$ vanishes.

The third virial coefficient has a rather simple form. It is measured experimentally (with an accuracy of about 10%) and evaluated theoretically; hence, it gives the possibility of studying the nonadditivity effects. The approximate contribution of the nonadditivity of long- and short-range forces to the third virial coefficient of the noble gases neon, argon, krypton and xenon has been studied in Reference [16]. It was found that the correction to $C^{add}$ due to the exchange repulsion has the opposite sign to that due to the three-particle dipole dispersion forces. As a whole, accounting for the nonadditivity improves the agreement of the calculated values of $C(T)$ with the experimental results. Barker and Pomp [17] estimated that $\Delta C$ accounts for up to 50% of the value of the experimental $C(T)$ for the noble gases.

[3] The term 'virial' originates from the Latin word 'vis' (pl. vires), which means 'force'. The virial coefficients account for the deviation from ideality due to intermolecular forces.

An essential contribution of the nonadditive corrections to the third virial coefficient was also obtained for the gases $N_2$ and $C_2H_4$ [18].

During the last decade, the spectroscopy of van der Waals molecules has became a very important tool in the study of many-body forces. The measurement of high-resolution spectra for van der Waals complexes formed in molecular beams, coupled with recent theoretical advances in the treatment with the associated quantum-dynamical problems, has led to the determination of anisotropic intermolecular pair potentials of high accuracy [19].

Many-body interactions play an essential role in the processes occurring on a surface. The forces between molecules adsorbed on a surface differ considerably from those between the molecules in the gas phase. In the case of physical adsorption, the surface acts as a macroscopic body. The electrostatic and fluctuation fields induce dipole moments in the adsorbed molecules. The latter leads to the appearance of repulsive forces. According to the evaluation of Sinanoğlu [20], adsorption on a metallic surface decreases the depth of the pair potential of rare-gas atoms, in comparison with the value in gas phase, by 20 to 40%.

Many-body interactions determine shifts of the absorption and luminescence bands of molecules in different solvents. The detailed discussion of the regularities observed are given in the review by Liptay [21]. Kestner and Sinanoğlu [22] considered the change of the pair potential due to the three-body dispersion interactions, observed in the transition from gas to condensed phase. The authors studied a liquid within the framework of the continuum model without taking into account the exchange effects. The London formula was obtained, but the dispersion coefficients had to be changed:

$$E_{disp} = -\frac{C_6'}{R_{AB}^6} \tag{4.34}$$

$$C_6' = C_6\,(1 - \overline{\alpha}_0 k) \tag{4.35}$$

where $C_6$ is the dispersion coefficient for the pair interaction in vacuum; $k$ is a constant, approximately equal to 7.1; $\overline{\alpha}_0$ is the polarizability of the unit volume of liquid: $\overline{\alpha}_0 = \alpha\,(0)\,\rho N_A/M$, where $\alpha\,(0)$ is the static polarizability of the molecules of the liquid, $\rho$ is the density, $M$ is the molecular mass and $N_A$ is the Avogadro constant. The evaluation, carried out via the Formula (4.35), demonstrates that the weakening of the pair dispersion interaction, caused by the three-body dispersion forces, accounts for ~9% in liquid helium, ~15% in liquid argon, ~15% in nitrogen, ~20% in methane ($CH_4$) and ~33% in carbon tetrachloride ($CCl_4$) [22].

Thus, in condensed matter, the form of the two-body dispersion term can be conserved if the dispersion coefficient is scaled. A similar situation also exist for the exchange term, whose exponent must also be scaled. This is the reason for the good agreement with experimental data obtained for condensed phases with the help of the empirical pair potentials of the Buckingham (exp–6) or the Lennard-Jones (12–6) types (see Chapter 5). The parameters of these potentials 'absorb' the nonadditivity effects. For example, the (12–6) potential for liquid argon, with the parameters $\epsilon/k \simeq 120$ K and $\sigma \simeq 3.4$ Å [23], corresponds to the dispersion

**Table 4.1** Comparative analysis of theoretical and experimental properties of $(HF)_2$, $(HF)_\infty$ and crystalline HF [24]

| Structure | | $R_{HF}$ (Å) | $R_{FF}$ (Å) | $f_{HF}$ (mdyne/Å$^{-1}$)$^a$ | $q_H$ | $\Delta E$ kcal mol$^{-1}$ |
|---|---|---|---|---|---|---|
| $(HF)_2$ | theor. | 0.919 | 2.704 | 9.23 | 0.512 | −7.49 |
| | expt. | – | $2.79 \pm 0.05$ | – | – | $-6 \pm 1$ |
| $(HF)_\infty$ | theor. | 0.937 | 2.49 | 6.85 | 0.437 | −11.79 |
| HF | crystal expt. | 0.950 | 2.49 | 6.52<br>5.24 | –<br>– | – |

$^a$ $f_{HF}$ is the force constant of the H–F bond; in the case of the crystal, the two values of $f_{HF}$ correspond to the asymmetric and symmetric vibrational modes, respectively; $q_H$ is the charge on H (in units of electron charge) evaluated via the Mulliken population analysis, $\Delta E$ for $(HF)_\infty$ is determined as the difference of the energies of the elementary cell of crystalline HF and the molecule HF.

interaction, which is twice as large as that in the argon dimer in gas phase. At the same time, this potential has a smaller depth than the $Ar_2$ potential [17]. The combination of such characteristics enables one to obtain the satisfactory description of liquid argon.

The influence of many-body forces on the structure and energetic characteristics of condensed media is clearly traced through the transition from dimers to polymers and crystals. The comparative theoretical and experimental data for $(HF)_2$, $(HF)_\infty$ and crystalline HF [24] are presented in Table 4.1. As follows from these data, the transition from dimer to polymer and crystal is accompanied by drastic changes of bond lengths and force constants. The distance $R_{FF}$ in the polymer and crystal is appreciably smaller than in the dimer $(HF)_2$, while for the intramonomer distance $R_{HF}$, the situation is contrary to the $R_{FF}$ behavior. The estimated stabilization energy $\Delta E$ per monomer in $(HF)_\infty$ increased by a factor 1.5 in comparison with the stabilization energy of $(HF)_2$ calculated at the same approximation.

The inadequacy of the additive pair potential description for crystals was demonstrated long ago by Löwdin [25]. In cubic lattices the pair potential model at zero pressure leads rigorously to the so-called Cauchi equality for the elastic constants: $c_{12} = c_{44}$ [26]. Löwdin showed that for alkali halide crystals, the Cauchi equality is not fulfilled. At present, it is well known that in transition metals and semiconductors, the Cauchi equality is often violated. According to Table 4.2, the ratio $c_{12}/c_{44}$ exceed 2 in most of metals, reaching 3.3 and 3.7 in platinum and gold, respectively.

Another manifestation of the inadequacy of the pair-potential model for metals is the magnitude of the vacancy formation energy $E_v$ [27, 28]. This energy is defined as the energy required to transfer an atom from its site (forming a vacancy) and placing it in the bulk of the crystal. In the pair-potential approximation, if the relaxation effects are neglected, the vacancy formation energy is equal to the cohesive energy per atom $E_c$. In metals, the vacancy relaxation energy $E_r$ is rather small, typically $E_r < 0.15E_c$. Whereas for most of metals, the measured $E_v$ is less than one third of $E_c$ (Table 4.2).

**Table 4.2** Indication of many-body effects in f.c.c. crystals [28]; the Lennard-Jones solid (LJ), as a typical two-body approximation, and the rare gas solids Ar and Kr are presented for comparison

| Material | $c_{12}/c_{44}$ | $E_v/E_c$ | $E_c/k_BT_m$ |
|---|---|---|---|
| LJ | 1.0 | 1.00 | 12.8 |
| Ar | 1.1 | 0.95 | 11.1 |
| Kr | 1.0 | 0.66 | 11.5 |
| Al | 2.0 | 0.23 | 41.5 |
| Pb | 2.3 | 0.24 | 39.4 |
| Ni | 1.2 | 0.31 | 29.8 |
| Pt | 3.3 | 0.26 | 33.3 |
| Cu | 1.5 | 0.33 | 29.9 |
| Ag | 1.9 | 0.36 | 27.8 |
| Au | 3.7 | 0.25 | 32.8 |

In additive approximation, the melting temperature $T_m$, as can be seen in Table 4.2 for the Lennard-Jones solid, is only one tenth of the cohesive energy per atoms, or $E_c/k_BT_m \simeq 10$. However, in real metals, this ratio is in three to four times larger.

Thus, for metal crystals, the adequate model potentials have to include the many-body terms (Section 5.1.12).

## 4.3 Perturbation Theory and Many-Body Decomposition

### 4.3.1 General formulae

At distances at which the magnitude of the interaction energy of a system under consideration can be considered small in comparison with the sum of the energies of the isolated subsystems, the interaction energy may be decomposed in a series over various orders of the perturbation theory (PT, see Section A3.3). Most of definitions of different kind of intermolecular interactions are based on the PT approach. In this section, physical terms of the PT series are considered from the viewpoint of the nonadditivity. Special attention is concentrated on the electron correlation contribution to the $m$-body interactions in different orders of PT.

In the general case, the PT series can be represented as:

$$E_{int} = \epsilon_{e\ell}^{(1)} + \epsilon_{exch}^{(1)} + \sum_{n=2}^{\infty}\left[\epsilon_{pol-exch}^{(n)} + \epsilon_{pol}^{(n)}\right] \tag{4.36}$$

The first-order energy, $\epsilon_{e\ell}^{(1)}$, is the electrostatic energy for an interaction of space-distributed charges:

$$\epsilon_{e\ell}^{(1)} = \left\langle \Psi_A^{(0)}\Psi_B^{(0)}\cdots\Psi_N^{(0)}\,|V|\,\Psi_A^{(0)}\Psi_B^{(0)}\cdots\Psi_N^{(0)}\right\rangle \tag{4.37}$$

It is evidently additive because it is calculated over unperturbed wave functions, which means that at this approximation the interacting charges are rigid, and as usual, the interaction operator $V$ is the sum of pair interactions:

$$V = \sum_{A<B} V_{AB} \tag{4.38}$$

The first-order exchange energy, $\epsilon_{exch}^{(1)}$, is expressed as:

$$\epsilon_{exch}^{(1)} = \left\langle \Psi_A^{(0)} \Psi_B^{(0)} \cdots \Psi_N^{(0)} \left| V \right| \sum_{Q \neq I} (-1)^q \, Q \Psi_A^{(0)} \Psi_B^{(0)} \cdots \Psi_N^{(0)} \right\rangle \tag{4.39}$$

where $Q$ are the permutations of electrons between subsystems and $q$ is the parity of permutation $Q$. It is nonadditive for all exchanges mixing electrons of three and more atoms.

As with exchange energy, polarization-exchange energy $\epsilon_{pol-exch}$ is also nonadditive. The standard PT cannot be applied to the calculation of the $\epsilon_{pol-exch}$. The reason is that the antisymmetrized functions of zeroth order $\widehat{A}\Psi_A^{(0)}\Psi_B^{(0)} \cdots \Psi_N^{(0)}$ are not eigenfunctions of the unperturbed Hamiltonian $H_0$, since the operator $H_0$ does not commute with the antisymmetrizer operator $\widehat{A}$. At this range, where the exchange is still important, the standard perturbation schemes have to be modified. The developed approaches are discussed in Section 3.2.1.

At large distances, at which the overlap of interacting charges becomes negligible, the expression for $E_{int}$ (Equation (4.36)) reduces to:

$$E_{int.asymp} = \epsilon_{e\ell}^{(1)} + \sum_{n=2}^{\infty} \epsilon_{pol}^{(n)} \equiv \epsilon_{e\ell}^{(1)} + E_{pol} \tag{4.40}$$

$\epsilon_{pol}^{(n)}$ is the polarization energy of the nth order of the standard PT, it comprises the induction, $\epsilon_{ind}^{(n)}$, and the dispersion, $\epsilon_{disp}^{(n)}$, energies. These energies are nonadditive in all orders of PT; the only exception is $\epsilon_{disp}^{(2)}$ (see the proof in Section 4.3.2). For a three-atom system, the additivity of $\epsilon_{disp}^{(2)}$ means:

$$\epsilon_{disp}^{(2)}(abc) = \epsilon_{disp}^{(2)}(ab) + \epsilon_{disp}^{(2)}(ac) + \epsilon_{disp}^{(2)}(bc) \tag{4.41}$$

Thus, in the energy partition (Equation (4.36)) only two terms are additive; $\epsilon_{e\ell}^{(1)}$ and $\epsilon_{disp}^{(2)}$. If we subtract them from $E_{int}$, the remaining part will contain only nonadditive contributions, $E_{int}^{nonadd}$ [29]:

$$E_{int}^{nonadd} = \epsilon_{exch}^{(1)} + E_{pol-exch} + E_{pol}^{nonadd} \tag{4.42}$$

where

$$E_{pol-exch} = \sum_{n=2}^{\infty} \epsilon_{pol-exch}^{(n)} \qquad E_{pol}^{nonadd} = E_{pol} - \epsilon_{disp}^{(2)} \tag{4.43}$$

Now we can decompose the $m$-body interaction energies defined in Section 4.1 into the PT series. For the two-body interaction energy, the expression directly follows from Equations (4.9) and (4.36):

$$E_2(N) = \sum_{a<b} \epsilon_{ab}$$

$$\epsilon_{ab} = \epsilon_{e\ell}^{(1)}(ab) + \epsilon_{exch}^{(1)}(ab) + \sum_{n=2}^{\infty}\left[\epsilon_{pol-exch}^{(n)}(ab) + \epsilon_{pol}^{(n)}(ab)\right] \tag{4.44}$$

For $m \geq 3$, the expressions for $m$-body energies contain only $E_{int}^{nonadd}$ as all additive terms are cancelled. So, $\epsilon_{abc}$ in Equation (4.12) is equal to:[4]

$$\epsilon_{abc} = E_{int}^{nonadd}(abc) - \sum_{a<b} E_{int}^{nonadd}(ab) \tag{4.45}$$

The sum in the right-hand side of Equation (4.45) contains $C_3^2 = 3$ terms and the sum in Equation (4.11) contains $C_N^3 = N!/(N-3)!3!$ terms. If we substitute Equation (4.45) into Equation (4.11) and subdivide $C_N^3 \cdot C_3^2$ two-body terms on $C_N^2$-membered sets, each corresponding to the nonadditive part of the two-body interaction energy of N-atom system [denoted as $\widetilde{E}_2(N)$], we obtain:

$$E_3(N) = \sum_{a<b<c} E_{int}^{nonadd}(abc) - (N-2)\widetilde{E}_2(N) \tag{4.46}$$

where

$$\widetilde{E}_2(N) = \sum_{a<b} E_{int}^{nonadd}(ab) \tag{4.47}$$

In Equations (4.44) and (4.45), the atom–atom distances in two- and three-atom clusters are taken the same as in the original N-atom clusters.

In the general case, the energy of $m$-body interactions in a PT approach is expressed through the energy of $k$-body ($k \leq m-1$) ones by the following recurrence formula [30]:

$$E_m(N) = \sum_{a<b<\cdots<m} E_{int}^{nonadd}(ab\ldots m) - a_{mN}^2 \widetilde{E}_2(N) - \sum_{k=3}^{m-1} a_{mN}^k E_k(N) \tag{4.48}$$

The term $\widetilde{E}_2(N)$ (Equation (4.47)) is separated, in so far as it differs from the energy of two-body interactions in an N-atom system, given by Equation (4.44). The expression for the coefficients $a_{mN}^k$ is giving by Equation (4.21). For clusters with number of monomers $N \geq 4$, Formula (4.48) is more convenient and efficient than the ordinary recurrence procedure.

Consider now the electron correlation energy. According to its definition [31], the correlation energy is the difference between the exact eigenvalue of the Hamiltonian

---

[4] In Equations (4.45) and (4.47) the energies of dimers, which cannot be nonadditive, are denoted by the general notation (4.42). Certainly, the nonadditivity appears in systems with $N \geq 3$.

describing the system under consideration and the Hartree–Fock (SCF) value, see Equation (3.157). However, except for the smallest systems, the exact energy cannot be obtained. So, the magnitude of the computed correlation energy depends upon the approximation used. For the energy of $m$-body interactions, the correlation contribution is defined as:

$$\Delta E_m^{corr}(N) = E_m(N) - E_m^{SCF}(N) \tag{4.49}$$

and depends upon the method which is applied for a calculation of $E_m(N)$.

The SCF interaction energy may be decomposed [32] as:

$$E_{int}^{SCF} = \epsilon_{el}^{(1)} + \epsilon_{exch}^{(1)} + \Delta E_{def}^{SCF} \tag{4.50}$$

The deformation energy, $E_{def}^{SCF}$, is the correction to the first order of PT due to the relaxation of orbitals in the self-consistent field. It originates from the induction–exchange interactions and can be denote as $E_{ind-exch}^{SCF}$. For the distances at which exchange effects may be neglected, $E_{def}^{SCF}$ is the classical induction energy.

Certainly, we can take into account only a finite number of terms in the perturbation series (4.36). Assume that calculations are performed to the $p$th order of PT. The expression for $\Delta E_2^{corr}(N)$ is easily obtained from Equations (4.44), (4.49) and (4.50):

$$\Delta E_2^{corr}(N) = \sum_{a<b}\left[\sum_{n=2}^{p}\left[\epsilon_{pol-exch}^{(n)}(ab) + \epsilon_{pol}^{(n)}(ab)\right] - \Delta E_{def}^{SCF}(ab)\right] \tag{4.51}$$

Equation (4.51) can be written in a more compact notation:

$$\Delta E_2^{corr}(N) = \sum_{a<b}\left[E_{pol-exch}(ab) + E_{pol}(ab) - \Delta E_{def}^{SCF}(ab)\right] \tag{4.52}$$

The contributions of the induction energy to the energy difference in brackets in Equation (4.52) are almost cancelled. To some degree, the same happens to the exchange energy. At large distances, the exchange terms can be totally neglected and the main contribution to $\Delta E_2^{corr}$ is given by the second-order dispersion energy:

$$\Delta E_{2,asymp}^{corr}(N) = \sum_{a<b}\epsilon_{disp}^{(2)}(ab) \tag{4.53}$$

The expression for the correlation contribution to the energy of three-body interactions can be represented as follows:

$$\Delta E_3^{corr}(N) = \sum_{a<b<c}\left[E_{pol-exch}(abc) + E_{pol}^{nonadd}(abc) - \Delta E_{def}^{SCF}(abc)\right] - (N-2)\sum_{a<b}\left[E_{pol-exch}(ab) + E_{pol}^{nonadd}(ab) - \Delta E_{def}^{SCF}(ab)\right] \tag{4.54}$$

Similar expressions can be obtained for the correlation contribution to the $m$-body energy for $m > 3$. At large distances, the exchange terms can be neglected and, as in the case of $\Delta E_{2,asymp}^{corr}$, the expression for $\Delta E_{3,asymp}^{corr}$ will contain only the dispersion terms in which the leading terms are the third-order dispersion energies: the Axilrod–Teller–Mutto dispersion energy, $\epsilon_{disp}^{(3)}(abc)$ (see Section 4.3.3), and the third-order dispersion energy for dimers, $\epsilon_{disp}^{(3)}(ab)$:

$$\Delta E_{3,asymp}^{corr}(N) = \sum_{a<b<c} \epsilon_{disp}^{(3)}(abc) - (N-2)\sum_{a<b} \epsilon_{disp}^{(3)}(ab) \tag{4.55}$$

The general formulae connecting the many-body interaction energies with physical contributions in the SAPT decomposition [33] are presented in Reference [34].

### 4.3.2 Proof of the additivity of the dispersion energy in the second order of PT

We will follow the proof presented in the author's book [35]. Consider the interaction energy of three atoms (molecules) in the second-order of PT, denoted as $E_{ABC}^{(2)}$. It is convenient to choose, as the initial expression for $E_{ABC}^{(2)}$, its representation in terms of the first-order correction $\Psi_{ABC}^{(1)}$ to the wave function (see Equation (A3.103)):

$$E_{ABC}^{(2)} = \left\langle \Psi_{ABC}^{(0)} \left| V_{ABC} \right| \Psi_{ABC}^{(1)} \right\rangle \tag{4.56}$$

The function $\Psi_{ABC}^{(1)}$ must satisfy the equation following from Equation (A3.149):

$$\left[\left(H_0^A + H_0^B + H_0^C\right) - \left(E_A^{(0)} + E_B^{(0)} + E_C^{(0)}\right)\right] \Psi_{ABC}^{(1)} = \left(E_{ABC}^{(1)} - V_{ABC}\right) \Psi_{ABC}^{(0)} \tag{4.57}$$

where the orthogonality condition:

$$\left\langle \Psi_{ABC}^{(0)} \mid \Psi_{ABC}^{(1)} \right\rangle = 0 \tag{4.58}$$

is assumed to be satisfied. It can be seen by the direct inspection that the solution of Equation (4.57) can be written in the form:

$$\Psi_{ABC}^{(1)} = \Psi_{AB}^{(1)}\Psi_C^{(0)} + \Psi_{AC}^{(1)}\Psi_B^{(0)} + \Psi_{BC}^{(1)}\Psi_A^{(0)} \tag{4.59}$$

where the first-order corrections to the wave function of each interacting pair satisfy an equation similar to Equation (4.57):

$$\left[\left(H_0^A + H_0^B\right) - \left(E_A^{(0)} + E_B^{(0)}\right)\right] \Psi_{AB}^{(1)} = \left(E_{AB}^{(1)} - V_{AB}\right) \Psi_{AB}^{(0)} \tag{4.60}$$

The following orthogonality relations:

$$\left\langle \Psi_{AB}^{(0)} \mid \Psi_{AB}^{(1)} \right\rangle = \left\langle \Psi_{AC}^{(0)} \mid \Psi_{AC}^{(1)} \right\rangle = \left\langle \Psi_{BC}^{(0)} \mid \Psi_{BC}^{(1)} \right\rangle = 0 \tag{4.61}$$

follow from those given by Equation (4.58). Substituting Equation (4.59) into Equation (4.56), one obtains, in the case of the dispersion forces,

$$\left\langle \Psi_{ABC}^{(0)} \left| V_{AB} \right| \Psi_{AC}^{(1)}\Psi_B^{(0)} \right\rangle = 0 \tag{4.62}$$

for all matrix elements of this type. In spite of the fact that Equation (4.61) holds in the case where the induction forces are also taken into account, the relation (Equation (4.62)) is not satisfied in this case, since the terms of the form $V_{00,n0}(AC)\,\Delta E_{on}^{-1}\Psi_n^{(0)}(A)\,\Psi_0^{(0)}(C)$ (see the derivation in Reference [36]), contribute to $\Psi_{AC}^{(1)}$. Taking into account that the matrix elements of the form of Equation (4.62) vanish one finds:

$$\epsilon_{disp}^{(2)}(abc) = \left\langle \Psi_{AB}^{(0)} \left| V_{AB} \right| \Psi_{AB}^{(1)} \right\rangle + \left\langle \Psi_{AC}^{(0)} \left| V_{AC} \right| \Psi_{AC}^{(1)} \right\rangle$$
$$+ \left\langle \Psi_{BC}^{(0)} \left| V_{BC} \right| \Psi_{BC}^{(1)} \right\rangle = \epsilon_{disp}^{2}(ab) + \epsilon_{disp}^{(2)}(ac) + \epsilon_{disp}^{(2)}(bc) \qquad (4.63)$$

That is, the dispersion energy is additive relative to the pair interactions within the framework of the second order of PT.

In contrast to the discussion given in the book by Margenau and Kestner [36], the derivation presented here is valid even if the multipole expansion is not used. The nonadditivity, mentioned in Reference [36], is due to the terms which contain the matrix elements $V_{00,0n}$ related formally to the induction energy but not to the dispersion energy.

### 4.3.3 The dispersion energies of higher orders

The expression for dispersion energy in the third order of PT for three $S$-state atoms was first studied by Axilrod and Teller [37] and independently by Mutto [38]. Using the dipole approximation, they obtained:

$$\epsilon_{disp}^{(3)}(ddd) = f^{ddd}(\theta_a, \theta_b, \theta_c)\,\frac{C_9(ABC)}{R_{ab}^3 R_{ac}^3 R_{bc}^3} \qquad (4.64)$$

It is often called as the ATM (Axilrod–Teller–Mutto) expression. The geometrical factor in Equation (4.64) is equal to:

$$f^{ddd}(\theta_a, \theta_b, \theta_c) = 3\cos\theta_a \cos\theta_b \cos\theta_c + 1 \qquad (4.65)$$

where $\theta_a, \theta_b$ and $\theta_c$ are the angles in a triangle formed by the interacting particles (Figure 4.3). The sign of the three-body dispersion energy is determined by the geometrical factor (Equation (4.65)). $\epsilon_{disp}^{(3)}$ is positive if all the angles $\theta < 117^\circ$, and negative if at least one of the angles $\theta > 126^\circ$. In the case of an equilateral

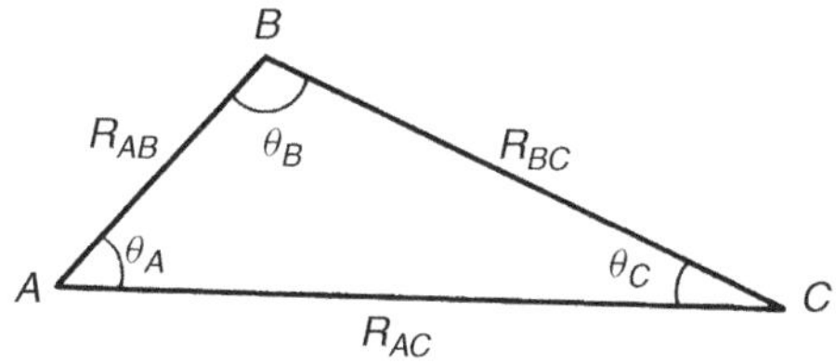


**Figure 4.3** Notations in a three-atomic system

triangle, the nonadditive contribution to the dispersion forces within the framework of the third order of the perturbation treatment leads to a repulsion and is equal to:

$$\epsilon_{disp}^{(3)} = \frac{11}{8} \frac{C_9(ABC)}{R_{ab}^9} \tag{4.66}$$

If the molecules are arranged linearly, it yields an attraction:

$$\epsilon_{disp}^{(3)} = -\frac{2C_9(ABC)}{R_{ab}^3 R_{ac}^3 R_{bc}^3} \tag{4.67}$$

that is, the contribution of the three-body forces stabilizes the linear system.

Aub and Zienau [39] obtained for $C_9(ABC)$ the integral expression similar to the Casimir–Polder formula for $C_6(AB)$ (Equation (2.69)), namely:

$$C_9(ABC) = \frac{3}{\pi} \int_0^{\infty} \overline{\alpha}_1^A(i\omega)\overline{\alpha}_1^B(i\omega)\overline{\alpha}_1^C(i\omega)\, d\omega \tag{4.68}$$

where $\overline{\alpha}_1^A(i\omega)$ is the average dipole dynamic polarizability as function of an imaginary variable.

For $C_9(ABC)$, as for the London formula (Equation (2.73)), an approximate expression via the dipole statical polarizabilities $\alpha_1(0)$ and the first ionization potentials $I_1$, can be derived, see Reference [35]:

$$C_9(ABC) = \frac{3}{2} \frac{\left(I_1^A + I_1^B + I_1^C\right) I_1^A I_1^B I_1^C}{\left(I_1^A + I_1^B\right)\left(I_1^A + I_1^C\right)\left(I_1^B + I_1^C\right)} \overline{\alpha}_1^A(0)\overline{\alpha}_1^B(0)\overline{\alpha}_1^C(0) \tag{4.69}$$

This expression is used for qualitative estimations. In the case of identical atoms (molecules), it becomes very simple:

$$C_9(AAA) = \frac{9}{16} I_1^A \left(\overline{\alpha}_1^A(0)\right)^3 \tag{4.70}$$

For $A = B$, the London formula (Equation (2.73)) reduces to:

$$C_6(AA) = \frac{3}{4} I_1^A \left(\overline{\alpha}_1^A(0)\right)^2 \tag{4.71}$$

Thus, the two dispersion constants are connected by the simple approximate relationship [40]:

$$C_9(AAA) = \frac{3}{4} \overline{\alpha}_1^A(0)\, C_6(AA) \tag{4.72}$$

Dalgarno and Davidson [41] have shown that for hydrogen and noble gas atoms, the relationship (Equation (4.72)) allows $C_9(AAA)$ to be evaluated accurately with the help of $C_6(AA)$. The values of $C_9(AAA)$ obtained seem to be more accurate than expected, taking into account the approximate nature of Equations (4.70) and (4.71).

The expressions for the higher orders of PT and subsequent terms of the multipole expansion were obtained by Bade [42]; see also Bell [43]. The second term, after the $ddd$ term in the multipole expansion of the dispersion energy of three interacting atoms, has the following form:

$$\varepsilon_{disp}^{(3)}(ddq) = f^{ddq}(\theta_a, \theta_b, \theta_c)\frac{C_{11}(ABC)}{R_{ab}^3 R_{ac}^4 R_{bc}^4} \tag{4.73}$$

where the quadrupole moment is located on the atom $C$, and the geometrical factor is equal to:

$$f^{ddq}(\theta_a, \theta_b, \theta_c) = 9\cos\theta_c - 25\cos 3\theta_c + 6\cos(\theta_a - \theta_b)[3 + 5\cos 2\theta_c] \tag{4.74}$$

The $ddd$-term in the fourth order of PT is equal to:

$$\varepsilon_{disp}^{(4)}(ddd) = \frac{45}{64}\left[\frac{1+\cos^2\theta_a}{R_{ab}^6 R_{ac}^6} + \frac{1+\cos^2\theta_b}{R_{ab}^6 R_{bc}^6} + \frac{1+\cos^2\theta_c}{R_{ac}^6 R_{bc}^6}\right] \tag{4.75}$$

At large distances, the energies (Equations (4.64), (4.73) and (4.75)) rapidly fall. At intermediate distances, which are most important when calculating the properties of clusters and condensed matter, the Formulas (4.64) to (4.75) must be corrected for the exchange and charge-overlap effects. As discussed in Chapter 3, the multipole expansion is valid only at large distances, at which the overlap of wave functions becomes negligible. O'Shea and Meath [44] investigated the effect of the charge overlap on the ATM dispersion term. They carried out a numerical study of the system: H(1$s$)–H(1$s$)–H(1$s$) and evaluated the partial energy $W^{(3)}(1, 1, 1)$ which transforms into the $\epsilon_{disp}^{(3)}(ddd)$ (Equation (4.64)) at $R \to \infty$. For distances at which the overlap effects are appreciable, $\epsilon_{disp}^{(3)}$ gives values that are high (more than four times at $R = 4$ bohrs) and falls off with the distance more sharply than $W^{(3)}(1, 1, 1)$ (Figure 4.4). The angular dependence of the interaction energy also undergoes essential changes with the distance. As $R$ decreases and the overlap of the wave functions increases, the range of angles $\theta$, where the three-body energy corresponds to an attraction, decreases. For example, if $\epsilon_{disp}^{(3)} < 0$ at $\theta > 117°$ for

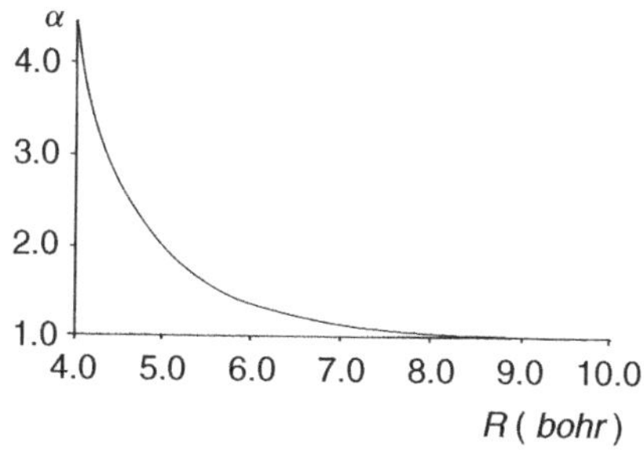


**Figure 4.4** Ratio of the three-body dispersion energy $\epsilon_{disp}^{(3)}$ to the partial energy $W^{(3)}(1, 1, 1)$ as a function of $R$ for the equilateral triangular conformation of H(1$s$)–H(1$s$)–H(1$s$) [44], $\alpha = \epsilon_{disp}^{(3)}/W^{(3)}(111)$

all distances; at $R = 6$ bohrs $W^{(3)}(1, 1, 1) < 0$ at $\theta > 121^\circ$; at $R = 4$ bohrs, for $\theta > 142^\circ$; and at $R \simeq 3.5$ bohrs, $W^{(3)}(1, 1, 1) > 0$ for all angles.

In the intermediate range of distances to the errors caused by the overlap effects and not the validity of the multipole expansion, the errors due to the neglect of the electron exchange effects in the ATM term (Equation (4.64)) and in the next terms (Equations (4.73) to (4.75)) are added. These expressions may be used only at distances that considerably exceed the magnitude of the van der Waals minimum of the dimer.

In a series of studies in the mid-1970s, Barker and collaborators showed that inclusion of the ATM three-body term to accurate pair potential allows a quantitative agreement with crystal properties of the rare-gas solids to be obtained. Barker summarized the remarkable success of the pair + ATM approach in his review [45]. Meath and Aziz [8] analyzed the different types of many-body contributions to the interaction energy and came to the conclusion that the satisfactory approximation of the many-body contribution in the bulk phase using only the ATM three-body term must be regarded as fortuitous. Barker [46] did not agree with their conclusion and emphasized that the existing calculations did not have the level of accuracy required for this purpose.

Meanwhile, the comparative ATM and *ab initio* calculations at the SCF level of the interaction energy in the $Ne_3$ and $Ar_3$ clusters by Wells and Wilson [47, 48] demonstrated the importance of both the ATM and exchange contribution to the three-body interaction energy, and showed that these contributions have opposite signs. Later on, Chałasinski *et al.* [32] using the much more precise SAPT approach also obtained the three-body exchange and dispersion contributions of the comparable magnitude and with opposite signs.

Thus, the theoretical results for small clusters show the importance of taking the three-body exchange forces into account. The success of the semiempirical pair + ATM potentials in reproducing some properties of rare-gas solids may be attributed to the optimized parameters that 'absorbed' other neglected many-body interactions, (compare with the discussion in Section 4.2). If a high level of accuracy is not needed, even pair semiempirical potentials satisfactory describe some bulk properties. However, it is reasonable to use a more reliable analytical form of model potentials.

## 4.4 Many-Body Effects in Atomic Clusters

### 4.4.1 Rare gas clusters

The first systems investigated with respect to the many-body interactions were rare gas clusters [36]. To study the nonadditive effects in these systems, the SCF methods [47–50] and then the more precise SAPT approach [32, 51, 52] were applied. It was shown that the many-body expansion converges rather well.

For trimers with the equilateral geometry, according to the *ab initio* calculations by the coupled cluster method (CCSD(T)), the many-body decomposition, performed according to modified Equation (4.27) in which the absolute value of $E_2(N)$ is used, is the following [52]:

$$
\begin{aligned}
E_{int}(\mathrm{He}_3, R = 5.6\ bohr) &= |E_2(3)|\,[-1 - 0.010] \\
E_{int}(\mathrm{Ne}_3, R = 6.0\ bohr) &= |E_2(3)|\,[-1 + 0.014] \\
E_{int}(\mathrm{Ar}_3, R = 7.0\ bohr) &= |E_2(3)|\,[-1 + 0.041]
\end{aligned}
\qquad (4.76)
$$

For all trimers, the $E_2(3)$ is negative and $E_3(3)$ is positive for $Ne_3$ and $Ar_3$ and negative for $He_3$.

As was found at the SCF level for the tetramer $Ne_4$ in $\mathbf{T}_d$ symmetry [50]:

$$E_{int}(\mathrm{Ne}_4, R = 5.0\ bohr) = |E_2(4)|\,[1 - 0.019 + 0.0006] \qquad (4.77)$$

The convergence of many-body expansion is very good, $E_4(4)/E_3(4) = 0.03$. However, it should be noted that the electron correlation corrections to many-body energies, in some cases, change not only a magnitude but even the sign of energy (compare with data for $Be_N$ and $Li_N$ in next section).

## 4.4.2 Metal clusters

It is instructive to make a comparative study of nonadditive effects in metal clusters with closed electronic subshell atoms (no valence electrons in ground state) and with open valence electronic subshell atoms. Such comparative studies have been carried out for $Be_N$ and $Li_N$ ($N = 2 - 4$) clusters in Reference [6]. Special attention was paid to estimating the electron correlation contribution to the many-body forces. The calculations were based on the formulae presented in Section 4.3.1 and carried out within the framework of 'supermolecular approach'. The Møller–Plesset perturbation theory (Section A3.3.2) up to the fourth order (MP4(SDTQ)), with the triply split valence basis set [6-311+G(3df)] and a correct account of the Basis Set Superposition Error (Section 3.2.2.2) was applied.

In Table 4.3, the many-body interaction energies for the $Be_N$ and $Li_N$ clusters are presented. For the $Be_N$ clusters the following decompositions, according to modified Equation (4.27), are obtained:

$$
\begin{aligned}
E_{int}(\mathrm{Be}_3) &= |E_2(\mathrm{Be}_3)|\,(-1 - 53.9) \\
E_{int}(\mathrm{Be}_4) &= |E_2(\mathrm{Be}_4)|\,(1 - 6.69 + 1.03)
\end{aligned}
\qquad (4.78)
$$

It is evident that for $Be_3$ and $Be_4$ clusters, the three-body energy is not only the dominant term of the many-body decomposition but it is the main (the single in the $Be_4$ case) stabilization factor. The extremely large magnitude of $\alpha_3(2,3) = E_3(\mathrm{Be}_3)/E_2(\mathrm{Be}_3)$ for $Be_3$ does not follow from physics of many-body interactions. It is due to the almost zero value of $E_2(\mathrm{Be}_3)$ because the equilibrium distance in the $Be_3$ triangle is located in the vicinity of the intersection of the $Be_2$ potential curve and the abscissa axis. The fact that the $Be_3$ and $Be_4$ clusters are

**Table 4.3** Many-body energy decomposition of the interaction energy of $Be_N$ and $Li_N$ clusters calculated at the MP4(SDTQ) level [6] for equilibrium geometries (in hartrees)

| $A_N$ | $E_{int}(N) = -E_{bind}$ | $E_2(N) = E_{add}$ | $E_3(N)$ | $E_4(N)$ |
|---|---|---|---|---|
| $Be_2$ | −0.00289 | −0.00289 | | |
| $Be_3$ | −0.04115 | −0.00075 | −0.04040 | |
| $Be_4$ | −0.14672 | −0.03150 | −0.21080 | 0.03258 |
| $Li_2$ | −0.03509 | −0.03509 | | |
| $Li_3$ | −0.04842 | −0.08846 | 0.04004 | |
| $Li_4$ | −0.10166 | −0.16342 | 0.17882 | −0.11706 |

stabilized by the three-body interactions and the flatness of the two-body potential explain the decrement in the interatomic distances from 4.2 bohr in $Be_3$ to 3.9 bohr in $Be_4$: the attractive three-body forces become larger with decreasing beryllium–beryllium distance, while the two-body forces undergo small changes until the beryllium–beryllium distance becomes smaller than 4 bohr.

For $Li_N$ clusters, the decompositions (modified Equation (4.27)) are the following:

$$E_{int}(Li_3) = |E_2(Li_3)|(-1 + 0.45)$$
$$E_{int}(Li_4) = |E_2(Li_4)|(-1 + 1.03 - 0.74) \quad (4.79)$$

The physical picture is opposite to the beryllium cluster case. The two-body interaction energies are large and stabilizing, while the three-body forces play a destabilizing role As a result, there is an increase of the equilibrium distances in the sequence $Li_2$, $Li_3$ and $Li_4$ in order to reduce the three-body repulsion: the latter diminishes more sharply with an increase in the interatomic distance than the two-body attraction does. But the most striking result is the important role of the four-body attraction. The four-body forces are decisive for the formation of the $Li_4$ cluster, since the two-body attraction in $Li_4$ is smaller than the three-body repulsion. It means that all terms in the many-body decomposition of the $Li_4$ interaction energy are important.

Calculations at the SCF and the MP4 levels allow the role of electron correlation in the cluster formation and in many-body interactions to be estimated. The results of such study [6] are presented in Table 4.4. It is well known that in the SCF approximation, the $Be_N$ clusters are stable only for $N \geq 4$. The inclusion of the electron correlation makes stable $Be_2$ $(E_b^{MP4} = 1.8 \text{ kcal mol}^{-1})$ and $Be_3$ $(E_b^{MP4} = 25.8 \text{ kcal mol}^{-1})$ and increases the binding energy of the $Be_4$ cluster from $E_b^{SCF} = 41.9 \text{ kcal mol}^{-1}$ to $E_b^{MP4} = 92.0 \text{ kcal mol}^{-1}$. Although, for $Li_N$ clusters, the inclusion of the electron correlation does not lead to qualitative changes, the quantitative changes are even larger than in the $Be_N$ case: the ratio $E_b^{MP4}/E_b^{SCF}$ is equal to 5.65 ($Li_2$), 2.55 ($Li_3$) and 3.26 ($Li_4$).

**Table 4.4** Influence of calculation method on the interaction energy and its many-body decompositions for the $Be_N$ and $Li_N$ clusters [6] (in hartree)

| $N$ | | $E(N)$ | $E_{int}(N) \equiv -E_{bind}$ | $E_2(N)$ | $E_3(N)$ | $E_4(N)$ | $E_{nonadd} = \sum_{m=3}^{4} E_m(N)$ | $\lvert E_{nonadd}/E_{add} \rvert$ |
|---|---|---|---|---|---|---|---|---|
| | a) $Be_N$ | | | | | | | |
| 2 | SCF | −29.13486 | 0.00905 | 0.00905 | | | | |
| | $MP_4$ | −29.22929 | −0.00289 | −0.00289 | | | | |
| 3 | SCF | −43.71494 | 0.00094 | 0.05706 | −0.05612 | | −0.05612 | 0.98 |
| | MP4 | −43.88126 | −0.04115 | −0.00075 | −0.04040 | | −0.04040 | 53.87 |
| 4 | SCF | −58.35469 | −0.06681 | 0.17244 | −0.32756 | 0.08831 | −0.23925 | 1.39 |
| | MP4 | −58.60040 | −0.14672 | 0.03150 | −0.21080 | 0.03258 | −0.17822 | 5.66 |
| | b) $Li_N$ | | | | | | | |
| 2 | SCF | −14.87028 | −0.00621 | −0.00621 | | | | |
| | MP4 | −14.89916 | −0.03509 | −0.03509 | | | | |
| 3 | SCF | −22.31507 | −0.01898 | −0.00406 | −0.01492 | | −0.01492 | 3.67 |
| | MP4 | −22.34452 | −0.04842 | −0.08846 | 0.04004 | | 0.04004 | 0.45 |
| 4 | SCF | −29.75927 | −0.03111 | 0.01725 | −0.03986 | −0.00850 | −0.04836 | 2.80 |
| | MP4 | −29.82983 | −0.10166 | −0.16342 | 0.17882 | −0.11706 | 0.06176 | 0.38 |

The electron correlation energy also has a very great influence on the many-body terms in the interaction energy decomposition. $E_2(N)$ even changes sign for $Be_2$ and $Be_3$, as well as for $Li_4$. The greatest differences are observed for the ratio $E_{nonadd}/E_{add}$ calculated at the SCF and the MP4 level. In Table 4.5 two kinds of the relative correlation contributions are presented. The relative correlation contribution to the many-body energy calculated at the MP4 level is defined as:

$$\beta_m(N) = \left| \frac{\Delta E_m^{corr}(N)}{E_m^{MP4}(N)} \right| \tag{4.80}$$

In those cases in which the correlation contributions are much larger than $E_m^{SCF}$, as for the $Li_4$ clusters, $\beta_m(N)$ will be close to 1, and it is more instructive to use the relative correlation contribution to the $E_m^{SCF}$:

$$\gamma_m(N) = \left| \frac{\Delta E_m^{corr}(N)}{E_m^{SCF}(N)} \right| \tag{4.81}$$

As follows from Table 4.5, the contribution of the electron correlation is essential for both $Be_N$ and $Li_N$ clusters. In particular, for the $Li_4$ cluster, its magnitude is surprisingly large.

As was shown by Habitz *et al.* [53] and confirmed later in more precise calculations [54, 55], to find the nonadditive contribution to the interaction energy of $(H_2O)_N$ clusters it is enough to perform calculations at the SCF level. The electron correlation corrections to many-body interaction energies of water clusters are very small. For example, according to the data [54], $\gamma_3((H_2O)_3) = 4\%$ and it becomes even smaller, $\sim 1\%$, if we use the data from Reference [55]. For small metal clusters, the situation is contrary to the $(H_2O)_N$ case; nonadditive interactions in $Be_N$ and $Li_N$ clusters cannot be studied without taking into account the electron correlation.

It is important to study the convergence of many-body decomposition for clusters with $N > 4$. In References [5 and 12], the many-body contributions to the interaction energy were calculated for the $Ag_N$ ($N = 4$–$6$) clusters. The calculations were performed by the all-electron nonlocal spin density (NLSD) method.

For all cases considered (excluding the linear geometries), the many-body decomposition is an alternating series: the two-body forces are attractive, the three-body

**Table 4.5** Relative contributions of the electron correlation into the many-body interaction energies for the $Be_N$ and $Li_N$ clusters; $\beta_m$ and $\gamma_m$ are defined by Equations (4.80) and (4.81), respectively

| $A_N$ | $\beta_3$, % | $\gamma_3$, % | $\beta_4$, % | $\gamma_4$, % |
|---|---|---|---|---|
| $Be_3$ | 40 | 28 | | |
| $Be_4$ | 55 | 36 | 171 | 63 |
| $Li_3$ | 63 | 168 | | |
| $Li_4$ | 122 | 549 | 93 | 1280 |

forces are repulsive and so on [5]. The convergence of the many-body series is poor or is absent entirely. For instance, for the tetragonal pyramidal pentamer, the decomposition (modified Equation (4.27)) is:

$$E_{int}\left(Ag_5; \mathbf{C}_{2v}, 3D\right) = |E_2(5)|\,[-1 + 1.28 - 0.87 + 0.19] \tag{4.82}$$

and for the tripyramidal hexamer, it is:

$$E_{int}\left(Ag_6; \mathbf{C}_{2v}, 3D\right) = |E_2(6)|\,[-1 + 1.54 - 1.60 + 0.82 - 0.17] \tag{4.83}$$

where by 3D is denoted three-dimensional confirmations.

According to Equations (4.82) and (4.83), the energy of the three-body interactions for the pentamer and hexamer are larger than the energy of the two-body ones; for the tripyramidal hexamer the four-body interactions are larger than the three-body and two-body ones. The contribution to the interaction energy from the five-body interactions also cannot be neglected. As mentioned in previous sections, it means that a good description of some clusters and solids by interatomic potentials, including only two- or two- and three-body terms has the single explanation that these potentials are effective ones. Their parameters (after fitting) depend implicitly upon the many-body interactions of higher orders.

In Table 4.6, the additive and nonadditive interaction energies for the silver cluster isomers with different point symmetry and space dimension are presented. For all considered conformations, the additive energy is negative and the total nonadditive energy is positive (one exception is for a one-dimensional case). Thus, the $Ag_N$ clusters are stabilized by the negative pairwise energy $E_{add}$ that is natural for atoms possessing valence electrons. The relative stability of different isomers is determined by the competition of $E_{add}$ and $E_{nonadd}$. For hexamers, the relative

**Table 4.6** Additive and nonadditive contributions to the interaction energy of $Ag_N$ clusters [12] (in hartree)

| $N(G, kD)^a$ | $E_{int}$ | $E_{add} \equiv E_2(N)$ | $E_{nonadd} = \sum_{m=3}^{N} E_m(N)$ | $\left\lvert \frac{E_{nonadd}}{E_{add}} \right\rvert$ % |
|---|---|---|---|---|
| 4($\mathbf{D}_{2h}$,2D) | −0.1409 | −0.2566 | 0.1156 | 45.0 |
| 4($\mathbf{C}_{2v}$,2D) | −0.1338 | −0.1936 | 0.0598 | 30.8 |
| 4($\mathbf{D}_{\infty h}$,1D) | −0.1233 | −0.1280 | 0.0047 | 3.7 |
| 5($\mathbf{C}_{2v}$,2D) | −0.1941 | −0.3507 | 0.1567 | 44.7 |
| 5($\mathbf{D}_{\infty h}$,1D) | −0.1539 | −0.1423 | −0.0116 | 8.1 |
| 6($\mathbf{C}_{2v}$,2D) | −0.2660 | −0.4355 | 0.1695 | 38.9 |
| 6($\mathbf{C}_{5v}$,3D) | −0.2606 | −0.5135 | 0.2529 | 49.2 |
| 6($\mathbf{C}_{2v}$,3D) | −0.2448 | −0.6018 | 0.3569 | 59.3 |

[a] The notations used for geometries: **G** is the point symmetry group; $kD$ is the dimensionality of a cluster. The parameters for the geometries are taken from Reference [56].

contribution of $E_{nonadd}$ is much greater in 3D conformations than in 2D ones. As a result, in spite of the greater values of the additive attractive energy, the 3D conformations of $Ag_N$ are less stable than the planar one.

## 4.4.3 Nature of binding in alkaline-earth clusters

### 4.4.3.1 Why the study of binding of alkaline-earth elements is important

Many definitions of the phenomenon of chemical bonding are used. One of the appropriate definitions can be formulated in the following way [57]:

> chemical bonding is the attraction between atoms due to the redistribution (collectivization, charge transfer) of their valence electron density.

Other types of binding can be attributed to physical binding. Examples of different types of binding are depicted in Figure 4.5. A two-atom molecule formed by atoms with an unclosed valence shell can be bound by a *covalent bond* (collectivization of valence electrons), by an *ionic bond* (charge transfer between atoms) and by a so-called *polar bond* (intermediate case: collectivization + charge transfer). These types of binding evidently belong to the definition of chemical binding. To chemical binding must be attributed the binding of molecules into stable associates, in which the hydrogen atom plays the role of a bridge connecting atoms of two similar or different molecules, a so-called *hydrogen bond*. The latter is formed by different types of intermolecular forces; a detailed discussion is given elsewhere [35, Chapter 1].

Compounds, constructed with atoms having closed electronic shells or subshells (that means they do not have valence electrons in the ground state), do not fulfill the chemical bond definition. A well known example is the rare gas atoms. They are stabilized by the van der Waals (dispersion) forces (Section 2.3.2). The *van der Waals bond* is caused by the quantum mechanical fluctuations in the electron density of the interacting atoms. On average, the atomic electron density is not changed. Thus, according to the aforementioned definition, the van der Waals bond cannot be attributed to chemical bonding but rather to physical binding.

On the other hand, in metals, which are traditional physical objects, the binding is realized by the electrostatic interaction between delocalized conducting electrons and positive ions localized at lattice sites. As there is delocalization of the valence atomic electronic density, the metallic binding should be attributed to a chemical type of binding. In this aspect, metals can be considered as giant molecules (graphite planes is the most evident case). This is one of the examples demonstrating that a line of demarcation between physics and chemistry does not exist.

Binding by van der Waals forces is very weak in comparison with chemical bonding. The weakest measured bond was found in $He_2$: the dissociation energy is $D_0 = 1.2$ mK or $2.38 \times 10^{-6}$ kcal mol$^{-1}$ [58] (the well depth is larger and equals 0.02 kcal mol$^{-1}$). Even in bulk, the rare-gas atoms have such small cohesive energy that they can form solids only at low temperature and helium remains liquid at all

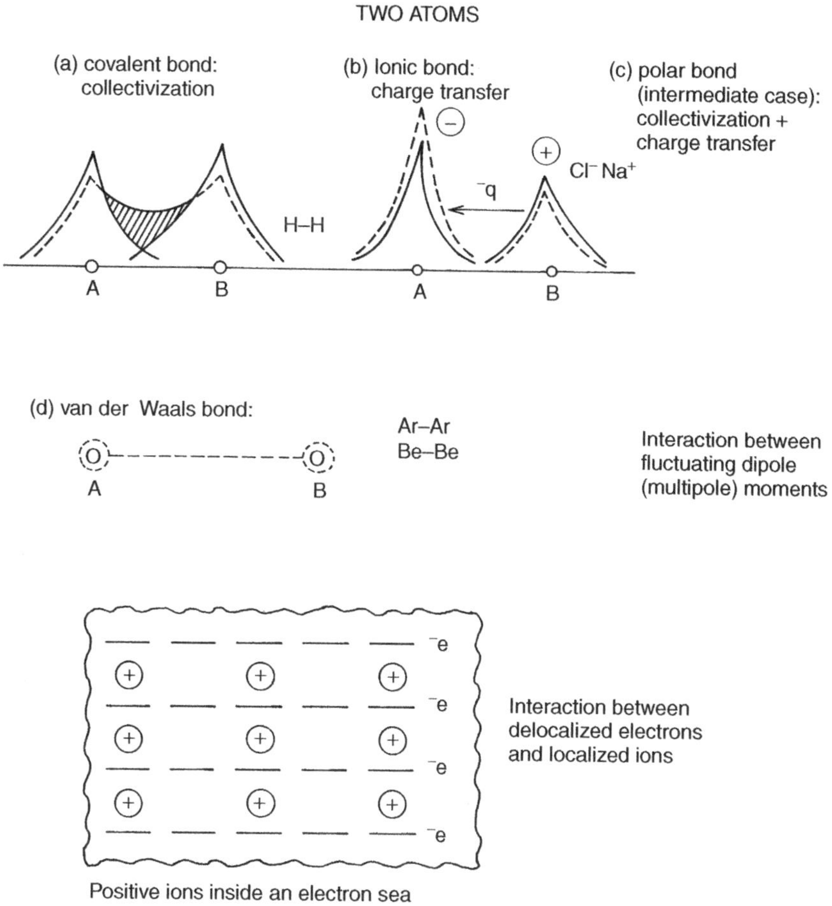


**Figure 4.5** Schematic picture of different types of binding

temperatures. This is a consequence of the closed-shell electronic structure of the rare gas elements and the high energy of their excited atomic states.

On the other hand, the alkaline-earth elements Be, Mg, Ca, etc. have a closed electronic subshell, $(ns)^2$, but form solids with quite large cohesive energy. The cohesive energy in solid beryllium equals 3.32 eV/atom, which is larger than that in solids of open one-valence *ns* shell atoms: Li (1.63 eV/atom) and Na (1.10 eV/atom).

The dimers of beryllium, magnesium and calcium are very weakly bound by the electron correlation effects, at the SCF level they are not stable. The binding energy of alkaline-earth dimers is only two to four times larger than that in krypton and xenon dimers. Thus, alkaline-earth dimers can be attributed to the van der Waals molecules. The situation is drastically changed in many-atom clusters, even in trimers (see Section 4.4.3.2). This is evidently a manifestation of the many-body effects. To reveal the details of binding in alkaline-earth clusters, it is important to carry out a comparative study of binding in clusters of beryllium, magnesium and calcium.

The analysis below is based on studies of the interaction energy and its many-body decomposition for $Be_n$, $Mg_n$ and $Ca_n$ ($n = 2$ and 3) clusters performed in References [59, 60]. All calculations were carried out at the SCF and Møller–Plesset MP4(SDTQ) levels, which allowed the SCF and electron correlation contributions to be studied separately and provided a physical analysis for each term in the dimer and trimer energy decomposition.

### 4.4.3.2 Nature of binding in dimers and trimers

At the SCF level, the interaction energy for the ground state of the beryllium, magnesium and calcium dimers is positive at all distances, that is, the potential curves do not have a minimum. It is the electron correlation energy that stabilizes the closed-subshell-atom dimers. The interaction energies at the equilibrium distances are presented in Table 4.7. The equilibrium distance rises from $R_0 = 2.56$ Å for $Be_2$ to $R_0 = 4.56$ Å for $Ca_2$. The increase in the equilibrium distance in the row $Be_2$, $Mg_2$ and $Ca_2$ is well correlated with an increase in the average radius of the atomic valence shell. However, the binding energy does not have such monotonic behavior. The decrease of $E_b$ from 1.83 kcal mol$^{-1}$ for $Be_2$ to 1.09 kcal mol$^{-1}$ for $Mg_2$ changes with an increase of $E_b$ to 2.14 kcal mol$^{-1}$ for $Ca_2$. The equilibrium distance for $Ca_2$ is very large (4.56 Å) and it could be expected that the Ca–Ca bond will be weaker than the bond in $Mg_2$. But this is not the case; the increase in the equilibrium distance in $Ca_2$ compared with that in $Mg_2$ does not lead to a weaker bond. It is explained by the smaller repulsive SCF energy and the larger correlation attraction at the equilibrium distances in the calcium dimer, with respect to the magnesium dimer (see Table 4.7). The same nonmonotonic behavior takes place for the binding energy of the trimers: $E_b =$ 25.9, 7.12 and 11.66 kcal mol$^{-1}$ for $Be_3$, $Mg_3$ and $Ca_3$, respectively (Table 4.8).

**Table 4.7** Interaction energy for the alkaline-earth dimers at the equilibrium distance at different levels of calculation [59] (in kcal mol$^{-1}$)

| Dimer | $E_{int}^{SCF}$ | $\epsilon_{MP}^{(2)}$ | $\epsilon_{MP}^{(3)}$ | $\epsilon_{MP}^{(4)}$ | $\Delta E^{corr}$ | $E_{int}^{MP4}$ |
|---|---|---|---|---|---|---|
| $Be_2$ $R_0 = 2.56$ Å | 6.12 | −6.71 | −0.81 | −0.43 | −7.94 | −1.83 |
| $Mg_2$ $R_0 = 3.92$ Å | 1.62 | −2.34 | −0.32 | −0.06 | −2.72 | −1.09 |
| $Ca_2$ $R_0 = 4.56$ Å | 1.05 | −2.56 | −0.46 | −0.17 | −3.19 | −2.14 |

**Table 4.8** Interaction energy and the many-body decomposition at the equilibrium geometry (the $\mathbf{C}_{3v}$ symmetry) of the alkaline-earth trimers [59] (in kcal mol$^{-1}$)

| Trimer | $E_{int}^{MP4}$ | $E_{int}^{SCF}$ | $\Delta E^{corr}$ | $E_2^{MP4}$ | $E_2^{SCF}$ | $\Delta E_2^{corr}$ | $E_3^{MP4}$ | $E_3^{SCF}$ | $\Delta E_3^{corr}$ | $E_3^{MP4}/E_2^{MP4}$ |
|---|---|---|---|---|---|---|---|---|---|---|
| $Be_3$ | | | | | | | | | | |
| $R_0 = 2.24$ Å | −25.90 | 0.60 | −26.50 | −0.79 | 35.45 | −36.24 | −25.11 | −34.80 | 9.75 | 31.8 |
| $Mg_3$ | | | | | | | | | | |
| $R_0 = 3.32$ Å | −7.12 | 8.06 | −15.18 | −0.15 | 15.00 | −15.15 | −6.97 | −6.94 | −0.03 | 46.5 |
| $Ca_3$ | | | | | | | | | | |
| $R_0 = 4.12$ Å | −11.66 | 2.16 | −13.82 | −4.44 | 9.15 | −13.59 | −7.22 | −6.98 | −0.23 | 1.6 |

It can be expected that this trend will be preserved in large clusters because, in solid alkaline earths, the cohesive energy and melting temperature show the similar behavior.

According to Equation (4.50), the SCF interaction energy can be presented as a sum of three components: $\epsilon_{e\ell}^{(1)}, \epsilon_{exch}^{(1)}$ and $\Delta E_{def}^{SCF}$. The electrostatic, $\epsilon_{e\ell}^{(1)}$, and exchange, $\epsilon_{exch}^{(1)}$, energies correspond to the first order of the perturbation theory; so, they are defined in the undisturbed atomic wave functions. The third term contains the induction interactions that cannot be separated from the exchange interactions. The induction forces polarize the SCF orbitals; accordingly $\Delta E_{def}^{SCF} \equiv \Delta E_{ind-exch}^{def}$ and is defined on the deformed (relaxed) orbitals.

Atoms with closed subshells have no multipole moments and their electrostatic and induction interactions have a pure overlap origin; from that follows their short-range character. The main contribution to $E_{int}^{SCF}$ gives the exchange interaction $\epsilon_{exch}^{(1)}$. Between atoms with closed subshells, it is repulsive (as in the noble-gas atom systems).[5] This determines the instability of the alkaline-earth dimers at the SCF approximation. They are stabilized by the attractive electron correlation forces.

At large distances, the electron correlation energy can be interpreted as the dispersion energy (see Table 3.7). At intermediate distances where the overlap of the atomic valence shells becomes essential, the dispersion forces cannot be defined without allowing for exchange effects.

As follows from the analysis given above, for a study of the stability of the closed-subshell atom clusters, it is necessary to consider the interplay of the two interaction: exchange and dispersion. In dimers the exchange interactions are repulsive. The main stabilization factor in dimers is the dispersion forces. Thus, the binding in the alkaline-earth dimers has a physical nature. To a great extent they can be attributed to van der Waals bonded molecules (as in the case of noble-gas atom clusters).

Let us now turn to trimers. The interaction energy at the SCF and MP4 levels for $Be_3$, $Mg_3$ and $Ca_3$ trimers in the equilateral triangle conformation is presented in Table 4.8. The trimers, as well as dimers, are not stable in the SCF approximation. According to Reference [59], the SCF energy is positive at all calculated distances. On the other hand, the electron correlation corrections are negative and lead to stabilization of the alkaline-earth trimers. The binding in trimers is much stronger than in dimers, especially in $Be_3$ where the value of $E_b = 25.9$ kcal mol$^{-1}$ is 14 times larger than the binding energy for $Be_2$, which has the van der Waals origin.

A more detailed analysis of the nature of binding is based on the many-body decomposition of the interaction energy:

$$E_{int}^{MP4}(A_3) = E_2^{MP4}(A_3) + E_3^{MP4}(A_3) \tag{4.84}$$

---

[5] $\epsilon_{exch}^{(1)}(ab)$ should not be mixed with $E_2(N)$, the sign of which was discussed in Section 4.4.2. Even at the SCF level, $E_2(N)$ contains $\epsilon_{e\ell}^{(1)}$ and $\Delta E_{def}^{SCF}$, in addition to $\epsilon_{exch}^{(1)}$; at the electron correlation level, it includes $\Delta E_2^{corr}$ also.

The results are presented in Table 4.8. The extremely large values of the ratio of the three-body to two-body energy for the equilibrium conformations of $Be_3$ and $Mg_3$ is connected with almost zero values of the two-body interaction energies. Thus, within the framework of the many-body decomposition of the interaction energy, it must be concluded that for beryllium and magnesium trimers the dominant factor in their stability is the three-body forces. For the calcium trimer, the two-body contribution to the interaction energy is essential and amounts to 38%, although the three-body interactions are still the main contributor to the stability of the cluster.

In the same manner as for the full interaction energy, the two- and three-body interaction energies can also be decomposed on the SCF and electron correlation part:

$$E_n^{MP4} = E_n^{SCF} + \Delta E_n^{corr} \qquad n = 2, 3 \tag{4.85}$$

The two-body SCF energy for an equilateral triangle is equal to:

$$E_2^{SCF}(A_3) = 3E_{int}^{SCF}(A_2) \tag{4.86}$$

It indicates that the physical sense of the two-body SCF energy in trimers is the same as that of the SCF interaction energy in dimers: it is predominantly the exchange interactions that are repulsive for two interacting atoms with closed subshells. The attractive contributions from the electrostatic and induction energies are less than the repulsive exchange contribution. This is the reason that $E_2^{SCF}(A_3)$ is positive for all studied alkaline trimers.

The situation is different in the case of the three-body SCF energy. The main contribution to $E_3^{SCF}(A_3)$ is given by the three-body exchange forces. These forces originate from the three-atomic electron exchange, which mixes electrons of all three atoms. In closed-shell atom systems, contrary to the two-body exchange forces, the three-body exchange forces are attractive and make a contribution to the stabilization of trimers.

The two-body electron correlation energy, $\Delta E_2^{corr}(A_3)$, as in the case of dimers, is reduced at large distances to the dispersion energy. At intermediate distances it contains both the exchange and dispersion contributions, which cannot be separated. The exchange effects decrease the dispersion attraction; nevertheless, the two-body electron correlation imports a significant contribution to the trimer formation, especially for the $Mg_3$ and $Ca_3$ trimers.

The three-body electron correlation energy, $\Delta E_3^{corr}(A_3)$, at large distances can be represented as the Axilrod–Teller–Mutto dispersion energy (Equation (4.64)). For an equilateral triangle, it is transformed to Equation (4.66). According to Equation (4.66), the three-body dispersion energy is positive as is $\Delta E_3^{corr}$ at large distances. At intermediate distances, the negative contributions from the three-body exchange and overlap effects can lead to negative values for $\Delta E_3^{corr}$. This is what is revealed for $Ca_3$ and, close to the equilibrium distances, for $Mg_3$ (Table 4.8).

The binding energy in the alkaline-earth trimers is much larger than that in dimers: in $Be_3$ it is 14 times larger, in $Mg_3$ and $Ca_3$ it is five to seven times larger than the corresponding dimer binding energies. The crucial role in trimer stability is played by the three-body forces. The reason is that at equilibrium distances, the attractive $\Delta E_2^{corr}$ interactions are almost compensated by the repulsive $\Delta E_2^{SCF}$ interactions. For the calcium trimer, the two-body contributions is nonnegligible (38%), although the three-body interactions are still the main contributor to the trimer stability. The decisive role in stabilization is played by three-body exchange forces. They cause the formation of the $\pi$-in-plane bond (see Section 4.4.3.3). Thus, in contrast to the van der Waals binding in the alkaline-earth dimers, the binding in the alkaline-earth trimers has a chemical nature.

#### 4.4.3.3 Population of vacant atomic orbitals

It is instructive to study the vacant atomic orbital population in dimers and trimers. In the 1980s, Bauschlicher *et al.* [61, 62] came to the conclusion that the promotion of $ns$-electrons to $np$-orbitals leading to $sp$-hybridization is the main mechanism responsible for binding in alkaline-earth clusters. This conclusion was based on a study of the SCF Mulliken population analysis for tetramers, which are stable at the SCF level. At present, a more precise, at the electron correlation level, natural bond orbital (NBO) analysis can be performed. It has been confirmed [59, 60] that the promotion of ns-electrons to np-orbitals in the alkaline-earth clusters takes place to a considerable extent, but it does not inevitably lead to binding.

The net population of valence orbitals in dimers and trimers is presented In Table 4.9. It can be seen that the $p$-population and even the $d$-population, especially for calcium clusters, are not negligible. The latter is correlated with the experimental atomic excitation energies $\Delta E_{at}$ [63]. The energy of the $4s \rightarrow 3d$ excitation in the calcium atom [63] is even smaller than the $ns \rightarrow np$ excitation energies in the beryllium and magnesium atoms. On the other hand, there is no quantitative relation between $\Delta E_{at}$ and the net population numbers $\Delta n_l$. The magnitude of $\Delta E_{at}$ $(ns \rightarrow np)$ in Be is larger than that in Ca; nevertheless, the $np$-population in the beryllium clusters is also larger than in the calcium clusters.

The NBO valence population at the MP4 level was calculated also for the isolated atoms. It could be expected that the inclusion of electron correlation effects leads to some population of vacant (in the SCF approximation) atomic orbitals. But the values obtained are surprisingly large. The $p$-population in the magnesium and calcium atoms are only slightly smaller than that in their dimers, and in Be the $p$-population is 0.7 of that in $Be_2$. Thus, to study the influence of interatomic interactions on the promotion of electrons to vacant orbitals, the atomic $p$-population has to be considered as a reference level. From this it also follows that the alkaline-earth atoms, assumed traditionally as the closed $ns$-subshell atoms, can to some extent manifest the anisotropic $p$-symmetry behavior.

One could ask about the meaning of fractional charges in Table 4.9. The answer is that these fractional numbers are the mean occupation numbers of the atomic states but not the real fractional charges. In the NBO population analysis, they appear as

**Table 4.9** The net valence population, $\Delta n_l$, for the isolated alkaline-earth atoms and its clusters at the equilibrium geometry, obtained by the NBO analysis at the SCF and MP4 levels [59][a]

| | SCF | | | | MP4 | | | |
|---|---|---|---|---|---|---|---|---|
| $A_n$ | $ns$ | $(n+1)s$ | $np$ | $nd$ | $ns$ | $(n+1)s$ | $np$ | $nd$ |
| Be | 0.000 | 0.000 | 0.000 | 0.000 | −0.135 | 0.004 | 0.130 | 0.001 |
| $Be_2$ | −0.044 | 0.006 | 0.037 | 0.001 | −0.199 | 0.008 | 0.185 | 0.006 |
| $Be_3$ | −0.257 | 0.005 | 0.246 | 0.005 | −0.315 | 0.009 | 0.288 | 0.016 |
| Mg | 0.000 | 0.000 | 0.000 | 0.000 | −0.112 | 0.004 | 0.105 | 0.003 |
| $Mg_2$ | −0.007 | 0.001 | 0.005 | 0.000 | −0.123 | 0.004 | 0.113 | 0.005 |
| $Mg_3$ | −0.045 | 0.002 | 0.040 | 0.002 | −0.173 | 0.005 | 0.154 | 0.012 |
| Ca | 0.000 | 0.000 | 0.000 | 0.000 | −0.138 | 0.003 | 0.124 | 0.011 |
| $Ca_2$ | −0.016 | 0.002 | 0.011 | 0.003 | −0.161 | 0.005 | 0.139 | 0.018 |
| $Ca_3$ | −0.074 | 0.003 | 0.057 | 0.015 | −0.299 | 0.006 | 0.184 | 0.039 |

[a]For atoms $\Delta n_l^{MP4}(A) = n_l^{MP4}(A) - n_l^{SCF}(A)$, for clusters $\Delta n_l^{MP4}(A_n) = n_l^{MP4}(A_n) - n_l^{SCF}(A)$; the similar definition is for $\Delta n_l^{SCF}$, therefore $\Delta n_l^{SCF}(A) = 0$.

the coefficients in the natural orbital expansion of the one-electron density matrix:

$$D(x \mid x) = \sum_{nl} c_{nl} \Phi_{nl}^2 (x) \tag{4.87}$$

For example, the ground state occupation numbers for beryllium at the MP4 level are:

$$K_0^{MP4} : 1s^2 2s^{1.87} 2p^{0.13}$$

In the MP perturbation theory, only double excited configurations are mixed. So, the ground state wave function of the beryllium atom can be presented with a good precision as a superposition of the two configurations:

$$\Psi_0^{MP4}(\mathrm{Be}) = a\Psi\left(1s^2 2s^2, {}^1S\right) + b\Psi\left(1s^2 2p^2, {}^1S\right) \tag{4.88}$$

with $b \ll a$. This leads to the fractional mean occupation numbers.

It is necessary to take into consideration that some atom–atom interactions, which enhance the excited orbital population, do not lead to a bound state. The last statement is confirmed by the valence orbital population in the alkaline-earth clusters at the SCF level. According to Table 4.9, at the SCF level there is a nonnegligible $p$-population, especially for trimers. But at the SCF approximation, the dimers and trimers are not stable. Thus, the repulsive, in total, SCF interactions also lead to the vacant orbital populations, although the hybridization caused by that does not lead to binding in the dimer and trimer cases.

The calculations in Reference [61] were performed at the SCF level. Within the framework of the latter, the isolated atoms do not have populated excited orbitals.

The authors [61] found the ratio of $p$-population in different tetramers (which are stable at the SCF level) proportional to the ratio of their dissociation energies. However, at an electron correlation level because of the nonzero $p$-population in the isolated atoms, such proportionality cannot be expected. It is evident for dimers: the $p$-population is largest in $Be_2$, although the bond strength is largest in $Ca_2$. For dimers, the $p$-population is not very different from that in the isolated atoms, whereas for trimers the increase of the $p$-population is rather large. It is not directly proportional to the bond strength, but its amount qualitatively reflects a trend toward a stronger bond formation by the $sp$-hybridization.

More insight into the bonding comes from the density difference maps. The total density difference is defined, like the interaction energy (Equation (4.6)), as the difference of the molecular electronic density and the atomic ones. The partition of the total density difference into two- and three-body terms was performed in the analogy to the interaction energy expressions (Equations (4.9)–(4.12)). In Figures 4.6 and 4.7 the total density difference maps are presented and its two- and

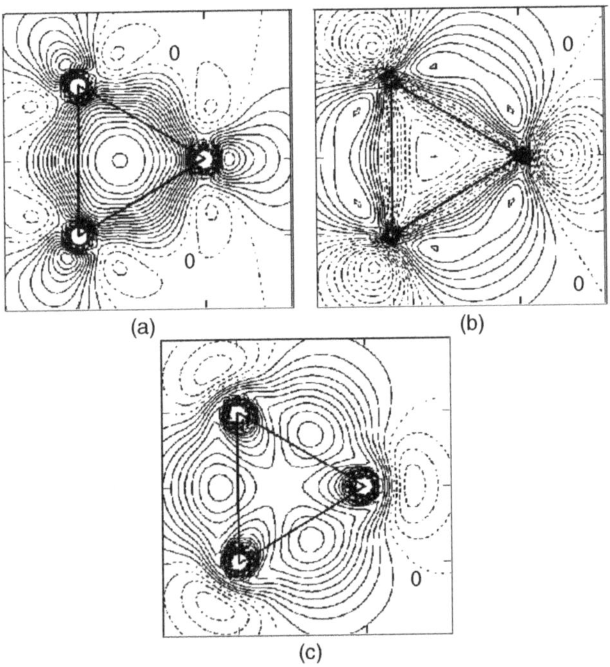


The plot is done in the plane of $Be_3$. The spacing between the contours is 0.001 electron/bohr. The contours with no charge density are labeled with zeros, while solid lines indicate the enhancement of electronic density.

**Figure 4.6** The electronic density difference maps for the $Be_3$ trimer partitioned for two-body (a), three-body (b), and the total difference density distribution (c)

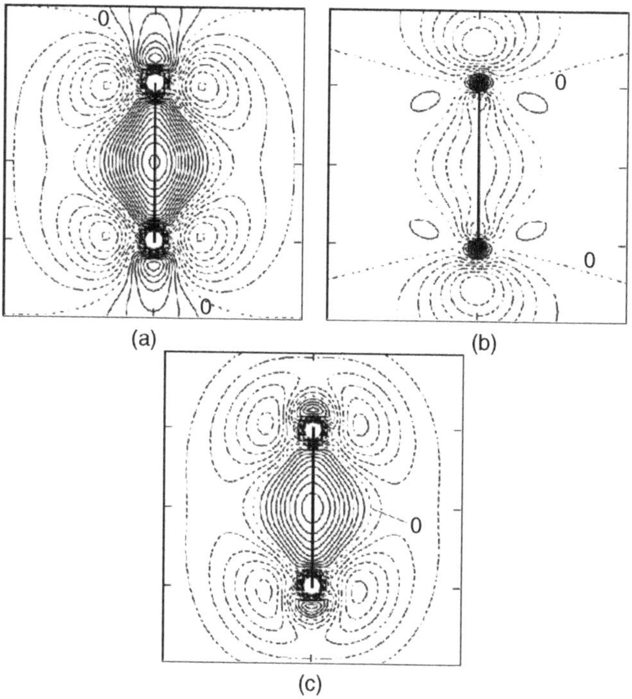


The plot is done in the plane perpendicular to the $Be_3$ plane and passing through the Be–Be bond. The spacing between the contours is 0.001 electron/bohr. The contours with no charge density are labeled with zeros, while solid lines indicate the enhancement of electronic density.

**Figure 4.7** The electronic density difference maps for the $Be_3$ trimer partitioned for two-body (a) and three-body (b) contributions, and the total difference density distribution (c)

three-body contributions in the $Be_3$ plane and in the perpendicular plane passing through the Be–Be bond. For two-body density difference, there are positive values along the Be–Be bond that can be attributed to the $\sigma$-type bonding (the latter is clearer from the plot in the plane perpendicular to the $Be_3$ plane, Figure 4.7a). The two-body dispersion interactions do not change the density distribution. Thus, the $\sigma$-type redistribution originates from the SCF two-body interactions, inside which there are two attractive interactions: electrostatic and induction. Because of the larger two-body exchange repulsion, these attractions are not sufficient for the stabilization of $Be_3$, and it must be concluded that the $\sigma$-type redistribution of the two-body density difference does not lead to a real bonding. This is analogical to the atomic $p$-population at the SCF level that also does not lead to trimer stability.

Figures 4.6a and 4.7a also reveal a positive density difference in the nonbonding region (the area outside the atoms attached to the vertices of the triangle

$Be_3$). This reflects the antibonding character of the two-body exchange interactions. The three-body interactions shift the electron density from nonbonding to bonding regions, because in the total density difference plots, the electronic density gain is observed only in the bonding area (Figure 4.6c).

According to Figures 4.6b and 4.7b, the three-body density difference is positive outside of the beryllium–beryllium line in a direction perpendicular to it, whereas it is negative beyond the $Be_3$ plane. So, the density difference has the $\pi$-in-plane character but is shifted outside the $Be_3$ triangle.

Let us finish at this stage the discussion of chemical bonding in compounds composed with atoms not having valence electrons in its ground state. In the last section of this chapter we discuss an approximate additive scheme permitting the study of large nonadditive systems: many-atomic molecular complexes (including biomolecules) and molecular crystals.

## 4.5 Atom–Atom Potential Scheme and Nonadditivity

It is very attractive, in the study of interactions between complex molecules, to represent the interaction energy as a sum of the pair interaction energies of atoms:

$$E_{int}(R) = \sum_{a<b} V_{ab}(R_{ab}) \tag{4.89}$$

in such a way that the pair potentials, $V_{ab}(R_{ab})$, possess some degree of universality, for instance, for a given homologous class. As discussed in the previous sections, the additivity is expected only in the case of rarefied gases. Hence, the representation (Equation (4.89)) contains effective, not real, pair interaction energies with the parameters scaled, with the help of some physical properties. The atom–atom potentials (AAPs) with such scaled parameters should be transferred, in order to evaluate the interactions between other molecules belonging to a given class.

Hill [64] was the first to propose that the interactions of atoms, or molecular groups, which are not bonded to each other, should be represented as a sum of semiempirical pair potentials. In that way, he obtained the conformations of molecular isomers and the potential barriers between them. Another atom–atom scheme was proposed independently by Kitaigorodsky [65] and developed by him and his coworkers for the study of molecular crystal properties [66–69]. Kitaigorodsky chose the Buckingham (exp–6) potential:

$$V_{ab}(R_{ab}) = \frac{\epsilon}{1-\frac{6}{\alpha}}\left\{\frac{6}{\alpha}\exp\left[\alpha\left(1-\frac{R_{ab}}{R_0}\right)\right] - \left(\frac{R_0}{R_{ab}}\right)^6\right\} \tag{4.90}$$

However, instead of using three parameters, such as the well depth $\epsilon$, the steepness $\alpha$ of the repulsive branch and the equilibrium distance $R_0$, he used only $R_0$. The other two parameters were fixed. The universal potential, proposed initially by

Kitaigorodsky [65], had the following form (with coefficients fitted to the interaction energy in kcal $mol^{-1}$):

$$V_{ab}(R_{ab}) = 3.5\left[8.6 \cdot 10^3 \exp(-13R_{ab}/R_0) - 0.04\,(R_0/R_{ab})^6\right] \tag{4.91}$$

Later, the numerical coefficients were revised [66, 67]. The potential, given by Equation (4.91), does not depend on the type of interacting molecules and has the same depth of the potential well for all pairs of atoms. Nevertheless, the application of the potential within the framework of the additive scheme (Equation (4.89)) has given the possibility of determining the mutual arrangement of molecules in the elementary cell and the sublimation heats of hydrocarbon crystals in a satisfactory agreement with the experimental data. Williams [70] evaluated all three parameters of the AAP (Equation (4.90)) by a least-squares method in such a manner that the sublimation heat, the elasticity constants and the crystal structure of the series of hydrocarbon molecular crystals were described with better accuracy. An analysis of the possibilities of the atom–atom additive scheme in molecular crystals has been presented in References [68, 69].

Besides the studies of molecular crystals, the atom–atom potential approach is widely used in conformational analyses, where the direct electrostatic interaction energy, the deformation energy of the valence angles and the so-called torsion energy, giving the change in the intramolecular interaction with respect to the reference conformation, have been added to the potential (4.90). The simulation of the interaction of nonbonded atoms by the AAPs yields the equilibrium conformations of complex molecules, the corresponding barriers and some other properties. It has been shown that different potentials can lead to an accuracy comparable to that of the experimental results. For example, the numerical study of the equilibrium conformations of 85 alkanes, carried out by Engler *et al.* [71] with the sets of the parameters, proposed by Allinger *et al.* [72] and by the authors [71],[6] led to results for a majority of molecules within the limits of the experimental error. With some exceptions, the bond lengths are reproduced with an accuracy of 0.01 Å, the angles with $1^\circ$ to $2^\circ$ and the heats of formation with $\sim$1 kcal $mol^{-1}$.

Ramachandran [73] carried out a calculation of conformations of peptides and sugars with the Kitaigorodsky [65], Scott–Scheraga [74] and Flory–Brant–Miller [75] potentials and showed that in the case of peptides all three potentials gave satisfactory results; although, these potentials behave quite differently in the range of 2.2–3.8 Å, which is important for the equilibrium conformations. It is interesting that in the case of sugars, the best agreement with the experimental results was obtained by the most idealized Kitaigorodsky potential. The recent achievement in the AAP calculations of polypeptides and proteins are presented in the review by Vásquez *et al.* [76].

The goal of conformation analysis is to find the stable conformations of molecules or clusters of molecules (see the discussion in Reference [77]). One

[6] It is instructive to stress that the parameters chosen in Reference [71] are completely different from those proposed in Reference [72].

of the preliminary conditions for a successful search is the selection of an appropriate molecular force field. The AAP method gives an approximate but one of the simplest expression for the molecular force field. In conformation analysis, the problem of the global minimum search is very important because of the multiple–minima behavior of the potential energy function [78–80]. We will return to this problem in Section 5.4.

The use of different experimental data and approaches leads to quite different potentials for the same pairs of atoms. Moreover, different sets of parameters can be obtained from the same experimental data due to the ambiguity of the fitting procedure. Burgos and Bonadeo [81] carried out an accurate fitting of the parameters of the Buckingham potential:

$$V_{ab}(r_{ab}) = -AR_{ab}^{-6} + B\exp(-CR_{ab}) \tag{4.92}$$

based on the sublimation heats, lattice equilibrium parameters and lattice vibrational frequencies for crystals of bromine-substituted benzenes. The authors [81] found two different sets of parameters (Table 4.10) describing the experimental data with the same accuracy.

It is not only the set of parameters but the analytical form of AAPs having the same accuracy that can be different. The behavior of AAP curves does not follow unambiguously from the system described. In Figure 4.8, the AAP curves for the pairs carbon–carbon and hydrogen–hydrogen found in References [66, 70, 82] are depicted. In spite of the differences for the position and depth of the minimum and the slope, all these potentials describe satisfactorily the geometry and the sublimation energy of the hydrocarbon crystalline lattice. It emphasizes the weak sensitivity of the crystal properties under study to the form of the potential and also the physical irrelevance of these potentials, whose role is reduced to that of subsidiary functions in the calculation (in this regard see References [7, 83]).

The real AAPs should depend on the structure of the molecule considered, since the atomic electron cloud is distorted in a different way in different molecules. The

**Table 4.10** Two equivalent sets of parameters for the atom–atom potentials for the bromine–benzene crystal [81]

| | | $A\left(\frac{\text{kcal}}{\text{mol}}\text{Å}^6\right)$ | $B\left(\frac{\text{kcal}}{\text{mol}}\right)$ | $C$ (Å$^{-1}$) |
|---|---|---|---|---|
| Br–C | I | 730 | 78 500 | 3.37 |
| | II | 421 | 118 000 | 3.57 |
| Br–Br | I | 4 580 | 149 000 | 3.14 |
| | II | 7 830 | 64 300 | 2.78 |
| Br–H | I | 555 | 18 050 | 3.44 |
| | II | 401 | 20 900 | 3.55 |

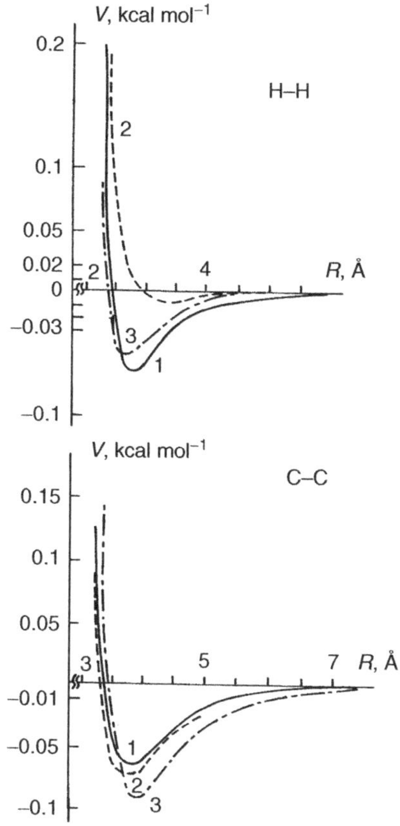


**Figure 4.8** Atom–atom potentials for H–H and C–C pairs

effect of chemical surrounding and conformation of the dimer under study on the parameters of the AAPs, determined by fitting the theoretical curve to experimental one, was investigated by the author and Rodimova [7]. As it should be expected, this effect appears to be rather essential. It means that the parameters of the AAPs cannot be directly related to a definite physical property and that it is a set of all the properties participating in the fitting procedure that influences the magnitudes of the parameters. For instance, the coefficient of the term $R^{-6}$ is not equal to the dispersion constant $C_6$ (compare this with the discussion of the empirical potential for argon–argon at the end of Section 4.2). That is the way it will be, if all the parameters of the AAP are adjustable. Some authors fix the coefficient of the term $R^{-6}$, evaluating it with the help of a semiempirical formula for $C_6$, in order to reduce the number of adjustable parameters. One should keep in mind in these cases that the value of this parameter is not the optimal one.

It should be stressed that a successful application of the AAP approach to determine the optimal packing of molecules in a crystal is possible, only if this application is based on a given crystal symmetry and size of the elementary cell. In such a case the equilibrium packing seems to show little sensitivity to the potential form; it is sensitive only to the location of the minimum of potential. For that reason, the universal one-parameter Kitaigorodsky potential (Equation (4.91)) manifested itself as quite satisfactory for the determination of the geometrical structure within a framework of given symmetry. On the other hand, the choice of the correct symmetry of a crystal is usually outside the limits of accuracy of the AAP method. For instance, the monoclinic allotrope of benzene found using the atom–atom scheme is more stable at all densities than the rhombic one, while it is known experimentally that the monoclinic allotrope exists only at pressures above 12 kbars [68].

The Buckingham potential (Equation (4.90)) does not contain a term corresponding to the direct electrostatic interactions. Such terms can be introduced into the atom–atom scheme as a sum of Coulomb interactions of charges located on the atoms. These charges may be determined with the help of any quantum-chemical approximation. According to Kitaigorodsky [68], inclusion of the Coulomb terms does not help, due to the small contribution of the electrostatic interactions to the energy of molecular crystals, at least for the hydrocarbon crystals, and the uncertainty of the concept of atomic charges. Originally, Williams [70] also did not consider the direct electrostatic interaction energy. But in a following publication [84], he gave arguments for the inclusion of the Coulomb energy in the AAPs.

If the inclusion of the Coulomb energy in the atom–atom scheme is questionable in the case of crystals, it is quite necessary for the study of molecular conformations of clusters [74, 76]. For instance, it is obvious in the case of dimers. It should be noted that the Buckingham or Lennard-Jones potentials, calibrated from crystalline or equilibrium molecular conformations, usually possess a minimum. It can be repulsive only if the dispersion term is very small. Therefore, if the electrostatic terms are not included, it is very difficult to obtain the repulsive potential curve for the dimer. Whereas, for some relative orientations of the monomers, the direct electrostatic interactions can lead to a repulsion and destabilizes the complex. For instance, the *ab initio* study of the ethylene dimer [85] has shown that the quadrupole–quadrupole and hexadecapole–hexadecapole interactions have positive sign for the parallel conformation. The latter determines the absence of a minimum in the potential curve for the parallel conformation of the dimer $(C_2H_4)_2$.

It is worthwhile to note that the inclusion of the electrostatic term in the Williams potentials [70], carried out by Basilevsky *et al.* [86], did not lead to a repulsive curve. A deep minimum for the parallel conformation of ethylene was obtained. This instructive result shows that the use of the atom–atom potentials, calibrated to the crystalline properties, for the study of dimers may lead to incorrect qualitative results. This statement is confirmed by studies in Reference [87] where was demonstrated that the Williams AAPs predict that the most stable conformation of the benzene dimer corresponds to the two parallel rings with the $\mathbf{D}_{4h}$ symmetry, in contradiction with the well-known perpendicular crystal arrangement of the benzene molecules in the elementary cell.

The discussed AAPs (Equations (4.89)–(4.91)) do not depend on the valence state of the atom. Therefore, the peculiarity of the different chemical environments of the atoms is not taken into account in these AAP schemes. The exchange interaction is essentially nonadditive. The partitioning of this nonadditive quantity into additive terms depends on the chemical structure of the interacting molecules. The electrostatic Coulomb term is determined by the atomic charges. The latter depend upon the valence state of atom. In References [74, 82] and many others, the AAPs were constructed accounting for the atomic valence state.

Clementi and coworkers [88, 89] proposed fitting parameters in AAPs not to experimental data but to theoretical potential surface. The Lennard-Jones, (12–6), potential modified by inclusion of a Coulomb term was chosen:

$$V_{ab}^{\alpha\beta}(r_{ab}) = -\frac{A_{ab}^{\alpha\beta}}{R_{ab}^{6}} + \frac{B_{ab}^{\alpha\beta}}{R_{ab}^{12}} + C_{ab}^{\alpha\beta}\frac{q_a^{\alpha} q_b^{\beta}}{R_{ab}} \tag{4.93}$$

where the indices $a$ and $b$ denote atoms of the molecules $A$ and $B$, respectively; $\alpha$ and $\beta$ indicate the type of atom; $q^{\alpha}$ are the charges on atoms determined in theoretical calculation; $A_{ab}^{\alpha\beta}$, $B_{ab}^{\alpha\beta}$ and $C_{ab}^{\alpha\beta}$ are the parameters to be determined by fitting. The atom classes were defined, taking into account the valence states of the atoms in the molecule and their charges $q$. In Reference [88], the potentials (Equation (4.93)) for the interactions of 21 amino acids with water, with consideration of 23 types of atoms in the amino acids, were determined. As a result, the authors constructed 46 AAPs to describe the potential curves for the amino acid–water interactions. The atom–atom potentials (Equation (4.93)) were also used to describe the interaction between the four DNA bases with water [89] and other systems of biological importance [90] (see calculation details in References [91, 92]). The method of transferable AAPs, which is based on approximating theoretical potential curves, was also developed in References [93, 94].

A final remark. Sometimes, in papers devoted to the basis of the AAP method, see for example Reference [68], it is claimed that quantum mechanics justifies the AAP approach due to the Born–Oppenheimer (adiabatic) approximation. This statement is based on a misunderstanding. The adiabatic approximation allows electron motion with fixed nuclei to be considered and, therefore, localizes only the nuclei that allow one to fix the atomic coordinates. This is a necessary condition for considering the interaction of individual atoms. However, the electrons are not localized; the electron density in molecules is delocalized and may not be represented as a sum of atomic electron densities. Therefore, it cannot be said that quantum mechanics justifies the AAP method. The justification of this method has an utilitarian character: the achieved agreement between calculated and measured properties.

# References


1. H.B. Casimir and D. Polder, *Phys. Rev.* **73**, 360 (1948).
2. V. Heine, I.G. Robertson and M.C. Payne, *Phil. Trans. Roy. Soc. (London)* A **334**, 393 (1991).
3. W. Kolos, F. Nieves and O. Novaro, *Chem. Phys. Lett.* **41**, 431 (1976).
4. J.P. Daudey, O. Novaro, W. Kołos and M. Berrondo, *J. Chem. Phys.* **71**, 4297 (1979).
5. I.G. Kaplan, R. Santamaría and O. Novaro, *Mol. Phys.* **84**, 105 (1995).
6. I.G. Kaplan, J. Hernández-Cobos, I. Ortega-Blake and O. Novaro, *Phys. Rev.* A **53**, 2493 (1996).
7. I.G. Kaplan and O.B. Rodimova, *Doklady Acad. Sci. USSR,* **256**, 1174, (1982).
8. W.J. Meath and R.A. Aziz, *Mol. Phys.* **52**, 225 (1984).
9. J. Hernández-Cobos, I.G. Kaplan and J.N. Murrell, *Mol. Phys.* **92**, 63 (1997).
10. I.G. Kaplan, J.N. Murrell, S. Roszak and J. Leszczynski, *Mol. Phys.* **100**, 843 (2002).
11. I.G. Kaplan, I. Garzón, R. Santamaría, B.S. Vaisberg and O. Novaro, *J. Mol. Structure (Theochem)* **398–399**, 333 (1997).
12. I.G. Kaplan, R. Santamaría and O. Novaro, *Int. J. Quantum Chem.* **55**, 237 (1995).
13. M. Ross and B. Alder, *J. Chem. Phys.* **46**, 4203 (1967).
14. G.C. Maithland, M. Rigby, E.B. Smith and W.A. Wakeham, *Intermolecular Forces*, Clarendon Press, Oxford (1987).
15. J.O. Hirschfelder, C.F. Curtiss and R.B. Bird, *Molecular Theory of Gases and Liquids*, John Wiley & Sons, Inc., New York (1954).
16. A.E. Sherwood, A.J. de Rocco and E.A. Masson, *J. Chem. Phys.* **44**, 2984 (1966).
17. J.A. Barker and A. Pompe, *Australian J. Chem.* **21**, 1683 (1968).
18. K. Datta and A.K. Barna, *J. Phys. B: Atom. Molec. Phys.* **5**, 1676 (1972).
19. M.J. Elrod and R.J. Saykally, *Chem. Rev.* **94**, 1975 (1994).
20. O. Sinanoğlu, in *Modern Quantum Chemistry*, O. Sinanoğlu (ed), Academic Press, New York (1965), Vol. 2.
21. W. Liptay, in *Modern Quantum Chemistry*, O. Sinanoğlu (ed), Academic Press, New York (1965), Vol. 2.
22. N.R. Kestner and O. Sinanoğlu, *J. Chem. Phys.* **38**, 1730 (1963).
23. J.A. Barker, D. Henderson and W.R. Smith, *Mol. Phys.* **17**, 579 (1969).
24. A. Karpfen and P. Schuster, *Chem. Phys. Lett.* **44**, 459 (1976).
25. P.O. Löwdin, *A Theoretical Investigation into some Properties of Ionic Crystal* (Thesis Uppsala University, Almquist and Wiksells, 1948).
26. M. Born and K. Huang, *Dynamical Theory of Crystal Lattices*, Clarendon Press, Oxford (1954).
27. A.E. Carlsson and N.W. Aschcroft, *Phys. Rev.* B **27**, 2101 (1983).
28. F. Ercolessi, M. Parrinello and E. Tosatti, *Phil. Mag.* A **58**, 213 (1988).
29. I.G. Kaplan, *Adv. Quant. Chem.* **31**, 137 (1999).
30. I.G. Kaplan, *CAM-94 Physics Meeting, AIP Conference Proceedings 342*, Woodbury, New York (1995), p. 154.
31. P.O. Löwdin, *Adv. Chem. Phys.* **2**, 207 (1959).
32. G. Chałasiński, M.M. Szcześniak and S.M. Cybulski, *J. Chem. Phys.* **92**, 2481 (1990).
33. M.M. Szcześniak, G. Chałasiński and P. Piecuch, *J. Chem. Phys.* **99**, 6732 (1993).
34. I.G. Kaplan, *Polish J. Chem.***72**, 1454 (1998).
35. I.G. Kaplan, *Theory of Molecular Interactions*, Elsevier, Amsterdam (1986).

36. H. Margenau and N.R. Kestner, *Theory of Intermolecular Forces*, Pergamon Press, New York (1971).
37. B.M. Axilrod and E. Teller, *J. Chem. Phys.* **11**, 299 (1943).
38. J. Mutto, *Proc. Phys. Math. Soc. Japan* **17**, 629 (1943)
39. M.R. Aub and S. Zienau, *Proc. Roy. Soc. (London)* A **257**, 464 (1960).
40. T. Kihara, *Adv. Chem. Phys.* **11**, 1045 (1956).
41. A. Dalgarno and W.D. Davidson, *Adv. Atom. Molec. Phys.* **2**, 1 (1966).
42. W.L. Bade, *J. Chem. Phys.* **27**, 1280 (1957); **28**, 282 (1958).
43. R.J. Bell, *J. Phys.* B **3**, 751 (1970).
44. S.F. O'Shea and W.J. Meath, *Mol. Phys.* **28**, 1431 (1974); **31**, 515 (1976).
45. J.A. Barker, In *Rare Gas Solids*, M.L. Klein and J.A. Venables (eds), Academic Press, London (1976), Chapter 4.
46. J.A. Barker, *Mol. Phys.* **57**, 155 (1986).
47. B.H. Wells and S. Wilson, *Mol. Phys.* **57**, 21 (1986).
48. B.M. Wells, *Mol. Phys.* **61**, 1283 (1987).
49. O. Novaro, *Kinam* **2**, 175 (1980).
50. B.H. Wells and S. Wilson, *Mol. Phys.* **66**, 457 (1989).
51. G. Chałasiński and M.M. Szcześniak, *Chem. Rev.***94**, 1723 (1994).
52. G.Chałasiński, M. Szcześniak and R.A. Kendall, *J. Chem. Phys.* **101**, 8860 (1994).
53. P. Habitz, P. Bagus, P. Siegbahn and E. Clementi, *Int. J. Quant. Chem.* **23**, 1803 (1983).
54. G. Chałasiński, M.M. Szcześniak, P. Cieplak and S. Scheiner, *J. Chem. Phys.* **94**, 2873 (1991).
55. W. Klopper, M. Schutz, H. Luthi and S. Leutwyler, *J. Chem. Phys.* **103**, 1 (1995).
56. R. Santamaría, I.G. Kaplan and O. Novaro, *Chem. Phys. Lett.* **218**, 395 (1994).
57. I.G. Kaplan, *Int. J. Quant. Chem.* **74**, 241 (1999).
58. F. Luo *et al.*, *J. Chem. Phys.* **98**, 9687, 10086 (1993).
59. I.G. Kaplan, S. Roszak and J. Leszczynski, *J. Chem. Phys.* **113**, 6245 (2000).
60. I.G. Kaplan, S. Roszak and J. Leszczynski, *Adv. Quant. Chem.* **40**, 257 (2001).
61. C.W. Bauschlicher, V. Bagus and B.N. Cox, *J. Chem. Phys.* **77**, 4032 (1982).
62. S.P. Walch and C.W. Bauschlicher, *J. Chem. Phys.* **83**, 5735 (1985).
63. C.E. Moore, *Atomic Energy Levels*, Circular of the NBS 467, Washington, DC (1949), Vol. 1.
64. T.L. Hill, *J. Chem. Phys.* **14**, 465 (1946); *Ibid.* **16**, 399 (1948).
65. A.I. Kitaigorodsky, *Tetrahedron* **14**, 230 (1961).
66. A.I. Kitaigorodsky and K.B. Mirskaya, *Sov. Phys.-Crystallography* **6**, 408 (1961); **9**, 137 (1964).
67. A.I. Kitaigorodsky, K.B. Mirskaya and A.B. Tovbis, *Sov. Phys.-Crystallography* **13**, 176 (1968).
68. A.I. Kitaigorodsky, *Chem. Soc. Rev.* **7**, 133 (1978).
69. A.J. Pertsin and A.I. Kitaigorodsky, *The Atom-Atom Potential Method. Applications to Organic Molecular Solids*, Springer-Verlag, Berlin (1987).
70. D.E. Williams, *J. Chem. Phys.* **45**, 3770 (1966); **47**, 4680 (1967).
71. E.M. Engler, J.D. Andose and P.V.R. Schleyer, *J. Am. Chem. Soc.* **95**, 8005 (1973).
72. N.L. Allinger, M.T. Tribble, M.A. Miller and D.H. Wertz, *J. Am. Chem. Soc.* **93**, 1637 (1971).
73. *Conformation of Biopolymers*, G.N. Ramachandran (ed), Academic Press, London (1967).

74. R.A. Scott and H.A. Scheraga, *J. Chem. Phys.* **45**, 2091 (1966).
75. P.J. Flory, D.A. Brant and W.J. Miller, *J. Mol. Biol.* **23**, 47 (1967).
76. M. Vásquez, G. Némethy and H.A. Scheraga, *Chem. Rev.* **94**, 2183 (1994).
77. I. Kolossváry and W.C. Guida, in *Encyclopedia of Computational Chemistry*, (P.v.R. Schleyer (ed)), John Wiley & Sons, Ltd, Chichester (1998), 513–520.
78. Z. Li and H.A. Scheraga, *Proc. Natl. Acad. Sci. USA* **84**, 6611 (1987).
79. L. Piela, J. Kostrowicki and H.A. Sheraga, *J. Phys. Chem.* **93**, 3339 (1989).
80. J.P.K. Doye and D.J. Wales, *Phys. Rev. Lett.* **80**, 1357 (1998).
81. E. Burgos and H. Bonadeo, *Chem. Phys. Lett.* **49**, 475 (1977).
82. V.I. Poltev, *Sov. Phys.-Crystallography* **22**, 259 (1977).
83. I.B. Golovanov and M.V. Vol'kenshtein, *Doklady Physical Chemistry* **236**, 279 (1977).
84. D.E. Williams, *Acta Cryst.* A **30**, 71 (1974).
85. P.E.S. Wormer and A. van der Avoird, *J. Chem. Phys.* **62**, 3326 (1975).
86. M.V. Basilevsky, S.N. Elovskii and V.A. Tikhomirov, *Teor. Eksp. Khimia* **14**, 156 (1978).
87. V.M. Promyslov, I.A. Misurkin and A.A. Ovchinnikov, *Theor. Eksp. Khimia* **12**, 591 (1976).
88. E. Clementi, F. Cavallone and R. Scordamaglia, *J. Am. Chem. Soc.* **99**, 5531 (1977).
89. R. Scordamaglia, F. Cavallone and E. Clementi, *J. Am. Chem. Soc.* **99**, 5545 (1977).
90. G. Corongiu and E. Clementi, *Gazzetta Chimica Italiana* **108**, 273 (1978).
91. E. Clementi, *Computational Aspects for Large Chemical Systems*. Lecture Notes in Chemistry, Vol. 19, Springer-Verlag, Berlin (1980).
92. E. Clementi, G. Corongiu and G. Ranghino, *J. Chem. Phys.* **74**, 578 (1981).
93. S. Lifson, A.T. Hagler and P.J. Dauber, *J. Am. Chem. Soc.* **101**, 5111 (1979).
94. W.L. Jorgensen, *J. Am. Chem. Soc.* **101**, 2011, 2016 (1979); **103**, 335 (1981).

# 5 Model Potentials

## 5.1 Semiempirical Model Potentials

Model potentials with experimentally determined parameters and simple analytical representations are widely used in different physical and chemical problems [1–3]. A simple potential form often allows the analytical solution of the problem under study. As a rule, an analytical representation of potential is based on the theoretical concepts of intermolecular theory selecting the most important contributions for different types of intermolecular interactions (Chapter 2). In the last decade, due to the development of high-speed computers, more complex analytical potential functions with a large number of parameters have been applied. The model potentials are widely used in the molecular dynamics and Monte Carlo simulation studies of clusters and condensed matter.

Below, some model potentials that have been extensively used are presented.

### 5.1.1 Hard-sphere model potentials

A rigid, impenetrable sphere is the simplest model for an atom. This sphere is described by the one-parameter potential function:

$$V(R) = \begin{cases} \infty & R \leq \sigma \\ 0 & R > \sigma \end{cases} \tag{5.1}$$

where $\sigma$ is the radius of the sphere (Figure 5.1(a)). This potential is widely used for those problems in which a qualitative result is sufficient. For example, in the study of atomic displacement caused by irradiation of solids, application of the potential (5.1) permits one to estimate the order of magnitude of the number of displaced atoms and their energy distribution. In that problem, the parameter $\sigma$ can be considered as a function of the energy of the particles bombarding a solid. In spite of its idealization, the potential (5.1) was found to be quite useful also in simulation of the liquid state.

An attractive term is often added to the hard-sphere potential. In the most simple form it is the rectangular well with a depth $\epsilon$ and a width $\sigma(a-1)$ (Figure 5.1(b)):

$$V(R) = \begin{cases} \infty & r \leq \sigma \\ -\epsilon & \sigma < r \leq a\sigma \\ 0 & r > a\sigma \end{cases} \tag{5.2}$$

*Intermolecular Interactions* I.G. Kaplan

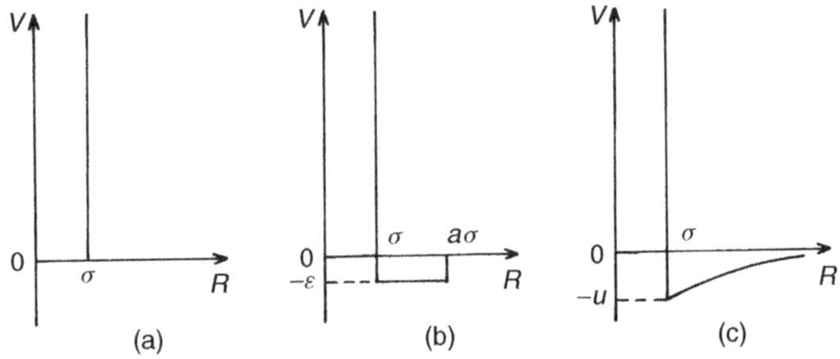


**Figure 5.1** Hard-sphere potentials

An advantage of the potential (5.2), in comparison with the potential (5.1), is the existence of two parameters, $a$ and $\sigma$, increasing its flexibility in the fitting procedure.

Another potential, which combines the hard-sphere model with an attraction in a more realistic way, is the so-called *Sutherland potential* $(\infty - 6)$ (Figure 5.1(c)):

$$V(R) = \begin{cases} \infty & R \leq \sigma \\ -u\,(\sigma/R)^6 & R > \sigma \end{cases} \tag{5.3}$$

where $u$ is the attraction value at $R = \sigma$.

### 5.1.2 Lennard–Jones potential

The general form of this potential is:

$$V(R) = \frac{\lambda_n}{R^n} - \frac{\lambda_m}{R^m} \tag{5.4}$$

It was initially proposed by Lennard-Jones [4] to study the thermodynamical properties of rare gases, in particular, their virial coefficients. Later, it has been used widely to study various systems. The so-called (12–6) potential, is written as:

$$V(R) = 4\epsilon\left[\left(\frac{\sigma}{R}\right)^{12} - \left(\frac{\sigma}{R}\right)^{6}\right] \tag{5.5}$$

where $\epsilon$ is the depth of the potential well with the minimum at $R_m = 2^{1/6}\sigma$ and $\sigma$ is the value of $R$ at $V(R) = 0$ (Figure 5.2). The attractive term corresponds to the dispersion dipole–dipole interaction. The repulsion is also approximated by the power term. The choice of order 12 is due to mathematical convenience. The magnitudes of $\sigma$ and $\epsilon/k$ for rare-gas atoms are given in Table 5.1.

The Buckingham (exp–6) potential (Section 5.1.4) is equivalent to the Lennard-Jones (12–6) potential and with a better theoretical justification than the latter. Nevertheless, because of its mathematical convenience, the Lennard-Jones potential has been widely used in many physical and chemical applications. The exponent $n$ of the repulsive term in Equation (5.4) was varied in the different problems in the range $n = 12$–25.

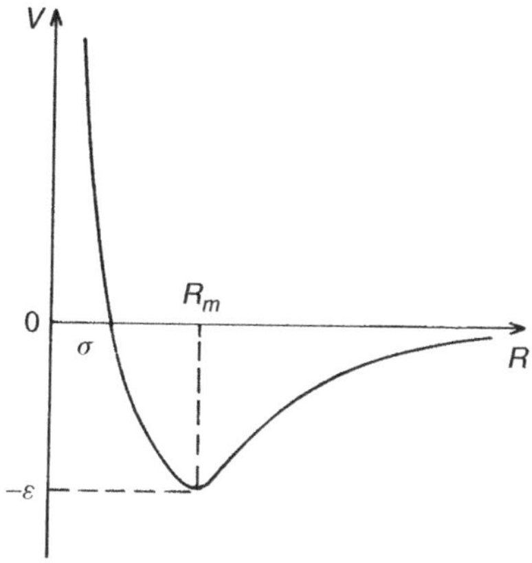


**Figure 5.2** The Lennard-Jones potential

**Table 5.1** Typical values of $\sigma$ and $\varepsilon/k$ for the Lennard-Jones potential for rare gas atoms [3]

| | He–He | Ne–Ne | Ar–Ar | Kr–Kr | Xe–Xe |
|---|---|---|---|---|---|
| $\sigma$ (Å) | 2.556 | 2.749 | 3.405 | 3.60 | 4.10 |
| $\varepsilon/k$ (K) | 10.22 | 35.60 | 119.8 | 171.0 | 221.0 |

## 5.1.3 Modifications of the Lennard–Jones potential

### 5.1.3.1 (12–6–4) potential

The term $R^{-6}$ makes the predominant contribution to the dispersion interaction energy of neutral systems. For an interaction between ions and neutral molecules or atoms, the induction energy is also important. The leading term of the induction energy in the considered case is proportional to $R^{-4}$. For this reason, Mason and Schamp [5] proposed the (12–6–4) potential, which describes the interaction of ions with neutral systems:

$$V(R) = 2\epsilon\left[(1+\gamma)\left(\frac{\sigma}{R}\right)^{12} - 2\gamma\left(\frac{\sigma}{R}\right)^{6} - 3(1-\gamma)\left(\frac{\sigma}{R}\right)^{4}\right] \qquad (5.6)$$

where the parameter $\gamma$ measures the relative influence of the term proportional to $R^{-6}$. If $\gamma = 1$, the (12–6–4) potential transforms into the (12–6) one, and if $\gamma = 0$ it transforms into the (12–4) potential.

### 5.1.3.2 (m–6–8) potential

Klein and Hanley [6] proposed adding the dipole–quadrupole term to the Lennard-Jones potential and considering the order $m$ in the repulsive term as a varying parameter. The *Klein–Hanley potential*:

$$V(R) = \frac{A}{R^m} - \frac{B}{R^6} - \frac{C}{R^8} \tag{5.7}$$

has four parameters: $m$, $A$, $B$, and $C$. The authors [6] also applied a particular form of the Potential (5.7):

$$V(R) = \epsilon \left[ \frac{6+2\gamma}{m-6} \left(\frac{R_m}{R}\right)^m - \frac{m-\gamma(m-8)}{m-6} \left(\frac{R_m}{R}\right)^6 - \gamma \left(\frac{R_m}{R}\right)^8 \right] \tag{5.8}$$

where $m, \gamma, \epsilon$ and $R_m$ are parameters. If $\gamma = 0$, potential (5.8) transforms to:

$$V(R) = \epsilon \left[ \frac{6}{m-6} \left(\frac{R_m}{R}\right)^m - \frac{m}{m-6} \left(\frac{R_m}{R}\right)^6 \right] \tag{5.9}$$

For $m = 12$ and $R_m = 2^{1/6}\sigma$, potential (5.9) transforms into the Lennard-Jones potential.

Potential (5.9) was further modified by Maitland and Smith [7]. They chose a potential with a form similar to potential (5.9), but in which the exponent in the repulsive term depends on $R$ as:

$$m = n + \gamma \left( \frac{R_m}{R} - 1 \right)$$

This modified potential was applied in order to describe the properties of gases in Reference [8].

Potential (5.8) proved to be flexible enough to describe the equilibrium properties and the transport coefficients in different monoatomic gases with the same set of parameters. Hanley and Klein [9] found that $m = 11$ and $\gamma = 3$ are the optimum parameters for rare gases. The different parameterizations of potential (5.8) has been discussed elsewhere [10], where it was shown that this potential adequately describes the heat conductivity data of noble gases.

#### 5.1.3.3 Kihara potential

In some problems, molecular size becomes important. The packing of molecules in a crystalline lattice and in a liquid, the reduced equation of state, and some others belong to this category. Kihara [11] proposed a modification to the Lennard-Jones potential that takes into account the size of the molecule under study. According to Kihara, each molecule is approximated by a convex rigid rotor and the distance used in the intermolecular potential is that between the surfaces of these bodies. The *Kihara potential* has the form:

$$V(\rho) = \epsilon \left[ \left(\frac{\rho_0}{\rho}\right)^{12} - 2\left(\frac{\rho_0}{\rho}\right)^6 \right] \tag{5.10}$$

where $\rho$ is the smallest distance between the 'surfaces' of the interacting molecules for a given conformation. It should be emphasized that $\rho$ is the distance between the nearest points of the molecules and, hence, depends on the distance $R$ between the molecular centers of mass and also on their sizes and relative arrangement. Averaging over the orientation of molecules gives:

$$\rho = R - 1/2\left(\overline{1}_a + \overline{1}_b\right) \quad (5.11)$$

where $\overline{1}_a$ and $\overline{1}_b$ are the mean diameters of the molecules, determined from experimental data for diffusion or other thermophysical properties [11]. The parameter $\rho_0$ corresponds to the minimum of the potential curve and $\epsilon$ determines the depth of the potential well.

The optimized values of the parameters for potential (5.10) for some molecules are presented in Table 5.2. The parameters of the interaction potential between different molecules can be obtained through the parameters of the pair potential for the similar molecules with the help of the following approximate relations:

$$\rho_0^{AB} = 1/2\left(\rho_0^{AA} + \rho_0^{BB}\right) \quad \epsilon^{AB} = \left[\epsilon^{AA}\epsilon^{BB}\right]^{1/2} \quad (5.12)$$

Potential (5.10) with the parameters given in Table 5.2 was applied in References [11, 12] to calculate the properties of molecular crystals and aggregates. Sinanoğlu [13] used the Kihara potential to study intermolecular interactions in liquids.

### 5.1.4 Buckingham potential

The model potential proposed by Buckingham [14] in 1938 includes attractive terms due to the dispersion dipole–dipole $\left(\sim R^{-6}\right)$ and dipole–quadrupole $\left(\sim R^{-8}\right)$

**Table 5.2** Parameters of the Kihara potential for homomolecular dimers [11]

| Molecule | $\rho_0$ (Å) | $\varepsilon/k$ (K) |
|---|---|---|
| $H_2$ | 2.62 | 48.5 |
| $N_2$ | 3.38 | 128 |
| $O_2$ | 3.12 | 160 |
| $F_2$ | 2.85 | 158 |
| $CO_2$ | 3.20 | 337 |
| $CH_4$ | 3.10 | 224 |
| $CF_4$ | 3.10 | 298 |
| $C(CH_3)_4$ | 3.40 | 691 |
| $SiF_4$ | 3.10 | 350 |
| $SF_6$ | 3.00 | 484 |
| $C_2H_6$ | 3.10 | 420 |
| $C_2H_2$ | 3.30 | 354 |
| *cyclo*–$C_3H_6$ | 3.00 | 630 |
| $C_6H_6$ | 3.42 | 850 |
| *cyclo*–$C_6H_{12}$ | 3.10 | 955 |

interactions; the repulsive term is approximated by an exponential function:

$$V(R) = Ae^{-\alpha R} - \frac{\lambda}{R^6} - \frac{\lambda'}{R^8} \tag{5.13}$$

By comparison with the Lennard-Jones potential, this potential is more difficult in practical calculations due to the exponential term. At the same time, it seems to be more realistic from physical point of view. Potential (5.13) contains four parameters.

The Buckingham (exp–6) potential has been widely used. It can be written in the form:

$$V(R) = \frac{\epsilon}{1 - 6/\alpha} \left\{ \frac{6}{\alpha} \exp\left[\alpha\left(1 - \frac{R}{R_m}\right)\right] - \left(\frac{R_m}{R}\right)^6 \right\} \tag{5.14}$$

where $\epsilon$ is the depth of the potential well, $R_m$ is the location of the minimum and $\alpha$ characterizes the steepness of the exponential repulsion. This potential has been used extensively in many applications; for example, in the atom–atom potential schemes [15–17] in which the intermolecular interaction potential is approximated by a sum of atom-atom interactions (Section 4.5).

Note that the Buckingham potential can not be applied at small $R$; it has a false maximum and approaches $-\infty$ at $R \to 0$ (Figure 5.3). Because $R_{max}$ is small, the simplest way to remove this defect is to introduce a hard sphere at $R \leq R_{max}$:

$$V(R) = \begin{cases} \infty & R \leq R_{max} \\ \text{Eq. (5.13) or (5.14)} & R > R_{max} \end{cases} \tag{5.15}$$

In most applications, however, the region of small $R$ is not essential. In such cases, it is valid to use the Buckingham potential in the simple (exp–6) form.

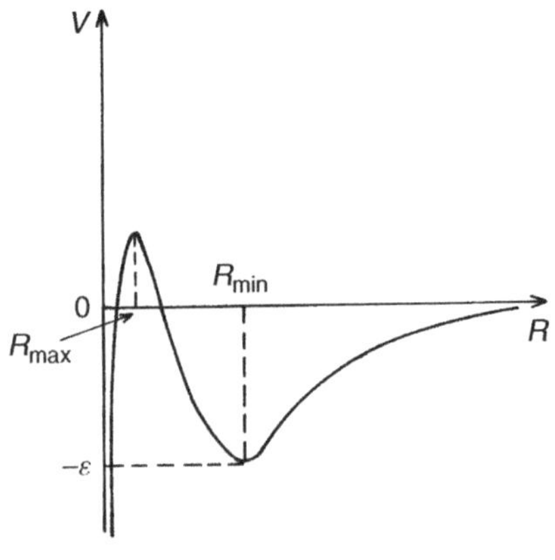


**Figure 5.3** The Buckingham potential (note: in all equations in text $R_{min}$ is denoted as $R_m$)

### 5.1.5 Modifications of the Buckingham potential

A large number of modifications of the Buckingham potential with correct behavior at small $R$ have been proposed in the literature. Let us consider some of them.

In the *Buckingham–Corner potential* [18], the incorrect behavior at small $R$ is removed by adding an exponential term to the attractive part of the Buckingham potential:

$$V(R) = \begin{cases} A\exp\left[-\alpha\dfrac{R}{R_m}\right] - \left(\dfrac{\lambda}{R^6} + \dfrac{\lambda'}{R^8}\right)\exp\left[-4\left(\dfrac{R_m}{R} - 1\right)^3\right], & R < R_m, \\ A\exp\left[-\alpha\dfrac{R}{R_m}\right] - \left(\dfrac{\lambda}{R^6} + \dfrac{\lambda'}{R^8}\right), & R > R_m; \end{cases} \tag{5.16}$$

where

$$A = \left[-\epsilon + (1+\beta)\left(\alpha/R_m^6\right)\right]e^{\alpha},$$

$$\lambda = \epsilon\alpha R_m^6/\left[\alpha(1+\beta) - 6 - 8\beta\right], \quad \lambda' = \beta R_m^2\lambda,$$

$$\beta = \left[\frac{\lambda'}{R^8}\Big/\frac{\lambda}{R^6}\right]_{R=R_m},$$

and $\alpha$, $\epsilon$ and $R_m$ have the same meaning as they do in Potential (5.14).

Ahlrichs *et al.* [19] proposed a modification of the Buckingham–Corner potential that involves the dispersion terms up to the power $R^{-10}$, with the dispersion coefficients $C_6, C_8, C_{10}$ chosen as parameters. In the case of mixtures of rare gas atoms, the *Ahlrichs–Penco–Scoles potential* has the form

$$V(r) = \begin{cases} A\exp(-\alpha R) - \left(\dfrac{C_6}{R^6} + \dfrac{C_8}{R^8} + \dfrac{C_{10}}{R^{10}}\right) \\ \mathrm{x}\exp\left[-\left(\dfrac{1.28R_m}{R} - 1\right)^2\right], & R < 1.28\ R_m, \\ A\exp(-\alpha R) - \left(\dfrac{C_6}{R^6} + \dfrac{C_8}{R^8} + \dfrac{C_{10}}{R^{10}}\right), & R \geq 1.28\ R_m. \end{cases} \tag{5.17}$$

At small $R$, the additional exponential term improves the potential; in fact, it is a damping function (Section 3.1.5). At large $R$, potential (5.17) transforms into the theoretical expression for the dispersion energy.

An analysis of the effect of the different parameterizations of potential (5.17) on the description of the thermodynamical properties of rare gas mixtures and a comparison with other empirical potentials was carried out in Reference [20].

Authors [21] suggested using the theoretical values for the dispersion part of Potential (5.17) and representing the exchange part via the electrostatic energy $E_{el}^{(1)}$, calculated in the first order of perturbation theory, as follows:

$$E_{ex} = -\gamma(1 + 0.1\ R)\,E_{el}^{(1)} \tag{5.18}$$

where $\gamma$ is a variational parameter. This potential contains a single parameter because all the other quantities can be evaluated. Such a one-parameter potential was used with success to describe mixtures of rare gas atoms.

In potentials (5.16) and (5.17), the incorrect behavior of $V(R)$ at $R \to 0$ was improved by introducing the exponential term in the dispersion part. Another way consists of multiplying the exponential repulsive term by $R^{-n}$. Then, the dispersion term will not be the leading one at small $R$. Such a procedure, applied to the Buckingham (exp–6) potential, leads to the *Karr–Konowalow potential* [22]:

$$V(r) = \frac{\epsilon}{\alpha}(\alpha+6)\left(\frac{R_m}{R}\right)^6 \left\{\frac{6}{\alpha+6}\exp\left[\alpha\left(1-\frac{R}{R_m}\right)\right]-1\right\} \tag{5.19}$$

There is one more way of removing the divergence of the dispersion terms at $R \to 0$. It consists of introducing a small constant term to the denominators of the dispersion part. This procedure was used in the construction of the *Barker–Pompe potential* [23]:

$$V(R) = \epsilon\left\{\exp\left[\alpha\left(1-\frac{R}{R_m}\right)\right]\sum_{i=0}^{n} A_i\left(\frac{R}{R_m}-1\right)^i - \left[\frac{C_6}{R^6+\delta}+\frac{C_8}{R^8+\delta}+\frac{C_{10}}{R^{10}+\delta}\right]\right\} \tag{5.20}$$

where the parameters $\alpha, \epsilon$ and $R_m$ have the usual meaning and the parameter $\delta$ is introduced to remove the divergence at $R \to 0$. The exponential repulsive term is multiplied by a polynomial of $R$, which is varied rather slowly and allows a better approximation of the repulsive part of potential (5.20) to be obtained.

Potential (5.20) is a multiparameter one. Hence, it requires the use of high-speed computers. The potential was successfully applied to the study of properties of rare gas atoms in different aggregate states [24].

### 5.1.6 Potentials describing spectroscopic properties of diatomic molecules

#### 5.1.6.1 Morse potential

This potential was proposed by Morse [25] in 1929 in order to evaluate the vibrational energy levels of diatomic molecules. Morse did not use a power series because in such a case it is not possible to reproduce the sequence of the energy levels observed experimentally and described by the formula:

$$E_v = -D + \hbar\omega_0\left[(v+1/2) - a(v+1/2)^2\right] \tag{5.21}$$

where the energy of nuclear vibrations is added to the electronic energy $D$, $\omega_0$ is the vibrational frequency, $v$ labels the vibrational levels and $a$ is a constant.

The *Morse potential* consists of two exponential terms:

$$V(R) = D\{\exp[-2\alpha(R-R_m)] - 2\exp[-\alpha(R-R_m)]\} \tag{5.22}$$

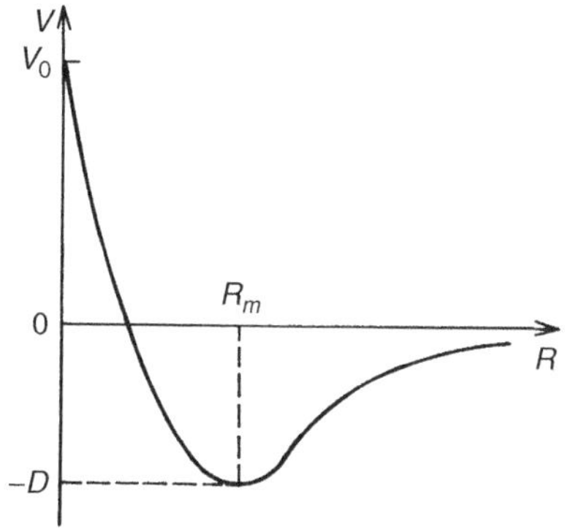


**Figure 5.4** The Morse potential

where $D$ is the well depth and $R_m$ defines the minimum. This potential is presented in Figure 5.4. At $R = 0$, Potential (5.22) becomes finite, namely, $V(0) = D\exp(\alpha R_m)\left[\exp(\alpha R_m) - 2\right]$.

The approximate solution of the Schrödinger equation for the nuclear motion with Potential (5.22) leads to:

$$E_v = -D + \hbar\omega_0\left[(v + 1/2) - \frac{\hbar\omega_0}{4D}(v + 1/2)^2\right] \tag{5.23}$$

that is, it reproduces with a good accuracy the $v$-dependence of the energy levels. Potential (5.22) contains three parameters: $\alpha$, $D$ and $R_m$. Morse fitted the potential parameters for a large number of different molecules on the basis of spectroscopic data.

The Morse potential does not behave so well at large separations, since in that range the inverse power dependence 'works' better than the exponential one. At $R = 0$ the Morse potential has a finite value, which is not correct either. Nevertheless, the Morse potential describes the vibrational levels quite satisfactorily, since it is only an interval near the minimum that is important in that connection.

The Morse potential has been used not only in molecular spectroscopy but also for the evaluation of kinetic properties of gases and, especially, in studies of various crystal properties. This last application is related to the fact that the crystal properties are most sensitive to that range of separations in which the Morse potential describes the real potential curve quite satisfactorily.

#### 5.1.6.2 Rydberg potential

This potential was proposed by Rydberg [26] to evaluate the vibrational spectrum of diatomic molecules, just as the Morse potential does. The *Rydberg potential* has the form:

$$V(R) = D\left[1 + \frac{b}{R_m}(R - R_m)\right]\exp\left[-\frac{b}{R_m}(R - R_m)\right] \tag{5.24}$$

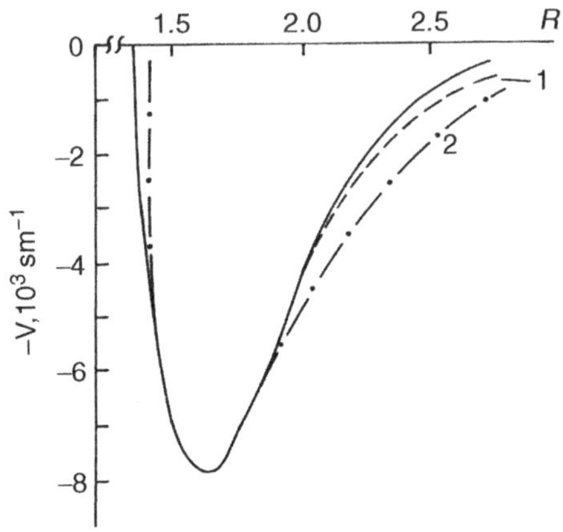


1 is the Rydberg potential; 2 is the Morse potential; the solid curve corresponds to the RKR experimental potential [26]

**Figure 5.5** Potential curves of the $^3\sum_u^-$ state of the $O_2$ molecule

It contains three parameters. At $R = 0$ the potential becomes finite and has the value $V(0) = D(1 - b)e^b$. This potential has the same qualitative behavior as the Morse potential (Figure 5.4). Nevertheless, in contrast with the latter, it is characterized by a smaller attraction at large $R$. As a result, the Rydberg potential approximates the potential curves of diatomic molecules better. The potential curves for the excited electronic state $^3\sum_u^-$ of the oxygen molecule approximated by means of the Rydberg and Morse potentials, are presented in Figure 5.5. Both potentials reproduce with a good accuracy the potential well, but at large $R$ the Rydberg potential agrees more closely with the reconstructed experimental curve.

The Rydberg potential was widely used in studies of thermodynamical properties of gases at high temperatures. Sinanoğlu and Pitzer [27] derived the expressions for the second virial coefficient and its derivatives on the basis of the Rydberg potential.

### 5.1.6.3 Pöschl–Teller potential

The potential constructed by Pöschl and Teller [28] in 1933 from hyperbolic functions has a great similarity with the Morse potential. The *Pöschl–Teller potential* has the form:

$$V(R) = D\left[\frac{sh^4\frac{\alpha R_m}{2}}{sh^2\frac{\alpha R}{2}} - \frac{ch^4\frac{\alpha R_m}{2}}{ch^2\frac{\alpha R}{2}}\right] \tag{5.25}$$

where the parameters $D, \alpha$ and $R_m$ have the same meaning as in the Morse potential. It is presented in Figure 5.6. The Pöschl–Teller potential describes the asymptotic behavior of the potential curve more correctly than the Morse potential.

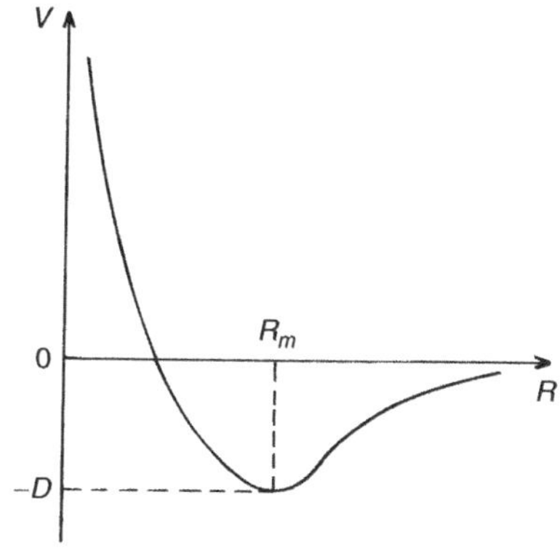


**Figure 5.6** The Pöschl–Teller potential

As $R \to 0$, potential (5.25) approaches infinity as $R^{-2}$, while the Morse potential has a finite value. As $R \to \infty$, the asymptotic behavior of the Pöschl–Teller potential differs of that for the Morse potential by the function $-2D \exp[-\alpha (R + R_m)]$. Inclusion of this function improves the asymptotic behavior, as was shown for hydrogen halides [29].

The Pöschl–Teller potential, just like the Morse potential, allows the Schrödinger equation to be solved with good accuracy. The energy spectrum obtained is similar to Equation (5.23).

A comparative analysis of different model potentials suggested for describing energy levels of diatomic molecules was performed in References [30, 31]. In Reference [31], comparison of some semiempirical model potentials with the empirical potentials $V_{RKR}$, reconstructed via the *RKR* procedure (see Section 5.3.1) for 19 states of different diatomic molecules, was carried out. The relative deviation $\overline{\Delta V}$ from the experimental potential average over all calculated states and over all $R$ is equal to 3.68% for the Morse potential, 3.48% for the Pöschl–Teller potential and 2.94% for the Rydberg potential.

These potentials have three fitted parameters. As was demonstrated in Reference [31] (Table 5.3), the mean deviation, $\overline{\Delta V}$, from the *RKR* experimental curve can be made smaller for five-parameter model potentials. It is of the order of 1.51% for the potential suggested by Hilburn and Hirschfelder [32] and 2.17% for the Lippincott potential [33, 34].

### 5.1.6.4 Kratzer potential

Kratzer [35] proposed his potential in 1920 (five years before the creation of modern quantum mechanics) to describe the vibrational–rotational energy levels in diatomic molecules. It can be presented in the following form [36, 37]:

$$V(x) = -\frac{a}{x + x_0} + \frac{b}{(x + x_0)^2} \tag{5.26}$$

**Table 5.3** Relative deviation, $(V_{RKR} - V)/V \cdot 100\,\%$, of different model potentials from the RKR curve evaluated on the basis of the data presented in Reference [31]

| $r$, Å | Morse | Hulburt-Hirschfelder | Rydberg | Pöschl-Teller | Varshni-III | Lippincott |
|---|---|---|---|---|---|---|
| | | Ground state, $X\ ^1\Sigma_g^+$, of molecule $N_2$ | | | | |
| 0.896 | 1.08 | 0.96 | 0.96 | 1.46 | 2.22 | 1.69 |
| 0.919 | 1.02 | 2.43 | 2.33 | 0.82 | 3.17 | 0.56 |
| 0.942 | 0.56 | 0.36 | 0.23 | 0.68 | 0.76 | 0.42 |
| 0.983 | 0.10 | 0.24 | 0.17 | 0.14 | 0.36 | 0.22 |
| 1.027 | 0.02 | 0.09 | 0.08 | 0.01 | 0.11 | 0.09 |
| 1.185 | 0.01 | 0.08 | 0.04 | 0 | 0.10 | 0.09 |
| 1.261 | 0.28 | 0.17 | 0.02 | 0.29 | 0.25 | 0.23 |
| 1.358 | 1.09 | 0.21 | 0.38 | 1.12 | 0.44 | 0.61 |
| 1.447 | 1.92 | 0.28 | 0.71 | 1.96 | 0.77 | 1.24 |
| 1.528 | 3.68 | 0.73 | 2.03 | 3.74 | 0.04 | 0.82 |
| | | Excited state, $B\ ^3\Sigma_u^-$, of molecule $O_2$ | | | | |
| 1.334 | 17.51 | 1.39 | 11.24 | 17.61 | 4.68 | 4.88 |
| 1.365 | 14.53 | 0.40 | 10.05 | 14.43 | 5.27 | 5.87 |
| 1.405 | 5.77 | 1.79 | 3.88 | 5.67 | 1.79 | 2.39 |
| 1.531 | 0.40 | 0.10 | 0.30 | 0.40 | 0.30 | 0.30 |
| 1.683 | 0.60 | 0.40 | 0.60 | 0.60 | 0.60 | 0.60 |
| 1.962 | 5.87 | 1.89 | 4.58 | 5.87 | 2.98 | 1.99 |
| 2.232 | 8.26 | 4.88 | 5.77 | 8.26 | 2.89 | 1.39 |
| 2.865 | 1.79 | 3.88 | 0 | 1.79 | 1.89 | 0.10 |

where $x_0$ is not a free parameter, it is expressed via $a$ and $b$, $x_0 = 2b/a$. The Kratzer potential is a two-parameter potential with a minimum at $x = 0$, $V(0) = -a^2/4b = -D$. It grows to infinity as $x \to -x_0$ and converges to zero, as $x \to \infty$ (Figure 5.7). In order to compare it with the potentials discussed above, the curve in Figure 5.7 has to be shifted by $x_0$ along the $x$-axis. For this, a new coordinate $R = x + x_0$ is introduced, then $V(R = 0) \equiv V(x = -x_0) = \infty$ and $V(R = x_0) \equiv V(x = 0) = -D$. The point $R = x_0$ corresponds to a minimum, denoted as $R_m$. Thus:

$$x = R - R_m$$

and the representation the Kratzer potential, as a series, has the following form:

$$V(R) = -D + \frac{1}{2}\omega_k^2 (R - R_m)^2 + c_3 (R - R_m)^3 + c_4 (R - R_m)^4 + \dots \tag{5.27}$$

where $D = a^2/4b$ and other coefficients are also functions of two parameters, $a$ and $b$.

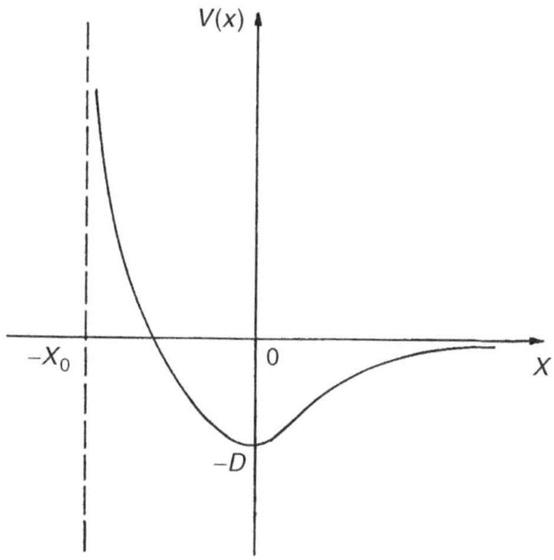


**Figure 5.7** The Kratzer potential

The Schrödinger equation with the Kratzer potential, as well as the Morse, Rydberg and Pöschl–Teller potentials, can be solved analytically. The first solution was derived by Fues [38] just after the development by Schrödinger of his equation (see also Reference [36]). It is the well-known problem in quantum mechanics of anharmonic oscillator. The general solution for the energy levels can be presented as [37]:

$$E_v = D'\left[1 - \frac{1}{1 + 2fv + f^2v^2}\right] \tag{5.28}$$

where $D' = 4b^2D/(D + 2b)^2$ and $f = 2D/(D + 2b)$.

The Taylor expansion of Equation (5.28) contains arbitrary high powers of $v$, whereas in the anharmonic Morse case[1] there are up to quadratic terms only (Equation (5.23)). This property of the Kratzer potential was recently used by Rushev and Moule [37] for applying the modified Kratzer potential to estimate the vibrational level density in polyatomic molecules in a given electronic state at very high vibrational excitation energies, up to the dissociation threshold. Their modification was in the introduction of a third variable parameter in the expression for the energy levels (Equation (5.28)) and conformation of this modified expression with the Morse expression for the energy levels at the lower $v$ values.

The Kratzer potentials played rather a large role in the early days of quantum mechanics. Later on, it was 'shielded' by the Morse potential, which was considered as more physical and has been widely used. The study [37] enlightened some beneficial sides of the Kratzer potential.

[1] The different signs in Equations (5.28) and (5.23) is caused by the different signs in the definition of the Kratzer and Morse potentials.

### 5.1.6.5 Dunham expansion and its modification

Dunham [39] considered a rotating vibrator and came to conclusion that, although the Morse potential describes the pure vibrational terms sufficiently well, the situation with the rotational terms is far from good. He suggested expanding the potential function in the Taylor series around the equilibrium point $R_m$. The *Dunham potential* can be represented in the form:

$$V(R) = a_0 \left( \frac{R - R_m}{R_m} \right)^2 \left[ 1 + \sum_n a_n \left( \frac{R - R_m}{R_m} \right)^n \right] \tag{5.29}$$

In his original paper [39], Dunham found, within the framework of semiclassical approximation, the relationships between the expansion coefficients and the energy levels. The Dunham expansion is of great flexibility due to the infinite number of parameters. However, evidently, the Dunham expansion diverges at $R > 2R_m$.

Simons *et al.* [40] proposed a modification of the Dunham expansion using $(R - R_m)/R$ instead of $(R - R_m)/R_m$ as the expansion parameter in Equation (5.29). It extends considerably the convergence region and accelerates the convergence. The *Simons–Parr–Finlan potential* (SPF) is written in the form [40]:

$$V(R) = b_0 \left( \frac{R - R_m}{R} \right)^2 \left[ 1 + \sum_{n=1}^{\infty} b_n \left( \frac{R - R_m}{R} \right)^n \right]. \tag{5.30}$$

The SPF potential was applied successfully in experimental data processing.

The further generalization was proposed by Thakkar [41]. The *Thakkar potential* has the following form:

$$V(R) = e_0(p)\, \lambda^2 \left( 1 + \sum_{n=1}^{\infty} e_n(p)\, \lambda^n \right) \tag{5.31}$$

where

$$\lambda(R, p) = sgn(p) \left[ 1 - \left( \frac{R_m}{R} \right)^p \right], \quad sgn(p) = \begin{cases} 1, & p > 0, \\ -1, & p < 0, \end{cases} \tag{5.32}$$

$p$ is a nonzero real number. If $p = -1$, the Thakkar potential transforms into the Dunham one, for $p = 1$, it transforms into the SPF potential. But this is not exhaust all particular cases of the Thakkar potential. If $e_n(p) = 0$ for all $n \geq 1$, it reduces to:

$$V(R) = e_0(p) + e_0(p) \left[ \left( \frac{R_m}{R} \right)^{2p} - 2 \left( \frac{R_m}{R} \right)^p \right] \tag{5.33}$$

Setting $p = 1$ in Equation (5.33) leads to the Kratzer potential and setting $p = 6$ leads to the familiar Lennard-Jones (12–6) potential, with its zero of energy shifted from the dissociation minimum to the minimum of the potential well.

Thakkar [41] formulated some criteria concerning the optimum choice for $p$. The use of the Thakkar expansion for the determination of characteristics of diatomic potential curves is discussed in References [42, 43].

## 5.1.7 Anisotropic potentials

### 5.1.7.1 Keesom potential

This potential is appropriate for systems that are characterized by a very small range of repulsive forces and interact via the dipole–dipole law. It has the form [44, 45]:

$$V(R, \theta_a, \theta_b, \xi_a - \xi_b) = \begin{cases} \infty, & R < \sigma, \\ -\dfrac{d_a d_b}{R_3} g(\theta_a, \theta_b, \xi_a - \xi_b), & R \geq \sigma, \end{cases} \quad (5.34)$$

where the function $g$ depends on the relative orientation of the dipoles. In spherical coordinates it is defined as:

$$g(\theta_a, \theta_b, \xi_a - \xi_b) = 2\cos\theta_a \cos\theta_b - \sin\theta_a \sin\theta_b \cos(\xi_a - \xi_b) \quad (5.35)$$

In fact, the Keesom potential corresponds to the interaction of two impenetrable spheres possessing the permanent dipole moments.

The Keesom potential was one of the first model potentials used for calculating the second virial coefficient [44].

### 5.1.7.2 Stockmayer potential

This potential was proposed by Stockmayer [46] in 1941 to describe the interaction of polar molecules with large dipole moments, for example, ammonia or water. The potential represents a superposition of the Lennard-Jones (12–6) potential and the potential of interaction of the two dipoles:

$$V(R, \theta_a, \theta_b, \xi_a - \xi_b) = 4\epsilon\left[\left(\frac{\sigma}{R}\right)^{12} - \left(\frac{\sigma}{R}\right)^{6}\right] - \frac{d_a d_b}{R^3} g(\theta_a, \theta_b, \xi_a - \xi_b) \quad (5.36)$$

where the function $g$ is given by Equation (5.35).

The Stockmayer potential rather satisfactorily describes the interaction between polar molecules at distances where the dipole–quadrupole and higher-order multipole–multipole interactions are not important. This potential was used for calculation of the second and third virial coefficients.

### 5.1.7.3 Atom–linear molecule interaction potentials

The interaction of arbitrary fixed anisotropic systems can not be described by a unique model potential. But it is possible for the simplest systems, such as an atom–diatomic molecule. The potential for that system depends on the distance $R$ and the polar angle $\theta$ (see Figure 5.8 for notations). In general, the potential $V(R, \theta)$ can be expanded in a series:

$$V(R, \theta) = \sum_{n=0}^{\infty} V_n(R) P_n(\cos\theta) \quad (5.37)$$

where $P_n(\cos\theta)$ is the Legendre polynomial. In Expansion (5.37) in the case of homonuclear molecules, only even $n$ appear. The interaction of $H_2$ with rare gas

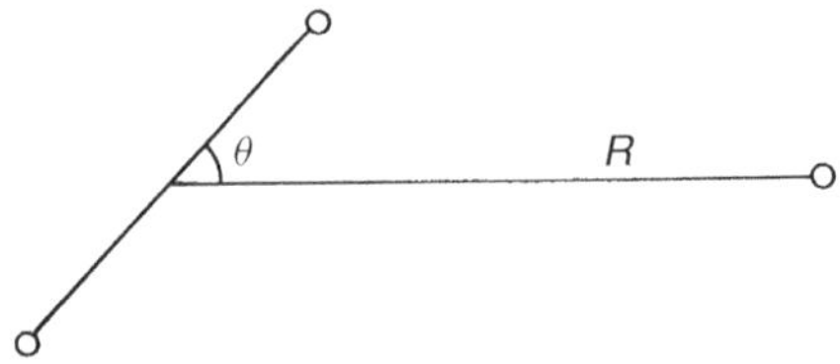


**Figure 5.8** Notations in the system: atom–linear molecule

atoms is the case most extensively studied, both experimentally and theoretically. An analysis of the infrared spectra of $H_2$–rare gas atom complexes confirms also the anisotropic nature of the potential, that is, the hindrance of the rotations in that complex. According to References [47, 48], it is sufficient to choose only two terms in Expansion (5.37):

$$V(R,\theta) = V_0(R) + V_2(R)\,P_2(\cos\theta) \tag{5.38}$$

to satisfactorily describe the experimental data. In Equation (5.38), the angular dependence is described by the Legendre polynomial $P_2(\cos\theta) = \frac{1}{2}\left(3\cos^2\theta - 1\right)$. The Lennard-Jones, Buckingham–Corner, and also the more flexible Morse–spline–van der Waals (MSV) potential (Section 5.1.11.2), were used as $V_0(R)$ and $V_2(R)$. The potential $V_2(r)$ involves anisotropic coefficients, which are usually different for the repulsion and attraction term.

For example, the anisotropic Lennard-Jones-type potential, used in Reference [49] for the description of the $H_2$–He interaction, has the form:

$$V(r,\theta) = \epsilon\left\{\left[\left(\frac{R_m}{R}\right)^{12} - 2\left(\frac{R_m}{R}\right)^{6}\right] + \left[\alpha_s\left(\frac{R_m}{R}\right)^{12} - 2\alpha_1\left(\frac{R_m}{R}\right)^{6}\right] P_2(\cos\theta)\right\} \tag{5.39}$$

The coefficients $\alpha_s$ and $\alpha_1$ characterize the relative contribution of the anisotropic component to the repulsive and the attractive terms, respectively (compare the coefficients $A_r$, $A_a$, and $A_d$ of the MSV potential in Equation (5.65)). Note that if $R_m = 2^{1/6}\sigma$ is substituted, the expression for the Lennard-Jones potential in Equation (5.39) becomes identical with Equation (5.5).

Pack [50] proposed another modification of the Lennard-Jones anisotropic potential. Instead of introducing anisotropic factors for the repulsive and attractive terms, Pack chose both parameters of the Lennard-Jones potential to be anisotropic, namely:

$$V(r,\theta) = \epsilon(\theta)\left\{\left[\frac{R_m(\theta)}{R}\right]^{12} - 2\left[\frac{R_m(\theta)}{R}\right]^{6}\right\}$$

$$\epsilon(\theta) = \overline{\epsilon}\left[1 + aP_2(\cos\theta)\right], \quad R_m(\theta) = \overline{R}_m\left[1 + bP_2(\cos\theta)\right] \tag{5.40}$$

The Potential (5.40), as Potential (5.39), involves four parameters. A small change of $b$ considerably affects Potential (5.40). Hence, the latter is high flexible and can be applied to describe the anisotropic potentials for such systems, as atom–linear triatomic molecule, for example Ar–$CO_2$ [50].

A detailed description of the applications of anisotropic model potentials to the study of the high-resolution vibrational–rotational–tunneling spectra of van der Waals molecules is available in the review by van der Avoird *et al.* [51].

#### 5.1.7.4 Model potentials applied in water and aqueous-system studies

One of the earliest model potentials for water was the *Rowlinson potential.* Rowlinson [52] proposed for study the $(H_2O)_2$ dimer using an anisotropic potential that consisted of a spherically symmetric part, given by the Lennard-Jones (12–6) potential, and an anisotropic component, chosen in the form of the Coulomb interaction of point charges:

$$V(R, \Omega_1, \Omega_2) = 4\epsilon\left[\left(\frac{\sigma}{R}\right)^{12} - \left(\frac{\sigma}{R}\right)^{6}\right] + \frac{1}{2}\sum_{\alpha,\beta}\frac{q_\alpha q_\beta}{R_{\alpha\beta}} \tag{5.41}$$

where $R$ is the distance between the oxygen atoms, $R_{\alpha\beta}$ is the distance between the $\alpha$th and $\beta$th point charges and $\Omega_1$ and $\Omega_2$ are the sets of Euler angles, which determine the orientations of the water molecules. The parameters were chosen in such a way that the second virial coefficient, dipole and quadrupole moments (for this the last term in Equation (5.41) was expanded into the dipole and quadrupole components), the energy and the geometry of the ice lattice were reproduced adequately. It was found that the following choice was optimum: the positive charges, $q = 0.32$ a.u., were located on the hydrogen atoms and the two equal negative charges, $-q$, on a line passing through the oxygen atom perpendicularly to the $H_2O$ plane, on both sides from the oxygen atom at a distance of 0.5 Å (Figure 5.9(a)).

In order to describe the water–water interaction when the water molecules approach to each other, Ben-Naim and Stillinger (BNS) [53] modified the Rowlinson potential. The *BNS potential* has the form of Equation (5.41), but the Coulomb

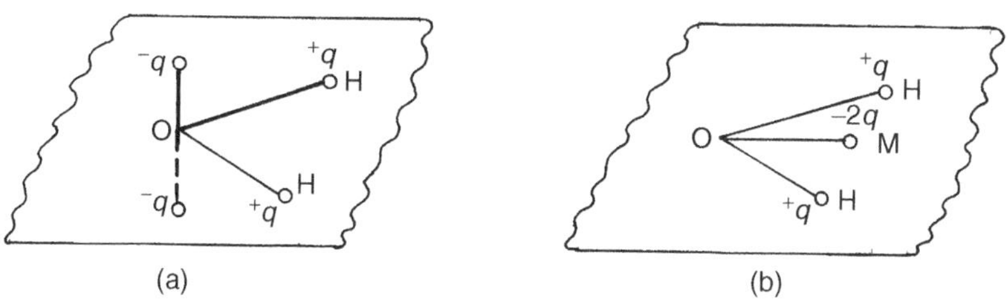


**Figure 5.9** Point charge model used in (a) the Rowlinson potential (b) the PIP4P potential

term is multiplied by:

$$f(R) = \begin{cases} 0 & 0 < R < R_1 \\ (R - R_1)^2 \dfrac{(3R_2 - R_1 - 2R)}{(R_2 - R_1)^3} & R_1 \le R < R_2 \\ 1 & R_2 \le R \end{cases} \tag{5.42}$$

It provides a smooth transition from 0 at $R < R_1$ to the Coulomb sum at $R \ge R_2$, where $R_1 = 2.0379$ Å and $R_2 = 3.1877$ Å. In 1972 Stillinger [54] modified the parameters of the BNS potential. The modified BNS potential is denoted in the literature as the *ST2 potential.*

These potentials, because of their simple analytical form, have been widely used in Monte Carlo simulations of the structure of liquid water [3, 55, 56]. The Rowlinson potential was applied up to $R > 2$ Å; at smaller $R$, it was replaced by a hard sphere. The parameters of the Rowlinson and BNS potentials are given elsewhere [56].

Later on, the so-called *TIP4P potential* suggested by Jorgensen *et al.* [57] has been widely used in studies of water systems. The TIP4P potential is a modification of a set of transferable intermolecular potential functions (TIPS) developed by Jorgensen [58, 59]. As with all other potentials discussed above, the TIP4P potential also has, for all monomer pairs, the analytical form Equation (5.41) with the Lennard-Jones potential modeling the interaction between oxygens and with the Coulomb interaction between all intermolecular pairs of charges. It includes three point charges: two positive equal charges $q$ on the hydrogen atoms and a negative charge $-2q$ at some point M on the bisector of the HOH angle (Figure 5.9(b)). The monomer geometry is fixed. The TIP4P potential was recently used in the global minimum search of $(H_2O)_N$ clusters, up to $N = 21$ by the basin-hopping algorithm [60] and by the evolution algorithm [61] (see Section 5.4).

It should be emphasized that all the water–water potentials considered above are based on a rigid geometry of point charges. In the *polarization model* proposed by Stillinger and Davis [62, 63], the polarization term was added to the pairwise atom–atom potential:

$$V^{SD} = \sum_{a<b} V_{ab} + V_{pol} \tag{5.43}$$

In the first term of Equation (5.43), the sum is taken over all pairs HH, HO, and OO; the second term is the nonadditive interaction of point charges and multipole moments with induced dipole moments on each of the oxygen atoms.

According to classical electrostatics, the induced dipole moment in atom $i$ can be determined by its polarizability, $\alpha_0$, and the external electric field, $E_i$, at that atom as a linear response to the electric field:

$$\boldsymbol{\mu}_i = \alpha_0 \mathbf{E}_i \tag{5.44}$$

The external sources creating the electric field can be both point charges, $q_i$, and the atomic or molecular multipole moments (including induced dipole moments).

At the approximation when induced dipole moments are only caused by the point charges:

$$V_{pol} = \sum_{i<j} \frac{q_i \left( \boldsymbol{\mu}_i \cdot \mathbf{R}_{ij} \right)}{R_{ij}^3} \left[ 1 - L \left( R_{ij} \right) \right] \tag{5.45}$$

where the function $L\left(R_{ij}\right)$ takes into account the delocalization of the electron density on oxygens. It differs from zero only at small distances around the oxygen nucleus. In the original model [62], the charges were located only at atoms and $q_{\mathrm{H}} = 1e$ and $q_{\mathrm{O}} = -2e$ were used.

The polarization model was further elaborated in References [64–70]. Niesar *et al.* [65, 66] in the potential called the *NCC potential* used the polarization scheme proposed by Stillinger and David [62] but assumed the induced dipole moment to be centered on the OH bonds and considered the polarization induced by the external field, $E_i$, due to both the point charges and the induced dipole moments. They used the point charge distribution as in the MCY potential [71] (it is similar to that used in the PIP4P potential, see Figure 5.9(b)) and considered the magnitudes of point charges and the location of the negative point charge on the bisector of the angle HOH as variable parameters. The pairwise atom–atom potentials were presented as a sum of exponential terms.

The specific feature of the NCC potential is its *ab initio* nature. It was fitted to the *ab initio* calculated interaction energies of 350 water dimer and 250 water trimer configurations (the *ab initio* model potential approach is discussed in Section 5.1.13). Corongiu [67] has modified the NCC potential adding the vibrational flexibility of monomer molecules, which includes the changes in the OH bond length and the HOH angle. This modified potential was named *NCC-vib potential* and used in the molecular dynamics simulation of liquid water [67] and hexagonal ice ($\mathbf{I}_h$) [72].

In References [69, 70], the so-called *polarization point-charge (PPC) potential* for water in condensed phase was elaborated. It consists of three terms:

$$V^{PPC} = V_q + V_{00}^{LJ} + V_{pol} \tag{5.46}$$

where the first two have an analytical form as in Equation (5.41), $V_q$ is the Coulomb interaction energy of point charges and $V_{00}^{LJ}$ is the Lennard-Jones oxygen–oxygen potential. The polarization term is expressed as:

$$V_{pol} = \frac{1}{2} \sum_{\tau,i} \frac{\left( d_\tau^i - d_\tau^{(0)} \right)^2}{\alpha_\tau} \tag{5.47}$$

where $d_\tau^i$ are the Cartesian components ($\tau = x, y, x$) of the dipole moment of the $i$th molecule in condensed phase and $d_\tau^{(0)}$ is the Cartesian component of the dipole moment of the isolated molecule. The PPC potential was successfully applied in the molecular dynamics simulation of liquid water under normal and extreme conditions [70].

Another form of the water–water model potential with the intramonomer flexibility was recently elaborated by Burnham and Xantheas [73]. Their model is based on smeared Coulomb and dipole–dipole interactions proposed by Thole [74]. Many-body effects are accounted via an induction scheme based on atomic site dipoles and polarizabilities. The important feature of this potential is that the charges are a function of the monomer geometry, and the potential energy of the water monomer is described at a high level *ab initio* approach [75]. The authors [73] named their potential as *TTM2-F (Thole-type model II, flexible) potential*. The TTM2-F potential predicts an increase in the HOH angle with cluster size. According to experimental data [76], the bonding angle in ice is almost tetrahedral and is in the range of 109.5 $\pm 0.3^\circ$. So, it increases on $5^\circ$ with respect to the gas-phase monomer for which the angle is equal to $104.52^\circ$. The other monomer flexible models, not using geometry-dependent charges [72, 77] failed to predict this increase; on the contrary, they predicted a contraction of the bending angle in the $\mathbf{I}_h$ ice. Note also that the quality of some recent water–water potentials is tested in Reference [78].

The TTM2-F potential was exploited by Hartke [79] in the global geometry optimization of the $(H_2O)_N$ clusters in the range $N = 2 - 30$. To realize such a time consuming calculation, Hartke elaborated a parallel implementation of the evolution algorithm. The global minimum structures for relatively small clusters with $N = 2 - 11$ found with this potential are qualitatively identical with those found earlier with the TIP4P potential [61], but for a larger cluster size, $N \geq 12$, serious qualitative differences were revealed. The latter can be considered as a manifestation of an inability of the simpler TIP4P potential to model large molecular clusters where the monomer changes become significant.

### 5.1.8 Screened Coulomb potential

In this section some model potentials which can be applied at short atom–atom distances are presented. These distances are important in the scattering processes. The physics of high-energy atom–molecule collisions are determined by the repulsive part of the interaction potential at very short distances (fractions of an Angström). In this case, the direct Coulomb repulsion of the nuclei, with corrections due to the screening of the electronic shell, makes a predominant contribution. Such a *screened Coulomb potential* has the general form:

$$V(R) = \frac{Z_1 Z_2 e^2}{R} f(R) \tag{5.48}$$

where $Z_1 e$ and $Z_2 e$ are the charges of the interacting nuclei and $f(R)$ is the screening function which satisfies the boundary conditions: $f(0) = 1$ and $f(\infty) = 0$.

The simplest *screened Coulomb potential* was proposed by Bohr [80]. The *Bohr screening function* has the form:

$$f(R) = e^{-R/a} \tag{5.49}$$

where the screening radius $a$ is expressed in terms of the Bohr radius $a_0$ as:

$$a = \frac{a_0}{\left(Z_1^{\frac{2}{3}} + Z_2^{\frac{2}{3}}\right)^{\frac{1}{2}}} \tag{5.50}$$

The Bohr potential falls down as distance increases very rapidly, and at a distance of some tenths of an Angström it becomes unreliable. This fact restricts its use to studies of only high-energy collisions with $E \gtrsim 100$ keV.

Brinkman [81] proposed using the screening function:

$$f(R) = \frac{a_1^2 e^{-R/a_2} - a_2^2 e^{-R/a_1}}{a_1^2 - a_2^2} \tag{5.51}$$

where $a_1$ and $a_2$ are the screening radii of the first and second atom, respectively. For identical atoms, in the limit $a_1 \to a_2 = a$:

$$f(R) = (1 - R/2a)\, e^{-R/a} \tag{5.52}$$

If $R \ll a$ Equation (5.52) transforms into the Bohr function. The Brinkman potential falls down with distance more rapidly than the Bohr potential and at $R > 2a$ it even leads to an attraction. However, at these distances, the screened Coulomb potential does not describe the real physical picture and the exchange and other types of interatomic interactions have to be taken into account.

Firsov [82] proposed another way of constructing the screened Coulomb potential. He showed that the dimensionless Thomas–Fermi function $\chi(x)$ can be used as screening function. The function $\chi(x)$ satisfies the Thomas–Fermi equation:

$$\frac{d^2\chi(x)}{dx^2} = x^{-\frac{1}{2}} \chi^{\frac{2}{3}}(x) \tag{5.53}$$

The *Firsov potential* involves the numerical function $\chi(R/a)$:

$$V(R) = \frac{Z_1 Z_2 e^2}{R} \chi(R/a) \tag{5.54}$$

with the screening parameter:

$$a = 0.885\, a_0 \left(Z_1^{\frac{1}{2}} + Z_2^{\frac{1}{2}}\right)^{-\frac{2}{3}} \tag{5.55}$$

The function $\chi(x)$ has been tabulated for a wide range of $x$. For that reason, the Firsov potential has been often used in numerical studies of the repulsion of atoms at small separations.

There is a large number of fairly accurate analytical approximations for the Thomas–Fermi screening function. They have been used as a basis for constructing some analytical potentials, based on the Thomas–Fermi model. Two widely-used potentials are presented below.

The Sommerfeld approximation [83] for the Thomas–Fermi screening function is the most well-known one:

$$\chi(x) = \left[1 + \left(12^{-\frac{2}{3}} x\right)^{\lambda}\right]^{-3/\lambda} \tag{5.56}$$

Sommerfeld chose $\lambda = 0.772$. Later, values of $\lambda$ that led to a better agreement with the numerical solution of the Thomas–Fermi equation, namely, $\lambda = 0.8034$ [84] and $\lambda = 0.8371$ [85], were obtained. The potential (5.54), with the function (5.56) as $\chi(R/a)$ where $a$ is given by Equation (5.55), is denoted as the *Sommerfeld potential.*

Moliere [86] proposed a three-term exponential approximation for the Thomas–Fermi function:

$$\chi(x) = 7pe^{-qx} + 11pe^{-4qx} + 2pe^{-20qx} \tag{5.57}$$

where $p = 0.05$ and $q = 0.3$. At large $x$, the Moliere function diverges from the exact solution. Nevertheless, the Moliere potential:

$$V(R) = \frac{Z_1 Z_2 e^2}{R}\left[0.35e^{-0.3R/a} + 0.55e^{-1.2R/a} + 0.10e^{-6R/a}\right] \tag{5.58}$$

behaves more correctly, as $R$ increases, than the Firsov potential based on the accurate numerical solution of the Thomas–Fermi equation, since the latter falls off more slowly as $R$ increases. A comparative analysis of the behavior of the different screened Coulomb potentials as $R$ increases is available in the book by Torrens [1].

### 5.1.9 Born–Mayer potential

This two-parameter potential was used by Born and Mayer [87] in their study of the properties of ionic crystals in order to describe the repulsion of closed shells of ions. It involves the single exponential term:

$$V(R) = A\exp[-BR] \tag{5.59}$$

For small separations (but not very small, since potential (5.59) becomes finite at $R = 0$), the Born–Mayer potential correctly describes the exchange repulsion (strictly speaking, it is correct only for closed electron shell systems). Because of its simplicity, this potential was widely used in calculations of crystalline properties, for example, the elastic constants [88].

In the study of crystal structure, a modification of potential (5.59), proposed by Hungtington [88], is often used. This potential involves the equilibrium distance $R_0$ between the nearest neighbors in the lattice:

$$V(R) = A'\exp\left[-\beta(R - R_0)/R_0\right] \tag{5.60}$$

### 5.1.10 Boys–Shavitt multi-parameter potential

Boys and Shavitt [89] proposed a potential with an infinite number of variable parameters:

$$V(R) = 4\epsilon\left[\left(\frac{R}{\sigma}\right)^2 + B^2\right]^{-3}\sum_{i=0}^{\infty} C_{2i}\left[\left(\frac{R}{\sigma}\right)^{2i}\exp\left\{A\left[1-\left(\frac{R}{\sigma}\right)^2\right]\right\} - 1\right] \tag{5.61}$$

where $\sigma$ is the value of $R$ for which $V(R) = 0$; the parameter $\epsilon$ can be equal to the depth of the potential well under the corresponding normalization of the parameters $C_{2i}$; $A$ and $B$ are usually fixed and $C_{2i}$ are the variable parameters. The parameter $B^2$ is introduced to simply the integration of the potential (5.61) near $R = 0$. Note that the authors proposed their potential before the widespread application of computers. The subsequent development of high-speed computers has made it possible to use such complicated potentials with a large number of parameters.

It should be mentioned that, taking the first term in the sum, $i = 0$, with $A = 4$, $B^2 = 0.1$ and $C_0 = 1$, leads to the potential:

$$V(R) = 4\epsilon\left[\left(\frac{R}{\sigma}\right)^2 + 0.1\right]^{-3}\left[\exp\left\{4\left[1-\left(\frac{R}{\sigma}\right)^2\right]\right\} - 1\right] \tag{5.62}$$

which, in fact, is indistinguishable from the Lennard-Jones (12–6) potential in a wide range of distances [89].

The Boys–Shavitt potentials, with the parameters given in Table 5.4, are presented in Figure 5.10. The constants $C_{2i}$ are normalized in such a way that the

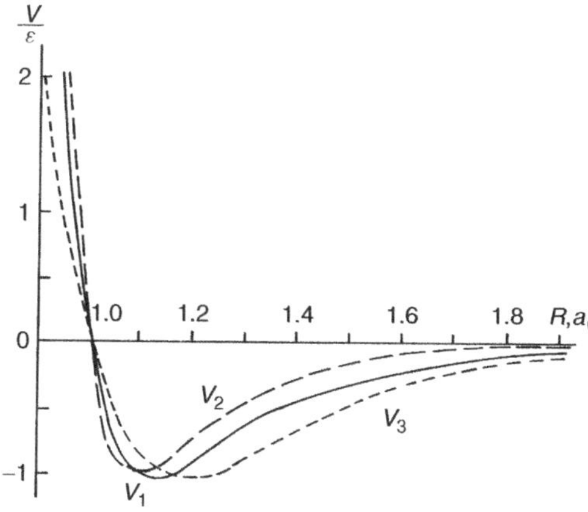


**Figure 5.10** The Boys–Shavitt potentials, $\varepsilon$ is the depth of the potential well after normalization

**Table 5.4** Parameters of the three Boys–Shavitt potential curves represented in Figure 5.10

| | $A$ | $B^2$ | $C_0$ | $C_2$ | $C_4$ | $C_6 = C_8 = \ldots$ | $R_m/\sigma$ |
|---|---|---|---|---|---|---|---|
| $V_1$ | 4 | 0.1 | 0.9725 | 0 | 0 | 0 | 1.123 |
| $V_2$ | 4 | 0.1 | 3.4934 | −2.9881 | 0 | 0 | 1.092 |
| $V_3$ | 4 | 0.1 | 1.3414 | −3.5452 | 3.8331 | 0 | 1.184 |

well depth is equal to $\epsilon$ for all these curves. The curves differ from each other by the value of $R_m$. These potentials were used by Boys and Shavitt [89] to evaluate the second, third and fourth virial coefficients over a wide range of temperatures in order to study the influence of the potential form on the behavior of the virial coefficients. The four-term Boys–Shavitt potential ($i = 0, 1, 2, 3$) was used by Munn [90] to study the behavior of the second virial coefficient of noble gases in the range of temperatures $T \gg T_B$, where $T_B$ is the Boyle temperature defined by the condition that the second virial coefficient is equal to zero.

### 5.1.11 Combined (piecewise) potentials

Each of the potentials mentioned above works rather well only in a definite range of distances. Therefore, in order to achieve a more adequate description, combined potentials are used. These potentials are formed from different pieces, each presented by some model potential. Such piecewise potentials are not appropriate for an analytical representation, although quite acceptable for computational studies without any additional inconvenient features. Rather complicated combinations of potentials can be used in high-speed computer simulations. Let us consider some of these potentials, which were used for the solution of various problems.

#### 5.1.11.1 Erginsoy–Vineyard–Englert potential

Erginsoy *et al.* [91] proposed a potential formed from four model potentials: the screened Coulomb, Born–Mayer, Morse and modified Morse potential. This piecewise potential was applied in numerical simulations of the dynamics of radiation defects in crystals, in particular, in the $\alpha$-iron lattice. The potential has the following form:

$$V(x) = \begin{cases} (0.7/x)\, A \exp[-\alpha x] & 0 < x \leq 0.7 \\ A \exp[-\alpha x] & 0.7 < x < 1.35 \\ D\,\{\exp[-2\beta(x - R_m)] - 2\exp[-\beta(x - R_m)]\} & 1.35 \leq x \leq 2.0 \\ f(x)\, D\,\{\exp[-2\beta(x - R_m)] & \\ -2\exp[-\beta(x - R_m)]\} & 2.0 < x \leq 2.5 \\ 0 & 2.5 < x < \infty \end{cases} \tag{5.63}$$

where all distances are measured in units of $R_0 = 2.48$ Å, $R_0$ being the equilibrium distance between the closest neighbors in the $\alpha$-iron lattice, $x = R/R_0$; $A = 8573$ eV, $\alpha = -6.547$, $D = 0.223$ eV, $\beta = 1.388$ Å$^{-1}$, $R_m = 2.845$ Å. The choice of the constant in the pre-exponential term of the Coulomb potential provides continuity with the Born–Mayer potential at $x = 0.7$. The parameters of the Born–Mayer potential were chosen in such a way that its coupling with the Morse potential occurs at $x = 1.35$; $f(x)$ is an arbitrary function that is equal to unity at $x = 2.0$ and decreases smoothly up to 0.1 at $x = 2.5$. At $x = 2.5$, $V = 0.014$ eV, causing a small discontinuity of the potential due to its cutting-off at $x > 2.5$.

### 5.1.11.2 ESMSV and MSV potentials

The *ESMSV potential* [92] was proposed in order to analyze the properties of rare gases and was widely applied later on [20, 93]. The repulsion is described by an exponential term, the potential well by the Morse function and the van der Waals long-range attraction by dispersion terms. These three potentials are joined smoothly with the help of the so-called spline functions.[2] The authors [92] denoted this potential as *ESMSV* (*exponent–spline-Morse-spline-van der Waals*). It has the form:

$$\frac{V(x)}{\epsilon} = \begin{cases} A\exp[-\alpha(x-1)] & 0 < x \le x_1 \\ \exp[a_1 + (x-x_1)\{a_2 + (x-x_2)[a_3 + (x-x_1)a_4]\}] & x_1 < x < x_2 \\ \exp[-2\beta(x-1)] - 2\exp[-\beta(x-1)] & x_2 \le x \le x_3 \\ b_1 + (x_1 - x_3)\{b_2 + (x-x_4)[b_3 + (x-x_3)b_4]\} & x_3 < x < x_4 \\ -c_6x^{-6} - c_8x^{-8} - c_{10}x^{-10} & x_4 \le x < \infty \end{cases} \tag{5.64}$$

where $x = R/R_m$. The coupling points $x_1$, $x_2$, $x_3$ and $x_4$ are varied in order to obtain the best agreement with experimental data. The constants $a_i, b_i$ $(i = 1,2,3,4)$ in the spline functions are chosen in such a manner that the transition between the potentials are smooth. The parameters $\epsilon$, $R_m$, $A, \alpha$ and $\beta$ have the usual meanings. The coefficients $c_6$, $c_8$ and $c_{10}$ are expressed in terms of the dispersion constants: $c_n = C_n/\epsilon R_m^n$. Sets of the parameters of the potential (5.64) for different pairs of rare gas atoms are tabulated elsewhere [93] (see also References [10, 20]).

If the behavior of the potential is not so essential at small separations, a simplified version of the potential ESMSV, without the exponential Born–Mayer potential, is used. The Morse potential is joined with the dispersion van der Waals potential by means of the spline function. This potential was denoted as the *MSV (Morse–spline-van der Waals) potential.* This potential was used to analyze the anisotropic interaction in the complexes $H_2$–rare gas atom [48]. In Reference [48],

[2] The splines are functions formed by combination of polynomials, which are coupled in such a way that the resulting function is continuous and differentiable [94, 95]. Cubic splines are commonly used, although the exponential ones are also used sometimes, as in Equation (5.64). The term 'spline' originates from the drawing technique in which thin strips, denoted as splines, were used to draw smooth curves passing through given points.

the MSV potential was chosen in the form (5.38), in which the following potentials were taken as the central potentials $V_0(R)$ and $V_2(R)$:

$$\left.\begin{aligned} V_0(R) &= \epsilon\{\exp[-2\beta(R-R_m)] - 2\exp[-\beta(R-R_m)]\} \\ V_2(R) &= \epsilon\left\{A_r\exp\left[-2\beta'(R-R_m)\right] - 2A_a\exp\left[-\beta'(R-R_m)\right]\right\} \end{aligned}\right\} \quad 0 < R \le R_m$$

$$\left.\begin{aligned} V_0(R) &= a_0 + a_1(R-R_m) + a_2(R-R_m)^2 + a_3(R-R_m)^2(R-R_v) \\ V_2(R) &= b_0 + b_1(R-R_m) + b_2(R-R_m)^2 + b_3(R-R_m)^2(R-R_v) \end{aligned}\right\} \quad R_m < R \le R_v$$

$$\left.\begin{aligned} V_0(R) &= -C_6R^{-6} - C_8R^{-8} \\ V_2(R) &= -A_d\left[C_6R^{-6} + C_8R^{-8}\right] \end{aligned}\right\} \quad R_v < R < \infty \tag{5.65}$$

where $A_r$ and $A_a$ are the anisotropic coefficients of the repulsion and attraction terms near the potential well; $A_d$ is the dispersion anisotropy coefficient; $C_6$ and $C_8$ are the dispersion constants obtained independently from calculations or experiments.

The two Morse functions, coupled at a point $\sigma$ where $V(\sigma) = 0$, are sometimes used instead of the single Morse function in the MSV potential. Such a potential is denoted as *MMSV* or $M^2SV$. Pack *et al.* [96] used the potential with three Morse functions, $M^3SV$, which contains 18 parameters. These parameters were fitted on the basis of some properties measured experimentally, such as the differential cross-section of scattering, viscosity and virial coefficients.

### 5.1.12 Model potentials applied in metal and semiconductor studies

As demonstrated in Section 4.2, the influence of many-body forces on some physical properties of metal clusters leads to the qualitative changes that cannot be taken into account by effective pairwise potentials. Below we present some model potentials for solids taking into account the many-body nonadditive effects.

#### 5.1.12.1 Glue models

This term was introduced by Ercolessi *et al.* [97, 98] for describing the 'gluing' effect of the conduction electrons in the metal bonding. Different glue models for the N-atom system may be set in the following form:

$$V = \frac{1}{2}\sum_{i,j}{}' V_{ij} + \sum_i V_i(\rho_i) \tag{5.66}$$

where $V_{ij}$ is an ordinary pairwise model potential and the prime means that $i \neq j$; $V_i(\rho_i)$ is a many-body 'glue' term and $\rho_i$ is the local electronic density at site $i$ due to the contributions from its surrounding:

$$\rho_i = \sum_j \phi(R_{ij}) \tag{5.67}$$

An analytical form of $V(\rho_i)$ and $\phi(R_{ij})$ depends upon approximations used, $R_{ij} = |\mathbf{R}_i - \mathbf{R}_j|$. The sum over $j$ may be limited by the nearest neighbors. Even in this case, the potential $V(\rho_i)$ has the many-body character, if the number of nearest neighbors $C \geq 2$.

Evidently, $\rho_i$ depends upon the coordination of atom $i$ in the lattice. At equilibrium distances, it can be expressed directly through the coordination number (see Equation (5.76) below). On the other hand, $\rho_i$ is the local electronic density at site $i$ set up by the surrounding atoms and, to the first approximation, is the same as if it would be in a homogeneous electron gas of that density. The last idea was assumed in the effective-medium theories [99–102], which were based on the density functional theory.

Daw and Baskes [101] substantiated the general Equation (5.66) for the potential energy representation and emphasized that Equation (5.66) can be applied to both an impurity in solids and a pure solid, because each atom can be viewed as an impurity embedded in a host comprising all the other atoms. They named their approach as the *embedded-atom method (EAM)* [101].[3] Later, Baskes *et al.* [105, 106] developed the *modified embedded-atom method (MEAM)* incorporating in the EAM angular terms and making some other improvements. For the embedding function ('glue' function) a logarithmic form was suggested [105, 107]:

$$V_i(\rho) = A_i E_i \rho \ln \rho \tag{5.68}$$

where $E_i^0$ is the sublimation energy at site $i$ and $A_i$ is a parameter to be determined. The application of the MEAM for some atomic solids and model potentials obtained are presented elsewhere [106, 108, 109].

Finnis and Sinclair [110] applied the second moment approximation to the tight-binding model [111], from which follows that the cohesive energy per atom varies as the square root of a site density $\rho_i$. In the original paper, the *Finnis–Sinclair (FS) potential* was presented as:

$$V = \frac{1}{2}\sum_{i,j}{}' V_{ij}(R_{ij}) - A\sum_i \left(\sum_j \phi(R_{ij})\right)^{1/2} \tag{5.69}$$

For functions $V_{ij}$ and $\phi$, the polynomial form was accepted. The repulsive pairwise potential has the form:

$$V_{ij}(R_{ij}) = \begin{cases} (R_{ij} - c)^2 \left(c_0 + c_1 R_{ij} + c_2 R_{ij}^2\right) & R \leq c \\ 0 & R > c \end{cases} \tag{5.70}$$

where $c$ is a disposable parameter lying between the second and third neighbors, and:

$$\phi(R_{ij}) = \begin{cases} (R_{ij} - d)^2 + \dfrac{\beta (R_{ij} - d)^3}{d} & R \leq d \\ 0 & R > d \end{cases} \tag{5.71}$$

[3] The embedded-atom method has not to be mixed with the embedded-cluster method [103, 104] developed for study of the local electronic structure in solids.

where the cutoff parameter $d$ also lies between the second and third neighbors, $\beta$ was chosen to introduce a maximum in $\phi$ within the first-neighbor distance. The total number of parameters in the FS potential equals seven: $A, c_0, c_1, c_2, c, d, \beta$. They are optimized by fitting the experimental equilibrium volume, cohesive energy and elastic constants with the parameters $c, d$ and $\beta$ fulfilling the conditions mentioned in the preceding text.

To the best of our knowledge, one of the first applications of tight-binding approximation to the model potential for transition metals was that of Gupta [112]. Gupta based his study, as later Finnis and Sinclair [110] did, on the tight-binding model. The $d$ band width was calculated from the second moment $\mu_2$ of the density of states and this gave for the cohesive energy, provided by the $d$ electron, the proportionality to $\sqrt{\mu_2}$. If the second and higher neighbor $d$–$d$ overlap and hopping integrals are neglected, $\mu_2$ can be expressed in terms of the average intersite hopping integrals $\beta\left(R_j\right)$ between the atom in the origin and atoms situated at sites $j$:

$$\mu_2 = \sum_j{}' \beta^2\left(R_j\right) \tag{5.72}$$

where the summation extends over the nearest neighbors only and the prime indicates that the $R_j = 0$ term has to be excluded. The cohesive energy is:

$$E_d \sim \left(\sum_j{}' \beta^2\left(R_j\right)\right)^{1/2} \tag{5.73}$$

In the neighborhood of the equilibrium distance $R_0$, the hopping integrals vary exponentially [111, 113]:

$$\beta\left(R\right) = \beta_0 e^{-q\left(R-R_0\right)} \tag{5.74}$$

Assuming that the repulsive pairwise potential has the same exponential behavior that coincides with the Born–Mayer potential (5.60), Gupta obtained one of the first potentials of the glue model family. The *Gupta potential* may be presented in the following form [114]:

$$V^G = u\sum_i{}' \left\{ A\sum_j \exp\left[-p\left(\frac{R_{ij}}{R_0}-1\right)\right] - \left(\sum_j \exp\left[-2q\left(\frac{R_{ij}}{R_0}-1\right)\right]\right)^{\frac{1}{2}} \right\} \tag{5.75}$$

where $R_0$ is the equilibrium distance between the nearest neighbors. At $R_{ij} = R_0$, the cohesive part of the Gupta potential is equal $\sqrt{C}$, where $C$ is the coordination number. The parameters $u$, $A$, $p$ and $q$ are obtained by fitting to the experimental data and depend upon the material. The Gupta potential has been extensively used

**Table 5.5** Calculated energies per atom and coordination numbers for the eighteen aluminium structures [116]

| Structure | Coordination number | Energy per atom (eV) |
|---|---|---|
| Atom | 0 | −54.95 |
| Dimer | 1 | −55.66 |
| Line | 2 | −56.28 |
| Graphite | 3 | −56.95 |
| Girder | 4 | −57.04 |
| Square layer | 4 | −57.29 |
| Diamond | 4 | −57.42 |
| Square slab | 5 | −57.64 |
| Close packed layer | 6 | −57.49 |
| Simple cubic | 6 | −57.91 |
| fcc 110 slab | 6 | −57.54 |
| Close packed slab | 7 | −57.89 |
| fcc 100 slab | 8 | −57.85 |
| Vacancy lattice | 8 | −58.10 |
| Simple hexagonal | 8 | −58.12 |
| bcc | 8 | −58.24 |
| fcc 111 slab | 9 | −57.97 |
| fcc | 12 | −58.31 |

for simulation studies of metals and clusters; recent application to the Ni, Ag and Au nanoclusters see in Reference [115].

The glue model has been tested [116] by comparison with the density functional theory (DFT) calculations of the total energies per atom for different aluminium structures. In Table 5.5 the total energies per atom, $E_a$, and the coordination numbers $C$ for the 18 equilibrium aluminium structures are presented. These structures range in coordination number from 0 to 12. The calculated energies were approximated by a simple analytical function of $C$:

$$E_a = E_0 + A\sqrt{C} + BC \tag{5.76}$$

The fit has a root-mean-square (rms) deviation of 0.20 eV with $E_0 = 54.74$ eV, $A = -1.41$ eV and $B = 0.09$ eV. The energy per bond $E_a/C$ decreases with increasing coordination number. Thus, even at this simplified level, the glue model is shown to be rather successful. The cohesive energies of this very wide variety of structures are determined to a large extent simply by the number of nearest neighbors.

Robertson *et al.* [116] have tested 25 different glue models to see how they reproduce the aluminium structures with variated distances and coordination number calculated at the DFT level. For all models tested, the errors obtained were in the range ±0.3 eV per atom. The authors came to conclusion that the minimum rms error of glue approach is no less than 0.1 eV per atom. In connection with this conclusion, it is worthwhile to note that the DFT method is far from precise, the error in DFT calculations can also be in the range 0.1 to 0.3 eV.

The method of fitting parameters in glue model potentials to theoretical data, produced by first principle calculations, was proposed by Ercolessi and Adams [117]. The method lies in fitting the glue potential to *ab initio* atomic forces of various atomic configurations. The analytical functions constituting the potential were defined as third order polynomials (cubic splines). This approach was applied to the construction of the model potentials for aluminium [117], magnesium [118] and some binary alloys with aluminium [119, 120].

The method [117] was also applied to constructing the model potential for silicon [121]. The authors [121] employed the MEAM approach of Baskes *et al.* [105], although the three-body potential was used in the form similar to the SW potential (see next section). All analytical functions in the model potential were represented as cubic splines each with 10 fitting parameters. The 49 independent parameters were fitted to a large theoretical database and to experimental values for elastic constants and phonon frequencies. The constructed potential accurately reproduces phonons and elastic constants, point defect energetics, formation energy and geometries of interstitial complexes, etc.

#### 5.1.12.2 Explicit inclusion of the three-body term in model potential

In several studies, the three-body potential was explicitly incorporated into model potentials describing semiconductors and metals.

Stillinger and Weber (SW) [122] included a three-body potential for simulating the solid and liquid forms of silicon. The *SW potential* comprises a five-parameter pairwise potential:

$$V_2^{SW} = \sum_{i<j} V_{ij}^{SW}\left(R_{ij}\right) \tag{5.77}$$

$$V_{ij}^{SW} = \begin{cases} A\left(BR_{ij}^{-p} - R_{ij}^{-q}\right)\exp\left[\left(R_{ij}-a\right)^{-1}\right] & R_{ij} < a \\ 0 & R_{ij} \geq a \end{cases}$$

and a three-body potential:

$$V_3^{SW} = \sum_{i<j<k} V_3^{SW}\left(\mathbf{R}_i, \mathbf{R}_j, \mathbf{R}_k\right) \tag{5.78}$$

$$V_3^{SW}\left(\mathbf{R}_i, \mathbf{R}_j, \mathbf{R}_k\right) = h\left(R_{ij}, R_{ik}, \theta_i\right) + h\left(R_{ji}, R_{jk}, \theta_j\right) + h\left(R_{ki}, R_{kj}, \theta_k\right)$$

where for the function $h$ was used the angular dependent two-parameter form with the same cutoff as in Equation (5.77):

$$h\left(R_{ij}, R_{ik}, \theta_i\right) = \begin{cases} \lambda \exp\left[\gamma\left(R_{ij}-a\right)^{-1} + \gamma\left(R_{ik}-a\right)^{-1}\right] & R_{ij} < a \\ \quad \times \left(\cos\theta_i + \frac{1}{3}\right)^2 & \\ 0 & R_{ij} \geq a \end{cases} \tag{5.79}$$

$\theta_i$ is the angle at vertex $i$. The three-body terms vanish for the tetrahedral structure in which $\cos\theta_t = -\frac{1}{3}$.

A limited search over seven parameters $(A, B, p, q, a, \lambda$ and $\gamma)$ was carried out to ensure that the diamond structure is slightly more stable than the cubic structure and that computer simulation gives a melting point and liquid structure in reasonable accord with the experiment. The particular feature of the SW potential is that two- and three-body terms have a cutoff at 3.77 Å, which is 1.6 times the interatomic spacing, so only nearest neighbors contribute to the energy.

Biswas and Hamann (BH) [123] developed a more elaborated and flexible scheme, which generalizes the SW potential. As seen above (Equation (5.79)), the three-body potential may be expressed as a function of two bond distances and the included angle between them. The authors [123] expanded the angular dependence of this potential in the set of Legendre polynomials and presented this expansion in a separable form with respect to the function of two distances:

$$V_3^{BH}\left(R_{ij}, R_{ik}, \theta_i\right) = \sum_{\ell} C_\ell \phi_\ell\left(R_{ij}\right) \phi_\ell\left(R_{ik}\right) P_\ell\left(\cos\theta_i\right) \tag{5.80}$$

Obviously, in applications, Expansion (5.80) has to be truncated. In the original paper [123], the following potential was used to study the silicon structures in bulk, surface and cluster form:

$$V_3^{BH}\left(R_{ij}, R_{ik}, \theta_i\right) = \Big[B_1\phi_1\left(R_{ij}\right)\phi_1\left(R_{ik}\right)\left(\cos\theta_i + \tfrac{1}{3}\right)^2 \\ + B_2\phi_2\left(R_{ij}\right)\phi_2\left(R_{ik}\right)\left(\cos\theta_i + \tfrac{1}{3}\right)^3\Big] f_c\left(R_{ij}\right) f_c\left(R_{ik}\right) \tag{5.81}$$

The best fit was obtained with the Gaussian functions $\phi_\ell(R) = \exp\left[-\alpha_\ell R^2\right]$; $f_c(R)$ is the cutoff function:

$$f_c(R) = \left[1 + \exp\left(\frac{R - R_c}{\mu}\right)\right]^{-1} \tag{5.82}$$

The two-body potential was also expressed through the Gaussian functions:

$$V_2^{BH}\left(R_{ij}\right) = \left[A_1 e^{-\lambda_1 R_{ij}^2} + A_2 e^{-\lambda_2 R_{ij}^2}\right] f_c\left(R_{ij}\right) \tag{5.83}$$

The comparative discussion of the SW and BH potentials has been presented in the review by Carlsson [124], in which he also discusses other approaches beyond the pair–potential description of elemental transition metals and semiconductors.

Murrell and Mottram (MM) [125] proposed, for atomic solids, another model potential with an original form of the three-body potential. In the latter, the symmetric coordinates (Q) were used. This distinguishes the *MM potential* from the SW and BH potential. The angular dependence was implicitly included into the polynomial in the symmetry coordinates. The MM potential is constructed as a sum of two-body and three-body terms:

$$V = \sum_{i<j} V_{ij}^{(2)} + \sum_{i<j<k} V_{ijk}^{(3)}$$

The two-body term expressed as the Rydberg potential (Equation (5.24)):

$$V_{ij} = -D\left(1 + a\rho_{ij}\right) e^{-a\rho_{ij}} \tag{5.84}$$

with

$$\rho_{ij} = \frac{R_{ij} - R_e}{R_e} \tag{5.85}$$

$D$ is the dissociation energy and $R_e$ is the equilibrium distance of the two-body potential. These are scaling factors that ensure that the cohesive (lattice) energy and lattice constants of the reference structure are reproduced exactly.

The three-body potential has been constructed as a totally symmetric to the exchange of atoms polynomial $P(Q_1, Q_2, Q_3)$ with an exponential factor:

$$V_{ijk}^{(3)} = DP\left(Q_1, Q_2, Q_3\right) \exp\left(-bQ_1\right) \tag{5.86}$$

where:

$$\begin{bmatrix} Q_1 \\ Q_2 \\ Q_3 \end{bmatrix} = \begin{bmatrix} \frac{1}{\sqrt{3}} & \frac{1}{\sqrt{3}} & \frac{1}{\sqrt{3}} \\ 0 & \frac{1}{\sqrt{2}} & -\frac{1}{\sqrt{2}} \\ \sqrt{\frac{2}{3}} & -\frac{1}{\sqrt{6}} & -\frac{1}{\sqrt{6}} \end{bmatrix} \begin{bmatrix} \rho_{ij} \\ \rho_{ik} \\ \rho_{jk} \end{bmatrix} \tag{5.87}$$

In the first studies [125–127], a cubic polynomial with 7 parameters was employed:

$$\begin{aligned} P\left(Q_1, Q_2, Q_3\right) &= c_0 + c_1 Q_1 + c_2 Q_1^2 + c_3\left(Q_2^2 + Q_3^2\right) \\ &\quad + c_4 Q_1^3 + c_5 Q_1\left(Q_2^2 + Q_3^2\right) + c_6\left(Q_3^3 - 3Q_3 Q_2^2\right) \end{aligned} \tag{5.88}$$

In subsequent studies, a polynomial with more parameters was employed. The parameters are derived by fitting the experimental phonon-dispersion curves and elastic constants by means of a least-square fit.

The MM potential reproduces lattices and some other properties of different atomic solids [125–127] quite well. It was also intensively used in cluster studies, see Reference [128] and references therein. Let us note that the application of the MM potential (and the others described above) to cluster studies cannot be rigorously validated. As with every semiempirical potential with parameters fitted to experimental data of crystal systems, the MM potential is not adequate when describing the properties of small finite systems. In molecules and clusters there is no translation symmetry; the other important difference with bulk is the discrete nature of electronic spectra. In the MM potential case, an additional specific reason exists that makes it inappropriate in the cluster studies – the absence of the dispersion terms. Hence, the MM potential describes the interactions at large distances rather poorly. The latter is not very important in crystals, in which the nearest neighbor approximation works quite well, but important in cluster studies. The procedure of constructing model potentials for clusters, based on molecular conception, is discussed in the following section.

### 5.1.13 Model potentials fitted to *ab initio* calculated potential surfaces

The parameters of almost all model potentials discussed in this Chapter are derived by fitting to experimental data. Although, two exceptions from this rule have been already discussed. The water–water potentials elaborated by Clementi and his group [65–67] have been fitted to points on an *ab initio* calculated potential energy surface (Section 5.1.7.4). In the approach developed by Ercolessi and Adams [117], the glue potential is fitted to an *ab initio* calculated database (Section 5.1.12.1). The trend of fitting parameters to *ab initio* calculated data looks quite promising, if the precision of calculation is high enough. At the present state-of-the-art of computational chemistry and physics, this condition can be easily fulfilled.

To the best of our knowledge, the first pairwise model potentials for molecular systems with parameters fitted to *ab initio* calculated potential surfaces were elaborated by Clementi and co-workers [129–131]. In these studies, the so-called atom–atom potential (AAP) approach was used (Section 4.5). In the AAP scheme, the interaction between molecules is represented as a sum of pair atom–atom potentials:

$$V(R, \Omega) = \sum_{a<b} V_{ab}(R_{ab}) \tag{5.89}$$

$R$ and $\Omega$ are sets of radial and angular coordinates determining the mutual arrangement of molecules.

For complex molecular systems, Clementi *et al.* [131] introduced an additional index for the same atom placed in a different chemical surrounding and used the Lennard-Jones potential modified by inclusion of a Coulomb term (Equation (4.93)). The atom classes were defined taking into account the valence states of the atoms in the molecule and their charges $q$. The AAPs for the interactions of 21 amino acids with water were determined, considering 23 types of atoms in the amino acids and two types of atoms in the water molecule. As a result, the authors constructed 46 AAPs to describe the potential curves for the amino acid–water interactions. In this procedure, $46 \times 3 = 138$ parameters were fitted. To decrease the linear dependence between the parameters, the authors chose for each parameter 15 to 20 points of the theoretical potential curves obtained for different conformations.

In Reference [132] the two-stage procedure of fitting parameters was proposed. The Clementi-type potential with the exponential form of repulsive part was used. In a first step, the fitting is carried out over the magnitudes of the interaction energy, obtained via the first- and second-order perturbation treatment. The parameters of the exchange term in the AAP are fitted by means of $E_{int}^{(1)}$, and the parameters of the dispersion term by means of $E_{int}^{(2)}$. It leads to a considerable reduction of the number of parameters, which are fitted simultaneously. The atomic charges, appearing in the Coulomb term, are not varied, being determined directly from the condition of the best agreement with the known data on the molecular multipole moments. The parameters of the AAPs obtained in such a way were used next as starting values in the adjustment of the *ab initio* potential curves.

The model potentials with parameters fitted to *ab initio* calculated potential surfaces are called *ab initio model potentials*, in contrast to semiempirical model potential. *Ab initio* model potentials taking into account many-body forces have been elaborated by several groups [133–139]. The *ab initio* model potential for $Ag_6$ [136, 137] includes two-body, three-body and four-body terms. Each consists of an exponential exchange term and dispersion terms. The analytical form of the latter was taken from perturbation theory up to the fourth order. The proposed model potential was used in molecular dynamics simulation of structural and dynamical properties of the $Ag_6$ cluster [136, 137, 140].

Note that in the case of open-shell atoms, as silver (or lithium, sodium, etc.), the exchange energy contributes significantly into the covalent bonding and is negative in the region of the potential minimum and larger distances [141]. The two-body model potential constructed as a sum of an exponential exchange term and appropriate dispersion terms (the Buckingham potential) allows a precise reproduction of an arbitrary two-atom potential curve to be achieved, but the fitted two-body exchange energy is positive for all distances. Thus, the two-body model potential in the Buckingham form cannot be used to analyze the nature of bonding created by open-shell atoms.

For atoms with closed electron shells (or subshells, as in the case of alkaline earths), the choice of the Buckingham-type potential for the two-body interaction can be substantiated. The reason for this is that atoms with closed shells (subshells) have no multipole moments and their electrostatic and induction interactions have a pure overlap origin, as in the exchange case. From that follows their short-range character. Thus, all these interactions can be approximated by the same analytical function with exponential behavior. At large distances, in which overlap between electron functions of interacting atoms becomes negligible, only the dispersion forces contribute to the interaction energy. Below we represent the *ab initio* model potentials elaborated for the alkaline-earth trimers $Be_3$, $Mg_3$ and $Ca_3$ [142].

For a three-atom system, the model potential is naturally decomposed into the two-body and three-body interaction terms:

$$V(ABC) = V_2 + V_3 = \sum_{a<b} V_{ab} + V_{abc} \tag{5.90}$$

The two-body potential was taken in the Buckingham form with dispersion terms up to $R^{-10}$ and the exchange term multiplied by a polynomial as follows:

$$V_{ab} = \left(a_0 + a_1 R_{ab} + a_2 R_{ab}^2 + a_3 R_{ab}^3\right) \exp\left(-\alpha R_{ab}\right) - \left(C_6 \frac{D_6(R_{ab})}{R_{ab}^6} + C_8 \frac{D_8(R_{ab})}{R_{ab}^8} + C_{10} \frac{D_{10}(R_{ab})}{R_{ab}^{10}}\right) \tag{5.91}$$

Thus, the dipole–dipole $(R^{-6})$, dipole–quadrupole $(R^{-8})$ and dipole–octopole plus quadrupole–quadrupole $(R^{-10})$ dispersion interactions were taken into account. The damping functions $D_n(R_{ab})$, which improve the behavior of the dispersion energy in the overlap region, were taken in the form:

$$D_n = \begin{cases} \exp\left[-\beta_n\left(\dfrac{k_n}{R_{ab}} - 1\right)^2\right] & R_{ab} < k_n \\ 1 & R_{ab} \geq k_n \end{cases} \tag{5.92}$$

The same value of $k_n$ was taken for each $D_n$; relaxing this condition did not significantly improve the fitting.

The reason for including the damping function in a model potential is the following. For the intermediate distances, the expressions for the dispersion energy given by standard perturbation theory are not valid. At these distances, the exchange effects must be taken into account. Also, the multipole expansion is not valid at distances where the overlap of the wave functions of interacting atoms becomes important. As shown in Reference [143], the corrections to the multipole expansion are large enough. The damping functions improve the behavior of dispersion terms for intermediate and short distances; see the discussion in Section 3.1.5.

The three-body potential was also presented as a sum of exchange and dispersion terms:

$$V_3 = V_3^{exch} + V_3^{disp} \tag{5.93}$$

The three-body exchange terms were based on the symmetry-adapted coordinates developed for solids [125] (Section 5.1.12.2). These coordinates are related to the three sides of the triangle by:

$$\begin{aligned} Q_1 &= (R_{ab} + R_{bc} + R_{ca}) / (3)^{\frac{1}{2}} \\ Q_2 &= (R_{bc} - R_{ca}) / (2)^{\frac{1}{2}} \\ Q_3 &= (2R_{ab} - R_{bc} - R_{ca}) / (6)^{\frac{1}{2}} \end{aligned} \tag{5.94}$$

The potential must be symmetric to the exchange of atoms and this is achieved by using only the totally symmetric combinations:

$$Q_1, \; Q_2^2 + Q_3^2 \quad \text{and} \quad Q_3^3 - 3Q_3Q_2^2 \tag{5.95}$$

For $C_{2v}$ geometries ($R_{bc} = R_{ca}$) $Q_2$ is zero but because there is only one quadratic and one cubic symmetry combination of $Q_2$ and $Q_3$, fitting points on a $Q$ surface, using only $C_{2v}$ geometries, will unambiguously determine an expansion in $Q_2$ and $Q_3$ up to fifth power; only for higher powers are data on $C_s$ geometries needed. The exchange part of the three-body term was defined by one exponent and the polynomial of the fifth order with 15 coefficients as follows:

$$V_3^{exch} = \left[\left(b_2 + b_3 Q_1 + b_4 Q_1^2\right) + \left(Q_2^2 + Q_3^2\right)\left(b_5 + b_6 Q_1 + b_7 Q_1^2\right) + \left(Q_3^3 - 3Q_3 Q_2^2\right)\left(b_8 + b_9 Q_1 + b_{10} Q_1^2\right) + \left(Q_2^2 + Q_3^2\right)^2\left(b_{11} + b_{12} Q_1 + b_{13} Q_1^2\right) + \left(Q_2^2 + Q_3^2\right)\left(Q_3^3 - 3Q_3 Q_2^2\right)\left(b_{14} + b_{15} Q_1 + b_{16} Q_1^2\right)\right]\exp(-b_1 Q_1) \quad (5.96)$$

For the three-body dispersion energy, the ATM expression (Equation (4.64)) with an added product of three two-body damping functions was used:

$$V_3^{disp} = C_9 \frac{D_3(R_{ab})\, D_3(R_{ac})\, D_3(R_{bc})}{R_{ab}^3 R_{ac}^3 R_{bc}^3}\left(1 + 3\cos\theta_a \cos\theta_b \cos\theta_c\right) \quad (5.97)$$

The damping function $D_3$ has the same form (5.92) as used in Equation (5.91) but with independent values for $k_3$ and $\beta_3$.

The parameters of $V_2$ and $V_3$ were fitted separately to the *ab initio* calculated potential curve for dimers and the three-body potential surface, respectively. The *ab initio* calculations were carried out at the Møller–Plesset perturbation level up

**Table 5.6** Fitted parameters for two-body potentials [142] (see text for definition of each parameter); distances in Å and energies in kcal mol$^{-1}$. The root-mean-square errors (rms) are also presented

| Parameters | $Be_2$ | $Mg_2$ | $Ca_2$ |
|---|---|---|---|
| $V_2$(exch) | | | |
| $a_0$ | 9.836 (3)[a] | −3.489 (3) | −4.718 (4) |
| $a_1$ | −9.021 (3) | 1.472 (4) | 5.102 (4) |
| $a_2$ | 2.906 (3) | −6.068 (3) | −1.339 (4) |
| $a_3$ | −2.960 (2) | 8.144 (2) | 1.074 (3) |
| $\alpha$ | 1.920 (0) | 2.230 (0) | 1.980 (0) |
| rms | 5.70 (−2) | 9.80 (−2) | 6.00 (−3) |
| $V_2$(disp) | | | |
| $C_6$ | 4.107 (3) | 1.041 (4) | 3.287 (4) |
| $C_8$ | 4.853 (4) | 1.158 (5) | 5.543 (5) |
| $C_{10}$ | 2.788 (4) | 2.835 (4) | 3.617 (5) |
| $\beta_6$ | 1.540 (0) | 4.980 (−1) | 7.310 (−1) |
| $\beta_8$ | 6.010 (−1) | 1.989 (0) | 2.494 (0) |
| $\beta_{10}$ | 2.770 (−1) | 2.610 (−1) | 3.730(−1) |
| $k_n$[b] | 6.5 | 7.0 | 8.0 |
| rms | 7.00 (−3) | 3.36 (−2) | 1.82 (−2) |

[a]The notation 9.836 (3) means 9.836 × $10^3$.
[b]The same value of $k_n$ was taken for each $D_n$ in the dispersion part of potential (5.91).

**Table 5.7** Fitted parameters for three-body potentials [142] (see text for definition of each parameter); distances in Å and energies in kcal mol$^{-1}$

| Parameter | $Be_3$ | $Mg_3$ | $Ca_3$ |
|---|---|---|---|
| $b_1$ | 2.187 (0)[a] | 1.016 (0) | 1.402 (0) |
| $b_2$ | −5.751 (5) | 2.692 (3) | −4.710 (4) |
| $b_3$ | 3.537 (5) | −2.193 (3) | 3.846 (4) |
| $b_4$ | −6.242 (4) | 2.241 (2) | −7.807 (3) |
| $b_5$ | −5.296 (4) | 2.179 (3) | −9.334 (4) |
| $b_6$ | 3.851 (4) | −1.698 (3) | −1.714 (4) |
| $b_7$ | −9.112 (3) | 3.781 (2) | 1.805 (4) |
| $b_8$ | −7.203 (4) | −3.499 (2) | 1.624 (5) |
| $b_9$ | 4.079 (4) | 2.062 (2) | −6.781 (4) |
| $b_{10}$ | −3.601 (3) | −9.744 (1) | 7.819 (2) |
| $b_{11}$ | −1.514 (4) | 2.347 (2) | −1.640 (4) |
| $b_{12}$ | 1.032 (4) | −1.574 (2) | −6.108 (3) |
| $b_{13}$ | −9.708 (2) | 4.954 (0) | 2.446 (2) |
| $b_{14}$ | 2.999 (2) | 1.739 (2) | 3.209 (4) |
| $b_{15}$ | −3.956 (3) | 1.663 (1) | −2.233 (3) |
| $b_{16}$ | 5.180 (2) | −7.840 (−1) | 4.961 (1) |
| $\beta_3$ | 4.000 (−1) | 1.200 (0) | 5.000 (−1) |
| $k_3$ | 1.5 | 6.0 | 6.0 |
| $C_9$ [b] | 4.1585 (3) | 2.27756 (4) | 2.206014 (5) |
| rms | 4.50 (−1) | 3.10 (−1) | 1.31 (−1) |

[a] The notation 2.187 (0) means $2.187 \times 10^0$.

[b] The values of $C_9$ are taken from calculations in Reference [144].

to the IV order (MP4(SDTQ)) in the frozen core approximation. The basis set used was a triply-split valence plus one diffuse $p$ function, three sets of five $d$ functions and one set of seven $f$ functions labelled as (6−311+G(3df)).

The optimum values for the parameters are presented in Tables 5.6 and 5.7. The units of the parameters are obvious from expressions for model potential; thus, $b_1$ is kcal mol$^{-1}$ and $b_5$ is kcal mol$^{-1}$ A$^{-2}$.

To the best of our knowledge, experimental data for the alkaline-earth trimers have not been reported. The quality of the model potentials obtained can be checked only by comparison with theoretical calculations. The model potential values of equilibrium geometry ($R_0$) and binding energy ($E_b$) for trimers with $D_{3h}$ symmetry and the reference *ab initio* calculations at the MP4 (SDTQ) level [145] are presented in Table 5.8. The agreement is quite satisfactory. The best published theoretical results for the trimers presented in Table 5.8 indicate that the choice of the MP4 (SDTQ) method as a reference *ab initio* approach for fitting parameters is well justified.

The analytical form used in the model potentials (5.90)–(5.97) is accurate over a broad range of distances. So after fitting, these electron correlated *ab initio* model potentials can be utilized in molecular dynamics simulations of metal clusters to

**Table 5.8** Comparison of model potential values of equilibrium geometries ($R_0$) and binding energies ($E_b$) for the trimers ($D_{3h}$ geometries) with the reference *ab initio* calculation at the MP4 (SDTQ) level and some published calculations (distances in Å, energies in kcal mol$^{-1}$)

| | $Be_3$ | | $Mg_3$ | | $Ca_3$ | |
|---|---|---|---|---|---|---|
| | $R_0$ | $E_b$ | $R_0$ | $E_b$ | $R_0$ | $E_b$ |
| Model potential | 2.21 | 25.68 | 3.39 | 7.21 | 4.17 | 11.05 |
| MP4 (SDTQ) [145] | 2.24 | 25.90 | 3.32 | 7.12 | 4.12 | 11.66 |
| MRCI [146] | 2.22 | 22.4 | 3.37 | 6.30 | | |
| Best estimations, | | | | | | |
| MP2–R12 [147] | 2.20 | 26.9 | 3.37 | 8.00 | | |
| CI [148] | | | | | 3.97 | 11.53 |
| MRCI [149] | | | | | 4.16 | 12.10 |

study their structural and dynamic properties, in collision dynamics and other physical and chemical applications.

## 5.2 Determination of Parameters in Model Potentials

The widely used way of obtaining an intermolecular potential from experimental data is based on evaluating a measured property using a given model potential, with subsequent fitting of the parameters in such a way that the best agreement of the measured values with the evaluated values is attained. Mathematical methods exist that permit standardization of the fitting procedure. The method of maximum probability and the least-square method are the most common [150–152].

It should be mentioned that information about the potential can be also obtained directly by solving the reverse problem, which consists of the direct reconstruction of the potential on the basis of the experimental values for a given property, without a preliminary representation of its analytical form. The solution of this problem offers difficulties of a basic and purely mathematical nature. Nevertheless, various effective procedures for reconstructing the potential on the basis of experimental data have been developed. Some of them are examined in Section 5.3.

A block-scheme of the procedure for the fitting of parameters is depicted in Figure 5.11. Let $n$ experimental values of the definite quantity $F$ be given:

$$F_1 \pm \sigma_1, \quad F_2 \pm \sigma_2, \ldots, \quad F_n \pm \sigma_n$$

where $\pm\sigma_i$ is the error in the $i$-th measurement. Assume that the measured quantity is related, directly or indirectly, through a functional dependence with the intermolecular potential. Therefore, the measured quantity can be represented as a function of the parameters. Substituting the initial set of parameters $\mathbf{p}^0\left(p_1^0, p_2^0, \ldots, p_m^0\right)$ into the potential, the quantity $F$ is evaluated at measured points, that is, one finds

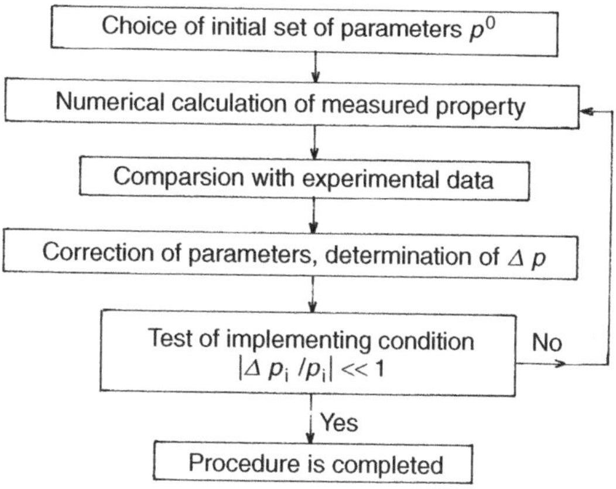


**Figure 5.11** Iteration cycle in the fitting parameter procedure

the function $f_j(\mathbf{q}_j, \mathbf{p}^0)$, where $\mathbf{q}_j(q_{j1}, q_{j2}, \ldots, q_{jr})$ are the coordinates of experimental points. Usually, the analytical expressions $f_j(\mathbf{q}_j, \mathbf{p}^0)$ are the same for all $j$, but they may also be different, in particular, when one set of parameters is fitted using different properties. Since the initial set of parameters is chosen on the basis of some physical arguments or even by intuition, the calculated and measured values of quantity $F$ are not equal. To determine the best set of parameters, the standard procedure of the least-square method (lsm) is often used. This procedure is now considered in detail.

In the lsm approach, the sum of the squares of the deviations, with weights that are inversely proportional to the squares of the errors, is chosen as the functional to be minimized:

$$S(\mathbf{p}) = \sum_{j=1}^{n} \frac{1}{\sigma_j^2} \left[ f_j(\mathbf{q}_j, \mathbf{p}) - F_j \right]^2 \tag{5.98}$$

The required set of parameters $\mathbf{p}$ must provide the minimum of Equation (5.98), and, therefore, it must satisfy the equations:

$$\partial S(\mathbf{p}) / \partial p_i = 0 \qquad i = 1, 2, \ldots, m \tag{5.99}$$

Solving Equations (5.99) gives a set of parameters $\mathbf{p}$ which differs from $\mathbf{p}^0$ by a correction:

$$\mathbf{p} = \mathbf{p}^0 + \Delta\mathbf{p}$$

The set of Equations (5.99) is solved via an iterative procedure under the assumption that the correction, $\Delta\mathbf{p}$, is small.[4] Then the left-hand side of Equation (5.99)

[4] If the function $f_j(\mathbf{q}_j, \mathbf{p})$ depends linearly on the parameters $\mathbf{p}$, the iterative procedure is not needed. In that case, the solutions of Equation (5.100) give the coordinates $\mathbf{p} + \Delta\mathbf{p}$ of the exact minimum.

can be approximated by the two first terms of a Taylor series at the point $\mathbf{p}^0$:

$$\left[\frac{\partial S(\mathbf{p})}{\partial p_i}\right]_{\mathbf{p}^0} + \sum_{k=1}^{m}\left[\frac{\partial^2 S(\mathbf{p})}{\partial p_i \partial p_k}\right]_{\mathbf{p}^0} \Delta p_k = 0 \qquad i = 1, \ldots, m \tag{5.100}$$

where the derivatives are obtained easily from Expression (5.98):

$$\frac{\partial S(\mathbf{p})}{\partial p_i} = 2\sum_{j=1}^{n}\frac{1}{\sigma_j^2}\left[f_j\left(\mathbf{q}_j, \mathbf{p}\right) - F_j\right]\frac{\partial f_j}{\partial p_i} \tag{5.101}$$

$$\frac{\partial^2 S(\mathbf{p})}{\partial p_i \partial p_k} = 2\sum_{j=1}^{n}\frac{1}{\sigma_j^2}\frac{\partial f_j}{\partial p_k}\frac{\partial f_j}{\partial p_i} + 2\sum_{j=1}^{n}\frac{1}{\sigma_j^2}\left[f_j\left(\mathbf{q}_j, \mathbf{p}\right) - F_j\right]\frac{\partial^2 f_j}{\partial p_i \partial p_k} \tag{5.102}$$

Within the framework of the lsm procedure, the approximate expression for the second derivatives:

$$Z_{ik} = 2\sum_{j=1}^{n}\frac{1}{\sigma_j^2}\frac{\partial f_j}{\partial p_k}\frac{\partial f_j}{\partial p_i} \tag{5.103}$$

is often used instead of the exact ones given by Equation (5.102). Application of the Formula (5.103) considerably simplifies the numerical procedure because it contains only the first derivatives of the function $f$.

Substituting Equations (5.101) and (5.103) into Equation (5.100), gives the set of equations:

$$\sum_{j=1}^{n}\frac{1}{\sigma_j^2}\left[f_j\left(\mathbf{q}_j, \mathbf{p}\right) - F_j\right]\left[\frac{\partial f_j}{\partial p_i}\right]_{\mathbf{p}^0} + \sum_{k=1}^{m}\Delta p_k \sum_{j=1}^{n}\frac{1}{\sigma_j^2}\left[\frac{\partial f_j}{\partial p_k}\frac{\partial f_j}{\partial p_i}\right]_{\mathbf{p}^0}, \qquad i = 1, 2, \ldots, m. \tag{5.104}$$

These equations are linear with respect to the correction $\Delta p_k$. Their solution is expressed in terms of a matrix inverse to that given by Equation (5.103). Let us rewrite the set of Equations (5.104), using Expression (5.101) and (5.103), as:

$$\left[\frac{\partial S(\mathbf{p})}{\partial p_i}\right]_{\mathbf{p}^0} + \sum_{k=1}^{m} Z_{ik}\left(\mathbf{p}^0\right)\Delta p_k = 0 \qquad i = 1, 2, \ldots, m \tag{5.105}$$

Then one obtains:

$$\Delta p_k = -\sum_{i=1}^{m}\left[\frac{\partial S(\mathbf{p})}{\partial p_i}\right]_{\mathbf{p}^0} Z_{ik}^{-1}\left(\mathbf{p}^0\right) \tag{5.106}$$

Thus, the solution is expressed by means of the matrix elements of the $Z^{-1}$. For its existence, it is necessary that the determinant of the matrix $Z$ be nonvanishing (the condition for the matrix to be nonsingular).[5] After the corrected set of parameters $\mathbf{p}^{(1)} = \mathbf{p}^0 + \Delta\mathbf{p}$ is determined, the new values $f_j\left(\mathbf{q}_j, \mathbf{p}^{(1)}\right)$ are evaluated and the procedure is repeated. Such an iterative procedure is continued until the relative correction $\left|\Delta p_i^{(n)}/\Delta p_i^{(n-1)}\right|$ becomes smaller than a given criterion of convergence. For instance, Le Roy and van Kranendonk [47] used the lsm procedure to fit seven parameters of the anisotropic potential of the system: rare gas atom–$H_2$. They used a unique convergence criterion for all the parameters, equal to 0.00001.

Other problems arise if a strong dependence exists between the parameters. The dependence of a given parameter on the others is characterized by a correlation factor, $R_i$, defined as:

$$R_i = Z_{ii}\left(Z^{-1}\right)_{ii} \qquad R_i \geq 1 \tag{5.107}$$

The matrix $Z$ is denoted as the *information matrix by Fisher*, and the matrix $Z^{-1}$ as the *dispersion matrix*. The diagonal elements of the dispersion matrix, $\left(Z^{-1}\right)_{ii} = \sigma_i^2$, characterize the dispersion $\sigma_i^2 = \left(p_i - \overline{p}_i\right)^2$ of the parameter $p_i$. If the dispersion matrix is normalized in such a way that its diagonal elements become equal to unity, that is, the matrix with elements $(1/\sigma_i\sigma_k)\left(Z^{-1}\right)_{ik}$ is chosen, then the matrix obtained is denoted as the *correlation matrix*. Its off-diagonal elements characterize the degree of correlation between parameters of the corresponding pairs. The degree of dependence of a pair of parameters is proportional to the corresponding off-diagonal element of the correlation matrix. The last statement means that a change in one parameter can be compensated for by the corresponding variation of the other. As an example, consider the correlation matrix for seven parameters derived by Le Roy and van Kranendonk [47]:

| | $\epsilon$ | $R_e$ | $S_1$ | $S_2$ | $a_{13}$ | $a_6$ | $S_3$ |
|---|---|---|---|---|---|---|---|
| $\varepsilon$ | 1 | | | | | | |
| $R_e$ | 0.856 | 1 | | | | | |
| $S_1$ | 0.157 | 0.030 | 1 | | | | |
| $S_2$ | −0.029 | −0.136 | 0.976 | 1 | | | |
| $a_{13}$ | −0.410 | −0.521 | −0.563 | −0.492 | 1 | | |
| $a_6$ | −0.415 | −0.507 | −0.570 | −0.495 | 0.996 | 1 | |
| $S_3$ | −0.034 | −0.119 | 0.326 | 0.260 | −0.125 | −0.193 | 1 |

As follows from the correlation matrix, the strongest dependence exists between parameters of the pairs $S_1, S_2$ and $a_{13}, a_6$. The reasons for such a dependence are discussed in Reference [47].

It should be remembered that consistency reached between a given empirical potential and experimental data is not a justification that an unique solution has

[5] In practice, the ill-conditioned matrices are often found. Determining the inverse matrix is then very difficult. In those cases, the special methods of regularization of matrices have to be applied [153].

been found, because experimental data can be often described by different analytical potentials. In addition, a potential, with parameters fitted to one property, may describe poorly another property because the different regions of the potential curve may be essential for different properties. For instance, the cross-section of high-energy collisions is determined by the short-range repulsive region of the potential curve, while the region of the potential curve near the well is crucial for the description of equilibrium thermophysical properties. To reproduce the potential curve more correctly, it is useful to fit the parameters to several physical properties. In those cases in which the number of physical properties described by the single potential is not smaller than the number of parameters, the latter can be determined by the solution of the corresponding set of equations without the need to use a fitting procedure.

The so-called *minimal maximum error* fitting scheme in conjunction with the lsm procedure to select the best individual fitting quantities (properties) was elaborated by Jellinek and López [154, 155]. The concept of the minimum maximum error was used as a guiding criterion for specifying the weights in the lsm procedure. It was implemented to obtain the optimal values of the parameters in the Gupta potential (Equation (5.75)) for nickel, see details in Reference [155].

In order that the potential with parameters fitted on the basis of experimental data be reliable, the following obvious criteria should be satisfied:

(i) existence a fairly rigorous theory relating the measured property with the intermolecular potential;

(ii) high sensitivity of the measured property to the analytical form of potential;

(iii) rather precise measurements, that is, a small experimental error.

## 5.3 Reconstructing Potentials on the Basis of Experimental Data

In previous section, the problems arising in the determination of model potentials by fitting their parameters were discussed. These problems are connected with an uncertainty in the choice of analytical form and an ambiguity of the fitting procedure. Therefore, the possibility of reconstructing a potential on the basis of experimental data without a preliminary assignment of its analytical form is of great importance. Such a procedure was rigorously elaborated in the theory of scattering in which it is known as the solution of the *inverse scattering problem*. But first, the procedure of reconstructing the potential was developed for potential curves of diatomics.

### 5.3.1 Rydberg–Klein–Rees method

On the basis of a semiclassical approach, Rydberg [26, 156] and Klein [157] developed a graphical method, which allows the experimentally observed vibrational-rotational energy levels to be related to the classical turning points in the movement of the nuclei. That method gives the possibility of reproducing the potential on the basis of spectroscopic data. This graphical procedure was later modified in an analytical form by Rees [158] and is usually designated as the *Rydberg–Klein–Rees (RKR) method.*

A typical potential curve of a diatomic molecule and some notations are presented in Figure 5.12. The effective potential energy has the form:[6]

$$V_{eff}(r) = V(r) + K/r^2 \tag{5.108}$$

where $K/r^2$ is the centrifugal energy, $K = \left(\hbar^2/2\mu\right) J(J+1)$, $J$ is the rotational quantum number and $\mu$ is a reduced mass of a given molecule. The area $P$, between a given vibrational–rotational energy level $E_m$ and the potential curve, is defined by the integral:

$$P = \int_{r_1}^{r_2} \left[E_m - V_{eff}(r)\right] dr \tag{5.109}$$

This integral depends only on two parameters, $E_m$ and $K$. The derivatives of $P$ with respect to these parameters are related in a simple way to the turning points $r_1$ and $r_2$:

$$\left(\frac{\partial P}{\partial E_m}\right)_K = \int_{r_1}^{r_2} dr = r_2 - r_1 \tag{5.110}$$

$$\left(\frac{\partial P}{\partial K}\right)_{E_m} = -\int_{r_1}^{r_2} \frac{dr}{r^2} = \frac{1}{r_2} - \frac{1}{r_1} \tag{5.111}$$

To obtain $r_1$ and $r_2$ by means of Equations (5.110) and (5.111), the area $P$ must be expressed in terms of energy values observed in experiments. Since the number of vibrational–rotational energy levels is very large, the form of the potential well can be reproduced quite satisfactorily.

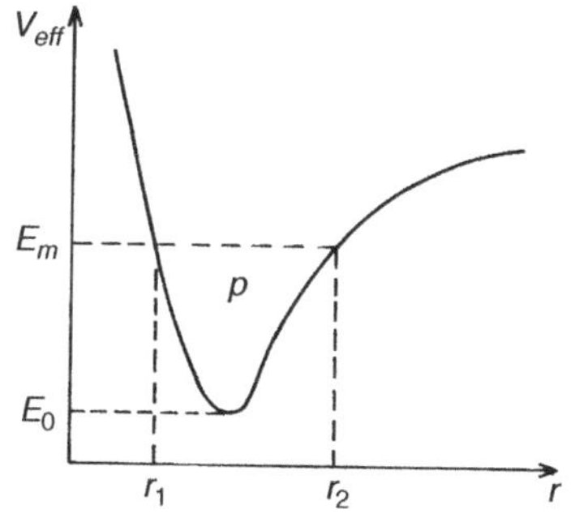


**Figure 5.12** Schematic diagram illustrating the RKR method of reconstruction of diatomic potential curve

[6] Here and below the internuclear distance is denoted by the lowercase letter $r$.

The integrand in Equation (5.109) can be represented as an Euler integral of the first kind:

$$E_m - V_{eff} = \frac{2}{\pi} \int_{V_{eff}}^{E_m} \left( \frac{E_m - E'}{E' - V_{eff}} \right)^{1/2} dE' \tag{5.112}$$

and Equation (5.109) is transformed into:

$$\begin{aligned} P &= \frac{2}{\pi} \int_{r_1}^{r_2} dr \int_{V_{eff}}^{E_m} \left( \frac{E_m - E'}{E' - V_{eff}} \right)^{1/2} dE' \\ &= \frac{1}{\pi} \int_{E_0}^{E_m} dE' \left(E_m - E'\right)^{1/2} \oint \frac{dr}{\left(E' - V_{eff}\right)^{1/2}} \end{aligned} \tag{5.113}$$

where $E_0$ is the value of the potential curve at the minimum.

Let us introduce now the phase integral that satisfies the *Bohr–Sommerfeld quantization rule*:

$$I = \oint P_q dq \equiv \oint \sqrt{2\mu \left(E - V_{eff}\right)} dr = \left( v + \frac{1}{2} \right) \hbar \tag{5.114}$$

It follows from Equation (5.114) that:

$$\frac{dI}{dE} = \sqrt{\frac{\mu}{2}} \oint \frac{dr}{\sqrt{E - V_{eff}}} \tag{5.115}$$

The relation (5.115) permits the unknown function $V_{eff}(r)$ in Equation (5.113) to be eliminated:

$$P = \sqrt{\frac{2}{\pi^2 \mu}} \int_{E_0}^{E_m} \left(E_m - E'\right)^{1/2} \frac{dI}{dE'} dE' \tag{5.116}$$

or

$$P = \sqrt{\frac{2}{\pi^2 \mu}} \int_{0}^{I} \left(E_m - E'\right)^{1/2} dI' \tag{5.117}$$

where $I$ is the magnitude of the phase integral corresponding to the energy $E_m$ and it is obtained from the condition $E'\left(I', K\right) = E_m$. The vibrational–rotational energy can be expressed as a polynomial in the rotational $(J)$ and vibrational $(v)$ quantum numbers:

$$E_{vJ} = \sum_{l,n} a_{ln} \left(v + \tfrac{1}{2}\right)^l \left[J\left(J+1\right)\right]^n \tag{5.118}$$

By introducing $I$ via Equation (5.114) and $K$ via its expression through $J$, Equation (5.118) is represented as:

$$E\left(I, K\right) = \sum_{l,n} b_{ln} I^l K^n \tag{5.119}$$

Substituting Equation (5.119) in place of $E'$ into Equation (5.117) and differentiating the obtained integrals with respect to $E_m$ and $K$, one determines the points

$r_1$ and $r_2$ of the potential curve that correspond to the experimental value $E_m$ of energy. The effective procedure of calculating the integrals, appearing in the RKR method, which removes their divergence at an upper bound, has been developed by Kaminsky [159]. The subsequent developments in the RKR procedure have been presented in a review [43].

The approximate procedure of reconstructing the potential surface of a polyatomic molecule on the basis of its vibrational–rotational spectrum, which is a generalization of the RKR method to the coupled modes of polyatomic molecules, was developed by Gerber *et al.* [160].

## 5.3.2 Inverse scattering problem

### 5.3.2.1 General statement of the problem

According to the quantum theory of scattering [161, 162], all the information concerning the scattering potential is contained in the phase shifts $\delta_1$. The scattering amplitude, which determines the scattering cross section measured in experiments, is expressed in terms of the phase shifts via the Faxén–Holtsmark relation. The integral relations, which connect directly the scattering amplitude with the scattering potential, can be obtained only in the two following extreme cases:

(i) the Born approximation, when $\delta_1 \ll 1$;

(ii) the semiclassical approximation, when $\delta_1 \gg 1$.

In all other cases, the scattering amplitude is related indirectly to the scattering potential by means of the phase shifts. The procedure of obtaining the scattering amplitude and the phase shifts from the experimental values of the scattering cross section is a complex problem [163, 164].

The inverse problem in the quantum theory of the scattering consists of the reconstruction of the potential from the S-matrix of scattering phases. It was first formulated by Frëdberg [165] and Hylleraas [166]. Bargmann [167] discussed the ambiguity of its solution and found a class of equivalent potentials which possess the same scattering phases and discrete spectra. As shown by Levinson [168], the existence of the bound states is one of the reasons of the lack of uniqueness in the determination of potential.

An original approach to the inverse problem was developed by Gel'fand and Levitan [169]. They solved analytically the problem of reproducing the potential from the phase shifts for states with a fixed angular momentum. This method is presented in detail in monographs [170, 171]. Below the general formulation will be restricted to the particular case with $l = 0$.

Instead of using phase shifts, the Gel'fand–Levitan method [169] is based on the so-called spectral function, $\rho(E)$. As a consequence of the completeness of the set of wave functions corresponding to the discrete and continuous spectrum of the radial Schrödinger equation with the potential $V(R)$, the following relationship of completeness exists:

$$\int \xi(E,r)\,\xi\left(E,r'\right)\frac{d\rho(E)}{dE}dE = \delta\left(r-r'\right) \tag{5.120}$$

where $d\rho/dE$ is a weight function needed for the condition (5.120) to hold. The spectral function $\rho(E)$ is determined in a unique way if the normalization constants of the wave functions of the bound states and the so-called Jost function, which is defined in terms of phase shifts and bound-state energies, are known. Assuming that $\rho(E)$ is known, Gel'fand and Levitan proved that the required potential $V(r)$ can be represented as:

$$V(r) = 2\frac{d}{dr}K\left(r,r'\right) + V_1(r) \tag{5.121}$$

where $V_1(r)$ is an arbitrary potential with the spectral function $\rho_1(E)$, the function $K\left(r,r'\right)$ is the solution of the Fredholm-type integral equation:

$$K\left(r,r'\right) = g\left(r,r'\right) + \int_0^r K(r,x)\,g\left(x,r'\right)dx \tag{5.122}$$

with the auxiliary function:

$$g\left(r,r'\right) = \int \xi_1(E,r)\,\xi_1\left(E,r'\right)\left[d\rho_1(E) - d\rho(E)\right] \tag{5.123}$$

determined from the known solution $\xi_1(E,r)$ of the radial Schrödinger equation with the potential $V_1(r)$. The solution obtained is unique. In addition, for each set of phase shifts $\delta_1$ and energies $E_m$ of bound states there exists a family of $n$-parameter, phase-equivalent potentials ($n$ is the number of bound states). The inverse scattering problem and the difficulties associated with it have been discussed extensively in the literature [172–174].

The inverse problem for a fixed value of the energy and an arbitrary value of the angular momentum was studied by De Alfaro and Regge [170] and Newton [171]. If the wave functions, corresponding to the states with a fixed value of the angular momentum and different energy, form a complete set; the wave functions of the states with a fixed value of the energy and all possible values for the angular momentum do not generate a complete set. This fact does not permit the theorem of completeness (Equation (5.120)) to be applied to the determination of the spectral function and hinders the solution of the inverse problem. The number of papers devoted to solving the inverse problem for a fixed energy value is quite large; however, the procedures developed are unstable with respect to a small dispersion of experimental data and give an infinite set of equivalent solutions [163, 164].

It should be mentioned that in spite of the fact that the inverse scattering problem is formulated in a rather elegant mathematical way, its practical solution offers great difficulties. On one hand, these procedures require completeness of the experimental data used. For instance, the Gel'fand–Levitan method is based on the knowledge of the phase shifts $\delta_l$ for the energies of all the bound states. The problem concerning the influence of incompleteness of information on the form of the solution remains open. On the other hand, a rather complicated integral equation must be solved.

And, finally, to remove the uncertainties connected with obtaining a family of equivalent potentials, additional information about the bound states is needed.

As a result of the problems mentioned above, in practice, the approximate methods of solving the inverse problem, based on semiclassical approaches, are applied.

#### 5.3.2.2 Quasi-classical treatment: the Firsov approach

Hoyt [175] was the first to solve the inverse problem for the classical scattering in the case of monotonic potential. Hoyt's approach is based on the knowledge of the differential cross section $\sigma(\theta)$ for different values of the energy $E$. However, Firsov [176] proposed a more convenient method, which requires a knowledge of $\sigma(\theta)$ for a single value of the energy $E$. Later on, many other authors studied the problem of reconstructing the potential using semiclassical and classical treatments. These practical methods are discussed in detail by Buck [163, 164].

Let us consider the Firsov approach. It is based on the classical treatment of the nuclear motion, valid if the angular momentum of the nuclear motion is $J \gg \hbar$. The scattering angle $\theta$ must be considerably greater than $\hbar/J$. These conditions are satisfied almost always in the scattering experiments of atoms and molecules.

In the quasi-classical approximation of the scattering theory [162], the scattering angle $\theta$ is connected by an integral relation with the central potential $V(r)$ and the sighting parameter $\rho$ (for its definition see in Figure 5.13). In the Firsov approach, this relation was chosen as the starting point. It can be written in the form:

$$\theta = \pi - 2\rho \int_{r_0}^{\infty} \frac{r^{-1}dr}{\sqrt{\left[1 - \frac{V(r)}{E}\right] r^2 - \rho^2}} \tag{5.124}$$

where $E = mv_{\infty}^2/2$ and $r_0$ is the smallest distance determined from the condition that the radicand vanish, that is, $\left[1 - V(r_0)/E\right] r_0^2 = \rho^2$. For convenience, a new

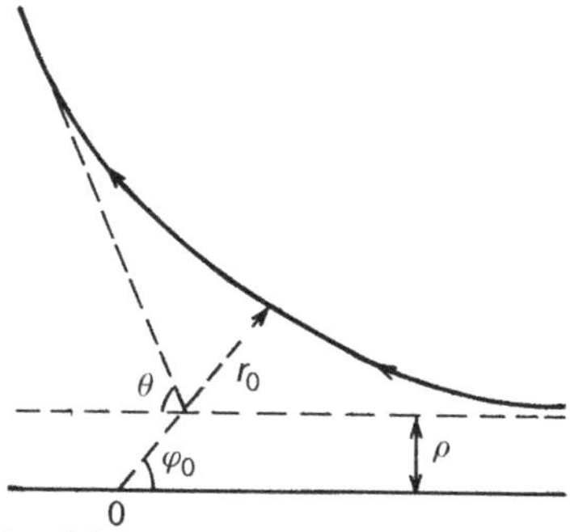


$\rho$ is the sighting parameter, $\theta$ is the scattering angle, and $r_0$ is the smallest distance between the center and particle

**Figure 5.13** Classical scattering of particle on the force center located at point $O$

variable can be introduced:

$$u = \left[1 - \frac{V(r)}{E}\right] r^2 \tag{5.125}$$

The change of variable corresponds to integration from $\rho^2$ to $\infty$. Replacing the variable in integral (5.124) gives:

$$\theta = \pi - 2\rho \int_{\rho^2}^{\infty} \frac{d\ln r}{du'} \frac{du'}{\sqrt{u' - \rho^2}} \tag{5.126}$$

If the following identity:

$$\rho \int_{\rho^2}^{\infty} \frac{d\ln u'}{du'} \frac{du'}{\sqrt{u' - \rho^2}} = \pi \tag{5.127}$$

which is proved fairly easy, is taken into account, Equation (5.126) can be rewritten:

$$\theta = \int_{\rho^2}^{\infty} \frac{d\ln\left(u'/r^2\right)}{du'} \frac{\rho du'}{\sqrt{u' - \rho^2}} \tag{5.128}$$

Operating on Equation (5.128) with $d\rho/\sqrt{\rho^2 - u}$ and integrating over $\rho$ from $\sqrt{u}$ to $\infty$ one obtains:

$$\int_{\sqrt{u}}^{\infty} \frac{\theta(r)\,d\rho}{\sqrt{\rho^2 - u}} = \int_{\sqrt{u}}^{\infty} \frac{\rho d\rho}{\sqrt{\rho^2 - u}} \int_{\rho^2}^{\infty} \frac{d\ln\left(u'/r^2\right)}{du'} \frac{du'}{\sqrt{u' - \rho^2}} \tag{5.129}$$

Using the Dirichlet theorem [177], it can be proven that the double integral on the right-hand side of Equation (5.129) is equal to $(\pi/2)\ln\left(r^2/u\right)$. Therefore:

$$\ln\frac{r^2}{u} = \frac{2}{\pi} \int_{\sqrt{u}}^{\infty} \frac{\theta(\rho)\,d\rho}{\sqrt{\rho^2 - u}} \tag{5.130}$$

or

$$r(u) = \sqrt{u}\exp\left[\frac{1}{\pi} \int_{\sqrt{u}}^{\infty} \frac{\theta(\rho)\,d\rho}{\sqrt{\rho^2 - u}}\right] \tag{5.131}$$

To obtain $r(u)$, then $u(r)$ and finally $V(r)$, the experimental dependence $\theta(\rho)$ must be known. In the classical theory of scattering, the sighting parameter is related to the differential cross-section of scattering $d\sigma/d\theta$, measured in the experiments, via the simple expression:

$$\rho^2 = \frac{1}{\pi} \int_{\theta}^{\pi} \frac{d\sigma(\theta)}{d\theta} d\theta \tag{5.132}$$

The relation (5.132) closes the chain of equalities that are required to reconstruct the potential.

Thus, Firsov's approach suggests the following order of operation for the reconstruction of the potential:

(i) the experimental dependence $\sigma(\theta)$ is substituted into Equation (5.132) and $\rho(\theta)$ and, after inversion, $\theta(\rho)$ are determined.

(ii) $\theta(\rho)$ is substituted into Equation (5.131) and $r(u)$ is obtained. Inverting $r(u)$, one determines the function $u(r)$.

(iii) $V(r)$ is found from $u(r)$ via Equation (5.125).

Firsov tested his procedure on the following example. He assumed that the experimental dependence is described by the Rutherford formula [161, 162]:

$$\sigma(\theta) = \frac{q^2}{16E^2}\frac{1}{\sin^4(\theta/2)} \tag{5.133}$$

It is well known that the Rutherford cross section corresponds to scattering in a Coulomb field. Let us determine the scattering field via the procedure described above. Substituting Equation (5.133) into Equation (5.132) one obtains:

$$\rho^2 = \frac{q^2}{8E^2}\int_0^{\pi}\frac{\sin\theta d\theta}{\sin^4(\theta/2)} = \frac{q^2}{4E^2 tg^2(\theta/2)}$$

or

$$\theta = \pm 2\arctan\frac{q}{2E\rho} \tag{5.134}$$

Then one evaluates the integral appearing in Equation (5.131) with function (5.134):[7]

$$\frac{1}{\pi}\int_{\sqrt{u}}^{\infty}\frac{\theta(\rho)\,d\rho}{\sqrt{\rho^2-u}} = \frac{1}{\pi}\int_{\sqrt{u}}^{\infty}\frac{\pm\arctan(q/2E\rho)}{\sqrt{\rho^2-u}}d\rho = \ln\frac{\sqrt{(q/2E)^2+u}\pm q/2E}{\sqrt{u}}$$

According to Equation (5.131):

$$r(u) = \sqrt{(q/2E)^2+u}\pm q/2E \tag{5.135}$$

Inverting $r(u)$ into $u(r)$ and substituting this result into the definition of $u$ (Equation (5.125)), gives finally the exact analytical potential:

$$V(r) = \pm\frac{q}{r} \tag{5.136}$$

It is a very impressive illustration, although the analytical result for the reconstructed potential follows from the possibility in this case of substituting in Equation (5.132) the analytical function for $\sigma(\theta)$.

Firsov's approach was applied with great success to the reconstruction of the scattering potential of atoms and ions of rare gases by Lane and Everhart [178]. For this procedure to be unique, the function $\theta(\rho)$ and the function $u(r)$, and therefore $V(r)$, have to be monotonic. Otherwise, it is impossible to determine in a unique way the inverse functions $\rho(\theta)$ and $r(u)$. Buck [163, 164, 179] modified Firsov's procedure for a class of oscillating deviation functions. The potential for the Cs–Hg system restored by Buck *et al.* [180] is presented in Figure 5.14. The authors [180] used the experimental cross sections for five different energy

[7] The integral is handled through differentiation on the parameter $q/2E$.

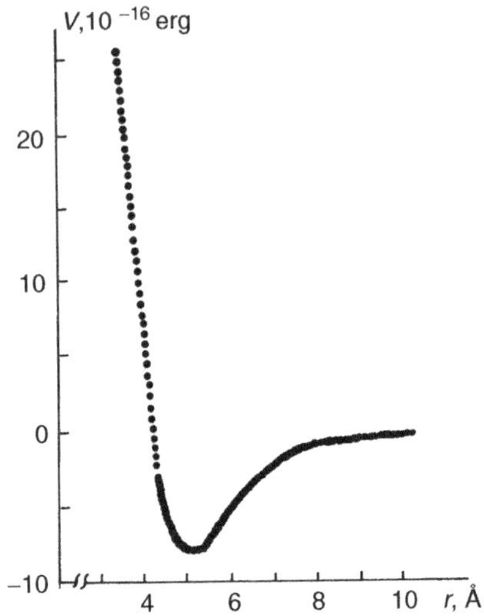


**Figure 5.14** Potential curve for the Cs–Hg system reconstructed via the modified Firsov procedure [180] using experimental cross sections for five different energy values

values. In fact, the potentials, reproduced from each cross section, overlap each other. The thickness of the line for the potential curve in the figure reflects this fact.

The procedure described above allows the reconstruction of the isotropic potentials, that is, the potentials that depend only on the distance. Buck and co-workers [181] developed a method of reconstructing the anisotropic potential of the interaction between an atom and a diatomic molecule, expressed as:

$$V(r, \theta) = V_0(r) + V_2(r)\ P_2(\cos\ \theta)$$

To reconstruct the potential (or the potential surface, in fact), the experimental data for the differential cross sections of elastic and inelastic scattering involving rotational molecular states are used. This method is applicable to molecules with large separations between their rotational levels ($H_2$, $D_2$, HD, and, to a considerable extent, $N_2$ also belong to this class). The suggested two-step procedure [181] can be depicted in the following form:

| Experimental differential cross sections | $\Longrightarrow$ | S-matrix elements | $\Longrightarrow$ | Potential being sought |
|---|---|---|---|---|

The use of the semiclassical approach and the method of exponentially-distorted waves makes it possible to express directly the S-matrix elements in terms of measured cross sections. The potential is then determined in the following way. The isotropic component of the potential $V_0(r)$ is determined by the total differential cross section and the anisotropic potential $V_2(r)$ by the ratio of the inelastic differential cross section to the elastic one. Some assumptions (the use of the exponentially-distorted wave approximation and consideration of a single channel of inelastic scattering) simplify considerably the numerical application of this method, although at the same time restricting its applicability. The procedure was tested for the Ne–$D_2$ system [181]; yielded the accuracy of the anisotropic component of the potential to within a few percentage points.

### 5.3.3 Reconstructing potentials from thermophysical data

For a long time, thermophysical data were traditionally used to determine the parameters of empirical potentials. The procedure of reconstructing potentials from thermophysical measurements has been initiated after Gough *et al.* [182] proposed an effective method of reconstructing a potential from viscosity data. Their idea is as follows.

The viscosity may be expressed as:

$$\eta \sim \sqrt{kT}\,\overline{\Omega}^{(2,2)}\left(T, V(r)\right) \tag{5.137}$$

where $\overline{\Omega}^{(2,2)}$ is the reduced collision integral, which is defined as a ratio of the collision integral $\Omega^{(2,2)}$, evaluated for the real potential, to that obtained for hard spheres with a certain diameter $d$ [45, 183]. The integral $\overline{\Omega}^{(2,2)}$ is connected with the interaction potential by means of a triple integration. Gough *et al.* showed that if it is assumed that at a fixed temperature the collision integral is equal to the square of an effective distance $\overline{r}$, that is:

$$\overline{\Omega}^{(2,2)}\left[T, V(r)\right] = \overline{r}^2 \tag{5.138}$$

the potential as a function of $\overline{r}$ can be represented in the form:

$$V(\overline{r}) = G\left(T^*\right)kT \tag{5.139}$$

where the function $G(T^*)$ depends, in fact, mainly on the single parameter $T^* = kT/\epsilon$, where $\epsilon$ is the depth of the potential well, and only very slightly on the form of the potential (a test, carried out for different potentials, yields a maximum dependence of about 10%).

Under the assumption that $\epsilon$ is evaluated independently, the relations (5.137) to (5.139) allow the following iterative procedure to be proposed to reconstruct the potential. As a first step, one chooses the trial potential $V_0(r)$ and evaluates the function $G(T^*) = V(\overline{r})/kT$. On the basis of the experimental magnitude for $\eta(T)$ and relationship (5.137), the experimental value of the collision integral $\overline{\Omega}^{(2,2)}_{\text{exp}}$ is obtained. Substituting then the set of values for $G_0(T^*)$ and $\overline{\Omega}^{(2,2)}_{\text{exp}}(T)$ for various $T$ into Equations (5.138) and (5.139), one obtains a set of values $(V_1, \overline{r})$, which gives the first-order approximation $V_1(\overline{r})$ to the potential. Each experimental point $\eta(T)$ corresponds to one point of the reconstructed potential. In a second step,

$G_1(T^*)$ for $V_1(\bar{r})$ is obtained via the relation (5.139) and the reduced collisions integral $\overline{\Omega}^{(2,2)}(T, V_1(\bar{r}))$ is evaluated. The new values for the collision integral and the function $G_1(T^*)$ allow the next approximation to the potential $V_2(\bar{r})$ to be found via relations (5.138) and (5.139). The procedure is continued until a given value of convergence is achieved.

Veihland *et al.* [184] developed a procedure for reconstructing the potential on the basis of experimental data on the mobility of ions. This procedure is similar to the one presented above for viscosity. The analysis of the inversion procedure for reconstructing model potentials in polyatomic gases and more details of reconstruction can be found in the book by Maitland *et al.* [2].

## 5.4 Global Optimization Methods

### 5.4.1 Introduction to the problem

To find the most stable conformation of a many-atomic cluster or macromolecule, such as proteins, one has to perform a global optimization [185, 186], that is, find the deepest minimum on the potential energy surface (PES). The PES is usually approximated by semiempirical model potentials, although it can be also calculated for each conformation by some effective quantum-mechanical computer program. For many-atomic systems, the PES is a multidimensional surface possessing many local minima.

From a mathematical viewpoint, the problem reduces to the minimization of the function of many variables having many minima, thus it is the multi-minima problem [187]. It cannot be solved by scanning all local minima, because the number of local minima increases exponentially with the number $N$ of atoms. For the PES of the 13-atom cluster described by the Lennard-Jones potential (denoted as $LJ_{13}$), Tsai and Jordan [188] enumerated about $10^3$ minima. The extrapolation of their approach to larger clusters leads to tremendous numbers: the $LJ_{55}$ cluster has at least $10^{12}$ minima [189] and the PES of the $LJ_{147}$ cluster possesses of the order of $10^{60}$ minima, or much more for another functional form [190]. Such problems are known in mathematics as *nondeterministic polynomial-time complete (NP-complete) problems* [191]. This class of problem may not be solved using polynomial-time algorithms. The optimization problems connected with solution of NP-complete problems are called *NP-hard*.

A well-known problem that belongs to the class of NP-hard problems is the *travelling salesman problem*. Its statement is the following. Given a list of $N$ cities and a means of calculating the cost of travelling between any two cities, one must compose the salesman's route, which has to through each city once and return finally to the initial point, minimizing the total cost. The main difficulty in finding an exact solution to this problem is that methods for determining an optimal route require a computing effort that increases exponentially with $N$.

The object of combinatorial optimization is to develop an efficient technique for finding the extreme values (maximum and minimum) of a function of very many independent variables [192, 193]. This function called the *cost function* or *objective function*. It represents a quantitative measure of the 'goodness' of some complex system and depends on the configuration of all constituent parts of the system. In the conformation analysis of clusters and macromolecules, the role of cost function is played by the PES, the global minimum of which has to be found. Wille and Vennik [194] proved that the global minimum search is an NP-hard problem. They showed that the global minimum problem contains the salesman travelling problem as a special case. This means that no polynomial-time algorithms solving this problem can be found. Thus, instead of searching direct algorithms, some heuristic approaches to the global minimum problem, giving a near-optimal solution in polynomial time, are needed.

There are two basic strategies applied in heuristic methods [195]: iterative improvement and 'divide-and-conquer' strategy. In the first, one starts with some initial configuration of the system and rearranges all parts of the system in turn until a rearranged configuration that improves the cost function is found. The rearranged configuration then becomes the starting one, and the process is continued until no further improvement can be found. There is a great probability that the process will stop in a local but not global minimum. Therefore, it is necessary to carry out this iterative procedure several times, starting from different randomly generated initial configurations, and save the best result.

In the 'divide-and-conquer' approach, one divides the problem into subproblems of manageable size and solves each subproblem separately. The solutions to the subproblems must then be patched back together. For this method to produce a good solution, the subproblems must be naturally disjointed, and the division made must be an appropriate one, so that errors made in patching do not offset the gains obtained in applying more powerful methods to the subproblems [196].

The global optimization methods for finding the most stable conformation deal with the PES. One can expect, and it has been really manifested, that the efficient route for the global minimum search depends upon the specific structure of the PES for the system under study. Successful heuristic methods applied to the minimization of PES's take into account its detailed geometry. On the other hand, most of these optimization methods are explicitly or implicitly based on one or both heuristic strategies described above.

Below we present a short description of the main families of global optimization methods: the simulation annealing and, related to it, quantum annealing, hypersurface deformation approaches and genetic algorithms.

### 5.4.2 Simulated annealing

This simulation method is based on the connection of statistical behavior of many-atomic system in the thermal equilibrium at a finite temperature with combinatorial optimization of many-variable function. It was developed by Kirkpatrick *et al.* [195], which started from the direct analogy with the experimental process of growing a single crystal from a melt. In this physical process, at first, the material is melted, then the temperature is slowly lowered, spending a long time in the vicinity of the freezing point in order to avoid the formation of thermodynamically metastable states (in mathematical language it is equivalent to avoiding the local minima on the PES).

The authors [195] exployed the Metropolis Monte Carlo algorithm [197] developed at the earliest time of computer calculations. In the Reference [197], the canonical ensemble of $N$ particles was considered at fixed temperature $T$. The average energy, or any quantity of interest, was defined using the Boltzmann $\exp(-E/kT)$ distribution, and the $3N$-dimensional integrals were calculated by the Monte Carlo method. At each step of this algorithm, the atoms undergo in succession small random displacements and the resulting change in the energy of the system $\Delta E = E_{i+1} - E_i$ is computed. If $\Delta E < 0$, i.e., if the move brings the system to a state of lower energy, the displacement is accepted and the configuration with the displaced atom is used as the starting point of the next step. If $\Delta E > 0$, this move is treated with the probability:

$$P(\Delta E) = \exp\left[-\Delta E/kT\right] \tag{5.140}$$

using the generator of random numbers in the interval between 0 and 1. If a random number $\xi_i < P(\Delta E)$, the particle is moved to its new position. If $\xi_i > P(\Delta E)$, it is returned to its old position. By repeating the basic step many times, one simulates the thermal motion of atoms at constant $T$ permitting the system to evolve into the Boltzmann distribution.

The concept of temperature is naturally introduced in a canonical ensemble of $N$ atoms when $N$ is a large number. In the case of an $N$-atomic cluster, the temperature can be also introduced, but it will be some effective temperature. As is well-known in classical statistics, the ideal gas has the permanent specific heat $c_v = \frac{1}{2}$ per each degree of freedom [198]. This is the so-called *uniform distribution law*. The energy of an ideal gas of $N$ atoms is equal to:

$$E = N\epsilon_0 + Nc_v kT \tag{5.141}$$

where $\epsilon_0$ is the energy of atom and $c_v = \frac{3}{2}$ is its specific heat (the atom has three translation degrees of freedom). If atoms are in a nondegenerate state, it is possible, by a simple energy shift, to put $\epsilon_0 = 0$. Then Equation (5.141) transforms into:

$$E = \frac{3}{2}NkT \tag{5.142}$$

The energy is distributed uniformly among $3N$ degrees of freedom, $\frac{1}{2}kT$ per degree. The effective temperature for an arbitrary many-atom cluster can be defined, if we assume the uniform distribution law for the internal cluster energy. An $N$-atom cluster has $3N - 6$ ($3N - 5$ for the linear geometry) internal degrees of freedom. Hence, its average energy can be represented as:

$$\langle E \rangle = \frac{3N - 6}{2} kT_{eff} \tag{5.143}$$

or

$$kT_{eff} = \frac{2\langle E \rangle}{3N - 6} \tag{5.144}$$

Using the Metropolis algorithm with the effective cluster temperature and an analogy with the physical annealing process, Kirkpatrick *et al.* [195] developed the optimization method called *simulated annealing* (SA). It starts at a high effective temperature sufficient for 'melting' the system being optimized. Then, the temperature is lowered by slow stages until the system 'freezes' and no further changes occur. At each temperature, the simulation must proceed long enough for the system to reach a steady state. The application of 'thermal fluctuations' help the system to avoid becoming trapped in local minima.

One of the first applications of the SA method was carried out by Biswas and Hamann [199]. They performed the SA of silicon clusters with a Langevin molecular-dynamic approach. The SA procedure is easily implemented into standard Monte Carlo or molecular dynamics codes [200–202]. For these reasons, the optimization procedure has been used for a large variety of systems; for recent applications see References [203, 204].

It was manifested that SA is effective in finding the global minimum of multidimensional function having large numbers of local minima. Although, its effectiveness depends to a great extent upon the structure of PES. In the case of several deep minima, the system processed by the SA procedure can be trapped in some deep local minimum but not global one.

To the family of SA methods, can be also attributed the so-called *quantum annealing* (QA) approach [205–208]. This approach is more robust than classical SA with respect to avoiding local minima because exploits the quantum-mechanical tunneling and delocalization. Amara *et al.* [205] presented a method for finding the global minimum via an approximate solution of the Schrödinger equation in imaginary time. The wave function was expressed as an antisymmetrized product of single-particle Gaussian wave packets. While evolving in time, the wave packets tunnel through barriers, expand and contract in search of the global minimum. The classical global minimum is then found by setting the Planck constant, $h$, equal to zero.

The method developed in References [206, 207] does not use the wave functions. It is based on the observation that the Schrödinger equation written in imaginary time is isomorphic to the diffusion equation with the growth/depletion term. The diffusion Monte Carlo method [209] is one, relatively simple technique for treating such problems.

Ma and Straub [210] developed the classical analog of the QA method of solving the Schrödinger equation in imaginary time [205], combining it with the simulated annealing of the classical density distribution. While the classical SA is based on the molecular dynamics of the single point particle, the classical density distribution provides information on an ensemble of systems. The density distribution is found from the solution of the Liouville equation using the Gaussian density annealing algorithm. The comparison of this approach performed in Reference [210] with the SA, based on molecular dynamics for LJ clusters, demonstrated that the probability of finding the global minimum is essentially enhanced.

Recently, Liu and Berne [211] suggested a new version of QA based on the Feynman path integral approach [212]. The latter allows one to employ the path integral Monte Carlo sampling, which efficiently samples barrier crossing events and avoids local trapping by means of local minimization of individual imaginary time slices. This method, named by authors [189] as *quantum path minimization* (QPM), demonstrated very high efficiency. It was able to locate the global minima of all LJ clusters of size up to $N = 100$, except for $N = 76, 77$ and 98. Thus, QPM can be comparable with the best global optimization methods (see Sections 5.4.3 and 5.4.4).

## 5.4.3 Hypersurface deformation methods

This family of global optimization methods solved the problem by an appropriate transformation of the cost function. The general strategy of this transformation was formulated by Stillinger and Weber [213]. In the case of some potential surface (hypersurface), it has to be a search for a continuous deformation of the hypersurface possessing the two following properties:

(a) shallow local minima are drastically reduced in size or are eliminated altogether;

(b) global minimum continuously transforms and, in an ideal case, the deformed hypersurface exhibits only a single minimum that can be traced back to the desired global minimum as the deformation continuously switches off.

There are various ways of realizing this strategy. Two rather different approaches are discussed below.

### 5.4.3.1 Diffusion equation method

The ideas outlined above were independently developed by Piela *et al.* [214] in a practical working method. They suggested an algorithm that was based on the deformation of the original hypersurface in such a way as to make shallow wells gradually disappear and, finally, the process ends up with a single potential well. In most cases, this single minimum arises from the one that corresponds to the original global minimum.

For simplicity, it is better to consider first a one-dimensional problem with a function $f(x)$. The transformation that destabilizes any potential well of $f(x)$ by decreasing its depth can be defined as:

$$f^{[1]}(x) = f(x) + \beta f''(x) \qquad \beta > 0 \tag{5.145}$$

where $f''(x)$ denotes $d^2 f(x)/dx^2$. The values of the inflection points do not undergo any change since they corresponds to $f'' = 0$, while the regions of the curve where the function is convex or concave go up or down, respectively. Minima become shallower because they correspond to $f'' > 0$.

The authors [214] illustrated their method on the numerical example:

$$f(x) = x^4 + 2x^3 + 0.9x^2 \tag{5.146}$$

Let us consider in detail the result of action of the transformation (5.145) on the function (5.146). The extrema of function (5.146) are easily found as roots of the equation:

$$f'(x) = x\left(4x^2 + 6x + 1.8\right) = 0$$

It has three roots: $x_1 = 0$, $x_2 = -0.415$, and $x_3 = -1.085$. The second derivative is equal at these points to:

$$f''(x_1) = 1.8 > 0, \quad f''(x_2) = -1.11 < 0, \quad \text{and} \quad f''(x_3) = 1.11 > 0 \tag{5.147}$$

Hence, at points $x_1$ and $x_3$, $f(x)$ has minima and at $x_2$ it has a maximum with the following values:

$$f(x_1) = 0, \quad f(x_2) = 0.042, \quad \text{and} \quad f(x_3) = -0.109 \tag{5.148}$$

The depth of a potential well can be defined as the difference between the values of the lowest neighboring maximum and minimum of the potential well. So, the depth of the shallower minimum at $x_1 = 0$ is equal to 0.042.

According to Equations (5.147) and (5.148), the transformed function at points $x_1$ and $x_2$ is equal to:

$$f^{[1]}(x_1) = 1.8\beta \quad \text{and} \quad f^{[1]}(x_2) = 0.042 - 1.114\beta$$

and the depth of the minimum of $f^{[1]}(x)$ at $x_1 = 0$ is equal to:

$$\Delta f^{[1]} = f^{[1]}(x_2) - f^{[1]}(x_1) = 0.042 - 2.914\beta$$

$\Delta f^{[1]}$ is equal to zero at $\beta = 0.014$. Hence, if the location of extrema of the potential curves $f^{[1]}(x)$ and $f(x)$ is not be significally changed, it can be expected that at values of parameter $\beta$ larger than 0.014, for instance at $\beta = 0.02$, the second shallower minimum will disappear. As follows from Figure 5.15, it is really the case.

The transformation (5.145) may be repeated for the new curve and, in the *Nth* iteration, it gives:

$$f^{[N]}(x) = \left(1 + \beta\frac{d^2}{dx^2}\right)^N f(x) \qquad \beta > 0 \tag{5.149}$$

The procedure is most effective when $\beta$ is small and $N$ is large. Taking $\beta = t/N$, with $t > 0$ being a parameter, Equation (5.149) may be transformed into:

$$F(x,t) = \lim_{N\to\infty}\left(1 + \frac{t}{N}\frac{d^2}{dx^2}\right)^N f(x) = \exp\left[t\frac{d^2}{dx^2}\right] f(x) \tag{5.150}$$

where the action of an exponential operator is defined, as usual, by expanding it into the Taylor series:

$$\exp T = 1 + T + \frac{1}{2!}T^2 + \frac{1}{3!}T^3 + \cdots \tag{5.151}$$

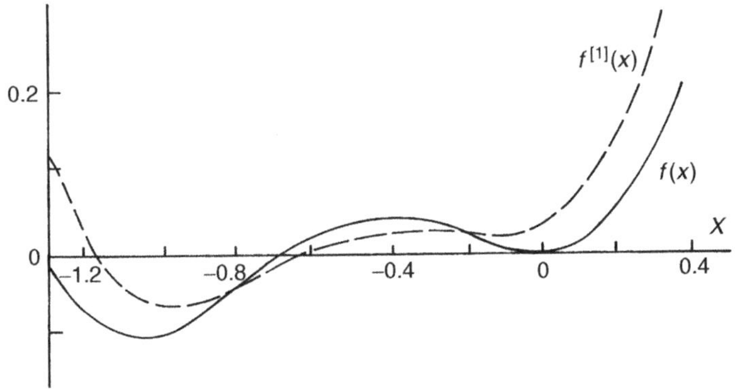


**Figure 5.15** Transformation (5.145) applied to the Function (5.146) [214]. For $\beta = 0.02$, the transformed function $f^{[1]}(x)$ exhibits only one minimum

As shown in Reference [214] by verifying the Taylor expansion of the right-hand side of Equation (5.150) term by term, its sum (when it does converge) is a solution of the diffusion or heat conductivity equation:

$$\frac{\partial^2 F}{\partial x^2} = \frac{\partial F}{\partial t} \tag{5.152}$$

with the initial condition $F(x, 0) = f(x)$. This is the reason that this method was called by the authors [214] the *diffusion equation method* (DEM). For the $n$-dimensional problem, the second derivative operator has to be replaced by the $n$-dimensional Laplacian:

$$\Delta = \sum_{i=1}^{n} \frac{\partial^2}{\partial x_i^2} \tag{5.153}$$

and the diffusion equation becomes:

$$\Delta F = \frac{\partial F}{\partial t} \tag{5.154}$$

The DEM procedure has been applied to a variety of mathematical functions [214, 215] and to $LJ_N$ clusters [216]. In the last case, the Lennard-Jones potential was approximated by a sum of Gaussians. The global minima were found for all studied clusters up to $N = 55$. For $LJ_{55}$, the global minimum corresponds to the Mackay icosahedron [217] (see Table A2.3 in Appendix 2).

In some particular cases, the global minimum may not be found. As mention by the authors [214], it may be when the global minimum belongs to a narrow potential well of large depth. This narrow potential well may disappear earlier than some wider originally shallower potential well. The smoothed surface must be mapped back to the real surface [218], and it can be a challenge. In Reference [219] it was shown that some of the smoothing procedures can change the global minimum. Hence, it is necessary to couple smoothing with an efficient local search procedure and apply it in mapping minima back from the deformed surface to the original one. To improve efficiency, more than one minimum of the smoothed surface must be tracked backwards [219, 220].

A discussion of the DEM approach and its modifications see in References [220–222]. The QA global minimization methods [205, 206, 210] discussed in the end of previous section can be also attributed to the DEM family.

#### 5.4.3.2 Basin-hopping algorithm

Before presenting this approach, it is necessary to describe the useful mapping from the continuous configuration space onto the discrete set of its local minima and give some definitions [223, 224].

The potential energy of the system composed of $N$ atoms, $V(\mathbf{r})$, is presented as a hypersurface in the $3N-6$ dimensional conformation space $\mathcal{R}^{3N-6}$ ($3N-5$ dimensions for a linear geometry) where only the internal degrees of freedom are taken into account and $\mathbf{r}$ denotes the set of $3N$ atomic coordinates. The potential hypersurface possesses a multitude of local minima which can be described as a discrete set indexed by $\alpha$. The map from the continuum $\mathcal{R}^{3N-6}$ space to the discrete set of minima $\{\alpha\}$:

$$V(\mathbf{r}) \Longrightarrow \widetilde{V}[\{\alpha\}] \tag{5.155}$$

is defined by direct minimization from any point $\mathbf{r}$ along a steepest-descent path to the nearest local minimum $\alpha$. Let $R(\alpha) \subset \mathcal{R}^{3N-6}$ denote the set of configurations, which map to a local minimum $\alpha$. $R(\alpha)$ is a connected set, since all $\mathbf{r} \in R(\alpha)$ are commuted by a path through $\alpha$, and the different $R(\alpha)$ are disjoint. The $R(\alpha)$ partition of the $\mathcal{R}^{3N-6}$ space naturally divides it into so-called *attraction basins* around each local minimum $\alpha$. All points of a given basin correspond to a constant energy equal to the energy of the minimum $\alpha$.

As a result of the mapping (5.155), the initial supersurface is mapped onto a set of plateaus, one for each basin (Figure 5.16). Together with the transition state regions, the mapping (5.155) removes the barriers between minima; on the other hand, the energies of local minima and the global minimum as well remain unchanged. The potential energy varies only in discrete steps when the geometry moves from one basin to another. It is useful to combine the groups of minima connected by low barriers (for instance, by barriers lower than $kT$) into *superbasins*. Very often the energy landscape of hypersurface is rugged with many deep valleys corresponding to local minima. Sometimes these valleys flow in a funnel toward a deep minimum. The term *funnel* is now accepted for describing this peculiarity of potential hypersurfaces [225, 226].

In the *basin-hopping algorithm* (BHA) [187, 190, 220], the transformation to attraction basins (Equations (5.155)) has been combined with the Monte Carlo search technique at constant temperature. The procedure starts from a given basin and after random displacement, the energy difference between the old and the new configuration is calculated:

$$\Delta V = V_{old} - V_{new} \tag{5.156}$$

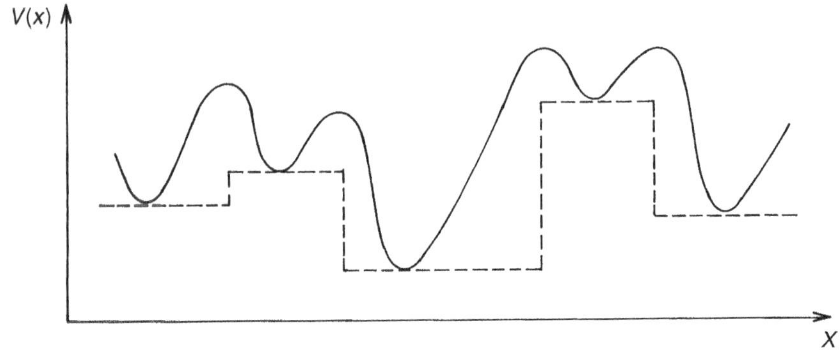


The solid line is the original surface and the dashed line is the surface transformed according to Equation (5.155)

**Figure 5.16** A schematic diagram illustrating the mapping an one-dimensional surface $V(x)$ onto the attraction basins

A step is accepted, if $\Delta V > 0$, i.e. the energy of the new minimum is lower than the starting point energy. If $\Delta V < 0$, then the step is accepted if:

$$\exp\left[\Delta V/kT\right] > \xi_i \tag{5.157}$$

where $\xi$ is a random number drawn from the interval [0, 1]. The reader can see that this is just the standard Metropolis Monte Carlo algorithm discussed in Section 5.4.2. The only difference is that $\Delta E$ in Equation (5.140) has the sign opposite to $\Delta V$ (Equation (5.156)), because it is defined as:

$$\Delta E = E_{new} - E_{old} \tag{5.158}$$

The conversion of potential hypersurface into the set of attraction basins dramatically reduces the mean time for interbasin motion [220]. On the original hypersurface, most classical trajectories that approach the barrier between two minima are reflected back due to the high potential energy; the passage proceeds only along the transition state valleys. On the contrary, on the transformed hypersurface it is feasible for the system to hop between basins at any point along the basin boundary.

The BHA has been successfully applied to the optimization of a large variety of systems. In an application to the $LJ_N$ clusters [190], all the lowest known minima were located up to $N = 110$, including some difficult cases for $N = 38, 69, 98, 103$ and 107; other applications are illustrated elsewhere [227–231]. For lead clusters, $Pb_N$, with the interatomic interactions approximated by the many-body glue potential (see Section 5.1.12.1), the global minima were located for all clusters up to $N = 160$ [231]. The data on global minima obtained for different systems by the BHA approach are presented on website [232].

### 5.4.4 Genetic algorithm

The genetic algorithm (GA) is based on the ideas of Darwin's evolution theory. Potential solutions of the problem are considered as individuals that compete for the right to reproduce themselves. The GA method was first proposed in 1962 by Holland [233, 234] for the study of adaptation in artificial systems. The publication of textbooks on this subject by Goldberg [235] and Koza [236] promoted a wide application of the GA approaches to different scientific and technical problems. A popular introduction into the GA and its application is presented in References [237–241].

In a typical implementation, points in the search space (such as particular sets of values for independent variables) are coded as fixed-length strings in binary code that are the analogue of the *chromosomes* in biology. At the beginning, a specified number of points in the search space are chosen randomly as the initial population. Each point is then evaluated for its suitability as a solution, called the *fitness*. If the problem is to find the best fit of experimental data, the fitness would be a function of the total difference between experimental data and the data to be fitted. If the problem is to find the minima of some function on a given interval, the fitness of a string is defined as a value of the function at a point corresponding to this string. The smaller is its value, the better is the fitness.

After all strings in the initial population are ranked according to their fitness, the best fraction is selected for breeding to produce a new population (offspring). This is performed by using three operations; in mathematical language, by applying three genetic operators: *faithful reproduction, crossover* (also called *recombination*) and *mutation.*

The faithful reproduction is applied to the strings of highest fitness which are placed directly into the next generation with no change. Thus, the reproduction causes the search to be concentrated in areas of the search space which tend to have high average fitness.

Crossover is a process in which some or all selected strings (including the strings subjected to the faithful reproduction) are grouped into pairs for mating. Then a random position is chosen at which each partner is cut into two pieces and exchanges a section of itself with its partner. The crossover is a fundamental operation in the GA method. It mimics the reproduction in real organisms where the chromosomes cross over and swap portions of their genetic code beyond the crossover point.

Point mutation is a process in which the value of a randomly chosen bit of a randomly chosen string is changed. The rate of mutation is small to avoid loosing good strings that have been created during the evolution. The mutation alone does not generally advance the search for a solution, but it does provide insurance against the development of a uniform population incapable of further evolution.

The strings obtained in the second generation are again ranked according to their fitness and the genetic operators described above are applied to the best fit part. The process is repeated until little improvement in the fitness is observed from one generation to the next. In some versions of GA, the search is carried out in parallel with several independent populations, and occasionally individuals are exchanged between population.

The remarkable feature of GA is to focus its attention on the most promising parts of the search space. The GA favors the fittest strings as parents, so these strings will have more offspring in the next generations. Another advantage of GA is that due to the application of the genetic operators, the system does not easily become stuck in some local minimum. All these features promote the good efficiency of the GA approaches in finding the global minimum.

***Example***

Consider, as an illustration of applying the genetic operators, a one-dimensional PES of a linear molecule with four atoms. The location of each atom is defined by its coordinate $x_i$ and the point in the search space is defined by a sequence of coordinates $\{x_1, x_2, x_3, x_4\}$.

Let the PES be defined by some pairwise approximation using a model potential $U_{ij}\left(\left|x_i - x_j\right|\right)$. The interaction energy of the system depends on four coordinates and is presented as:

$$U\left(x_1, x_2, x_3, x_4\right) = \sum_{i<j} U_{ij}\left(\left|x_i - x_j\right|\right) \tag{5.159}$$

This function determines the fitness of the points in the search space. For the problem of finding the global minimum, the smaller the magnitude of $U\left(x_1, x_2, x_3, x_4\right)$ at some point, the larger the fitness.

It is necessary to limit the search space. We will consider atomic coordinates in the interval:

$$0 \leq x_i \leq 15.0 \text{ Å}$$

If only integral numbers are used, the binary code for each coordinate will contain no more than four bits.

Let the fittest points, selected from a randomly chosen initial set of points, corresponds to the following sets:

$$\{x_1, x_2, x_3, x_4\} : \{1, 3, 5, 7\}, \{2, 3, 5, 6, \}, \{1, 2, 5, 8\} \quad \text{and} \quad \{1, 3, 4, 6\}$$

In binary code they are represented as:[8]

[8] Note for those not acquainted with binary calculus. If in the decimal calculus one has ten digits: $0, 1, 2, \ldots, 9$, and the order of the number corresponds to the power of ten; in the binary calculus there are only two digits: 0 and 1 and the order of the number corresponds to the power of two. Then, there is the following correspondence:

| | | | | | | | | | |
|---|---|---|---|---|---|---|---|---|---|
| 1 | 2 | 3 | 4 | 5 | 6 | 7 | 8 | 9 | 10 |
| 1 | 10 | 11 | 100 | 101 | 110 | 111 | 1000 | 1001 | 1010 |

| | | | | |
|---|---|---|---|---|
| 1. | 0001 | 0011 | 0101 | 0111 |
| 2. | 0010 | 0011 | 0101 | 0110 |
| 3. | 0001 | 0010 | 0101 | 1000 |
| 4. | 0001 | 0011 | 0100 | 0110 |

(5.160)

These four strings participate in the breeding to create the second generation. Assume the following distribution: strings 1 and 2, having the highest rank of fitness, are allowed to reproduce themselves without change (faithful reproduction). Three pairs, namely: 1–2, 1–3 and 2–4 participate in the crossover. For the pair 1–2 the cut takes place between the third and fourth bit positions. The products of their recombination are:

| parents | | | | | children | | | |
|---|---|---|---|---|---|---|---|---|
| 000 ⋮ 1 | 0011 | 0101 | 0111 | → | 0000 | 0011 | 0101 | 0110 |
| 001 ⋮ 0 | 0011 | 0101 | 0110 | | 0011 | 0011 | 0101 | 0111 |

(5.161)

The pairs 1–3 and 2–4 are cut between the eighth and ninth bit positions and create children by exchanging their cutted parts, similar to Equation (5.161).

Let one of the children of the mating (Equation (5.161)) undergoes a point mutation at the thirteenth bit position:

| | | | | | | | | | |
|---|---|---|---|---|---|---|---|---|---|
| $\widehat{M}(13)$ | 0000 | 0011 | 0101 | 0110 | → | 0000 | 0011 | 0101 | 1110 |

(5.162)

where $\widehat{M}(13)$ denotes the operator of mutation.

Thus, there are 9 strings for the second generation: two from the first generation (Equation (5.160)), six children created by the crossover of three chosen pairs, and one mutated strings (Equation (5.162)):

| | | | | | | | | | |
|---|---|---|---|---|---|---|---|---|---|
| 1. | 0001 | 0011 | 0101 | 0111 | 2. | 0010 | 0011 | 0101 | 0110 |
| 3. | 0000 | 0011 | 0101 | 0110 | 4. | 0011 | 0011 | 0101 | 0111 |
| 5. | 0001 | 0011 | 0101 | 1000 | 6. | 0001 | 0010 | 0101 | 0111 |

| | | | | |
|---|---|---|---|---|
| 7. | 0010 | 0011 | 0100 | 0110 |
| 8. | 0001 | 0011 | 0101 | 0110 |
| 9. | 0000 | 0011 | 0101 | 1110 |

(5.163)

These strings have now to be ranked according to their fitness. The GA procedure is constrained to keep the gene pool population as well as the total population constant, so the best four strings from the population (Equation (5.163)) have to be chosen for the second generation from which the third generation is created and so on until little improvement in the fitness is observed in the next generations. In real computer algorithms, the number of genetic operations is chosen in a manner that allows the total population in each generation to remain constant.

The first application of the GA approach to optimize cluster structures was performed by Hudson *et al.* [242] and independently by Hartke [243] and Xiao and Williams [244]. After these works, many different modifications, improving the efficiency of application the GA to cluster structural optimization, were suggested [245–252].

An important development was made by Deaven and Ho [245, 246], who suggested acting the genetic operators not on a string representation of the points on the cluster potential surface, but directly on the cluster configuration space. Each child cluster is relaxed to the nearest local minimum using a conjugate–gradient minimization before determining its fitness. This approach is sometimes designated as *DH–GA*. Note that a similar idea was used in the basin-hopping algorithm [190] (Section 5.4.3.2), where instead of the GA approach, the Monte Carlo technique was applied.

The DH–GA method was further modified in several directions by Wolf and Landman [249]. These modifications allowed the authors to obtain all the global minima for the $LJ_N$ clusters up to $N = 100$. Further development of the DH–GA approach was elaborated by Hartke and named by him as *phenotype algorithm*; it is described in detail in the References [251].

As mentioned by Michaelian [250], for large clusters or molecules, the interatomic potential is often short range in comparison to the size of the system, and the contribution to the binding energy due to interactions between nearest neighbors is significally larger than that for interactions between atoms separated by larger distances. This means that, to a first approximation, the global problem can be reduced to a linear combination of various local problems of fewer variables. Following these ideas, Michaelian [250] proposed a *symbiotic algorithm* in which the cluster is studied by parts, and the evolution in overlapping cells inside the cluster is considered separately. Apart from reducing the complexity of the PES, this approach makes optimal use of the crossover operation, constraining it to act locally, only within each cell. The symbiotic algorithm was implemented in Reference [115] for the study of large noble metal clusters, up to $N = 75$, bound by the many-body Gupta potential (Section 5.1.12.1, Equation (5.75)).

In the preceding text, the implementation of different GA approaches for the atomic clusters was discussed. For molecular clusters, the number of independent variables is increased because of the additional orientational degrees of freedom. The problem becomes much more complicated in that it causes a great increase in the computational time for the global optimization of molecular clusters in comparison with atomic clusters of the same size. According to an estimation by Hartke [79], the computational time required for a global optimization of the water cluster with 20 monomers, bound by the rather simple TIP4P model potential (Section 5.1.7.4) is approximately equal to the time needed for treating the $LJ_N$ cluster with $N = 250$.

In Reference [61], Hartke extended his phenotype algorithm elaborated for the global optimization of the atomic clusters to a more difficult case of the molecular clusters. For $(H_2O)_N$ clusters bound by the TIP4P potential, all the global minima up to $N = 21$ were found. But for $N > 20$, a serial approach becomes very complicated and time-consuming. In his next study, Hartke [79] realized a parallel implementation of his algorithm. As a result, all the global minima for the water clusters in the range $n = 2$ to 30, bound by the more reliable and complex TTM2-F potential (Section 5.1.7.4), have been found.

Lai *et al.* [253] applied both the GA technique and the BHA approach [190] to a search of the global energy minima in Na, K, Rb, Cs and Pb clusters, bound by the Gupta potential, in the range $N = 3 - 56$. For all clusters studied, the global minima found by both approaches were in very good agreement. The complete coincidence was obtained in the global minimum search of the TIP4P $(H_2O)_N$ clusters ($N$= 2–20) performed by both the BHA approach [60] and the phenotype version of the GA [61]. This is the manifestation of comparable capacities of both approaches. However, for very large systems, the phenotype version of GA [251] looks more appropriate. Using the $(LJ)_N$ clusters as a benchmark system, Hartke [251] found the global minima for all considered clusters, up to $N = 150$. The scaling of computer time with cluster size was estimated as cubic, which makes this method promising for application to larger clusters.

# References


1. I.M. Torrens, *Interatomic Potentials*, Academic Press, New York (1972).
2. G.C. Maitland, M. Rigby, E.B. Smith and W.A. Wakeham, *Intermolecular Forces*, Clarendon Press, Oxford (1987).
3. R.O. Watts and I.G. McGee, *Liquid State Chemical Physics*, John Wiley & Sons, Inc., New York (1976).
4. J.E. Lennard-Jones, *Proc. Roy. Soc. (London)* A **106**, 463 (1924).
5. E.A. Mason and H.W. Schamp, *Ann. Phys.* (USA) **4**, 233 (1958).
6. M. Klein and H.J.M. Hanley, *J. Chem. Phys.* **53**, 4722 (1970).
7. G.C. Maitland and E.B. Smith, *Chem. Phys. Lett.* **22**, 443 (1973).
8. R.A. Aziz and W.Tokay, *Mol. Phys.* **30**, 857 (1975).
9. H.J.M. Hanley and M. Klein, *J. Phys. Chem.* **76**, 1743 (1972).
10. V.P.S. Nain, R.A. Aziz, P.C. Jain and S.C. Saxena, *J. Chem. Phys.* **65**, 3242 (1976).

11. T. Kihara, *Adv. Chem. Phys.* **5**, 147 (1964).
12. K. Kobashi and T. Kihara, *J. Chem. Phys.* **72**, 378 (1980).
13. P. Sinanoğlu, *Adv. Chem. Phys.* **12**, 283 (1967).
14. R.A. Buckingham, *Proc. Roy. Soc. (London)* A **168**, 264 (1938).
15. I.G. Kaplan, *Theory of Molecular Interactions*, Elsevier, Amsterdam (1986).
16. A.J. Pertsin and A.I. Kitaigorodsky, *The Atom–Atom Potential Method. Applications to Organic Molecular Solids*, Springer–Verlag, Berlin (1987).
17. O. Schnepp, *Adv. Atom. Molec. Phys.* **5**, 155 (1969).
18. R.A. Buckingham, *J. Plan. Space Sci.* **3**, 205 (1961).
19. R. Alrichs, R. Penco and G. Scoles, *Chem. Phys.* **19**, 119 (1977).
20. K.M. Smith *et al.*, *J. Chem. Phys.* **67**, 152 (1977).
21. K. Ng, W.J. Meath and A.R. Allnatt, *Chem. Phys.* **32**, 175 (1978); *Mol. Phys.* **37**, 237 (1979).
22. S. Karr and D.D. Konowalow, *Nuovo Cim*ento **34**, 205 (1964).
23. J.A. Barker and A. Pompe, *Australian J. Chem.* **21**, 1683 (1968).
24. J.A. Barker *et al.*, *J. Chem. Phys.*, **61**, 3081 (1974).
25. P.M. Morse, *Phys. Rev.* **34**, 57 (1929).
26. R. Rydberg, *Zs. f. Phys.* **73**, 376 (1932).
27. O. Sinanoğlu and K.S. Pitzer, *J. Chem. Phys.* **31**, 960 (1959).
28. G. Pöschl and E. Teller, *Zs. f. Phys.* **83**, 143 (1933).
29. M. Davies, *J. Chem. Phys.* **17**, 374 (1949).
30. Y.P. Varshni, *Rev. Mod. Phys.* **29**, 664 (1957).
31. D. Steele, E.R. Lippincott and J.T. Vanderslice, *Rev. Mod. Phys.* **34**, 239 (1962).
32. H.M. Hulburt and J.O. Hirschfelder, *J. Chem. Phys.* **9**, 61 (1941).
33. E.R. Lippincott, *J. Chem. Phys.* **21**, 2070 (1953).
34. E.R. Lippincott and D. Steele, *J. Chem. Phys.* **35**, 2065 (1961).
35. A. Kratzer, *Zs. f. Phys.* **3**, 289 (1920).
36. S. Flügge, *Practical Quantum Mechanics*, Springer–Verlag, Berlin (1971), Vol. 1, p. 173.
37. S. Rashev and D.C. Moule, *Chem. Phys.* **295**, 109 (2003).
38. E. Fues, *Ann. d. Phys. (Leipzig)* **80**, 367 (1929).
39. J.L. Dunham, *Phys. Rev.* **41**, 721 (1932).
40. G. Simons, R.G. Parr and J.M. Finlan, *J. Chem. Phys.* **59**, 3229 (1973).
41. A.J. Thakkar, *J. Chem. Phys.* **62**, 1693 (1975).
42. J.N. Goble and J.S. Winn, *J. Chem. Phys.* **70**, 2058 (1979); *Chem. Phys. Lett.* **77**, 168 (1981).
43. E.S. Kryachko and T. Koga, *Adv. Quant. Chem.* **17**, 97 (1985).
44. W.N. Keesom, *Comm. Phys. Lab. Leiden*, Suppl. **246**, Section 6 (1912).
45. J.O. Hirschfelder, C.F. Curtiss and R.B. Bird, *Molecular Theory of Gases and Liquids*, John Wiley & Sons, Inc., New York (1954).
46. W.N. Stockmayer, *J. Chem. Phys.* **9**, 398 (1941).
47. R.J. Le Roy and J. van Kranendonk, *J. Chem. Phys.* **61**, 4750 (1974).
48. A.M. Duncker and R.G. Gordon, *J. Chem. Phys.* **68**, 700 (1978).
49. K.R. Foster and J.H. Rugheimer, *J. Chem. Phys.* **56**, 2632 (1972).
50. R.T. Pack, *Chem. Phys. Lett.* **55**, 197 (1978).
51. A. van der Avoird, P.E.S. Wormer and R. Moszynski, *Chem. Rev.* **94**, 1931 (1994).
52. J.S. Rowlinson, *Trans. Faraday Soc.* **47**, 120 (1951).

53. A. Ben-Naim and F.H. Stillinger, in *Structure and Transport Processes in Water and Aqueous Solutions*, R.A. Horne (ed.), Interscience, New York (1972).
54. F.H. Stillinger, *J. Chem. Phys.* **57**, 1780 (1972).
55. A. Rahman and F.H. Stillinger, *J. Chem. Phys.* **57**, 1281 (1972).
56. R.O. Watts, *Mol. Phys.* **28**, 1069 (1974).
57. W.L. Jorgensen *et al.*, *J. Chem. Phys.* **79**, 926 (1983).
58. W.L. Jorgensen, *J. Amer. Chem. Soc.* **103**, 335 (1981).
59. W.L. Jorgensen, *J. Chem. Phys.* **77**, 4156 (1982).
60. D.J. Wales and M.P. Hudges, *Chem. Phys. Lett.* **286**, 65 (1998).
61. B. Hartke, *Z. Phys. Chem.* **214**, 1251 (2000).
62. F.H. Stillinger and C.W. David, *J. Chem. Phys.* **69**, 1473 (1978).
63. F.H. Stillinger, *J. Chem. Phys.* **71**, 1647 (1979).
64. P. Ahlström, A. Walcuist, S. Engström and B. Jönsson, *Mol. Phys.* **68**, 563 (1989).
65. U. Niesar *et al.*, *Int. J. Quant. Chem. Symp.* **23**, 421 (1989).
66. U. Niesar *et al.*, *J. Phys. Chem.* **94**, 7949 (1990).
67. G. Corongiu, *Int. J. Quant. Chem.* **42**, 1209 (1992).
68. D.N. Bernardo, Y. Ding, K. Krogh-Jespersen and R.M. Levy, *J. Phys. Chem.* **98**, 4180 (1994).
69. P.G. Kusalik, F. Liden and I.M. Svishchev, *J. Chem. Phys.* **103**, 10169 (1995).
70. I.M. Svishchev, P.G. Kusalik, J. Wang and R.J. Boyd, *J. Chem. Phys.* **105**, 4742 (1996).
71. O. Matsuoka, E. Clementi and M. Yoshimine, *J. Chem. Phys.* **64**, 1351 (1976).
72. F. Sciortino and G. Corongiu, *J. Chem. Phys.* **98**, 5694 (1993).
73. C.J. Burnham and S.S. Xantheas, *J. Chem. Phys.* **116**, 5115 (2002).
74. B.T. Thole, *Chem. Phys.* **59**, 341 (1981).
75. H. Patridge and D.W. Schwenke, *J. Chem. Phys.* **106**, 4618 (1997).
76. W.F. Kuhn and M.S. Lehman, *Water Sci. Rev.* **2**, 1 (1986).
77. P.W. Deutsch, B.N. Hale, R.C. Ward and D.A. Reago, Jr., *J. Chem. Phys.* **78**, 5103 (1983).
78. J.G.C.M. van Duijneveldt-van de Rijdt, W.T.M. Mooij, and F.B. van Duijneveldt, *Phys. Chem. Chem. Phys.* **5**, 1169 (2003).
79. B. Hartke, *Phys. Chem. Chem. Phys.* **5**, 275 (2003).
80. N. Bohr, *Kgl. Danske Vid. Selsk. Mat.- Fys. Medd.* **18**, 144 (1948).
81. J.A. Brinkman, *J. Appl. Phys.* **25**, 961 (1954).
82. O.B. Firsov, *Doklady Acad. Sci. USSR* **91**, 515 (1953); *Zh. Eksp. Teor. Fiz.* **33**, 696 (1957).
83. A. Sommerfeld, *Zs. f. Phys.* **78**, 283 (1932).
84. N.H. March, *Proc. Cambridge Phil. Soc.* **46**, 356 (1950).
85. K. Umeda, *J. Phys. Soc. Japan* **9**, 290 (1954).
86. G. Moliere, *Zs. Naturforsch.* **2a**, 133 (1947).
87. M. Born and J.E. Mayer, *Zs. f. Phys.* **75**, 1 (1932).
88. H.B. Hungtington, *Solid State Phys.* **7,** 213 (1958).
89. S.F. Boys and I. Shavitt, *Proc. Roy. Soc. (London)* A **254**, 499 (1960).
90. R.J. Munn, *J. Chem. Phys.* **40**, 1439 (1969).
91. C. Erginsoy, G.H. Vineyard and A. Englert, *Phys. Rev.* **133A**, 595 (1964).
92. P.E. Siska, J.H. Parson, T.P. Schafer and Y.T. Lee, *J. Chem. Phys.* **55**, 5762 (1971).
93. C.H. Chen., P.E. Siska and Y.T. Lee, *J. Chem. Phys.* **59**, 601 (1973).

94. J.H. Ahlberg, E.N. Nilson and J.L. Walsh, *The Theory of Splines and Their Applications*, Academic Press, New York (1967).
95. W.H. Press, S.A. Teukolsky, W.T. Vetterling and B.P. Flannery, *Numerical Recipes in C: The Art of Scientific Computing* (2nd edn), Cambridge University Press, Cambridge (1992).
96. R.T. Pack *et al.*, *J. Chem. Phys.* **77**, 5475 (1982).
97. F. Ercolessi, E. Tosatti and M. Parrinello, *Phys. Rev. Lett.* **57**, 719 (1986).
98. F. Ercolessi, M. Parrinello and E. Tosatti, *Philos. Mag.* A **58**, 213 (1988).
99. M.J. Stott and E. Zaremba, *Phys. Rev.* B **22**, 1564 (1980).
100. J.K. Nørskov, *Phys. Rev.* B **26**, 2875 (1982).
101. M.S. Daw and M.J. Baskes, *Phys. Rev. Lett.* **50**, 1285 (1983); *Phys. Rev.* B **29**, 6443 (1984).
102. K.W. Jacobsen, J.K. Nørskov and M.J. Puska, *Phys. Rev.* B **35**, 7423 (1987).
103. J. Vail, R. Pandey and A.B. Kuntz, *Rev. Solid State Sci.* **5**, 181 (1991).
104. I.G. Kaplan, J. Soullard, J. Hernández-Cobos and R. Pandey, *J. Phys.: Condens. Matter* **11**, 1049 (1999).
105. M.J. Baskes, J.S. Nelson and A.F. Wright, *Phys. Rev.* B **40**, 6085 (1989).
106. M.J. Baskes, *Phys. Rev.* B **46**, 2727 (1992).
107. M.J. Baskes, *Phys. Rev. Lett.* **59**, 2666 (1987).
108. M.J. Baskes and R.A. Johnson, *Modelling Simul. Mater. Sci. Eng.* **2**, 147 (1994).
109. M.J. Baskes, *Mater. Sci. Eng.* A **261**, 165 (1999).
110. M.W. Finnis and J.E. Sinclair, *Philos. Mag.* A **50**, 45 (1984).
111. J. Friedel, in *The Physics of Metals*, J. Ziman (ed), Cambridge University Press, Cambridge (1969), Chapter 8.
112. R. Gupta, *Phys. Rev.* B **23**, 6265 (1981).
113. F. Ducastelle, *J. Phys. (Paris)* **31**, 1055 (1970).
114. A. Posada-Amarillas and I.L. Garzón, *Phys. Rev.* B **53**, 8363 (1996).
115. K. Michaelian, N. Rendón and I.L. Garzón, *Phys. Rev.* B **60**, 2000 (1999).
116. I.J. Robertson, V. Heine and M.C. Payne, *Phys. Rev. Lett.* **70**, 1944 (1993).
117. F. Ercolessi and J.B. Adams, *Europhys. Lett.* **26**, 583 (1994).
118. X-Y. Liu, J.B. Adams, F. Ercolessi and J.A. Moriarty, *Modelling Simul. Mater. Sci. Eng.* **4**, 293 (1994).
119. X-Y. Liu and J.B. Adams, *Acta Mater.* **46**, 3467 (1998).
120. X-Y. Liu, W. Xu, S.M. Foiles and J.B. Adams, *Appl. Phys. Lett.* **72**, 1578 (1998).
121. T.J. Lenosky *et al.*, *Modelling Simul. Mater. Sci. Eng.* **8**, 825 (2000).
122. F.H. Stillinger and T.A. Weber, *Phys. Rev.* A **31**, 5262 (1985).
123. R. Biswas and D.R. Hamann, *Phys. Rev. Lett.* **55**, 2001 (1985); *Phys. Rev.* B **36**, 6434 (1987).
124. A.E. Carlsson, *Solid State Phys.* **43**, 1 (1990).
125. J.N. Murrell and R.E. Mottram, *Mol. Phys.* **69**, 571 (1990).
126. J.N. Murrell and J.A. Rodríguez-Ruiz, *Mol. Phys.* **71**, 823 (1990).
127. R.L. Johnston and J.Y. Fang, *J. Chem. Phys.* **97**, 7809 (1992).
128. J.E. Hearn and R.L. Johnston, *J. Chem. Phys.* **107**, 4674 (1997).
129. H. Popkie, H. Kistenmacher and E. Clementi, *J. Chem. Phys.* **59**, 1325 (1973).
130. G.C. Lie, E. Clementi and M. Yoshimine, *J. Chem. Phys.* **64**, 2314 (1976).
131. E. Clementi, F. Cavallone and R. Scordamaglia, *J. Amer. Chem. Soc.* **99**, 5531 (1977).
132. P.H. Smith, J.L. Derissen and F.B. Duijneveldt, *Mol. Phys.* **37**, 521 (1979).
133. E. Clementi and G. Corongiu, *Int. J. Quant. Chem.* Symp. **10**, 31 (1983).

134. J.N. Detrich, G. Corongiu and E. Clementi, *Chem. Phys. Lett.* **112**, 426 (1984).
135. E. Blaisten-Borojas and S.N. Khanna, *Phys. Rev. Lett.* **61**, 1477 (1988).
136. I.G. Kaplan *et al.*, *J. Mol. Str.* (Theochem) **398–399**, 333 (1997).
137. I.L. Garzón *et al.*, *Z. Phys.* D **40**, 202 (1997).
138. J. Hernández-Cobos, I.G. Kaplan and J.N. Murrell, *Mol. Phys.* **92**, 63 (1997).
139. I.G. Kaplan, *Int. J. Quant. Chem.* **74**, 241 (1999).
140. I.L. Garzón, I.G. Kaplan, R. Santamaría and O. Novaro, *J. Chem. Phys.* **109**, 2176 (1998).
141. S. Fraga and R.S. Mulliken, *Rev. Mod. Phys.* **32**, 254 (1960).
142. I.G. Kaplan, J.N. Murrell, S. Roszak and J. Leszczynski, *Mol. Phys.* **100**, 843 (2002).
143. S.F. O'Shea and M.J. Meath, *Mol. Phys.* **28**, 1431 (1974); **31**, 515 (1976).
144. J.M. Standart and P.R. Certain, *J. Chem. Phys.* **83**, 3002 (1985).
145. I.G. Kaplan, S. Roszak and J. Leszczynski, *J. Chem. Phys.* **113**, 6245 (2000).
146. T.J. Lee, A.P. Rendel and P.R. Taylor, *J. Chem. Phys.* **92**, 489 (1990); **93**, 6636 (1990).
147. W. Klopper and J. Almlöf, *J. Chem. Phys.* **99**, 5167 (1993).
148. S.P. Walch and C.W. Bauschlicher, Jr., *J. Chem. Phys.* **83**, 5735 (1985).
149. T.J. Lee, A.P. Rendel and P.R. Taylor, *Theoret. Chim. Acta* **83**, 165 (1992).
150. W.T. Eadie, D. Dryard, F.E. James, M. Roos and B. Sadoulet, *Statistical Methods in Experimental Physics*, North-Holland, Amsterdam (1971).
151. A.K. Jain and R.C. Dubes, *Algorithms for Clustering Data*, Prentice Hall, Englewood Cliffs, New Jersey (1988).
152. W.H. Press, B.P. Flannery, S.A. Teukolsky and V.T. Vetterling, *Numerical Recipes*, Cambridge University Press, Cambridge (1986).
153. A. Albert, *Regression and the Moor-Penrose Pseudoinverse*, Academic Press, New York (1972).
154. J. Jellinek and M.J. López, in *Fashioning a Model: Optimization Methods in Chemical Physics*, A. Ernesti, J.M. Hutson and N.J. Wright (eds), CCP6, Daresbury (1988), p. 12.
155. M.J. López and J. Jellinek, *J. Chem. Phys.* **110**, 8899 (1999).
156. R. Rydberg, *Zs. f. Phys.* **80**, 514 (1933).
157. O. Klein, *Zs. f. Phys.* **76**, 226 (1932).
158. A.L.G. Rees, *Proc. Roy. Soc. (London)* A **59**, 998 (1947).
159. M.E. Kaminsky, *J. Chem. Phys.* **66**, 4951 (1977).
160. R.B. Gerber, R.M. Roth and M.A. Rather, *Mol. Phys.* **44**, 1335 (1981).
161. N.F. Mott and H.S.W. Massey, *The Theory of Atomic Collisions*, Clarendon Press, Oxford (1965).
162. L.D. Landau and E.M. Lifshitz, *Quantum Mechanics (Nonrelativistic Theory)*, 3rd edn, Pergamon Press, Oxford (1977).
163. U. Buck, *Rev. Mod. Phys.* **46**, 369 (1974).
164. U. Buck, *Adv. Chem. Phys.* **30**, 313 (1975).
165. C.E. Frëdberg, *Phys. Rev.* **72**, 519 (1947).
166. E.A. Hylleraas, *Phys. Rev.* **74**, 48 (1948).
167. V. Bargmann, *Rev. Mod. Phys.* **21**, 488 (1949).
168. N. Levinson, *Phys. Rev.* **89**, 755 (1953).
169. I.M. Gel'fand and B.M. Levitan, *Izv. Acad. Sci. USSR, math. ser.* **15**, 309 (1951).
170. V. De Alfaro and T. Regge, *Potential Scattering*, North-Holland, Amsterdam (1965).

171. R.G. Newton, *Scattering Theory of Waves and Particles*, McGraw-Hill, New York (1966).
172. M.G. Krein, *Uspechi Math. Nauk* **13**, N5, 83 (1958).
173. Z.S. Agranovich and V.A. Marchenko, *Inverse Scattering Problem (Russ.), Nauka, Moscow* (1963).
174. K. Chadan and P.C. Sabatier, *Inverse Problems in Quantum Scattering Theory*, Springer–Verlag, New York (1977).
175. F.G. Hoyt, *Phys. Rev.* **55**, 664 (1939).
176. O.B. Firsov, *Zh. Eks. Theor. Fiz.* **24**, 279 (1953).
177. E.T. Whittaker and G.N. Watson, *A Course of Modern Analysis*, Cambridge University Press, Cambridge (1927).
178. G.H. Lane and E. Everhart, *Phys. Rev.* **120**, 2064 (1960).
179. U. Buck, *J. Chem. Phys.* **54**, 1923 (1971).
180. U. Buck, M. Kick and H. Pauly, *J. Chem. Phys.* **56**, 3391 (1972).
181. R.B. Gerber, V. Buck and U. Buck, *J. Chem. Phys.* **72**, 3596 (1980).
182. D.W. Gough, E.B. Smith and G.C. Maitland, *Mol. Phys.* **24**, 151 (1972).
183. S. Chapman and T.G. Cowling, *The Mathematical Theory of Nonuniform Gases*, Cambridge University Press, London (1952).
184. L.A. Veihland, M.M. Harrington, and E.A. Mason, *Chem. Phys.* **17**, 433 (1976).
185. R. Horst, P.M. Pardalos and N.V. Thoai, *Introduction to Global Optimization*, Academic Press, Dordrecht, Netherlands (1995).
186. F. Baletto and R. Ferrando, *Rev. Mod. Phys.* **77**, 371 (2005).
187. Z. Li and H.A. Scheraga, *Proc. Natl. Acad. Sci. USA* **84**, 6611 (1987).
188. C.J. Tsai and K.D. Jordan, *J. Phys. Chem.* **97**, 11227 (1993).
189. J.P.K. Doye and D.J. Wales, *J. Chem. Phys.* **112**, 9659 (1995).
190. D.J. Wales and J.P.K. Doye, *J. Phys. Chem.* **101**, 5111 (1997).
191. M.R. Garey and D.S. Johnson, *Computers and Interactability: A Guide to the Theory of NP-Completeness*, Freeman, San Francisco (1979).
192. E.L. Lawlor, *Combinatorial Optimization*, Halt, Rinehart & Winston, New York (1976).
193. A.V. Aho, J.E. Hopcroft and J.D. Ullman, *The Design and Analysis of Computer Algorithms,* Addison–Wesley, Reading, Massachusetts (1974).
194. L.T. Wille and J. Vennik, *J. Phys. A: Math. Gen.* **18**, L419 (1985).
195. S. Kirkpatrick, C.D. Gelatt, Jr. and M.P. Vecchi, *Science* **220**, 671 (1983).
196. R. Karp, *Math. Oper. Res.* **2**, 209 (1977).
197. N. Metropolis *et al.*, *J. Chem. Phys.* **21**, 1087 (1953).
198. L.D. Landau and E.M. Lifshitz, *Statistical Physics,* Part 1, Pergamon Press, Oxford (1986).
199. R. Biswas and D.R. Hamann, *Phys. Rev.* B **34**, 895 (1986).
200. W.H. Press, S.A. Teukolsky, W.T. Vetterling and B.P. Flannery, *Numerical Recipes in FORTRAN,* 2nd edn, Cambridge University Press, Cambridge (1992).
201. *Monte Carlo Methods,* 2nd edn, K. Binder (ed), Springer–Verlag, New York (1986).
202. M.P. Allen and D.J. Tildesley, *Computer Simulation of Liquids,* Clarendon Press, Oxford (1987).
203. R. Ahlrichs and S.D. Elliott, *Phys. Chem. Chem. Phys.* **1**, 13 (1999).
204. D.E. Bacelo, R.C. Binning, Jr. and Y. Ishikawa, *J. Phys. Chem.* A **103**, 4631 (1999).
205. P. Amara, D. Hsu and J.E. Straub, *J. Phys. Chem.* **97**, 6715 (1993).
206. A.B. Finnila *et al.*, *Chem. Phys. Lett.* **219**, 343 (1994).

207. J.D. Doll and D.L. Freeman, *IEEE Comp. Sci. Eng.* **1**, 22 (1994).
208. D.L. Freeman and J.D. Doll, *Ann. Rev. Phys. Chem.* **47**, 43 (1996).
209. D. Ceperly and B. Alder, *Science* **231**, 555 (1986).
210. J. Ma and J.E. Straub, *J. Chem. Phys.* **101**, 533 (1994).
211. P. Liu and B.J. Berne, *J. Chem. Phys.* **118**, 2999 (2003).
212. R.P. Feynman and A.R. Hibbs, *Quantum Mechanics and Path Integrals*, McGraw–Hill Co., New York (1965).
213. F.H. Stillinger and T.A. Weber, *J. Stat. Phys.* **52**, 1429 (1988).
214. L. Piela, J. Kostrowicki and H.A. Scheraga, *J. Phys. Chem.* **93**, 3339 (1989).
215. J. Kostrowicki and L. Piela, *J. Optim. Theory and Appl.* **69**, 269 (1991).
216. J. Kostrowicki, L. Piela, B.J. Cherayil and H.A. Scheraga, *J. Phys. Chem.* **95**, 4113 (1991).
217. A.L. Mackay, *Acta Crystallogr.* **15**, 916 (1962).
218. H.A. Scheraga, *Int. J. Quant. Chem.* **42**, 1529 (1992).
219. R.I. Wawak *et al., J. Phys. Chem.* A **102**, 2904 (1998).
220. D.J. Wales and H.A. Scheraga, *Science* **285**, 1368 (1999).
221. L. Piela, K.A. Olszewski and J. Pillardy, *J. Mol. Str. (Theochem)* **308**, 229 (1994).
222. J. Pillardy and L. Piela, *Polish J. Chem.* **72**, 1849 (1998).
223. F.H. Stillinger and T.A. Weber, *Phys. Rev.* A **25**, 978 (1982); *Science* **225**, 983 (1984).
224. O.M. Becker and M. Karplus, *J. Chem. Phys.* **106**, 1495 (1997).
225. P.E. Leopold, M. Montal and J.N. Onuchic, *Proc. Natl. Acad. Sci. USA* **89**, 8721 (1992).
226. P.G. Wolynes, J.N. Onuchic and D. Thirumalal, *Science* **267**, 1619 (1995).
227. D.J. Wales and J.P.K. Doye, *Phys. Rev. Lett.* **80**, 1357 (1998).
228. J.P.K. Doye and D.J. Wales, *Phys. Rev. Lett.* **86**, 5719 (2001).
229. J.P.K. Doye, D.J. Wales, W. Branz and P. Calvo, *Phys. Rev.* B **64**, 235409 (2001).
230. J.P.K. Doye, *J. Chem. Phys.* **119**, 1136 (2003).
231. J.P.K. Doye and S.C. Hendy, *Eur. Phys. J.* D **22**, 99 (2003).
232. D.J. Wales *et al., The Cambridge Cluster Database,* URL. http://www-wales.ch.cam.ac.uk/CCD.html
233. J.H. Holland, in *Selforganizing Systems*, M.C. Yovits, G.T. Jacobi and G.D. Goldstein (eds), Spartan Books, Washington (1962), pp. 215–230.
234. J.H. Holland, *Adaptation in Natural and Artificial Systems,* University of Michigan Press, Ann Arbor (1975).
235. D.E. Goldberg, *Genetic Algorithm in Search, Optimization and Machine Learning,* Addison–Wesley, Reading, MA (1989).
236. J.R. Koza, *Genetic Programming: On the Programming of Computers by Means of Natural Selection,* MIT, Cambridge, MA (1992).
237. J.H. Holland, *Sci. Am.* **267**, July, 66 (1992).
238. D.B. McGarrah and R.S. Judson, *J. Comp. Chem.* **14**, 1385 (1993).
239. P. Sutton and S. Boyden, *Am. J. Phys.* **62**, 549 (1994).
240. D.A. Coley, *Contemp. Phys.* **37**, 145 (1996).
241. K. Michaelian, *Am. J. Phys.* **66**, 231 (1998).
242. R.S. Judson *et al., Int. J. Quant. Chem.* **44**, 277 (1992).
243. B. Hartke, *J. Phys. Chem.* **97**, 9973 (1993).
244. Y. Xiao and D.E. Williams, *Chem. Phys. Lett.* **215**, 17 (1993).
245. D.M. Deaven and K.M. Ho, *Phys. Rev. Lett.* **75**, 288 (1995).
246. D.M. Deaven, N. Tit, J.R. Morris and K.M. Ho, *Chem. Phys. Lett.* **256**, 195 (1996).

247. S.K. Gregarick, M.H. Alexander and B. Hartke, *J. Chem. Phys.* **104**, 2684 (1996).
248. J.A. Niesse and H.R. Mayne, *J. Chem. Phys.* **105**, 4700 (1996); *J. Comput. Chem.* **18**, 1233 (1997).
249. M.D. Wolf and U. Landman, *J. Phys. Chem.* A **102**, 6129 (1998).
250. K. Michaelian, *Chem. Phys. Lett.* **293**, 202 (1998).
251. B. Hartke, *J. Comput. Chem.* **20**, 1752 (1999).
252. I. Rata *et al., Phys. Rev. Lett.* **85**, 546 (2000).
253. S.K. Lai *et al., J. Chem. Phys.* **117**, 10715 (2002).

# Appendix 1

# Fundamental Physical Constants and Conversion Table of Physical Units

**Table A1.1** Fundamental physical constants[a]

| Physical constant | Symbol | Value | Units |
|---|---|---|---|
| Speed of light in vacuum | $c$ | $2.99792458 \times 10^{10}$ | cm $s^{-1}$ |
| Planck constant | $\hbar = h/2\pi$ | $1.0545716 \times 10^{-34}$ | Js |
| Avogadro constant | $N_A$ | $6.022142 \times 10^{23}$ | $mol^{-1}$ |
| Boltzmann constant | $k$ | $1.380650 \times 10^{-23}$ | $JK^{-1}$ |
| Elementary charge | $e$ | $1.602176 \times 10^{-19}$ | C |
| Electron mass | $m_e$ | $9.109382 \times 10^{-28}$ | g |
| Proton mass | $m_p$ | $1.672622 \times 10^{-24}$ | g |
| Proton–electron mass ratio | $m_p/m_e$ | 1836.152 667 | |
| Bohr radius | $a_0$ | $0.529177 \times 10^{-8}$ | cm |
| Fine structure constant | $\alpha$ | $7.297352 \times 10^{-3}$ =1/137.036 | |
| Hartree energy | $E_h$ | 27.211 383 | eV |
| Rydberg constant | $R_\infty$ | $1.097373 \times 10^5$ | $cm^{-1}$ |
| | $R_\infty hc/e$ | 13.605 692 | eV |

[a]P.J. Mohr and B.N. Taylor, CODATA Recommended Values of the Fundamental Physical Constants–1998, *Physics Today*, August (2000).

**Table A1.2** Conversion table of energy units

| | J | eV | kcal/mol | $cm^{-1}$ | K | Hz |
|---|---|---|---|---|---|---|
| $J$ | 1 | $6.2415 \times 10^{18}$ | $1.4383 \times 10^{20}$ | $5.0341 \times 10^{22}$ | $7.2430 \times 10^{22}$ | $1.5092 \times 10^{33}$ |
| $eV$ | $1.6022 \times 10^{-19}$ | 1 | 23.0420 | $8.0655 \times 10^{3}$ | $1.1604 \times 10^{4}$ | $2.4180 \times 10^{14}$ |
| $kcal/mol$ | $6.9524 \times 10^{-21}$ | $4.3399 \times 10^{-2}$ | 1 | $0.3499 \times 10^{3}$ | $0.5035 \times 10^{3}$ | $1.0494 \times 10^{13}$ |
| $cm^{-1}$ | $1.9864 \times 10^{-23}$ | $1.2398 \times 10^{-4}$ | $2.8580 \times 10^{-3}$ | 1 | 1.4388 | $2.9979 \times 10^{10}$ |
| $K$ | $1.3806 \times 10^{-23}$ | $8.6173 \times 10^{-5}$ | $1.9861 \times 10^{-3}$ | 0.6950 | 1 | $2.0837 \times 10^{10}$ |
| $Hz$ | $6.6261 \times 10^{-34}$ | $4.1357 \times 10^{-15}$ | $9.5292 \times 10^{-14}$ | $3.3356 \times 10^{-11}$ | $4.7992 \times 10^{-11}$ | 1 |

# Appendix 2

# Some Necessary Mathematical Apparatus

## A2.1 Vector and Tensor Calculus

### A2.1.1 Definition of vector; the addition law

If *scalar* is characterized by one quantity – its *magnitude*, vector is characterized by $n$ quantities, where n is the dimension of the space. These quantities are called the *components* of the vector and can be interpreted as projections of the vector onto the coordinate axes. In the three-dimensional space and the Cartesian coordinate frame, a vector **A** is characterized by 3 components: $A_x$, $A_y$ and $A_z$, which completely determine the vector. This is an algebraic representation of the vector.

In the geometrical representation that is visual only in space with $n \leq 3$, the vector is characterized by its magnitude and direction. It may be conveniently represented by an arrow whose length is proportional to its magnitude in the direction of the vector. In this representation, the vector addition:

$$\mathbf{A} + \mathbf{B} = \mathbf{C} \tag{A2.1}$$

places the rear end of vector **B** at the point of vector **A**. Vector **C** is represented by an arrow drawn from the rear of **A** to the point of **B** (Figure A2.1).

The vector subtraction can be easily understand from Figure A2.1, if we take into account that:

$$\mathbf{B} = \mathbf{C} - \mathbf{A} \tag{A2.2}$$

For the subtraction of two vectors (**C**–**A**), they are drawn from one common point. The result of the subtraction, the vector **B**, is the vector drawn from the point of **A** to the point of **C**. Thus, it is directed to the point of the vector-minuend.

The connection between a vector and its component is shown in Figure A2.2, where the unit basis vectors along each of coordinate axes of the Cartesian frame are introduced. Using the basis vectors and the vector addition rule, the vector **A** can be presented as the vector sum of its components:

$$\mathbf{A} = A_x\mathbf{e}_x + A_y\mathbf{e}_y + A_z\mathbf{e}_z \tag{A2.3}$$

where $A_x\mathbf{e}_x$ is a vector directed in the positive x-direction with the magnitude equal to $A_x$. By successively applying the Pythagorean theorem, the magnitude of the vector **A** is:

$$A = \left(A_x^2 + A_y^2 + A_z^2\right)^{\frac{1}{2}} \tag{A2.4}$$

*Intermolecular Interactions* I.G. Kaplan

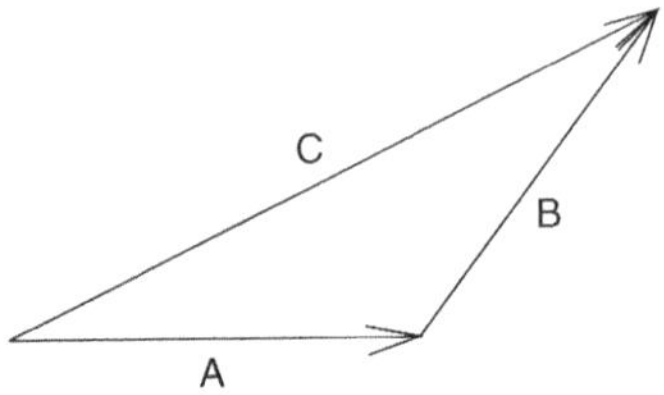


**Figure A2.1** Triangle law of the vector addition

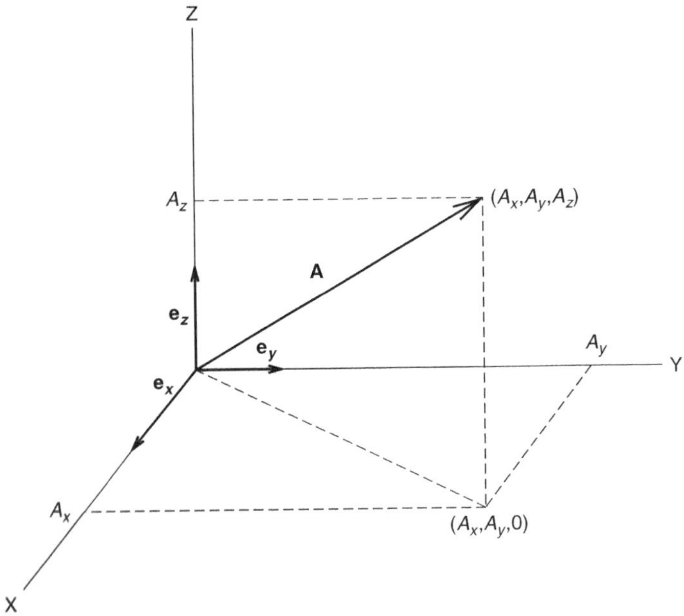


**Figure A2.2** Cartesian components of vector **A**

With the aid of Equation (A2.3), addition and subtraction of vectors may be carried out in terms of their components:

$$\mathbf{C} = \mathbf{A} \pm \mathbf{B} = (A_x \pm B_x)\,\mathbf{e}_x + \left(A_y \pm B_y\right)\mathbf{e}_y + (A_z \pm B_z)\,\mathbf{e}_z \tag{A2.5}$$

Thus, the components of the resulting vector are:

$$C_x = A_x \pm B_x, \quad C_y = A_y \pm B_y, \quad C_z = A_z \pm B_z. \tag{A2.6}$$

It can be proved that the following properties hold:

1. Addition of vectors is commutative:

$$\mathbf{A} + \mathbf{B} = \mathbf{B} + \mathbf{A}$$

2. Addition of vectors is associative:

$$(\mathbf{A} + \mathbf{B}) + \mathbf{C} = \mathbf{A} + (\mathbf{B} + \mathbf{C})$$

3. Scalar multiplication is distributive:

$$a\,(\mathbf{A}+\mathbf{B}) = a\mathbf{A} + a\mathbf{B}$$

$$(a+b)\,\mathbf{A} = a\mathbf{A} + b\mathbf{A}$$

4. Scalar multiplication is associative:

$$(ab)\,\mathbf{A} = a\,(b\mathbf{A})$$

### A2.1.2 Scalar and vector products; triple scalar product

There are two types of multiplication product of two vectors: a *scalar* or *dot* product and a *vector* or *cross* product.

The projection of a vector **A** onto a coordinate axes, which define its Cartesian components, is a special case of the scalar product of **A** and the basis unit vectors:

$$A_x = \mathbf{A}\cdot\mathbf{e}_x, \quad A_y = \mathbf{A}\cdot\mathbf{e}_y, \quad A_z = \mathbf{A}\cdot\mathbf{e}_z \tag{A2.7}$$

In the general case:

$$\mathbf{A}\cdot\mathbf{B} = A_x B_x + A_y B_y + A_z B_z \tag{A2.8}$$

This formula is easily obtained by applying Equation (A2.3) for the vector **B** and Equation (A2.7) for the vector **A**.

The scalar product is commutative:

$$\mathbf{A}\cdot\mathbf{B} = \mathbf{B}\cdot\mathbf{A} \tag{A2.9}$$

it is distributive:

$$\mathbf{A}\cdot(\mathbf{B}+\mathbf{C}) = \mathbf{A}\cdot\mathbf{B} + \mathbf{A}\cdot\mathbf{C}, \tag{A2.10}$$

and it is associative:

$$\mathbf{A}\cdot(a\mathbf{B}) = (a\mathbf{A})\cdot\mathbf{B} = a\,(\mathbf{A}\cdot\mathbf{B}) \tag{A2.11}$$

where $a$ is a number.

Vector **A** can be projected onto the direction of vector **B** or vice versa. This leads to the definition:

$$\mathbf{A}\cdot\mathbf{B} = AB\cos\theta \tag{A2.12}$$

where $\theta$ is the angle between the directions of vectors **A** and **B**. From Equation (A2.12) follows:

$$\mathbf{A}\cdot\mathbf{B} = 0 \quad \text{if} \quad \mathbf{A}\perp\mathbf{B} \tag{A2.13}$$

The components of vector **A** (Equation (A2.7)) are expressed via the direction cosines:

$$A_x = A\ cos\,(\mathbf{A},\mathbf{e}_x)\,, \quad A_y = A\ cos(\mathbf{A},\mathbf{e}_y), \quad A_z = A\ cos\,(\mathbf{A},\mathbf{e}_z)\,. \tag{A2.7a}$$

Equation (A2.12) is labelled a geometrical definition of the scalar product, then Equation (A2.8) can be considered as an algebraic definition of the scalar product.

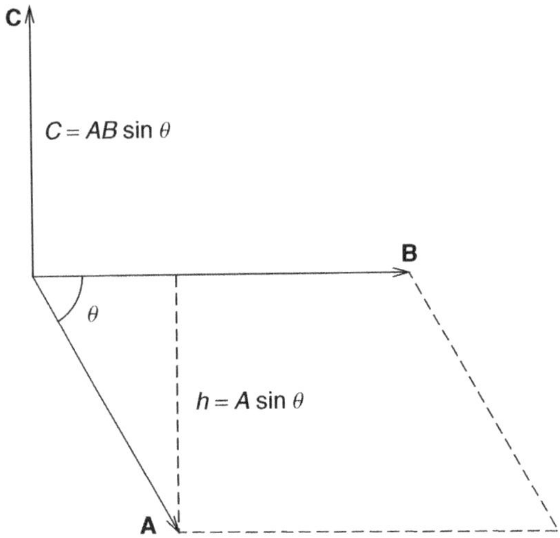


**Figure A2.3** The area of a parallelogram constructed on vector **A** and **B** is equal to the magnitude of the vector product

Unlike the scalar product, the *vector* or *cross* product:

$$\mathbf{C} = \mathbf{A} \times \mathbf{B} \tag{A2.14}$$

is a vector directed perpendicular to the plane of **A** and **B**, so that **A**, **B**, and **C** form a right-handed system, with the magnitude:

$$C = AB\ \sin\theta \tag{A2.15}$$

Equation (A2.15) can be interpreted as the area of parallelogram defined by **A** and **B** (Figure A2.3). The vector product is anticommutative:

$$\mathbf{A} \times \mathbf{B} = -\mathbf{B} \times \mathbf{A} \tag{A2.16}$$

The component of the vector product can be found using the decomposition (A2.3). Namely:

$$C_x = A_y B_z - A_z B_y, \quad C_y = A_z B_x - A_x B_z,$$
$$C_z = A_x B_y - A_y B_x. \tag{A2.17}$$

The vector product may be conveniently presented by a *determinant* (see next section):

$$\mathbf{C} = \begin{vmatrix} \mathbf{e}_x & \mathbf{e}_y & \mathbf{e}_z \\ A_x & A_y & A_z \\ B_x & B_y & B_z \end{vmatrix} \tag{A2.18}$$

Expansion of this determinant across the top row gives exactly the components (A2.17).

The *triple scalar* product is defined as:

$$\mathbf{A} \cdot (\mathbf{B} \times \mathbf{C}) \tag{A2.19}$$

and is a scalar. It has a simple geometric interpretation. The magnitude of the triple scalar product (A2.19) is equal to the volume of the parallelepiped defined by **A**, **B** and **C**. It is easy to understand, if we remember that the vector product $\mathbf{B} \times \mathbf{C}$ is a vector perpendicular to the plane of **B** and **C** with a magnitude equal to the area of the parallelogram defined by **B** and **C**. Then forming the scalar product of this vector and the vector **A**, we obtain the volume of the parallelepiped.

It can be proved that the dot and the cross operation in Equation (A2.19) may be interchanged:

$$\mathbf{A} \cdot (\mathbf{B} \times \mathbf{C}) = (\mathbf{A} \times \mathbf{B}) \cdot \mathbf{C} \tag{A2.20}$$

The explicit form of the triple scalar product can be directly obtained from the definition of the scalar product (A2.8) using the expressions for the components of the vector product (A2.17):

$$A \cdot (B \times C) = A_x \left(B_y C_z - B_z C_y\right) + A_y \left(B_z C_x - B_x C_z\right) + A_z \left(B_x C_y - B_y C_z\right) \tag{A2.21}$$

This rather cumbersome expression can be written in a compact form using the determinant notation:

$$A \cdot (B \times C) = \begin{vmatrix} A_x & A_y & A_z \\ B_x & B_y & B_z \\ C_x & C_y & C_z \end{vmatrix} \tag{A2.22}$$

In the next section, we briefly discuss the definition and properties of determinants.

### A2.1.3 Determinants

The *determinant* is defined as a square array of numbers (or functions):

$$D_n = \begin{vmatrix} a_1 & a_2 & \cdots & a_n \\ b_1 & b_2 & \cdots & b_n \\ \cdot & \cdot & \cdots & \cdot \\ q_1 & q_2 & \cdots & q_n \end{vmatrix} \tag{A2.23}$$

The number of columns (rows) in the array is called the *order* of the determinant. The determinant is a scalar quantity and its magnitude is equal to:

$$D_n = \sum_P (-1)^p \, P a_1 b_2 c_3 \ldots \tag{A2.24}$$

where the sum is taken over all $n!$ permutations $P$ of the indices in the product of main diagonal elements and $p$ is the parity of the permutation $P$. The parity of permutation is equal to an even number for an even $P$ and to an odd number for an odd $P$. Thus:

$$(-1)^p = \begin{cases} 1 & \text{for even} \quad P \\ -1 & \text{for odd} \quad P \end{cases} \tag{A2.25}$$

A permutation is even if it involves an even number of transpositions of adjacent indices, and it is odd if it involves an odd number of transpositions, see Section A2.2.3.

The most simple examples are:

$$D_2 = a_1 b_2 - a_2 b_1 \tag{A2.26}$$

$$D_3 = a_1 b_2 c_3 - a_2 b_1 c_3 - a_3 b_2 c_1 - a_1 b_3 c_2 + a_2 b_3 c_1 + a_3 b_1 c_2 \tag{A2.27}$$

Equation (A2.27) for the third-order determinant may be written in another form:

$$D_3 = a_1 \begin{vmatrix} b_2 & b_3 \\ c_2 & c_3 \end{vmatrix} - a_2 \begin{vmatrix} b_1 & b_3 \\ c_1 & c_3 \end{vmatrix} + a_3 \begin{vmatrix} b_1 & b_2 \\ c_1 & c_2 \end{vmatrix} \tag{A2.28}$$

This is a particular case of the Laplace expansion. The nth-order determinant may be expanded as a linear combination of the products of the elements of any row (or any column) and the $(n-1)$-order determinants formed by deleting the row and column in which the element is located. This $(n-1)$-order determinate is called a *minor* and denoted as $M_{ij}$, if the element is in the *ith* row and the *jth* column. The sign of each term in the expansion is $(-1)^{i+j}$. If the element located in the *ith* row and the *jth* column is designated as $a_{ij}$, the Laplace expansion across the ith row may be represented as:

$$D_n = \sum_{j=1}^{n} (-1)^{i+j} a_{ij} M_{ij} \tag{A2.29}$$

Equation (A2.28) is easily obtained from the general formula (Equation (A2.29)) as an expansion across the first row in the case $n = 3$. The Laplace expansion is useful in the evaluation of high-order determinants in which a lot of the elements are zero.

The determinant has an important property of antisymmetry: it changes sign if any two rows (or two columns) are interchanged. This property follows directly from Equation (A2.24) and can be obtained also from the Laplace expansion (A2.29).

### A2.1.4 Vector analysis; gradient, divergence and curl

Let us consider a scalar function in the three-dimensional (3D) space, $\varphi(\mathbf{r}) = \varphi(x, y, z)$. The derivative of this function with respect to some direction **s** (**s** is an unit vector) is defined as a limit:

$$\frac{\partial \varphi(\mathbf{r})}{\partial s} = \lim_{\epsilon \to 0} \frac{\varphi(\mathbf{r} + \epsilon \mathbf{s}) - \varphi(\mathbf{r})}{\epsilon} \tag{A2.30}$$

It is clear that this derivative depends upon the direction chosen. The Cartesian components of the unit vector **s** (see Equation (A2.7a)) are:

$$s_x = cos\,(\mathbf{s}, \mathbf{e}_x)\,, \quad s_y = cos(\mathbf{s}, \mathbf{e}_y), \quad s_z = cos\,(\mathbf{s}, \mathbf{e}_z) \tag{A2.31}$$

Therefore, in the coordinate representation, the nominator in Equation (A.30) is rewritten as:

$$\varphi\,(\mathbf{r} + \epsilon\mathbf{s}) - \varphi\,(\mathbf{r})$$
$$= \varphi(x + \epsilon\; cos\,(\mathbf{s}, \mathbf{e}_x)\,,\; y + \epsilon\; cos(\mathbf{s}, \mathbf{e}_y),\; z + \epsilon\; cos\,(\mathbf{s}, \mathbf{e}_z)) - \varphi\,(x, y, z)$$

After expansion of this difference into the Taylor series up to the first degree of $\epsilon$:

$$\varphi\,(\mathbf{r} + \epsilon\mathbf{s}) - \varphi\,(\mathbf{r}) = \epsilon\left[\frac{\partial\varphi}{\partial x} cos\,(\mathbf{s}, \mathbf{e}_x) + \frac{\partial\varphi}{\partial y} cos(\mathbf{s}, \mathbf{e}_y) + \frac{\partial\varphi}{\partial z} cos\,(\mathbf{s}, \mathbf{e}_z)\right]$$

and substitution of it in the definition of the derivative (Equation (A2.30)), we obtain:

$$\frac{\partial\varphi}{\partial s} = \frac{\partial\varphi}{\partial x} cos\,(\mathbf{s}, \mathbf{e}_x) + \frac{\partial\varphi}{\partial y} cos(\mathbf{s}, \mathbf{e}_y) + \frac{\partial\varphi}{\partial z} cos\,(\mathbf{s}, \mathbf{e}_z) \tag{A2.32}$$

This expression can be interpreted as a scalar product of the unit vector **s** on a vector with components $\left(\frac{\partial\varphi}{\partial x}, \frac{\partial\varphi}{\partial y}, \frac{\partial\varphi}{\partial z}\right)$. The latter is called the *gradient* of $\varphi$ or *del* $\varphi$ and is designated as:

$$grad\;\varphi = \nabla\varphi = \frac{\partial\varphi}{\partial x}\mathbf{e}_x + \frac{\partial\varphi}{\partial y}\mathbf{e}_y + \frac{\partial\varphi}{\partial z}\mathbf{e}_z \tag{A2.33}$$

where $\nabla(del)$ is the vector differential operator:

$$\nabla = \mathbf{e}_x\frac{\partial}{\partial x} + \mathbf{e}_y\frac{\partial}{\partial y} + \mathbf{e}_z\frac{\partial}{\partial z} \tag{A2.34}$$

Thus:

$$\frac{\partial\varphi}{\partial s} = \mathbf{s}\cdot\nabla\varphi = \mathbf{s}\cdot grad\;\varphi = |grad\;\varphi|\;\; cos\;(grad\;\varphi, \mathbf{s}) \tag{A2.35}$$

From Equation (A2.35) it follows that the derivative of $\varphi\,(\mathbf{r})$ with respect to some direction **s** is maximum in the direction of the $grad\;\varphi$ (in this case $cos(grad\;\varphi, \mathbf{s}) = 1$). This allows the geometric definition of the gradient to be given:

> grad $\varphi$ is a vector having the direction of the maximum space rate of change of $\varphi$ and the magnitude equal to the derivative of $\varphi$ with respect to this direction.

The gradient possesses the following properties:

$$grad\,(\varphi + \psi) = grad\ \varphi + grad\ \psi \tag{A2.36}$$

$$grad\,(\varphi\psi) = \varphi\ grad\ \psi + \psi\ grad\ \varphi \tag{A2.37}$$

$$grad\ F\,(\varphi) = \frac{dF}{d\varphi}\ grad\ \varphi \tag{A2.38}$$

In physics there is often a need to calculate a particular case of Equation (A2.38) with $\varphi = r$, where $r = \left(x^2 + y^2 + z^2\right)^{\frac{1}{2}}$ is the magnitude of the radius-vector $\mathbf{r}$:

$$grad\ F\,(r) = \frac{dF}{dr} grad\ r \tag{A2.38a}$$

Direct application of Expression (A2.34) for $\nabla$ to $r$ results in:

$$grad\ r = \nabla r = \left(\mathbf{e}_x \frac{\partial}{\partial x} + \mathbf{e}_y \frac{\partial}{\partial y} + \mathbf{e}_z \frac{\partial}{\partial z}\right)\left(x^2 + y^2 + z^2\right)^{\frac{1}{2}}$$

$$= \frac{1}{2r}\left(2x\mathbf{e}_x + 2y\mathbf{e}_y + 2z\mathbf{e}_z\right) = \frac{\mathbf{r}}{r} \tag{A2.39}$$

The gradient of the magnitude of the radius-vector is the unit vector in the radial direction. It is now easy to calculate the gradient of an arbitrary scalar function of $r$. For instance:

$$grad\ \frac{1}{r} = -\frac{\mathbf{r}}{r^3} \tag{A2.40}$$

$$grad\ r^n = nr^{n-2}\mathbf{r} \tag{A2.41}$$

Application of $\nabla$ to a scalar product of two vectors is performed employing Equations (A2.34) and (A2.8). In the case when a radius-vector is multiplied by a permanent vector, the result is:

$$grad\,(\mathbf{A}\cdot\mathbf{r}) = \mathbf{A} \quad for\ \mathbf{A} = \ const \tag{A2.42}$$

As we discussed above, $\nabla$ is the vector differential operator, that is, it has both the vector and differential properties. The gradient was defined as an operation of $\nabla$ on a scalar function. Consider the operation of $\nabla$ on a vector. There are two kinds of vector multiplication: dot and cross products.

The dot product $\nabla \cdot \mathbf{A}$, where $\mathbf{A}$ is an arbitrary vector, is called the *divergence* of $\mathbf{A}$ and can be found using the algebraic definition of dot (scalar) product (A2.8) and taking into account the differential properties of the operator $\nabla$:

$$div\ \mathbf{A} = \nabla\cdot\mathbf{A} = \frac{\partial A_x}{\partial x} + \frac{\partial A_y}{\partial y} + \frac{\partial A_z}{\partial z} \tag{A2.43}$$

The physical and geometrical sense of the divergence of $\mathbf{A}$ is connected with a flow of the vector $\mathbf{A}$ over some arbitrary surfaces. The divergence is define as a limit:

$$div\ \mathbf{A} = \lim_{\Delta V \to 0} \frac{\oint_S A_n\,dS}{\Delta V} \tag{A2.44}$$

where $A_n$ is the component of **A** normal to the surface and $\Delta V$ is a volume confined by this surface. According to the *Gauss theorem*:

$$\oint\limits_S A_n \, dS = \int\limits_V div\mathbf{A} \, dV \tag{A2.45}$$

that is, the flow of vector **A** over an arbitrary closed surface is equal to the volume integral of the divergence of the vector over the volume $V$ confined by the surface $S$.

The vector product, or cross product, $\nabla \times \mathbf{A}$ is called the *curl* of **A**. According to Equation (A2.18):

$$curl\ \mathbf{A} = \nabla \times \mathbf{A} = \begin{vmatrix} \mathbf{e}_x & \mathbf{e}_y & \mathbf{e}_z \\ \frac{\partial}{\partial x} & \frac{\partial}{\partial y} & \frac{\partial}{\partial z} \\ A_x & A_y & A_z \end{vmatrix} \tag{A2.46}$$

or:

$$\nabla \times \mathbf{A} = \mathbf{e}_x \left( \frac{\partial A_z}{\partial y} - \frac{\partial A_y}{\partial z} \right) + \mathbf{e}_y \left( \frac{\partial A_x}{\partial z} - \frac{\partial A_z}{\partial x} \right) + \mathbf{e}_z \left( \frac{\partial A_y}{\partial x} - \frac{\partial A_x}{\partial y} \right) \tag{A2.46a}$$

From Equation (A2.46a) follows:

$$curl\ \mathbf{r} = \nabla \times \mathbf{r} = 0 \tag{A2.47}$$

The divergence of **A** is connected with the flow of the vector **A** over some surface (Equation (A2.44)), whereas the curl of **A** is connected with the line integral describing the circulation of the vector **A** around some contour. Any component of the curl can be define as a limit:

$$curl_n\ \mathbf{A} = \lim_{\Delta S \to 0} \frac{\oint\limits_C A_r \, d\mathbf{r}}{\Delta S} \tag{A2.48}$$

The component of the *curl* **A** in the arbitrary direction **n** is a limit per unit area of the circulation of the vector **A** around a contour $C$ that confines some plane surface perpendicular to **n**. According to the *Stokes theorem*:

$$\oint\limits_C A_r d\mathbf{r} = \int\limits_S curl_n\ \mathbf{A} \, dS \tag{A2.49}$$

The circulation of **A** around closed contour $C$ is equal to the flow of the *curl* **A** over the surface $S$ confined by this contour.

Employing the properties of scalar and vector products one obtains:

$$div\,(grad\ \varphi) = \nabla \cdot \nabla\varphi = \nabla^2\varphi = \frac{\partial^2\varphi}{\partial x^2} + \frac{\partial^2\varphi}{\partial y^2} + \frac{\partial^2\varphi}{\partial z^2} \tag{A2.50}$$

$$div\,(curl\ \mathbf{A}) = \nabla \cdot (\nabla \times \mathbf{A}) = 0 \tag{A2.51}$$

$$curl\,(grad\ \varphi) = \nabla \times \nabla\varphi = 0 \tag{A2.52}$$

$$curl\,(curl\ \mathbf{A}) = \nabla \times (\nabla \times A) = grad\ div\ \mathbf{A} - \nabla^2\mathbf{A} \tag{A2.53}$$

The operator $\nabla \cdot \nabla = \nabla^2$ is named the *Laplacian* and designated as $\Delta$:

$$\Delta = \nabla^2 = \frac{\partial^2}{\partial x^2} + \frac{\partial^2}{\partial y^2} + \frac{\partial^2}{\partial z^2} \tag{A2.54}$$

Below some useful formulae for different types of products are presented:

$$div\,(\varphi \mathbf{A}) = \varphi\; div\; \mathbf{A} + \mathbf{A}\; grad\; \varphi \tag{A2.55}$$

$$div\;(\mathbf{A} \times \mathbf{B}) = \mathbf{B}\; div\; \mathbf{A} - \mathbf{A}\; div\; \mathbf{B} \tag{A2.56}$$

$$curl\;(\varphi \mathbf{A}) = \varphi\; curl\; \mathbf{A} + (grad\; \varphi) \times \mathbf{A} \tag{A2.57}$$

$$curl\;(\mathbf{A} \times \mathbf{B}) = (\mathbf{B} \cdot \nabla)\,\mathbf{A} + (\mathbf{A} \cdot \nabla)\,\mathbf{B} + \mathbf{A}\; div\; \mathbf{B} - \mathbf{B}\; div\; \mathbf{A} \tag{A2.58}$$

$$grad\;(\mathbf{A} \cdot \mathbf{B}) = (\mathbf{B} \cdot \nabla)\,\mathbf{A} + (\mathbf{A} \cdot \nabla)\,\mathbf{B} + \mathbf{B} \times curl\; \mathbf{A} + \mathbf{A} \times curl\; \mathbf{B} \tag{A2.59}$$

## A2.1.5 Vector spaces and matrices

A vector space of $n$ dimensions consists of the set of all the vectors $\mathbf{x}$ which can be obtained by forming all possible linear combinations of $n$ linearly independent basis vectors $\mathbf{e}_i$:

$$\mathbf{x} = \sum_{i=1}^{n} x_i \mathbf{e}_i \tag{A2.60}$$

As we know, $x_i$ is called the *component* of the vector $\mathbf{x}$ in the directions $\mathbf{e}_i$; and $x_i$ may be a complex quantity, in which case the vector space is complex. In the general case, the basis vectors $\mathbf{e}_i$ may not be unit and orthogonal, that is, do not form an orthonormal set.

Under a linear transformation from one system of basis vectors to another, the components of a vector in the old system transform linearly into the components in the new system. Thus, if:

$$\mathbf{e}'_k = \sum_i s_{ik} \mathbf{e}_i \tag{A2.61}$$

then from:

$$\mathbf{x} = \sum_i x_i \mathbf{e}_i = \sum_k x'_k \mathbf{e}'_k = \sum_{k,i} x'_k s_{ik} \mathbf{e}_i$$

it follows that:

$$x_i = \sum_k s_{ik} x'_k \tag{A2.62}$$

An array of numbers $s_{ik}$ in Equation (A2.62) forms a *matrix* $S$. In the case considered, it is a square array, but matrices may be also rectangular. The matrix may be defined as a square or rectangular array of numbers or functions that obeys certain laws (see below). An arbitrary matrix with $m$ rows and $n$ columns is represented as:

$$A = \begin{pmatrix} a_{11} & a_{12} & \cdots & a_{1n} \\ a_{21} & a_{22} & \cdots & a_{2n} \\ \cdot & \cdot & \cdots & \cdot \\ a_{m1} & a_{m2} & \cdots & a_{mn} \end{pmatrix} \tag{A2.63}$$

where $a_{ij}$ is the matrix element in the *ith* row and *jth* column. Contrary to a determinant, a matrix is not a single number; it is an array of numbers (functions). Matrix (A2.63) is designated as an $m \times n$ matrix.

The addition law for matrices is as for numbers: $A + B = C$ if and only if $a_{ij} + b_{ij} = c_{ij}$ for all matrix elements. This means that for matrices the commutation and associative laws are satisfied.

The multiplication of matrix $A$ by the scalar quantity $\alpha$ produces a new matrix $\alpha A$, in which each element of matrix $A$ is multiplied by this scalar factor. So, the matrix $\alpha A$ has elements $\alpha a_{ij}$.

Two different kind of products of two matrices may be defined: the *inner* product and the *direct* product. The inner product is defined as a matrix with elements:

$$C = AB \quad with \quad c_{ij} = \sum_k a_{ik} b_{kj} \tag{A2.64}$$

The matrix element $c_{ij}$ can be interpreted as a scalar product of the *ith* row of $A$ with the *jth* column of $B$. For instance:

$$\begin{pmatrix} a_{11} & a_{12} \\ a_{21} & a_{22} \end{pmatrix} \begin{pmatrix} b_{11} & b_{12} \\ b_{21} & b_{22} \end{pmatrix} = \begin{pmatrix} a_{11}b_{11} + a_{12}b_{21} & a_{11}b_{12} + a_{12}b_{22} \\ a_{21}b_{11} + a_{22}b_{21} & a_{21}b_{12} + a_{22}b_{22} \end{pmatrix}$$

The multiplication law (A2.64) demands that the number of columns in $A$ must be equal to the number of rows in $B$. For square matrices it means that the inner product may be defined only for matrices of the same order $m \times m$. For rectangular matrices, as follows from Equation (A2.64):

$$(m_1 \times n_1)(n_1 \times n_2) = (m_1 \times n_2)\,. \tag{A2.65}$$

The resulting matrix has the number of rows as in the first matrix of the product and the number of columns as in the second matrix of the product.

It can be easily checked that the matrix multiplication is not commutative:

$$AB \neq BA \tag{A2.66}$$

but it is associative.

The *direct* (*Kronecker*) product:

$$C = A \otimes B \tag{A2.67}$$

of the square matrices $m \times m$ $A$ and $n \times n$ $B$ is a $mn \times mn$ matrix with elements:

$$c_{ij,kl} = a_{ij} b_{kl} \tag{A2.68}$$

For instance, if $A$ and $B$ are both $2 \times 2$ matrices:

$$A \otimes B = \begin{pmatrix} a_{11}B & a_{12}B \\ a_{21}B & a_{22}B \end{pmatrix} = \begin{pmatrix} a_{11}b_{11} & a_{11}b_{12} & a_{12}b_{11} & a_{12}b_{12} \\ a_{11}b_{21} & a_{11}b_{22} & a_{12}b_{21} & a_{12}b_{22} \\ a_{21}b_{11} & a_{21}b_{12} & a_{22}b_{11} & a_{22}b_{12} \\ a_{21}b_{21} & a_{21}b_{22} & a_{22}b_{21} & a_{22}b_{22} \end{pmatrix} \tag{A2.69}$$

The vector in n-dimensional space is characterized by its $n$ components and can be represented as a $n \times 1$ matrix, that is the one-column matrix:

$$\underline{x} = \begin{pmatrix} x_1 \\ x_2 \\ \vdots \\ x_n \end{pmatrix} \tag{A2.70}$$

This allows transformation (A2.62) to be written in the matrix form:

$$X = SX' \tag{A2.71}$$

where $S$ is the $n \times n$ matrix. The product of a square matrix with a column matrix is a column matrix, in accordance with Equation (A2.65). From Equation (A2.71) follows:

$$X' = S^{-1}X \tag{A2.72}$$

where $S^{-1}$ is the matrix of the inverse transformation. According to the definition of the *inverse* matrix:

$$S^{-1}S = SS^{-1} = 1 \tag{A2.73}$$

Instead of transforming the basis vectors, we can carry out a transformation of the vector space, by which we associate with each vector $\mathbf{x}$ a vector $\mathbf{y}$:

$$\mathbf{y} = A\mathbf{x} \tag{A2.74}$$

The quantity $A$ can be regarded as an operator that carries the vector $\mathbf{x}$ into the vector $\mathbf{y}$. The analytical form of $A$ is an $n \times n$ matrix which connects the components of the vectors:

$$y_i = \sum_k a_{ik}x_k \tag{A2.75}$$

or in matrix notation:

$$Y = AX \tag{A2.76}$$

Let the vectors $\mathbf{y}$ and $\mathbf{x}$ be connected to each other by the relationship (A2.74) and carry out the transformation (A2.61) of the basis vectors. Then in the new coordinate system:

$$Y' = A'X' \tag{A2.77}$$

To find the relation between $A'$ and $A$, it is necessary to pass to the old coordinate system, carry out in it the transformation (A2.75) and return to the new system:

$$Y' = S^{-1}Y, \quad Y = AX, \quad X = SX'$$

or

$$Y' = S^{-1}ASX' \tag{A2.78}$$

On comparing Equations (A2.77) and (A2.78), we obtain:

$$A' = S^{-1}AS \tag{A2.79}$$

If the basis vectors form an orthonormal set

$$\mathbf{e}_i \cdot \mathbf{e}_j = \delta_{ij} \tag{A2.80}$$

then the scalar product of two vectors is given by:

$$\mathbf{x} \cdot \mathbf{y} = \sum_i x_i^* y_i, \tag{A2.81}$$

and the square of the length of a vector is equal to the sum of the squared moduli of its components:

$$\mathbf{x} \cdot \mathbf{x} = |\mathbf{x}|^2 = \sum_i |x_i|^2 \tag{A2.82}$$

In this section, the general case of the complex vector space is considered, as it takes place in quantum mechanics. So, the components of vectors are complex numbers; $x_i^*$ designates the complex conjugated of $x_i$, and $|x_i|^2$ has to be written instead of $x_i^2$. This is the reason for the difference between Equations (A2.81) and (A2.82) and corresponding Equations (A2.8) and (A2.4).

A linear transformation is said to be *unitary* if it leaves the scalar product of two vectors unchanged. It follows from this definition of a unitary transformation that:

$$\sum_k x_k'^* y_k' = \sum_i x_i^* y_i = \sum_{i,k,l} s_{ik}^* s_{il} x_k'^* y_l' \tag{A2.83}$$

Equation (A2.83) holds if the condition:

$$\sum_i s_{ik}^* s_{il} = \delta_{kl} \tag{A2.84}$$

is fulfilled. However, according to the definition of the inverse of a matrix (Equation (A2.73)):

$$\sum_i s_{ki}^{-1} s_{il} = \delta_{kl} \tag{A2.85}$$

and it therefore follows from a comparison of Equations (A2.84) and (A2.85) that $s_{ki}^{-1} = s_{ik}^*$. This can be written symbolically in the form:

$$S^{-1} = \widetilde{S}^* \tag{A2.86}$$

or

$$S\widetilde{S}^* = \widetilde{S}^* S = I \tag{A2.87}$$

$\widetilde{S}$ denotes the *transpose* of the matrix $S$. Its elements are related to those of $S$ by:

$$\widetilde{s}_{ik} = s_{ki} \tag{A2.88}$$

Matrices that satisfy the conditions of Equations (A2.86) or (A2.87) are called *unitary*. From Equation (A2.87) it can be seen that in addition to the condition of Equation (A2.84) the elements of a unitary matrix also obey the relationship:

$$\sum_k s_{ik}^* s_{lk} = \delta_{il} \tag{A2.89}$$

In the case of real matrices, one has instead of Equation (A2.87):

$$S\widetilde{S} = \widetilde{S}S = I \tag{A2.90}$$

Matrices that satisfy Equation (A2.90) are said to be *orthogonal*. Their elements satisfy the following orthogonality conditions:

$$\sum_i s_{ik} s_{il} = \delta_{kl}, \quad \sum_k s_{ik} s_{lk} = \delta_{il} \tag{A2.91}$$

### A2.1.6 Tensors

A simple product of two vectors **xy** defined in the n-dimensional space (not the scalar or vector product) is called the *diad*. Under a linear transformation from one system of basis vectors to another, the components of vectors transform linearly, according to Equation (A2.62). For the components of the diad the following is obtained:

$$x_i y_j = \sum_{k,l} s_{ik} s_{jl} x'_k y'_l \tag{A2.92}$$

The matrix of the diad transformation is the direct product of the transformation matrices $S \otimes S$ (Equation (A2.68)) and is the $(n^2 \times n^2)$ matrix. The $n^2$ quantities $x_i y_j$ constitute a *second rank tensor* defined in the n-dimensional vector space (nD space).

Any tensor of rank two can be decomposed into linear combinations of the diads formed from the basis vectors. In the 3D space with the aid of decomposition (A2.3), the diad **AB** can be represented as:

$$\mathbf{AB} = \sum_{ij} a_i b_j \mathbf{e}_i \mathbf{e}_j \tag{A2.93}$$

Let $\mathbf{T}^{(2)}$ designate a second rank tensor, Equation (A2.93) can be written as:

$$\mathbf{T}^{(2)} = \sum_{i,j} T_{ij} \mathbf{e}_i \mathbf{e}_j \tag{A2.94}$$

The $3^2$ Cartesian components of the tensor $\mathbf{T}^{(2)}$ form a $3 \times 3$ matrix:

$$\begin{pmatrix} T_{xx} & T_{xy} & T_{xz} \\ T_{yx} & T_{yy} & T_{yz} \\ T_{zx} & T_{zy} & T_{zz} \end{pmatrix} \tag{A2.95}$$

that is called the *matrix of the tensor* $\mathbf{T}_2$. As in the case of matrices, a tensor is *symmetric* if:

$$T_{ij} = T_{ji} \tag{A2.96}$$

a tensor is *antisymmetric* if:

$$T_{ij} = -T_{ji} \tag{A2.97}$$

and a tensor is orthogonal if:

$$\sum_k T_{ik} T_{jk} = \sum_k T_{ki} T_{kj} = \delta_{ij} \tag{A2.98}$$

An arbitrary tensor of the second rank can be presented as a sum of symmetric and antisymmetric tensors:

$$\mathbf{T}^{(2)} = \mathbf{S}^{(2)} + \mathbf{A}^{(2)} \tag{A2.99}$$

where:

$$S_{ij} = \tfrac{1}{2}\left(T_{ij} + T_{ji}\right) \textit{ and } A_{ij} = \tfrac{1}{2}\left(T_{ij} - T_{ji}\right) \tag{A2.100}$$

It can always be done because of the evident identity:

$$T_{ij} = \tfrac{1}{2}\left(T_{ij} + T_{ji}\right) + \tfrac{1}{2}\left(T_{ij} - T_{ji}\right)$$

The *Nth rank tensor* in the nD space is defined as the set of $n^N$ components:

$$T^{(N)}_{i_1 i_2 \ldots i_N} = x_{i_1} x_{i_2} \ldots x_{i_N} \tag{A2.101}$$

which under transformation of vector space transforms according to the rule:

$$T^{(N)}_{i_1 i_2 \ldots i_N} = \sum_{k_1, k_2, \ldots k_N} s_{i_1 k_1} s_{i_2 k_2} \ldots s_{i_N k_N} T^{(N)'}_{k_1 k_2 \ldots k_N} \tag{A2.102}$$

A transformation $S$ of the nD space induces a transformation:

$$\Pi_N(S) = \underbrace{S \otimes S \otimes \ldots \otimes S}_{N} \tag{A2.103}$$

in the $n^N$-dimensional space of Nth rank tensors.

Let $\mathbf{T}^{(N)}$ designate a Nth rank tensor, it can be decomposed as a linear combinations of the basis vector products:

$$\mathbf{T}^{(N)} = \sum_{i_1, i_2, \ldots, i_N} T^{(N)}_{i_1 i_2 \ldots i_N} \mathbf{e}_{i_1} \mathbf{e}_{i_2} \ldots \mathbf{e}_{i_N} \tag{A2.104}$$

This decomposition is a generalization of Equation (A2.94).

From the general definition, Equations (A2.101) and (A2.102), it follows that vectors are tensors of rank one and scalars are tensors of rank zero.

The tensors defined by Equation (A2.102) are called the *affine* tensors. If the basis vectors of the vector space form an orthonormal set, the tensors (A2.102) are called the *affine orthogonal*, or *Cartesian*, tensors. Another type of tensor, the *irreducible* tensor, is discussed in Section A2.2.6.

For the affine tensors three types of operation can be defined:

1. Addition of tensors of identical rank:
$$V^{(N)}_{i_1 i_2 \ldots i_N} = T^{(N)}_{i_1 i_2 \ldots i_N} + U^{(N)}_{i_1 i_2 \ldots i_N} \tag{A2.105}$$

2. Tensor multiplication that leads to a tensor whose rank is equal to the sum of the ranks of the tensors forming the product. For instance:
$$V^{(5)}_{ijklm} = T^{(2)}_{ij} U^{(3)}_{klm} \tag{A2.106}$$

3. Tensor multiplication with *contraction* over a pair of indices that leads to a lowering of the rank of the resulting tensor on two:
$$V^{(2N-2)}_{i_2 \ldots i_N j_2 \ldots j_N} = \sum_{i_1} T^{(N)}_{i_1 i_2 \ldots i_N} T^{(N)}_{i_1 j_2 \ldots j_N} \tag{A2.107}$$

This type of multiplication is called the *scalar product of tensors*. If it is applied $N$ times to a product of Nth rank tensors, it gives rise to a scalar, as it takes place in the scalar product of two vectors (Equation (A2.8)).

More details on the vector and tensor calculus can be found in References [1–5].

# A2.2 Group Theory

## A2.2.1 Properties of group operations

### Group postulates

A set of elements $A$, $B$,..., is said to form a group, **G**, if it satisfies the following four conditions:

1. A 'multiplication law' for the elements is specified, i.e., a rule is prescribed according to which each pair of elements $P$, $Q$ is placed into correspondence with some element $R$, which is also contained in the same set. Element $R$ is termed the *product of the elements $P$ and $Q$*, and is written in the form:
$$R = PQ \tag{A2.108}$$

2. The product of the factors is associative:
$$P\,(QR) = (PQ)\,R \tag{A2.109}$$
i.e., in order to specify a product uniquely it is sufficient to specify the order of the factors.

3. Among the elements of the group there is an unit element, $E$, possessing the property:

$$EQ = QE = Q \qquad \text{(A2.110)}$$

for any $Q$ belonging to the group (the condition that the element $Q$ belongs to the group $\mathbf{G}$ is denoted symbolically as follows: $Q \in \mathbf{G}$).

4. For every element $Q \in \mathbf{G}$ there exists an inverse element $Q^{-1} \in \mathbf{G}$ that satisfies the equation:

$$Q^{-1}Q = QQ^{-1} = E \qquad \text{(A2.111)}$$

The element, which is the inverse of a product of elements, is given by:

$$(PQ)^{-1} = Q^{-1}P^{-1} \qquad \text{(A2.112)}$$

as can easily be seen on multiplying $PQ$ by $Q^{-1}P^{-1}$ and using the associative rule for products.

The product of elements in a group is generally noncommutative:

$$PQ \neq QP$$

If all elements of a group satisfy the equation $PQ = QP$, the group is said to be *Abelian*. The *cyclic* groups are a particular case of Abelian groups in which all the elements are obtained by successively raising the power of one of the elements, i.e., the $n$ elements of a cyclic group can be represented as follows:

$$A, A^2, A^3, \ldots, A^n \equiv E \qquad \text{(A2.113)}$$

The following theorem is valid for the elements of a group:

> If $G_a$ runs through all the elements of a group $\mathbf{G}$ and $G_0$ is some fixed element of $\mathbf{G}$, then the product $G_0G_a$ (or $G_aG_0$) also runs through all the elements of the group and, moreover, does so once only.

Any element $G_b$ of the group can, in fact, be obtained by multiplying $G_0$ from the right by $G_a = G_0^{-1}G_b$. Furthermore, the product $G_0G_a$ cannot occur more than once since if $G_0G_a = G_0G_b$, then by multiplying this equation from the left by $G_0^{-1}$ gives $G_a = G_b$. Hence for different $G_a$ all $G_0G_a$ are different.

From this theorem it follows that an arbitrary function of the group elements when summed over all the elements of the group is an invariant:

$$\sum_{G_a} f(G_a) = \sum_{G_a} f(G_0G_a) = \sum_{G_b} f(G_b) \qquad \text{(A2.114)}$$

**Examples of groups**

1. The set of all vectors in three-dimensional space constitutes a group with respect to the operation of addition. In this case the operation of multiplying

the elements of the group is vector addition. Vector addition possesses the associative property, and the unit element is represented by a vector of zero length. Mutually inverse elements of the group are represented by vectors equal in magnitude and opposite to one another in direction.

2. As a more complicated example, we consider the permutations of $N$ objects. The objects are numbered by the integers 1 to $N$. As is well known, from $N$ numbers $N!$ different permutations are formed, which can be represented by the symbol:

$$P = \begin{pmatrix} 1 & 2 & 3 & \cdots & N \\ i_1 & i_2 & i_3 & \cdots & i_N \end{pmatrix} \tag{A2.115}$$

where $i_k$ stands for the number, which, as a result of the permutation, takes the place of the number $k$. The product of two permutations $P_2P_1$ is also a permutation, the effect of which is equivalent to the action of $P_1$ followed by that of $P_2$.

One can associate with every permutation (A2.115) an inverse permutation:

$$P^{-1} = \begin{pmatrix} i_1 & i_2 & i_3 & \cdots & i_N \\ 1 & 2 & 3 & \cdots & N \end{pmatrix} \tag{A2.116}$$

The effect of applying successively a permutation and its inverse is to leave the objects in their original positions, i.e., this forms the identical permutation, which we denote by $I$.

The $N!$ permutations of $N$ objects thus form a group which is called the *permutation group* or the *symmetric group*, and which we denote by $\pi_N$. It is convenient to decompose the elements of a permutation group into products of *cycles*. A permutation is called a cycle if it can be written in the form:

$$P_{i_1 i_2 \ldots i_k} = \begin{pmatrix} i_1 & i_2 & i_3 & \cdots & i_k & i_{k+1} & \cdots & i_N \\ i_2 & i_3 & i_4 & \cdots & i_1 & i_{k+1} & \cdots & i_N \end{pmatrix} \tag{A2.117}$$

It is convenient to write the cycle in a more compact form:

$$P_{i_1 i_2 \ldots i_k} = (i_1 i_2 \ldots i_k) \tag{A2.117a}$$

Thus the six elements of the group $\pi_3$ can be written as the cyclic permutations:

$$I,\ P_{12},\ P_{13},\ P_{23},\ P_{123},\ P_{132} \tag{A2.118}$$

3. Another example of a group is formed by the set of spatial transformations of an equilateral triangle under which the triangle is sent into itself. Such transformations are called symmetry operations (asymmetric bodies cannot be made to coincide with themselves under any such transformations apart from the identity). There are six distinct nonequivalent symmetry operations for an equilateral triangle and for these we choose the following (see Figure A2.4):

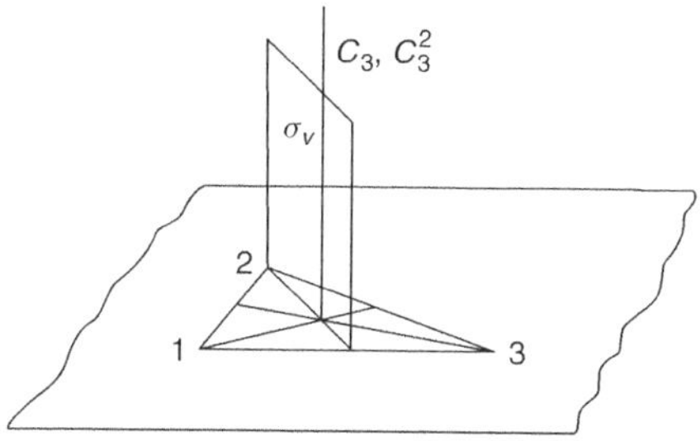


**Figure A2.4** Symmetry elements of the equilateral triangle

$E$–The identity operation, which leaves the triangle unchanged.

$C_3, C_3^2$–Clockwise rotations by 120° and 240°, respectively, about an axis perpendicular to the plane of the triangle and passing through its center of gravity.

$\sigma_v^{(1)}, \sigma_v^{(2)}, \sigma_v^{(3)}$–Reflections in planes perpendicular to the plane of the triangle and passing through its medians.

All other symmetry operations are equivalent to one of those just enumerated. Thus, for example, a rotation by 180° about an axis passing through a median is equivalent to a reflection $\sigma_v$.

Let us now determine the products of all possible pairs of the six symmetry operations of the equilateral triangle given in Figure A2.4 and put the results in the form of a table (Table A2.1). The operation, which acts on the triangle first, is placed on the top line of the table. From this it can be seen that the successive application of any two operations is equivalent to a single operation from the same set. For every operation, $Q$, there is an inverse operation, $Q^{-1}$, which leads to the identity transformation, $E$. By using Table A2.1, it is easy to verify that the associative rule is obeyed. The symmetry operations of a triangle therefore form a group, which is usually denoted by $\mathbf{C}_{3v}$. The group, $\mathbf{C}_{3v}$, is one of the so-called *point groups* (see Section A2.2.5).

**Table A2.1** Multiplication table for the group $\mathbf{C}_{3v}$

| | $E$ | $C_3$ | $C_3^2$ | $\sigma_v^{(1)}$ | $\sigma_v^{(2)}$ | $\sigma_v^{(3)}$ |
|---|---|---|---|---|---|---|
| $E$ | $E$ | $C_3$ | $C_3^2$ | $\sigma_v^{(1)}$ | $\sigma_v^{(2)}$ | $\sigma_v^{(3)}$ |
| $C_3$ | $C_3$ | $C_3^2$ | $E$ | $\sigma_v^{(2)}$ | $\sigma_v^{(3)}$ | $\sigma_v^{(1)}$ |
| $C_3^2$ | $C_3^2$ | $E$ | $C_3$ | $\sigma_v^{(3)}$ | $\sigma_v^{(1)}$ | $\sigma_v^{(2)}$ |
| $\sigma_v^{(1)}$ | $\sigma_v^{(1)}$ | $\sigma_v^{(3)}$ | $\sigma_v^{(2)}$ | $E$ | $C_3^2$ | $C_3$ |
| $\sigma_v^{(2)}$ | $\sigma_v^{(2)}$ | $\sigma_v^{(1)}$ | $\sigma_v^{(3)}$ | $C_3$ | $E$ | $C_3^2$ |
| $\sigma_v^{(3)}$ | $\sigma_v^{(3)}$ | $\sigma_v^{(2)}$ | $\sigma_v^{(1)}$ | $C_3^2$ | $C_3$ | $E$ |

**Isomorphism and homomorphism**

Groups are classified either as *finite* or *infinite* depending upon whether they contain a finite or infinite number of elements. The number of elements in a group is termed its *order*. Two groups $\mathbf{G}$ and $\mathbf{G}'$ of the same order are termed *isomorphic* if a one-to-one correspondence between their elements can be defined such that if the element $A' \in \mathbf{G}'$ corresponds to $A \in \mathbf{G}$ and $B' \in \mathbf{G}'$ corresponds to $B \in \mathbf{G}$, then the element $C' = A'B'$ corresponds to $C = AB$. Isomorphic groups are identical as far as their abstract group properties are concerned, although the actual meaning of their respective elements may be quite different.

The groups $\mathbf{C}_{3v}$ and $\boldsymbol{\pi}_3$ introduced in the previous section form an example of isomorphous groups. Thus we can regard each symmetry operation of the triangle as the corresponding permutation of its vertices, and so derive a one-to-one correspondence between the elements of the two groups:

$$\begin{array}{lcccccc} \mathbf{C}_{3v}: & E & C_3 & C_3^2 & \sigma_v^{(1)} & \sigma_v^{(2)} & \sigma_v^{(3)} \\ \boldsymbol{\pi}_3: & I & P_{132} & P_{123} & P_{23} & P_{13} & P_{12} \end{array}$$

This is a particular case of a more general theorem (*Cayley's theorem*) which states that any finite group of order $n$ is isomorphous with a subgroup (see below) of the permutation group $\boldsymbol{\pi}_n$.

If for each element $A$ of a group $\mathbf{G}$ there is a corresponding element $A'$ of a group $\mathbf{G}'$ so that if $AB = C$, then $A'B' = C'$, the two groups are said to be *homomorphic*. In contrast to isomorphism, homomorphism does not require a one-to-one correspondence between the elements of the groups, and one element of the group $\mathbf{G}'$ may correspond to several elements of $\mathbf{G}$.

**Subgroups and cosets**

If it is possible to select from a group $\mathbf{G}$ a set of elements, which form a group $\mathbf{H}$ among themselves with the same multiplication law, then $\mathbf{H}$ is called a *subgroup* of $\mathbf{G}$. Every group possesses a trivial subgroup that consists of the one unit element of the group. Henceforth when referring to a subgroup, a nontrivial subgroup will always be implied.

Let the element $G_1$ belongs to a finite group $\mathbf{G}$, but not to a subgroup $\mathbf{H}$ of $\mathbf{G}$, which consists of $h$ elements. Multiplying all the elements of $\mathbf{H}$ (e.g. from the left) by $G_1$ gives a set of $h$ elements denoted by $G_1\mathbf{H}$. No element of $G_1\mathbf{H}$ belongs to $\mathbf{H}$. Otherwise this would give for two elements $H_a, H_b \in \mathbf{H}$ the equation $G_1 H_a = H_b$, or $G_1 = H_b H_a^{-1}$, i.e., $G_1 \in \mathbf{H}$, which contradicts the initial assumption. Now take element $G_2 \in \mathbf{G}$ that does not belong to either $\mathbf{H}$ or $G_1\mathbf{H}$ and form the set $G_2\mathbf{H}$. In an analogous way it can be shown that $G_1\mathbf{H}$ and $G_2\mathbf{H}$ do not possess any elements in common. If $\mathbf{H}$, $G_1\mathbf{H}$ and $G_2\mathbf{H}$ do not exhaust all the elements of the group, form $G_3\mathbf{H}$, ... etc., until all the elements of $\mathbf{G}$ are divided into $m$ sets:

$$\mathbf{H}, G_1\mathbf{H}, G_2\mathbf{H}, \ldots, G_{m-1}\mathbf{H} \tag{A2.119}$$

The order of the group is therefore given by $g = mh$. As a result, we arrive at a theorem known as *Lagrange's theorem:*

> the order of a subgroup of a finite group is a divisor of the order of the group.

From this theorem it obviously follows that a group whose order is a prime number does not possess any subgroups.

The division of the group elements into sets (Equation (A2.119)) is uniquely determined on specifying the subgroup $\mathbf{H}$, since any element of the set $G_k\mathbf{H}$ may play the role of $G_k$. Thus let $G'_k = G_k H_a$, where $H_a$ is an arbitrary element of the subgroup $\mathbf{H}$. The set of elements $G'_k\mathbf{H} \equiv G_k H_a \mathbf{H} = G_k\mathbf{H}$, the last equality following from the theorem in the beginning of this section.

The sets (Equation (A2.119)) are called the *left cosets* of the group $\mathbf{H}$. *Right cosets* are defined in an analogous manner. The number of cosets of a subgroup is called the *index* of the subgroup. With the exception of $\mathbf{H}$, none of the cosets $G_k\mathbf{H}$ forms a subgroup since they do not contain the unit element.

As an example, consider the set of permutations $I$ and $P_{12}$ from the group $\pi_3$ (see Equation (A2.118)). This set forms a subgroup, which is denoted by $\pi_2$. The index of the subgroup, $m$ is equal to $g/h = 6/2 = 3$. There are therefore three cosets. The left cosets of the subgroup $\pi_2$ are given by:

$$\begin{aligned} \pi_2 &: \quad I, P_{12} \\ P_{13}\pi_2 &: \quad P_{13}, P_{123} \\ P_{23}\pi_2 &: \quad P_{23}, P_{132} \end{aligned}$$

### Conjugate elements. Classes

Two elements $A$ and $B$ are said to be *conjugate* if $A = QBQ^{-1}$, where $Q$ is an element of the same group. If two elements $A$ and $B$ are each conjugate to a third element $C$, then they are also conjugate to each other. Thus it follows from $A = QCQ^{-1}$ and $B = RCR^{-1}$ that $C = Q^{-1}AQ$, whence $B = RQ^{-1}AQR^{-1} = \left(RQ^{-1}\right) A \left(RQ^{-1}\right)^{-1}$. The elements of a group therefore fall into sets in each of which the elements are conjugate to one another. Such sets are called *classes* of the group. A class is determined by specifying one of its elements; thus, knowing $A$, the other elements can be obtained by forming $G_b A G_b^{-1}$, where $G_b$ runs through all the elements of the group (some elements of the class, however, may be repeated by this). By using the multiplication table for the group $\mathbf{C}_{3v}$ it is not difficult to confirm that its six elements fall into three classes: $E$; $C_3$, $C_3^2$; and $\sigma_v^{(1)}$, $\sigma_v^{(2)}$ and $\sigma_v^{(3)}$.

We quote without proof the following further properties of classes:

(a) The unit element forms a class by itself.

(b) Except for the unit class, classes do not form subgroups as they do not contain the unit element.

(c) Every element of an Abelian group forms a class by itself.

(d) The number of elements in a class is a divisor of the order of the group.

**Direct products of groups**

Let two finite groups, $\mathbf{G}_1$ and $\mathbf{G}_2$, possess no elements in common apart from the unit element. In particular, these groups might be subgroups of some larger group. Denote their order by $g_1$ and $g_2$, respectively. If the elements of the group $\mathbf{G}_1$ commute with those of $\mathbf{G}_2$, then by forming all possible products of the elements of these groups a set of $g_1 g_2$ elements is obtained which also form a group. The commutation condition ensures that the product of two elements of the set obtained in this way gives, in fact, an element belonging to the same set:

$$A_1 A_2 \bullet B_1 B_2 = A_1 B_1 \bullet A_2 B_2 = C_1 C_2$$

The satisfaction of the remaining group postulates is evident. The group which is thus obtained is called the *direct product* of the groups $\mathbf{G}_1$ and $\mathbf{G}_2$ and is denoted by $\mathbf{G}_1 \times \mathbf{G}_2$.

The simplest example of commuting groups are permutation groups whose elements act upon different sets of objects. Thus, for example, the direct product of two $\pi_2$ permutation groups may be formed. Let the permutations of the first group act on the numbers 1 and 2, and the permutations of the second upon the numbers 3 and 4. The direct product of the groups $\pi_2 \times \pi_2$ therefore contains the four elements:

$$\pi_2 \times \pi_2 : I,\ P_{12},\ P_{34},\ \ P_{12} \bullet P_{34} \tag{A2.120}$$

## A2.2.2 Representations of groups

**Definition**

A group of square matrices that is homomorphic with a given group is said to form a *representation* of the group. The number of rows or columns of the matrices is called the *dimension* of the representation.

It follows from this definition that one can associate with every element $Q$ of a group a matrix $\mathbf{\Gamma}(Q)$, such that corresponding to the product $QP = R$ of the group elements is the matrix product:

$$\mathbf{\Gamma}(Q)\,\mathbf{\Gamma}(P) = \mathbf{\Gamma}(R) \tag{A2.121}$$

Representations defined by Equation (A2.121) are called *vector representations*. A more general type of representation is possible, the matrices of which satisfy the relation:

$$\Gamma(Q)\,\Gamma(P) = \epsilon\Gamma(R) \quad |\epsilon| = 1 \tag{A2.121a}$$

Representations such as these are known as *projective or ray* representations.

Multiplication of the matrices on the left-hand side of Equations (A2.121) and (A2.121a) is carried out according to the usual rules of matrix algebra (see Equation

(A2.64)). If the matrices of a representation are all distinct, the representation is isomorphic with the group. Such a representation is then said to be *faithful*.

In physical applications representations usually arise as the result of applying the elements of a group to functions of some coordinates. The groups that occur in physics and chemistry are either groups of linear spatial transformations or of permutations of particle coordinates. When elements of these groups are applied to functions of the coordinates their effect is to generate a set of new functions that transform linearly into one another under the action of the group elements. We now examine this process in somewhat greater detail.

Let $\psi_0$ be a coordinate function defined in the configuration space of the system. The application of an element $Q$ of a group $\mathbf{G}$ to $\psi_0$ converts this function into another function which we denote by $\psi_Q \equiv Q\psi_0$. If $Q$ runs through all the elements of the group, we obtain $g$ functions. These functions may not be linearly independent. Let the number of linearly independent functions $\psi_i$ be equal to $f$ $(f \leq g)$. Under the action of the group elements the functions $\psi_i$ transform only into each other, since by virtue of the properties of a group (see the theorem in the beginning of this section) no other functions may appear in this set. We thus have:

$$Q\psi_k = \sum_{i=1}^{f} \mathbf{\Gamma}_{ik}(Q)\,\psi_i \tag{A2.122}$$

The coefficients $\mathbf{\Gamma}_{ik}(Q)$ form a square matrix of order $f$. It follows from the definition of the matrix $\mathbf{\Gamma}(Q)$ that under the action of the group element $Q$ the function $\psi_k$ transforms according to the $k$th column of the corresponding matrix $\mathbf{\Gamma}(Q)$. Corresponding to the product of two elements $QP$ is a matrix which is the product of the matrices $\mathbf{\Gamma}(Q)$ and $\mathbf{\Gamma}(P)$, for:

$$\begin{aligned} QP\psi_k &= Q\sum_i \mathbf{\Gamma}_{ik}(P)\,\psi_i = \sum_l \left[\sum_i \mathbf{\Gamma}_{li}(Q)\,\mathbf{\Gamma}_{ik}(P)\right]\psi_l \\ &= \sum_l [\mathbf{\Gamma}(Q)\,\mathbf{\Gamma}(P)]_{lk}\,\psi_l \end{aligned}$$

The matrices $\mathbf{\Gamma}(Q)$ consequently form an $f$-dimensional representation of the group $\mathbf{G}$. The set of $f$ functions $\psi_i$ which define the $\mathbf{\Gamma}(Q)$ matrices are said to form a *basis* for the representation.

### Reducibility of representations

The $f$ basis functions for a representation $\mathbf{\Gamma}$ can be regarded as basis vectors in an $f$-dimensional vector space. This space is denoted by $\mathfrak{R}$ and it is said that it transforms according to the representation $\mathbf{\Gamma}$. According to Equation (A2.122), the effect of an element $Q$ of a group $\mathbf{G}$ upon a basis vector $\psi_k$ is to generate a vector $Q\psi_k$, which also lies in $\mathfrak{R}$. Since any vector in a vector space can be expressed in the form of a linear combination of the basis vectors, the effect of an element of the group is to convert every vector of the space into another vector, which also belongs to $\mathfrak{R}$. The space $\mathfrak{R}$ is therefore invariant under the transformations of $\mathbf{G}$.

According to Equation (A2.79), a linear transformation of the basis vectors:

$$\psi_k' = \sum_{i=1}^{f} s_{ik}\psi_i \tag{A2.123}$$

converts the matrices of a representation $\mathbf{\Gamma}$ into the matrices:

$$\mathbf{\Gamma}'(Q) = S^{-1}\mathbf{\Gamma}(Q)\,S \tag{A2.124}$$

Two representations that are connected by a relation such as Equation (A2.124) are said to be *equivalent*. It is clear that there is an infinite number of equivalent representations. A transformation which converts a representation into an equivalent one is called a *similarity transformation*.

If a similarity transformation can be found that converts all the matrices of a representation into block-diagonal form:

$$\mathbf{\Gamma}' = S^{-1}\mathbf{\Gamma}S = \begin{vmatrix} \boxed{\mathbf{\Gamma}^{(1)}} & & & \\ & \boxed{\mathbf{\Gamma}^{(2)}} & & \\ & & \cdots & \\ & & & \boxed{\mathbf{\Gamma}^{(m)}} \end{vmatrix} \tag{A2.125}$$

then the representation is said to be *reducible*. As a result of such a similarity transformation, the representation is decomposed into $m$ representations of smaller dimensions. This is written as:

$$\mathbf{\Gamma} \doteq \mathbf{\Gamma}^{(1)} + \mathbf{\Gamma}^{(2)} + \ldots + \mathbf{\Gamma}^{(m)} \tag{A2.126}$$

The reducibility of a representation implies that by means of a linear transformation of the basis vectors, the space $\Re$ can be divided into a number of invariant subspaces $\Re^{(\alpha)}$, each of which transforms according to a representation $\mathbf{\Gamma}^{(\alpha)}$. The operations of $\mathbf{G}$ transform the vectors of each subspace among themselves without mixing vectors from different subspaces.

If there is no transformation that reduces the representation matrices to the block-diagonal form (Equation (A2.125)), then the representation is said to be *irreducible*. It should be noted that an irreducible representation of a group obviously forms a representation of a subgroup of the group. However, as far as the subgroup is concerned, this representation may be reducible and may be decomposed into irreducible representations of the subgroup. This process is called *reduction with respect to a subgroup*.

The following set of matrices offers an example of a reducible representation of the group $\mathbf{C}_{3v}$:

$$\mathbf{\Gamma}(E) = \begin{vmatrix} 1 & 0 & 0 \\ 0 & 1 & 0 \\ 0 & 0 & 1 \end{vmatrix} \qquad \mathbf{\Gamma}(C_3) = \begin{vmatrix} 0 & 0 & 1 \\ 1 & 0 & 0 \\ 0 & 1 & 0 \end{vmatrix}$$

$$\mathbf{\Gamma}\left(C_3^2\right) = \begin{vmatrix} 0 & 1 & 0 \\ 0 & 0 & 1 \\ 1 & 0 & 0 \end{vmatrix} \qquad \mathbf{\Gamma}\left(\sigma_v^{(1)}\right) = \begin{vmatrix} 1 & 0 & 0 \\ 0 & 0 & 1 \\ 0 & 1 & 0 \end{vmatrix}$$

$$\mathbf{\Gamma}\left(\sigma_v^{(2)}\right) = \begin{vmatrix} 0 & 0 & 1 \\ 0 & 1 & 0 \\ 1 & 0 & 0 \end{vmatrix} \qquad \mathbf{\Gamma}\left(\sigma_v^{(3)}\right) = \begin{vmatrix} 0 & 1 & 0 \\ 1 & 0 & 0 \\ 0 & 0 & 1 \end{vmatrix}$$

It is easily verified, by direct multiplication, that these matrices conform to the multiplication table for the group (see Table A2.1). By means of a transformation such as Equation (A2.125), it can be decomposed into two irreducible representations, one of which is one dimensional and the other two dimensional. Methods for obtaining the irreducible representations, which occur in the decomposition (A2.126), are described in the following subsections.

**Properties of irreducible representations**

Let quote without proof the following important properties of irreducible representations of finite groups. The proofs of these statements can be found in any textbook on group theory (for example [6, 7]).

1. The number of nonequivalent irreducible representations of a group is equal to the number of classes in the group.

2. The sum of the squares of the dimensions of the nonequivalent irreducible representations is equal to the order of the group:

$$f_1^2 + f_2^2 + \ldots + f_r^2 = g \tag{A2.127}$$

   where $f_\alpha$ denotes the dimension of the $\alpha$th irreducible representation. It follows from this that all the irreducible representations of an Abelian group are one-dimensional, since the number of its irreducible representations is equal to the number of elements in the group.

3. The dimension of an irreducible representation of a finite group is a divisor of the order of the group.

4. The following orthogonality relations hold for the matrix elements of irreducible representations:

$$\sum_R \mathbf{\Gamma}_{ik}^{(\alpha)}(R)^* \, \mathbf{\Gamma}_{lm}^{(\beta)}(R) = (g/f_\alpha)\,\delta_{\alpha\beta}\delta_{il}\delta_{km} \tag{A2.128}$$

$$\sum_{\alpha,i,k} (f_\alpha/g)\,\mathbf{\Gamma}_{ik}^{(\alpha)}(R)^* \, \mathbf{\Gamma}_{ik}^{(\alpha)}(Q) = \delta_{RQ} \tag{A2.129}$$

   The summation in Equation (A2.128) is taken over all the $g$ elements of the group. In Equation (A2.129) all $f_\alpha^2$ elements of the matrix $\mathbf{\Gamma}^{(\alpha)}(R)$ are summed for each irreducible representation $\mathbf{\Gamma}^{(\alpha)}$. The number of terms in the sum (A2.129) is equal to the order of the group in accordance with Equation (A2.127).

### Characters

Let the matrices $\mathbf{\Gamma}^{(\alpha)}(R)$ form a representation of a group **G**. The sum of the diagonal elements of a matrix $\mathbf{\Gamma}^{(\alpha)}(R)$ is called the *character* of the element $R$ in the representation $\mathbf{\Gamma}^{(\alpha)}$ and is denoted by $\chi^{(\alpha)}(R)$:

$$\chi^{(\alpha)}(R) = \sum_i \mathbf{\Gamma}_{ii}^{(\alpha)}(R) \tag{A2.130}$$

Each representation is typified by a set of $g$ characters.

The characters of equivalent representations are identical, for in accordance with the rule for matrix multiplication:

$$\sum_i \left(S^{-1}\mathbf{\Gamma}(R)\,S\right)_{ii} = \sum_{i,l,m} s_{il}^{-1}\mathbf{\Gamma}_{lm}(R)\,s_{mi}$$

$$= \sum_{l,m}\left(\sum_i s_{mi}s_{il}^{-1}\right)\mathbf{\Gamma}_{lm}(R) = \sum_{l,m}\delta_{ml}\mathbf{\Gamma}_{lm}(R) = \sum_m \mathbf{\Gamma}_{mm}(R)$$

Hence, the specification of a representation by means of a set of characters does not distinguish between equivalent representations. This is extremely useful, since in physical applications only inequivalent representations are important. By specifying the characters of a representation, nonequivalent representations can be distinguished.

The elements of a group belonging to a given class are connected among themselves by relationships analogous to Equation (A2.124). Their characters must therefore all be identical. Consequently, a class can be denoted by specifying the character of just one of its members. Furthermore, the number of distinct characters of a representation cannot exceed the number of classes in the group.

The characters of irreducible representations satisfy the following orthogonality relation:

$$\sum_R \chi^{(\alpha)}(R)^* \chi^{(\beta)}(R) = g\delta_{\alpha\beta} \tag{A2.131}$$

This relation is obtained from Equation (A2.128) by putting $i$ equal to $k$, $l$ equal to $m$, and summing over $k$ and $m$ on both sides.

Since the characters of all the elements belonging to a given class are equal, Equation (A2.131) can be written in the form:

$$\sum_{\mathbf{C}} g_{\mathbf{C}}\chi^{(\alpha)}(\mathbf{C})^* \chi^{(\beta)}(\mathbf{C}) = g\delta_{\alpha\beta} \tag{A2.132}$$

where the sum is taken over all classes **C** of the group, and $g_{\mathbf{C}}$ denotes the number of elements in $\mathbf{C}\left(\sum_{\mathbf{C}} g_{\mathbf{C}} = g\right)$.

If $(g_{\mathbf{C}}/g)^{1/2}\chi^{(\alpha)}$ is written as $a_{\mathbf{C}_\alpha}$, then since the number of classes is equal to the number of irreducible representations, the quantities $a_{\mathbf{C}_\alpha}$ form a square matrix. The orthogonality relation (A2.132), are then identical for the $a_{\mathbf{C}_\alpha}$, with the unitarity condition for the first subscript. The matrix $\|a_{\mathbf{C}_\alpha}\|$ is therefore unitary, with the unitarity condition also applying to its elements. From this we obtain a second orthogonality relation for the characters of irreducible representations:

$$\sum_{\alpha} \chi^{(\alpha)}(\mathbf{C}_i)^* \chi^{(\alpha)}(\mathbf{C}_k) = \left(g/g_{\mathbf{C}_i}\right) \delta_{\mathbf{C}_i \mathbf{C}_k} \qquad \text{(A2.133)}$$

Among the irreducible representations of a group there is always a one-dimensional representation which is generated by a basis function that is invariant under all the operations of the group. All the characters of this representation are equal to unity. This representation is usually called the *totally symmetric* or *unit* representation and is denoted by the symbol $A_1$.

To decompose a reducible representation into its irreducible components, it suffices to know just the characters of the representations. A reducible representation is either given directly by its matrices (in which case the determination of its characters is trivial) or its basis functions are given. In the latter case, the characters are obtained by eliciting the effect of the operations of the group on the basis functions and then summing the diagonal elements. For irreducible representations there are a number of methods for calculating the characters. These are based upon the properties of finite groups, and a knowledge of the basis functions is unnecessary (see References [6–8]).

## Decomposition of a reducible representation

Consider some reducible representation $\mathbf{\Gamma}$. By definition, one can always find a unitary transformation that will bring this to the block-diagonal form (Equation (A2.125)). We assume that no further reduction is possible, i.e., that the representations $\mathbf{\Gamma}^{(\alpha)}$ which appear in the decomposition (Equation (A2.126)) are irreducible. A given irreducible representation, however, may occur several times in the decomposition. Thus, in general:

$$\mathbf{\Gamma} \doteq \sum_{\beta} a^{(\beta)} \mathbf{\Gamma}^{(\beta)} \qquad \text{(A2.134)}$$

where $a^{(\beta)}$ denotes the number of times the representation $\mathbf{\Gamma}^{(\beta)}$ occurs in the decomposition of $\mathbf{\Gamma}$. Since a similarity transformation does not alter the characters of a representation, the characters of the representation $\mathbf{\Gamma}$ are given by:

$$\chi^{(\Gamma)}(R) = \sum_{\beta} a^{(\beta)} \chi^{(\beta)}(R) \qquad \text{(A2.135)}$$

We multiply Equation (A2.135) by $\chi^{(\alpha)}(R)^*$ and sum over all the elements in the group. By virtue of the orthogonality relation (A2.131), we obtain:

$$a^{(\alpha)} = (1/g) \sum_{R} \chi^{(\Gamma)}(R) \chi^{(\alpha)}(R)^* \qquad \text{(A2.136)}$$

For actual applications, it is convenient to rewrite this relation in the form:

$$a^{(\alpha)} = (1/g) \sum_{\mathbf{C}} g_{\mathbf{C}} \chi^{(\Gamma)}(\mathbf{C}) \chi^{(\alpha)}(\mathbf{C})^* \qquad \text{(A2.137)}$$

**Table A2.2** Characters of the irreducible representations of $\mathbf{C}_{3v}^{a}$

| | $E$ | $2C_3$ | $3\sigma_v$ |
|---|---|---|---|
| $\chi^{(1)}$ | 1 | 1 | 1 |
| $\chi^{(2)}$ | 1 | 1 | $-1$ |
| $\chi^{(3)}$ | 2 | $-1$ | 0 |

[a] Here the irreducible representations are enumerated. Usually, the irreducible representations of the point group are denoted by special symbols, the Schönflies notation is often used (see Section A2.2.5 and References [7, 8]).

If the characters of the representations are known, Equation (A2.137) enables the irreducible representations which occur in the decomposition of a given reducible representation to be determined easily.

As an example, let us determine the irreducible representations of $\mathbf{C}_{3v}$ which occur in the decomposition of the representation whose matrices are presented on p. 280. The characters of this representation are:

| | $E$ | $2C_3$ | $3\sigma_v$ |
|---|---|---|---|
| $\chi^{(\Gamma)}$ | 3 | 0 | 1 |

Taking the characters of the irreducible representations of $\mathbf{C}_{3v}$ from Table A2.2, we find from formula (A2.137) that:

$$a^{(1)} = \tfrac{1}{6}(3+3) = 1 \qquad a^{(2)} = \tfrac{1}{6}(3-3) = 0 \qquad a^{(3)} = \tfrac{1}{6}6 = 1$$

Thus, this representation consequently breaks down into two irreducible representations: the one-dimensional representation $\mathbf{\Gamma}^{(1)}$ and the two-dimensional representation $\mathbf{\Gamma}^{(3)}$.

### The direct product of representations

Let us consider two irreducible representations $\mathbf{\Gamma}^{(\alpha)}$ and $\mathbf{\Gamma}^{(\beta)}$ of a group **G**, each of which is defined by a set of basis functions $\psi^{(\alpha)}$ $(i = 1, 2, \ldots, f_\alpha)$ and $\psi_k^{(\beta)}$ $(k = 1, 2, \ldots, f_\beta)$, respectively. If all possible products of the form $\psi_i^{(\alpha)}\psi_k^{(\beta)}$ are constructed, an $f_\alpha f_\beta$ -dimensional basis for a representation of the group is obtained. This representation is known as the *direct product* of the representations $\mathbf{\Gamma}^{(\alpha)}$ and $\mathbf{\Gamma}^{(\beta)}$ and is denoted by $\mathbf{\Gamma}^{(a)} \times \mathbf{\Gamma}^{(\beta)}$. The matrix elements of the direct product of the two representations are of the form of products of the matrix elements of $\mathbf{\Gamma}^{(\alpha)}$ and $\mathbf{\Gamma}^{(\beta)}$:

$$R\psi_i^{(\alpha)}\psi_k^{(\beta)} = \left(R\psi_i^{(\alpha)}\right)\left(R\psi_k^{(\beta)}\right) = \sum_{l,m} \mathbf{\Gamma}_{li}^{(\alpha)}(R)\,\mathbf{\Gamma}_{mk}^{(\beta)}(R)\,\psi_l^{(\alpha)}\psi_m^{(\beta)} \qquad \text{(A2.138)}$$

and the characters of this representation are equal to the product of the constituent characters:

$$\chi^{(\alpha\times\beta)}(R) = \sum_{i,k} \mathbf{\Gamma}_{ii}^{(\alpha)}(R)\,\mathbf{\Gamma}_{kk}^{(\beta)}(R) = \chi^{(\alpha)}(R)\,\chi^{(\beta)}(R) \qquad \text{(A2.139)}$$

The matrix $\mathbf{\Gamma}^{(\alpha\times\beta)}(R)$, with matrix elements given by Equation (A2.138), is called the *direct product* of the matrices $\mathbf{\Gamma}^{(\alpha)}(R)$ and $\mathbf{\Gamma}^{(\beta)}(R)$ (see Section A2.1.5).

It is clear that although $A \times B \neq B \times A$, the two can be converted into each other by an appropriate permutation of the rows and columns. The characters of a direct product are therefore independent of the order in which the representations are multiplied. This also follows from formula (A2.139).

If the two representations that are multiplied together are identical but possess different bases, then:

$$\chi^{(\alpha\times\alpha)}(R) = \left[\chi^{(\alpha)}(R)\right]^2 \qquad \text{(A2.140)}$$

Thus, the characters of the direct product of the representations $\mathbf{\Gamma}^{(3)} \times \mathbf{\Gamma}^{(3)}$ of $\mathbf{C}_{3v}$ are (see Table A2.2):

| | $E$ | $2C_3$ | $3\sigma_v$ |
|---|---|---|---|
| $\left[\chi^{(3)}(R)\right]^2$ | 4 | 1 | 0 |

This representation can be reduced. Its decomposition into irreducible components is carried out according to formula (A2.137) and contains each irreducible representation of $\mathbf{C}_{3v}$ once:

$$\mathbf{\Gamma}^{(3)} \times \mathbf{\Gamma}^{(3)} \doteq \mathbf{\Gamma}^{(1)} + \mathbf{\Gamma}^{(2)} + \mathbf{\Gamma}^{(3)}$$

If in the direct product $\mathbf{\Gamma}^{(\alpha)} \times \mathbf{\Gamma}^{(\alpha)}$ the bases of the two representations $\mathbf{\Gamma}^{(\alpha)}$ coincide, then the direct product basis is symmetric with respect to a permutation of its two factors. A direct product of this kind is called the *symmetric product* of a representation with itself; it is denoted by $\left[\mathbf{\Gamma}^{(\alpha)}\right]^2$. Its dimension is less than $f_\alpha^2$ and is equal to $f_\alpha(f_\alpha+1)/2$; the characters of a symmetric product are not given by a product of characters (Equation (A2.140)), but by an equation which is a special case of more general formula (see Equation (4.18) in Reference [8]):

$$\left[\chi^{(\alpha)}\right]^2(R) = \tfrac{1}{2}\chi^{(\alpha)}\left(R^2\right) + \tfrac{1}{2}\left[\chi^{(\alpha)}(R)\right]^2 \qquad \text{(A2.141)}$$

For example, the characters of the symmetric product $\left[\mathbf{\Gamma}^{(3)}\right]^2$ of $\mathbf{C}_{3v}$ are:

| | $E$ | $2C_3$ | $3\sigma_v$ |
|---|---|---|---|
| $\left[\chi^{(3)}\right]^2(R)$ | 3 | 0 | 1 |

Decomposing this representation into its irreducible components, we obtain:

$$\left[\boldsymbol{\Gamma}^{(3)}\right]^2 \doteq \boldsymbol{\Gamma}^{(1)} + \boldsymbol{\Gamma}^{(3)}$$

The direct product of two irreducible representations is always reducible except when one of the representations is one dimensional. The decomposition of a direct product into irreducible representations:

$$\boldsymbol{\Gamma}^{(\alpha)} \times \boldsymbol{\Gamma}^{(\beta)} \doteq \sum_{\tau} a^{(\tau)} \boldsymbol{\Gamma}^{(\tau)} \tag{A2.142}$$

is conventionally called the *Clebsch–Gordan series*. The coefficients in the decomposition are found from the general formula (A2.136):

$$a^{(\tau)} = (1/g) \sum_{R} \chi^{(\alpha\times\beta)}(R)\, \chi^{(\tau)}(R)^* = (1/g) \sum_{R} \chi^{(\alpha)}(R)\, \chi^{(\beta)}(R)\, \chi^{(\tau)}(R)^* \tag{A2.143}$$

Determine the condition that the totally symmetric representation $A_1$ appears in the decomposition (A2.142). The characters of this representation $\chi^{(A_1)}(R) = 1$ for all elements in the group, and hence by virtue of the orthogonality relation (A2.131), we obtain:

$$a^{(A_1)} = (1/g) \sum_{R} \chi^{(\alpha)}(R)\, \chi^{(\beta)}(R) = \delta_{\beta\alpha^*} \tag{A2.143a}$$

where $\boldsymbol{\Gamma}^{(\alpha)*}$ denotes the representation whose matrix elements are the complex conjugates of the matrix elements of $\boldsymbol{\Gamma}^{(\alpha)}$. Such representations are called *complex conjugate representations*. Consequently, the totally symmetric representation occurs in the decomposition of a direct product of irreducible representations if, and only if, these representations are the complex conjugates of one another. In the case of real matrices, the totally symmetric representation occurs only in the direct product of an irreducible representation with itself.

In analogy with the direct product of two irreducible representations of a group, a direct product of an arbitrary number of irreducible representations may be defined. The characters of such a direct product are equal to the product of the characters of the representations that are multiplied:

$$\chi^{(\alpha\times\beta\times\ldots\times\omega)}(R) = \chi^{(\alpha)}(R)\, \chi^{(\beta)}(R) \ldots \chi^{(\omega)}(R) \tag{A2.144}$$

The direct products of representations that have been considered so far concern one and the same group. If $\mathbf{G}_1 \times \mathbf{G}_2$ is the direct product of two groups and the irreducible representations $\boldsymbol{\Gamma}^{(\alpha)} \in \mathbf{G}_1$, $\boldsymbol{\Gamma}^{(\beta)} \in \mathbf{G}_2$, then the direct product of the representations $\boldsymbol{\Gamma}^{(\alpha)} \times \boldsymbol{\Gamma}^{(\beta)}$ is an irreducible representation of the group $\mathbf{G}_1 \times \mathbf{G}_2$. By a derivation similar to that which led to Equation (A2.139), it can be shown that the element $R = R_1 R_2$ of the group $\mathbf{G}_1 \times \mathbf{G}_2$ possesses the character:

$$\chi^{(\alpha\times\beta)}(R) = \chi^{(\alpha)}(R_1)\, \chi^{(\beta)}(R_2) \tag{A2.145}$$

The different irreducible representations of $\mathbf{G}_1 \times \mathbf{G}_2$ are obtained by combining in pairs the irreducible representations of $\mathbf{G}_1$ and $\mathbf{G}_2$.

**Clebsch–Gordan coefficients**

The reduction of a direct product representation into its irreducible components (Equation (A2.142)) is accomplished by means of a linear transformation of the basis functions $\psi_i^{(\alpha)}\psi_k^{(\beta)}$ to a set of functions $\psi_t^{(\tau)}$ that do not mix under the operation of the group elements:

$$\psi_t^{(a\tau)} = \sum_{i,k} \psi_i^{(\alpha)}\psi_k^{(\beta)} \langle \alpha i, \beta k \mid a\tau t\rangle \tag{A2.146}$$

The index $a$ distinguishes the representations $\tau$ should they occur more than once in the decomposition (A2.142). The coefficients $\langle \alpha i, \beta k \mid a\tau t\rangle$ appearing in this equation are known as *Clebsch–Gordan coefficients*. Since the basis functions are normally chosen to be orthogonal, the Clebsch–Gordan coefficients constitute a unitary matrix [1], which is denoted by $C_{\alpha\beta}$. This matrix reduces the direct product. Thus, in agreement with Equation (A2.125):

$$C_{\alpha\beta}^{-1}\left(\mathbf{\Gamma}^{(\alpha)} \times \mathbf{\Gamma}^{(\beta)}\right) C_{\alpha\beta} = \left|\begin{array}{cccc} \boxed{\mathbf{\Gamma}^{(\tau_1)}} & & & \\ & \boxed{\mathbf{\Gamma}^{(\tau_2)}} & & \\ & & \boxed{\cdots} & \\ & & & \boxed{\mathbf{\Gamma}^{(\tau_m)}} \end{array}\right| \tag{A2.147}$$

The rows of the $C_{\alpha\beta}$ matrix are numbered by the indices $i, k$, which may take on a total of $f_\alpha f_\beta$ values, and the columns by the indices $a, \tau$ and $t$. For most of the groups that occur in physics it is possible to choose the matrix of Clebsch–Gordan coefficients to be real, so that the orthogonality relationships (A2.91) apply to its elements:

$$\sum_{a,\tau,t} \langle \alpha i, \beta k \mid a\tau t\rangle \left\langle \alpha\bar{i}, \beta\bar{k} \mid a\tau t\right\rangle = \delta_{i\bar{i}}\delta_{k\bar{k}}$$

$$\sum_{i,k} \langle \alpha i, \beta k \mid a\tau t\rangle \left\langle \alpha i, \beta k \mid \overline{a\tau t}\right\rangle = \delta_{a\bar{a}}\delta_{\tau\bar{\tau}}\delta_{t\bar{t}} \tag{A2.148}$$

In an orthogonal transformation, the inverse of a matrix is equal to its transpose, and therefore the inverse transformation to Equation (A2.146) is:

$$\psi_i^{(\alpha)}\psi_k^{(\beta)} = \sum_{a,\tau,t} \psi_t^{(a\tau)} \langle a\tau t \mid \alpha i.\beta k\rangle \tag{A2.146a}$$

or in matrix form:

$$\mathbf{\Gamma}^{(\alpha)} \times \mathbf{\Gamma}^{(\beta)} = \mathbf{C}_{\alpha\beta}\left(\sum_{a,\tau} \mathbf{\Gamma}^{(a\tau)}\right)\mathbf{C}_{\alpha\beta}^{-1} \tag{A2.149}$$

---

[1] Basis functions $\psi_t^{(a\tau)}$ that belong to different irreducible representations are automatically orthogonal to one another [8]. Even when an irreducible representation is repeated in Decomposition (A2.142), its bases can always be orthogonalized to one another.

Equation (A2.149) is equivalent to an equation between matrix elements:

$$\mathbf{\Gamma}^{(\alpha)}_{i\bar{i}}(R)\,\mathbf{\Gamma}^{(\beta)}_{k\bar{k}}(R) = \sum_{a,\tau}\sum_{t,\bar{t}} \langle \alpha i, \beta k \mid a\tau t\rangle\, \mathbf{\Gamma}^{(a\tau)}_{t\bar{t}}(R) \langle a\tau\bar{t} \mid \alpha\bar{i}, \beta\bar{k}\rangle \qquad \text{(A2.150)}$$

We multiply Equation (A2.150) by $\mathbf{\Gamma}^{(\tau_0)}_{t_0\bar{t}_0}(R)^*$ and sum over all elements in the group. If the representation $\mathbf{\Gamma}^{(\tau_0)}$ occurs only once in the decomposition (A2.142), so that the sum over $a\tau$ in Equation (A2.150) consists of just a single term with $\tau_0$, use of the orthogonality relation (A2.128) leads to the equation:

$$\langle \alpha i, \beta k \mid \tau_0 t_0\rangle \langle \tau_0\bar{t}_0 \mid \alpha\bar{i}, \beta\bar{k}\rangle = \left(f_{\tau_0}/g\right) \sum_R \mathbf{\Gamma}^{(\alpha)}_{i\bar{i}}(R)\,\mathbf{\Gamma}^{(\beta)}_{k\bar{k}}(R)\,\mathbf{\Gamma}^{(\tau_0)}_{t_0\bar{t}_0}(R)^* \qquad \text{(A2.151)}$$

Equation (A2.151) can be used to calculate the Clebsch–Gordan coefficients if the matrix elements of the irreducible representations of the group are known. If irreducible representations occur more than once in the decomposition (A2.142), the matrix of Clebsch–Gordan coefficients is defined only to within a unitary transformation.

From the definition of Clebsch–Gordan coefficients it follows that:

$$\langle \alpha i, \beta k \mid \tau t\rangle \equiv 0$$

for all $\mathbf{\Gamma}^{(\tau)}$ that do not appear in the decomposition of the direct product $\mathbf{\Gamma}^{(\alpha)} \times \mathbf{\Gamma}^{(\beta)}$.

**The regular representation**

Let $\psi_0$ be some coordinate function that does not possess any symmetry properties with respect to the operations of a group **G**. The operation of the $g$ elements of the group upon $\psi_0$ generates $g$ linearly independent functions:

$$\psi_R = R\psi_0 \qquad \text{(A2.152)}$$

The functions $\psi_R$ are converted into one another under the operations of the group:

$$Q\psi_R = QR\psi_0 = P\psi_0 \equiv \psi_P \qquad \text{(A2.153)}$$

The functions (A2.152) therefore constitute a $g$-dimensional basis for a representation of **G**. A representation such as this is called a *regular representation*. According to Equation (A2.153), the matrices of the regular representation, with the exception of the matrix of the identity transformation $E$, possess zero diagonal elements. The identity matrix is always a diagonal unit matrix. From this it follows that the characters of the regular representation are given by:

$$\chi(E) = g, \quad \chi(R) = 0 \quad \text{for} \quad R \neq E \qquad \text{(A2.154)}$$

Substituting Equation (A2.154) into Equation (A2.136) and remembering that for any representation $\chi^{(\alpha)}(E) = f_\alpha$, it is found that:

$$a^{(\alpha)} = (1/g)(gf_\alpha) = f_\alpha$$

i.e., the number of times each irreducible representation occurs in the decomposition of the regular representation is equal to its dimension.

Let us write formula (A2.135), which is an expression for the character of a reducible representation in terms of the irreducible representations occurring in it, for the case $R = E$:

$$\sum_{\beta} a^{(\beta)} f_{\beta} = f_{\Gamma}$$

If the reducible representation is the regular representation, $f_{\Gamma} = g$ and $a^{(\beta)} = f_{\beta}$. Hence we arrive at the result that:

$$\sum_{\beta} f_{\beta}^2 = g \tag{A2.155}$$

We have thus proved a property of which we have already made use:

> The sum of the squares of the dimensions of all the irreducible representations of a group is equal to the order of the group.

It should be noted that to construct the regular representation, it is not necessary to make use of the basis functions (Equation (A2.152)). The elements of the group themselves can be employed as a basis. A knowledge of the multiplication table of the group is therefore sufficient to write down all the matrices of the regular representation.

## Construction of basis functions for irreducible representations

The regular representation considered in the previous section is generated by the $g$ linearly independent functions (Equation (A2.152)). The decomposition of this representation into its irreducible components contains all the nonequivalent irreducible representations of the group. It can be shown that this decomposition can be accomplished by constructing the following linear combinations of the basis functions $\psi_R$:

$$\psi_{ik}^{(\alpha)} = (f_{\alpha}/g) \sum_{R} \mathbf{\Gamma}_{ik}^{(\alpha)}(R)^* \psi_R \tag{A2.156}$$

where $\mathbf{\Gamma}_{ik}^{(\alpha)}(R)$ is a matrix element of the irreducible representation $\mathbf{\Gamma}^{(\alpha)}$ corresponding to the operation $R$. The summation is taken over all $g$ operations in the group.

Let us apply an arbitrary operation $Q$ of the group to Function (A2.156):

$$Q\psi_{ik}^{(\alpha)} = (f_{\alpha}/g) \sum_{R} \mathbf{\Gamma}_{ik}^{(\alpha)}(R)^* QR\psi_0 = (f_{\alpha}/g) \sum_{P} \mathbf{\Gamma}_{ik}^{(\alpha)}\left(Q^{-1}P\right)^* P\psi_0 \tag{A2.157}$$

In this equation we have denoted the operation $QR$ by $P$ and made use of the invariance property of a sum over the group (Equation (A2.114)). Furthermore, we write the matrix element of the product as products of matrix elements and make use of the property (Equation (A2.86)) of unitary matrices:

$$\mathbf{\Gamma}_{ik}^{(\alpha)}\left(Q^{-1}P\right)^{*} = \sum_{l}\mathbf{\Gamma}_{il}^{(\alpha)}\left(Q^{-1}\right)^{*}\mathbf{\Gamma}_{lk}^{(\alpha)}(P)^{*} = \sum_{l}\mathbf{\Gamma}_{li}^{(\alpha)}(Q)\,\mathbf{\Gamma}_{lk}^{(\alpha)}(P)^{*} \quad \text{(A2.158)}$$

Substituting Equation (A2.158) into Equation (A2.157) we obtain finally:

$$Q\psi_{ik}^{(\alpha)} = \sum_{l}\mathbf{\Gamma}_{li}^{(\alpha)}(Q)\,\psi_{lk}^{(\alpha)} \quad \text{(A2.159)}$$

The function $\psi_{ik}^{(\alpha)}$ therefore transforms as the $i$th column of the irreducible representation $\mathbf{\Gamma}^{(\alpha)}$, and the set of $f_\alpha$ functions $\psi_{ik}^{(\alpha)}$ with fixed second index $k$ forms a basis for the irreducible representation $\mathbf{\Gamma}^{(\alpha)}$. Altogether it is possible to form $f_\alpha$ independent bases corresponding to the number of different values for $k$. This is to be expected, since in the decomposition of the regular representation, each irreducible representation occurs as many times as its dimension.

From the form of Equation (A2.156) it follows that to obtain a basis function for a representation $\mathbf{\Gamma}^{(\alpha)}$, it is sufficient to apply the operator:

$$\epsilon_{ik}^{(\alpha)} = (f_\alpha/g)\sum_{R}\mathbf{\Gamma}_{ik}^{(\alpha)}(R)^{*}R \quad \text{(A2.160)}$$

to some arbitrary function $\psi_0$. There are $f_\alpha^2$ such operators for every irreducible representation. They form $f_\alpha$ sets, which differ from one another in the second index. Each of these sets can be used to obtain basis functions for an irreducible representation.

Should the function $\psi_0$ possess specific symmetry properties with respect to the operations of the group $\mathbf{G}$, then in some cases the application of the operator $\epsilon_{ik}^{(\alpha)}$ to it may give zero. Let us determine the result of applying $\epsilon_{iK}^{(\alpha)}$ to a function of the form down in Equation (A2.156):

$$\begin{aligned}\epsilon_{ik}^{(\alpha)}\psi_{mn}^{(\beta)} &= (f_\alpha/g)\sum_{R}\mathbf{\Gamma}_{ik}^{(\alpha)}(R)^{*}R\psi_{mn}^{(\beta)} \\ &= (f_\alpha/g)\sum_{l}\sum_{R}\mathbf{\Gamma}_{ik}^{(\alpha)}(R)^{*}\mathbf{\Gamma}_{lm}^{(\beta)}(R)\,\psi_{ln}^{(\beta)} = \delta_{\alpha\beta}\delta_{km}\psi_{in}^{(\alpha)}\end{aligned} \quad \text{(A2.161)}$$

In this equation we made use of Equation (A2.159) and the orthogonality relations (A2.128). The application of $\epsilon_{ik}^{(\alpha)}$ to a basis function of an irreducible representation therefore either gives zero or another basis function belonging to the same irreducible representation. When $i = k$, the application of $\epsilon_{ii}^{(\alpha)}$ to a basis function $\psi_{in}^{(\alpha)}$ gives just the same function once more. Operators that possess such properties are known as *projection operators*. They satisfy the operator equation:

$$\epsilon_{ii}^{(\alpha)}\epsilon_{ii}^{(\alpha)} = \epsilon_{ii}^{(\alpha)} \quad \text{(A2.162)}$$

An arbitrary function $\psi$ can be represented in the form of a sum of functions each of which transforms according to an irreducible representation of a group:

$$\psi = \sum_{\alpha,i} \psi_{ii}^{(\alpha)} \tag{A2.163}$$

The proof of Equation (A2.163) is given in Reference [8]. The summation over $\alpha$ in this equation is taken over all the irreducible representations of the group, and the summation $i$ over all the independent bases for the representation $\mathbf{\Gamma}^{(\alpha)}$, $i = 1, 2, \ldots, f_\alpha$. The functions $\psi_{ii}^{(\alpha)}$ are given by formula (A2.156) with $i = k$.

It can be proved [8] that functions which transform according to different irreducible representations, or according to different columns of the same representation, are orthogonal to one another. An arbitrary function can thus be expanded in a set of orthogonal functions $\psi_{ii}^{(\alpha)}$. The $\psi_{ii}^{(\alpha)}$ can be visualized as the components of a vector $\psi$ in the space of the basis vectors for the irreducible representations of a group. The operator $\epsilon_{ii}^{(\alpha)}$ projects the vector $\psi$ onto the direction $(\alpha i)$, i.e., it picks out the component of $\psi$ in this direction. If there is no such component of $\psi$, then the result of operating with $\epsilon_{ii}^{(\alpha)}$ upon $\psi$ is of course zero. The geometric interpretation of the effect of $\epsilon_{ik}^{(\alpha)}$ upon $\psi$ is a little more complicated. It may be regarded as the projection of a previously rotated vector $\psi$ onto the direction $(\alpha i)$. This rotation orients the component $\psi_{kk}^{(\alpha)}$ in the direction $(\alpha i)$. The action of $\epsilon_{ik}^{(\alpha)}$ upon $\psi$ therefore gives zero if there is no component of $\psi$ in the direction $(\alpha k)$.

## A2.2.3 The permutation group

### Operations with permutations

Consider permutations of the $N$ integers $1, 2, \ldots, N$. There are $N!$ such permutations altogether, each of which is denoted by:

$$P = \begin{pmatrix} 1 & 2 & 3 & \cdots & N \\ i_1 & i_2 & i_3 & \cdots & i_N \end{pmatrix} \tag{A2.164}$$

Every permutation can be represented in the form of a product of commuting cycles (Equation (A2.117a)). For this purpose, one and $i_1$ in Equation (A2.164) are taken to form the first two elements of a cycle. These are followed by the number that replaces $i_1$, the process being continued until the number that replaces one is reached. A similar procedure is followed for the remaining elements. For example:

$$\begin{pmatrix} 1 & 2 & 3 & 4 & 5 & 6 \\ 2 & 4 & 5 & 1 & 3 & 6 \end{pmatrix} = (124)\,(35)\,(6)$$

A cycle, by definition, is invariant under a cyclic permutation of its elements:

$$(i_1 i_2 i_3 \ldots i_k) = (i_2 i_3 \ldots i_k i_1) = (i_3 \ldots i_k i_1 i_2) = \ldots$$

The number of elements in a cycle is called its *length*. It follows from the definition that if a cycle is raised to a power equal to its length, the identity permutation obtained is:

$$(i_1 i_2 \ldots i_k)^k = I \tag{A2.165}$$

From this it is seen that:

$$(i_1 i_2 \ldots i_k)^{-1} = (i_1 i_2 \ldots i_k)^{k-1} \tag{A2.166}$$

A cycle of two elements is commonly known as a *transposition*. It is obvious that:

$$(i_1 i_2) = (i_1 i_2)^{-1} \tag{A2.167}$$

The following rules are useful in manipulating permutations:

1. The product $QPQ^{-1}$, where $Q$ and $P$ are arbitrary permutations, is equal to the permutation that is obtained by letting the permutation $Q$ act on $P$ (in the sense of a permutation acting upon the arguments of a function). For example:

$$(13)\,(35)\,(13) = (15) \qquad (123)\,(23)\,(123)^{-1} = (13)$$

2. The product of two permutations is independent of their order if the cycles that constitute the permutations do not contain any common elements.

3. Two cycles that possess a common element can be combined by placing this element at the end of one cycle and at the beginning of the other, i.e.:

$$(ik \ldots lm)\,(mn \ldots q) = (ik \ldots lmn \ldots q) \tag{A2.168}$$

For example:

$$(1245)\,(346) = (5124)\,(463) = (512463) = (124635)$$

4. A cycle that results from multiplying other cycles may contain a number of identical elements. It is then useful to reduce this to cycles of the form:

$$(ik \ldots lmi) \equiv (k \ldots lm) \tag{A2.169}$$

Every cycle can always be written in the form of a product of transpositions. Such a representation, however, is not unique. Thus, for example:

$$\begin{aligned}(123 \ldots k) &= (12)\,(23) \ldots (k-1k)\\ &= (1k)\,(1k-1)\,(1k-2) \ldots (12)\end{aligned}$$

Nevertheless, the number of transpositions which form a given permutation always possess a unique parity. A permutation is classed as *even* or *odd* according to whether the number of its transpositions is even or odd.

It is not difficult to show that any permutation can be represented as a product of transpositions of the form $(i-1, i)$, where $i-1$ and $i$ are two consecutive numbers. Thus, for example:

$$(245) = (24)\,(45) = (23)\,(34)\,(23)\,(45)$$

This plays an important role in the determination of explicit forms for the matrices of the irreducible representations of the permutation group, see Reference [8].

### Classes

In Section A2.2.1, it was shown that the $N!$ permutations of $N$ objects satisfy the four group postulates and therefore form a group, which was denoted by the symbol $\pi_N$. The even permutations in $\pi_N$ form a group by themselves known as the *alternating group*, and constitute a subgroup of $\pi_N$. In addition to this, $\pi_N$ possess $N-1$ obvious further subgroups: $\pi_{N-1}, \pi_{N-2}, \ldots$ and $\pi_1$.

All permutations that are related to one another by the equation $P_i = QP_jQ^{-1}$, where $Q$ is any permutation of $\pi_N$, are, by definition, members of the same class. Since the permutation $QP_jQ^{-1}$ is obtained by the action of $Q$ upon $P_j$ (see the previous subsection), the cyclic structures of $P_i$ and $P_j$ are identical, i.e. the number and lengths of the cycles in the two permutations coincide. The permutations $P_i$ and $P_j$ differ only in the elements forming the cycles. Each class of $\pi_N$ is therefore characterized by a particular partition of the $N$ elements into cycles. The number of different classes is determined by the number of different ways of partitioning the number $N$ into positive integral components, i.e., it is equal to the number of different integral solutions (with the exception of zero) of the equation:

$$1\nu_1 + 2\nu_2 + \ldots + N\nu_N = N \tag{A2.170}$$

A set numbers $\nu_1, \nu_2, \ldots \nu_N$, which satisfies Equation (A2.170), uniquely defines a class of $\pi_N$. A class is designated by the symbol $\{1^{\nu_1}2^{\nu_2}\ldots m^{\nu_m}\}$, where $\nu_k$ is the number of cycles of length $k$ that appear in the permutations in the class. The class $\{1^N\}$ corresponds to the identity permutation. Thus, the six permutations of the group $\pi_3$ are divided into three classes:

$$\{1^3\} : (1)\,(2)\,(3) \equiv I$$

$$\{12\} : (1)\,(23) \quad (2)\,(13) \quad (3)\,(12)$$

$$\{3\} : (123) \quad (132)$$

The group $\pi_4$ contains five classes: $\{1^4\}, \{1^2 2\}, \{2^2\}, \{13\}$ and $\{4\}$.

It can be proved [7, 8] that the order of the class is given by:

$$g_{\{1^{\nu_1}2^{\nu_2}\ldots m^{\nu_m}\}} = \left(N!/\nu_1!\nu_2!2^{\nu_2}\ldots\nu_m!m^{\nu_m}\right) \tag{A2.171}$$

For example, the class $\{1^2 2\}$ of the group $\pi_4$ contains $4!/(2!2) = 6$ elements, and the class $\{13\}$ contains $4!/3 = 8$ elements.

### Young diagrams and irreducible representations

Since the number of nonequivalent irreducible representations of a group is equal to the number of its classes, the nonequivalent irreducible representations of $\pi_N$ are defined, as are the classes, by the different partitions of the number $N$ into positive integral components. Each irreducible representation is typified by one such partition. The partitions are usually written in order of decreasing components $\lambda^{(i)}$:

$$\lambda^{(i)} + \lambda^{(2)} + \ldots + \lambda^{(m)} = N \quad \lambda^{(1)} \geq \lambda^{(2)} \geq \ldots \geq \lambda^{(m)} \tag{A2.172}$$

Some of the $\lambda^{(i)}$ in Equation (A2.172) may coincide. It is clear that $m$ cannot exceed $N$. Equation (A2.172) can be regarded as an equation for finding the possible partitions of $N$, and in this sense is entirely equivalent to Equation (A2.170). The partitions (A2.172) can be depicted graphically by means of diagrams known as *Young diagrams*, in which each number $\lambda^{(i)}$ is represented by a row of $\lambda^{(i)}$ cells. Young diagrams will be subsequently denoted by the symbol $[\lambda] \equiv \left[\lambda^{(1)}\lambda^{(2)} \ldots \lambda^{(m)}\right]$. The presence of several rows of identical length $\lambda^{(i)}$ will be indicated by a power of $\lambda^{(i)}$. For example:

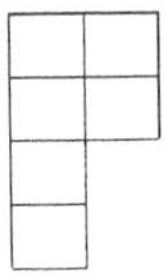

$[\lambda] = \left[2^2 1^2\right]$

It is obvious that from two cells only two Young diagrams can be formed:

(A2.173)

[2] $\left[1^2\right]$

For the permutation group of three elements, $\pi_3$, from three cells three Young diagrams can be formed:

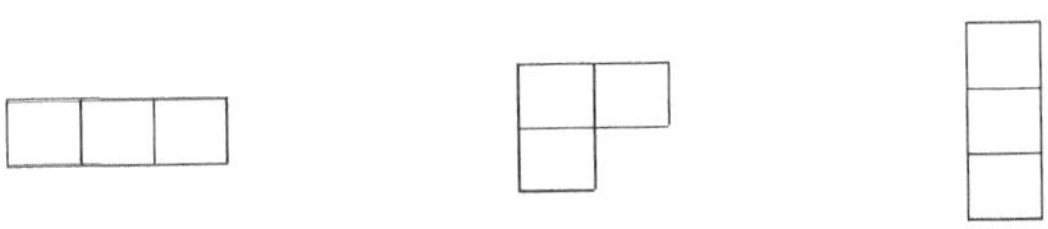

(A2.174)

[3] [21] $\left[1^3\right]$

The group $\pi_4$ has five Young diagrams:

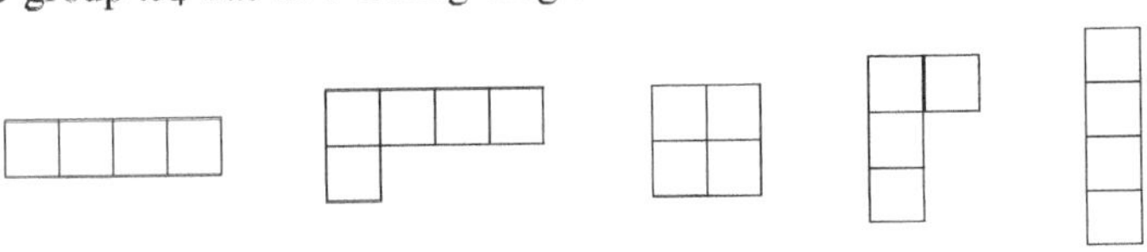

(A2.175)

[4] [31] $\left[2^2\right]$ $\left[21^2\right]$ $\left[1^4\right]$

Each Young diagram $[\lambda]$ uniquely corresponds to a specific irreducible representation $\Gamma^{[\lambda]}$ of the group $\pi_N$. The assignment of a Young diagram determines the permutation symmetry of the basis functions for an irreducible representation, i.e., it determines the behavior of the basis functions under permutations of their arguments. A Young diagram with only one row corresponds to a function symmetric in all its arguments. A Young diagram with one column corresponds to a completely antisymmetric function. All other types of diagrams correspond to intermediate types of symmetry. There are certain rules that make it possible to find the matrices of irreducible representations of the permutations group from the form of the corresponding Young diagram. Such rules are especially simple in the case of the so-called standard orthogonal representation (the Young–Yamanouchi representation), see References [7, 8].

For every standard representation $\mathbf{\Gamma}^{[\lambda]}$, a *conjugate* (or *associated*) representation $\mathbf{\Gamma}^{[\tilde{\lambda}]}$ of the same dimension can be constructed. The matrices of the representation $\mathbf{\Gamma}^{[\tilde{\lambda}]}(P)$ differ from those of the standard representation $\mathbf{\Gamma}^{[\lambda]}(P)$ by a factor $(-1)^p$, where $p$ is the parity of the permutation; namely, $\mathbf{\Gamma}_{rt}^{[\tilde{\lambda}]}(P) = (-1)^p\,\mathbf{\Gamma}_{rt}^{[\lambda]}(P)$.

The symmetric and antisymmetric representations defined above form the simplest example of conjugate representations. The first representation corresponds to the Young diagram $[N]$ and the second to the diagram $\left[1^N\right]$. The Young diagram $\left[1^N\right]$ is obtained from $[N]$ by changing the row into a column. In the general case, it can be shown that for a representation characterized by the Young diagram $[\lambda]$, the conjugate representation is characterized by a diagram known as the dual Young diagram $\left[\tilde{\lambda}\right]$, which is obtained from $[\lambda]$ by interchanging its rows and columns. For example:

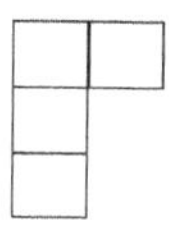

$[\lambda]$

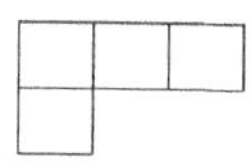

$\left[\tilde{\lambda}\right]$

**Construction of the permutation symmetric functions**

In Section A2.2.2 it was shown that a set of basis functions for an irreducible representation of any finite group can be obtained by applying the operators (A2.160) to some arbitrary function. The basis functions for an irreducible representation $\mathbf{\Gamma}^{[\lambda]}$ of the permutation group $\pi_N$ can be constructed by means of the so-called normalized Young operators[2] [8, 9]:

$$\omega_{rt}^{[\lambda]} = \sqrt{\frac{f_\lambda}{N!}} \sum_P \mathbf{\Gamma}_{rt}^{[\lambda]}(P)\,P \tag{A2.176}$$

where the summation over $P$ runs over all the $N!$ permutations in the group $\pi_N$, $\mathbf{\Gamma}_{rt}^{[\lambda]}(P)$ are the matrix elements and $f_\lambda$ is the dimension of the irreducible representation $\mathbf{\Gamma}^{[\lambda]}$.

The operator $\omega_{rt}^{[\lambda]}$ differs from the $\epsilon_{ik}^{(\alpha)}$ by the absence of the sign for complex conjugation on the matrix elements (the matrices in a representation $\mathbf{\Gamma}^{[\lambda]}$ are all

[2] Operators (A2.176) should not be mixed up with the operator that symmetries the rows and antisymmetrizes the columns in Young diagram, which is also often referred to as the Young operator [7].

real), and also in the factor in front of the summation. This factor is chosen so that application of the operator (A2.176) to a nonsymmetrized product of orthonormal one-particle functions $\varphi_a$:

$$\Phi_0 = \varphi_1(1)\varphi_2(2)\ldots\varphi_N(N) \tag{A2.177}$$

produces a normalized function:

$$\Phi_{rt}^{\lambda} = \omega_{rt}^{[\lambda]}\Phi_0 = \sqrt{\frac{f_\lambda}{N!}}\sum_P \mathbf{\Gamma}_{rt}^{[\lambda]}(P)\,P\Phi_0 \tag{A2.178}$$

transforming in accordance with the representation $\mathbf{\Gamma}^{[\lambda]}$. The function $\Phi_{rt}^{[\lambda]}$ transforms as the rth column of the irreducible representation $\mathbf{\Gamma}^{[\lambda]}$, and the set of $f_\lambda$ functions $\Phi_{rt}^{[\lambda]}$ with fixed second index t forms a basis for the irreducible representation $\mathbf{\Gamma}^{[\lambda]}$. Altogether it is possible to form $f_\lambda$ independent bases corresponding to the number of different values of t. This is natural, since $N!$ functions $P\Phi_0$ form a basis for the regular representation of $\pi_N$, and in the decomposition of the regular representation, each irreducible representation occurs as many times as its dimension. The first index, $r$, characterizes the symmetry of the function $\Phi_{rt}^{[\lambda]}$ under permutation of the arguments. The second index, $t$, enumerating the different bases of $\mathbf{\Gamma}^{[\lambda]}$, characterizes the symmetry of $\Phi_{rt}^{[\lambda]}$ under permutations of the one-particle functions $\varphi_a$ [8].

Let us write down, as an example, the basis functions for all irreducible representations of $\pi_3$ using Equation (A2.178). The group $\pi_3$ has two one-dimensional representations: the symmetric one with $\mathbf{\Gamma}^{[3]}(P) = 1$ for all $P$ and the antisymmetric one with $\mathbf{\Gamma}^{[1^3]}(P) = (-1)^p$ where $p$ is the parity of the permutation $P$. The representation $\mathbf{\Gamma}^{[21]}$ is two-dimensional and is given by the following matrices [8]:

$$\overset{I}{\begin{pmatrix} 1 & \\ & 1 \end{pmatrix}} \quad \overset{P_{12}}{\begin{pmatrix} 1 & \\ & -1 \end{pmatrix}} \quad \overset{P_{13}}{\begin{pmatrix} -\frac{1}{2} & -\frac{\sqrt{3}}{2} \\ -\frac{\sqrt{3}}{2} & \frac{1}{2} \end{pmatrix}}$$

$$\overset{P_{23}}{\begin{pmatrix} -\frac{1}{2} & \frac{\sqrt{3}}{2} \\ \frac{\sqrt{3}}{2} & \frac{1}{2} \end{pmatrix}} \quad \overset{P_{123}}{\begin{pmatrix} -\frac{1}{2} & \frac{\sqrt{3}}{2} \\ -\frac{\sqrt{3}}{2} & -\frac{1}{2} \end{pmatrix}} \quad \overset{P_{132}}{\begin{pmatrix} -\frac{1}{2} & -\frac{\sqrt{3}}{2} \\ \frac{\sqrt{3}}{2} & -\frac{1}{2} \end{pmatrix}}$$

Six independent basis functions can be constructed:

$$\Phi^{[3]} = \frac{1}{\sqrt{6}}\{\varphi_a(1)\varphi_b(2)\varphi_c(3) + \varphi_a(2)\varphi_b(1)\varphi_c(3) + \varphi_a(3)\varphi_b(2)\varphi_c(1) + \varphi_a(1)\varphi_b(3)\varphi_c(2) + \varphi_a(2)\varphi_b(3)\varphi_c(1) + \varphi_a(3)\varphi_b(1)\varphi_c(2)\}$$

$$\Phi^{[1^3]} = \frac{1}{\sqrt{6}}\{\varphi_a(1)\varphi_b(2)\varphi_c(3) - \varphi_a(2)\varphi_b(1)\varphi_c(3) - \varphi_a(3)\varphi_b(2)\varphi_c(1) - \varphi_a(1)\varphi_b(3)\varphi_c(2) + \varphi_a(2)\varphi_b(3)\varphi_c(1) + \varphi_a(3)\varphi_b(1)\varphi_c(2)\}$$

$$\Phi_{11}^{[21]} = \frac{1}{\sqrt{12}} \{2\varphi_a(1)\varphi_b(2)\varphi_c(3) + 2\varphi_a(2)\varphi_b(1)\varphi_c(3) - \varphi_a(3)\varphi_b(2)\varphi_c(1)$$
$$-\varphi_a(1)\varphi_b(3)\varphi_c(2) - \varphi_a(2)\varphi_b(3)\varphi_c(1) - \varphi_a(3)\varphi_b(1)\varphi_c(2)\}$$

$$\Phi_{21}^{[21]} = \tfrac{1}{2}\{-\varphi_a(3)\varphi_b(2)\varphi_c(1) + \varphi_a(1)\varphi_b(3)\varphi_c(2) - \varphi_a(2)\varphi_b(3)\varphi_c(1)$$
$$+\varphi_a(3)\varphi_b(1)\varphi_c(2)\}$$

$$\Phi_{12}^{[21]} = \tfrac{1}{2}\{-\varphi_a(3)\varphi_b(2)\varphi_c(1) - \varphi_a(1)\varphi_b(3)\varphi_c(2) + \varphi_a(2)\varphi_b(3)\varphi_c(1)$$
$$-\varphi_a(3)\varphi_b(1)\varphi_c(2)\}$$

$$\Phi_{22}^{[21]} = \frac{1}{\sqrt{12}} \{2\varphi_a(1)\varphi_b(2)\varphi_c(3) - 2\varphi_a(2)\varphi_b(1)\varphi_c(3) + \varphi_a(3)\varphi_b(2)\varphi_c(1)$$
$$+\varphi_a(1)\varphi_b(3)\varphi_c(2) - \varphi_a(2)\varphi_b(3)\varphi_c(1) - \varphi_a(3)\varphi_b(1)\varphi_c(2)\}$$

For the representation $\mathbf{\Gamma}^{[21]}$, two independent bases are constructed: $\left(\Phi_{11}^{[21]}, \Phi_{21}^{[21]}\right)$ and $\left(\Phi_{12}^{[21]}, \Phi_{22}^{[21]}\right)$. This should be expected, since the dimension of $\mathbf{\Gamma}^{[21]}$ is equal to two and we realize the decomposition of the regular representation.

The character tables for the permutation groups $\pi_3$ to $\pi_8$ and matrices of the orthogonal irreducible representations for groups $\pi_3$ to $\pi_6$ are presented in the book [8], Appendices 4 and 5.

### A2.2.4 The linear groups. The three-dimensional rotation group

**Definition**

In the preceding sections, *discrete groups* were considered, i.e., groups whose elements form a discrete set. A discrete set of elements can always be enumerated by positive integers. However, there is a large class of groups whose elements form a *continuous set*. Each element is characterized by a number of parameters that can vary continuously. Such groups are known as *continuous groups*.

We can also formally associate a value of some parameter $a$ with every element $G_a$ of a discrete group. The parameter corresponding to the product of two elements, $G_c = G_a G_b$, is equal to $c$. The discrete function:

$$c = \psi(a, b) \tag{A2.179}$$

which places the value of the parameter $c$ of the product into correspondence with the values of the parameters $a$ and $b$, determines the multiplication table for the group. Thus, all the elements of a discrete group can be described by specifying a single parameter that can assume $g$ discrete values. In the case of continuous groups, the function (A2.179) becomes a continuous function of its arguments, the number of parameters which characterize an element of the group being arbitrary, and even infinite.

In what follows, we consider groups of linear transformations in an $n$-dimensional vector space, the transformation being characterized by a finite number of parameters. These groups form a particular case of *Lie groups*. The group of transformations:

$$x_i' = f_i\left(x_1, \ldots, x_n; a_1, \ldots, a_r\right) \quad i = 1, 2, \ldots, n \tag{A2.180}$$

in which the functions are analytic in the parameters $a_k$, is known as a *r-parameter Lie group*.

The application of two such transformation in succession:

$$x_i' = f_i\left(x_1 \ldots, x_n; a_1, \ldots, a_r\right)$$

and:

$$x_i'' = f_i\left(x_1', \ldots, x_n'; b_1, \ldots, b_r\right)$$

is equivalent to a third transformation:

$$x_i''' = f_i\left(x_1, \ldots, x_n; c_1, \ldots, c_r\right)$$

in which the parameters are functions of the parameters of the first two transformations:

$$c_k = \psi_k\left(a_1, \ldots, a_r; b_1, \ldots, b_r\right) \quad k = 1, \ldots, r \tag{A2.181}$$

The transformations (Equation (A2.180)) also have to satisfy the remaining three group postulates.

The basic concepts and theorems that were established in Sections A2.2.1 and A2.2.2 for discrete groups can immediately be generalized to continuous groups. Only those statements that are based upon the finiteness of the order of the group cease to have any meaning (for example, the statements that the dimensions of the irreducible representations are divisors of the order of the group). Since there is a continuous set of elements in a group, each irreducible representations consists of a continuous set of matrices. The number of nonequivalent irreducible representations is infinite, though these form a discrete series in which the dimensions of all the irreducible representations are finite, i.e. the number of basis functions that transform into each other under the operations of the group is finite. [3]

For continuous groups a summation over the elements of a group is replaced by an integration over the domain over which the parameters can vary. The volume element in the integration is chosen so that the integral of an arbitrary continuous function of the group parameters is invariant under any transformation of the group. This is known as *invariant integration.*

Let us consider a vector space of $n$ dimensions. We subject the space to a linear transformation, with a matrix $A$. Every vector $\mathbf{x}$ in the space is transformed into a

[3] This is true only of groups known as *compact* continuous groups [10] to which practically all the continuous groups that are used in physical applications belong (the Lorentz group is an exception).

new vector $\mathbf{x}'$ with components:

$$x_l' = \sum_{k=1}^{n} a_{ik} x_k \tag{A2.182}$$

A linear transformation is said to be *nonsingular* if the determinant of the transformation matrix is nonzero. The set of all nonsingular linear transformations in an $n$-dimensional space forms a group known as the *general linear group*, and which is denoted by $\mathbf{GL}_n$. The product of two linear transformations is, of course, also a linear transformation, the matrix of which is found by multiplying the matrices of the individual transformations. Since the determinant of a transformation is nonzero, there exists for every transformation $A$ an inverse transformation $A^{-1}$, the successive application of the two transformations leading to the identity transformation. Finally, the associative rule is satisfied.

Since each linear transformation is determined by its matrix, these matrices form an $n$-dimensional representation of the group $\mathbf{GL}_n$. Generally speaking, the transformation matrices are complex, and therefore to characterize a transformation, $2n^2$ real parameters must be specified.

Restricting the linear transformations to unitary transformations, gives the *group of unitary transformations* in an $n$-dimensional space, which is denoted by the symbol $\mathbf{U}_n$. The elements of the transformation matrices satisfy the unitary conditions (Equation (A2.84)) that gives $n^2$ equations among $2n^2$ parameters, and so the number of independent parameters is equal to $n^2$. Every element of the unitary group is therefore characterized by specifying $n^2$ real parameters.

Unitary transformations with the determinant equal to unity form the *group of unitary unimodular transformations*, which is denoted by $\mathbf{SU}_n$ (the symbol is derived from the other name for this group: *the special unitary group)*. Every transformation in the group $\mathbf{SU}_n$ is characterized by specifying $n^2 - 1$ real parameters.

The set of all real unitary transformations, i.e., of all orthogonal transformations, forms a subgroup of the unitary group called the *orthogonal group* $\mathbf{O}_n$. The orthogonality conditions (see Equation (A2.91)) give $n + \frac{1}{2}n\,(n-1)$ equations among the $n^2$ parameters of a matrix. The elements of the orthogonal group are therefore characterized by specifying $n\,(n-1)\,/2$ parameters. Since the determinant of a matrix is equal to that of its transpose, it follows from Equation (A2.90) that the square of the determinant of an orthogonal transformation is equal to one, and so the determinant itself can only assume two values, plus or minus one. An orthogonal transformation with the determinant equal to plus one corresponds to a *rotation of the space* about the origin of coordinates. An orthogonal transformation with the determinant equal to minus one constitutes a combination of a rotation and an *inversion of the space* with respect to the origin of coordinates.

If the orthogonal transformations are restricted to those whose determinant is equal to plus one, the *rotation group in n-dimensional space*, $\mathbf{R}_n$, is obtained. For $n = 3$, $\mathbf{R}_n$ becomes the rotation group in three-dimensional space $\mathbf{R}_3$, which is widely used in physics.

Every subgroup of the orthogonal group is called a *point group*. The rotation group $\mathbf{R}_3$ is a continuous point group. Finite point groups, the elements of which consist of combinations of rotations through definite angles and reflections in planes, are used in the theory of molecules and crystals. The various point groups are classified in Section A2.2.5.

If the groups of linear transformations described above are placed in the order in which they are contained in each other, we obtain the following scheme:

$$\mathbf{GL}_n \supset \mathbf{U}_n \supset \begin{pmatrix} \mathbf{O}_n \\ \mathbf{SU}_n \end{pmatrix} \supset \mathbf{R}_n$$

**Irreducible representations of the $\mathbf{R}_3$ group**

As is established in quantum mechanics (see, for instance, Reference [11], Chapter 4), the eigenfunctions of the angular momentum operators satisfy the following eigenvalue equations:

$$\begin{aligned} J^2 \psi_m^{(j)} &= j\,(j+1)\,\psi_m^{(j)} \\ J_z \psi_m^{(j)} &= m \psi_m^{(j)} \\ \left(J_x + iJ_y\right)\psi_m^{(j)} &= \left[(j-m)\,(j+m+1)\right]^{1/2}\psi_{m+1}^{(j)} \\ \left(J_x - iJ_y\right)\psi_m^{(j)} &= \left[(j+m)\,(j-m+)\right]^{1/2}\psi_{m-1}^{(j)} \end{aligned} \qquad \text{(A2.183)}$$

where $j$ can assume integer and half-integer values only,[4] and for a given $j$, $m$ can assume the $2j+1$ values within the bounds $|m| \leq j$:

$$m = j, \quad j-1, \ldots, \quad -j \qquad \text{(A2.184)}$$

So, there are $2j+1$ independent functions $\psi_m^{(j)}$.

When $j$ is an integral number, $j = \ell$, the eigenfunctions of Equations (A2.183) are the spherical harmonics $Y_{\ell m}\left(\theta,\varphi\right)$, well known in mathematical physics [12]:

$$\psi_m^{(l)} = Y_{\ell m}\left(\theta,\varphi\right) = (-1)^m i^l \left[\frac{(2l+1)\,(l-m)!}{2\,(l+m)!}\right]^{1/2} P_l^m\,(\cos\theta)\, e^{im\varphi} \qquad \text{(A2.185)}$$

where $P_l^m\,(\cos\theta)$ is the associated Legendre functions:

$$P_l^m\,(\cos\theta) = \frac{1}{2^l l!}\sin^m\theta \frac{d^{l+m}}{d\,(\cos\theta)^{l+m}}\left(\cos^2\theta - 1\right)^l \qquad \text{(A2.186)}$$

For negative values of $m$:

$$Y_{l,-|m|} = (-1)^m\, Y_{l|m|}^* \qquad \text{(A2.187)}$$

Operators of angular momentum determine the change in a function due to an infinitely small rotations. It follows from Equations (A2.183) that functions $\psi_m^{(j)}$

---

4 This is correct only in the three-dimensional space in which the $\mathbf{R}_3$ group is defined. If $n = 2$ or 1, the restrictions on the value of $j$ are not valid, and the angular momentum can possess arbitrary values [13–15].

with fixed $j$ transform into one another under infinitely small rotations. An operator for a finite rotation through an angle $\alpha$ about an axis in the direction of an arbitrary unit vector $\mathbf{n}$ can also be represented via an angular momentum operator [16]:

$$R_{\mathbf{n},\alpha} = \exp(i\alpha\mathbf{n}\cdot\mathbf{J}) \tag{A2.188}$$

In order to find the effect of the operator (A2.188) upon a function of the coordinates, one must write the operator in the form of a power series and determine the effect of each term of the series upon the function. Since the operator for a finite rotation can be expressed in terms of angular momentum operators, the $\psi_m^{(j)}$ also transform among themselves under finite rotations:

$$R_{\mathbf{n},\alpha}\psi_m^{(j)} = \sum_{m'} \mathbf{D}_{m'm}^{(j)}\left(R_{\mathbf{n},\alpha}\right)\psi_{m'}^{(j)} \tag{A2.189}$$

where the coefficients $\mathbf{D}_{m'm}^{(j)}\left(R_{\mathbf{n},\alpha}\right)$ form a matrix that corresponds to a rotation through an angle $\alpha$ about an axis $\mathbf{n}$. The $2j+1$ functions $\psi_m^{(j)}$ therefore form a $(2j+1)$-dimensional basis for a representation of the rotation group $\mathbf{R}_3$. It can be shown that this representation, which is usually denoted by $D^{(j)}$, is irreducible [16]. Its matrix elements satisfy orthogonality relationships in which instead of summation, as in the case of finite group, it has to be invariant integration. The volume element for $\mathbf{R}_3$ can be written as $d\tau_R = \sin\beta\, d\beta\, d\alpha\, d\gamma$, where $\alpha$, $\beta$ and $\gamma$ are the Euler angles [16]. As a result we obtain:

$$\int \mathbf{D}_{m\mu}^{(j)}(R)^* \mathbf{D}_{m'\mu'}^{(j')}(R)\, d\tau_R = \left[8\pi^2/(2j+1)\right]\delta_{jj'}\delta_{mm'}\delta_{\mu\mu'}$$

An explicit form for the representation matrices $\mathbf{D}^{(j)}(\alpha,\beta,\gamma)$, expressed in terms of the Euler angles, has been obtained by Wigner [17].

It is not difficult to find the character of the class that corresponds to rotation through an angle $\alpha$. Since the classes of $\mathbf{R}_3$ are typified by the angle of rotation only and are independent of the direction of the axis of rotation, the $z$ axis is taken to be the axis of rotation. The rotation operator (A2.188) therefore assumes the form:

$$R_{z.\alpha} = e^{i\alpha J_z} \tag{A2.188a}$$

In order to find the effect of this operator on a function $\psi_m^{(j)}$, we expand the exponent and note that:

$$\mathbf{J}_z^k\psi_m^{(j)} = m^k\psi_m^{(j)}$$

i.e., in every term in the series the operator is replaced by its eigenvalue. This gives:

$$R_{z,\alpha}\psi_m^{(j)} = e^{im\alpha}\psi_m^{(j)} \tag{A2.188b}$$

The matrix of the rotation $R_{z,\alpha}$ is thus diagonal and its character in the representation $D^{(j)}$ is given by:

$$\chi^{(j)}(\alpha) = \sum_{m=-j}^{j} e^{im\alpha} = \frac{e^{i(j+1)\alpha} - e^{-ij\alpha}}{e^{i\alpha} - 1}$$
$$= \frac{\sin\left(j + \frac{1}{2}\right)\alpha}{\sin(\alpha/2)} \quad \text{(A2.190)}$$

From this formula it follows that:

$$\chi^{(j)}(-\alpha) = \chi^{(j)}(\alpha)$$

Two rotations through identical angles but in opposite directions thus belong to the same class; i.e., a class is characterized by the absolute value of the angle of rotation.

The set of rotations about a fixed axis forms the two-dimensional rotation group $\mathbf{R}_2$. This group is Abelian, and all of its irreducible representations are one dimensional. The $2j+1$ basis functions for an irreducible representation $\mathbf{D}^{(j)}$ of the group $\mathbf{R}_3$ belong to different irreducible representations of $\mathbf{R}_2$, each of which is characterized by a particular value of the number $m$.

It can be shown [16] that the matrix elements of the representations $\mathbf{D}^{(j)}$ for integral $j$ are proportional to the spherical harmonics:

$$D^{(l)}_{m'0}(\alpha, \beta, 0) = \left(\frac{4\pi}{2l+1}\right)^{\frac{1}{2}} Y_{lm'}(\beta, \alpha)$$

$$D^{(l)}_{0m}(0, \beta, \gamma) = (-1)^m \left(\frac{4\pi}{2l+1}\right)^{\frac{1}{2}} Y_{lm}(\beta, \gamma) \quad \text{(A2.191)}$$

where $\alpha, \beta$ and $\gamma$ are the Euler angles. The matrix elements $\mathbf{D}^{(j)}_{m'm}(\alpha, \beta, \gamma)$ are therefore known as *generalized spherical harmonics* of the $j$th order.

### Reduction of the direct product of two irreducible representations of $\mathbf{R}_3$

Let us construct the direct product of two irreducible representations, $\mathbf{D}^{(j_1)} \times \mathbf{D}^{(j_2)}$. The representation that is obtained is of dimension $(2j_1+1)(2j_2+1)$ and is reducible. Products of the basis functions for the individual representations, $\psi^{(j_1)}_{m_1}\psi^{(j_2)}_{m_2}$, form a basis for this representation. It can be proved that the decomposition has the following form [11, 16]:

$$\mathbf{D}^{(j_1)} \times \mathbf{D}^{(j_2)} \doteq \mathbf{D}^{(j_1+j_2)} + \mathbf{D}^{(j_1+j_2-1)} + \cdots + \mathbf{D}^{(|j_1-j_2|)} \quad \text{(A2.192)}$$

Each irreducible representation $D^{(j)}$ occurs once in the decomposition, the values of $j$ are located in the interval:

$$|j_1 - j_2| \leq j \leq j_1 + j_2, \quad j_1 + j_2 + j \text{ is integral} \quad \text{(A2.193)}$$

If one constructs a triangle with integral perimeter and sides of lengths $j_1$, $j_2$ and $j$, the condition that the sides must obey is just Equation (A2.193). Equation (A2.193) is called the triangle rule with integral perimeter; it is denoted by $\Delta\,(j_1 j_2 j)$.

The linear transformation from the basis functions of the reducible representation to basis functions for the irreducible representations which occur in the decomposition (A2.192) is carried out by means of the matrix of Clebsch–Gordan coefficients (see Equation A2.146):

$$\psi_m^{(j)} = \sum_{m_1, m_2} \psi_{m_1}^{(j_1)} \psi_{m_2}^{(j_2)} \langle j_1 m_1, j_2 m_2 \mid jm \rangle \tag{A2.194}$$

The Clebsch–Gordan coefficients $\langle j_1 m_1, j_2 m_2 \mid jm \rangle$ are nonzero only if $m_1 + m_2 = m$ and condition (A2.193) is fulfilled. The summation over $m_2$ in Equation (A2.194) is therefore purely formal, since $m_2 = m - m_1$. The $\langle j_1 m_1, j_2 m_2 \mid jm \rangle$ coefficients satisfy the orthogonal relations (A2.148). A number of different explicit expressions for the Clebsch–Gordan coefficients in terms of the parameters $j_1$, $j_2$, $j$, $m_1$ and $m_2$ may be found elsewhere [18–21].

Instead of the Clebsch–Gordan coefficients, one often makes use of more symmetric coefficients known as $3j$ *symbols*. The $3j$ symbols are written in the form of two-row matrices:

$$\begin{pmatrix} j_1 & j_2 & j_3 \\ m_1 & m_2 & m_3 \end{pmatrix} \tag{A2.195}$$

They are related to the Clebsch–Gordan coefficients by the equation:

$$\langle j_1 m_1, j_2 m_2 \mid jm \rangle = (-1)^{-j_1 + j_2 - m} (2j + 1)^{\frac{1}{2}} \begin{pmatrix} j_1 & j_2 & j \\ m_1 & m_2 & -m \end{pmatrix} \tag{A2.196}$$

The $3j$ symbols are invariant under any even permutation of their columns, and are multiplied by $(-1)^{j_1 + j_2 + j_3}$ under any odd permutation of their columns.

## A2.2.5 Point Groups

### Symmetry elements and symmetry operations

According to our classification of groups of linear transformations in a vector space (Section A2.2.4), every subgroup of the group of orthogonal transformations is called a *point* group. The group of rotation in three-dimensional space, which was considered in the previous section, is therefore a point group. These groups are referred to as 'point' groups because their transformations leave fixed at least one point of the space (the origin of coordinates). In reality, all orthogonal transformations reduce to combinations of two types of transformation: to rotations and to reflections in planes that pass through the origin of coordinates. Hence, the point in the space corresponding to the origin of coordinates remains fixed under these transformations.

A body is said to possess the symmetry of a point group if it is sent into itself by the transformations of this group. In such a case, the transformations of the

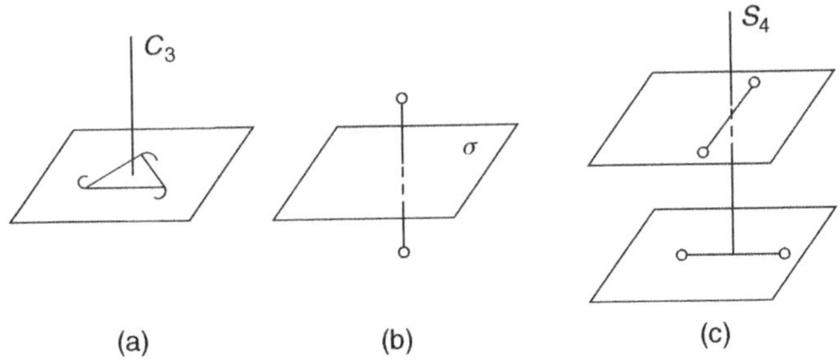


**Figure A2.5** Symmetry elements

point group coincide with the symmetry transformations of the body. A necessary condition for the body to be symmetric is that it possess axes and planes such that transformations with respect to these send the body into itself. Such axes and planes are usually referred to as *symmetry elements*. Each symmetry element gives rise to corresponding symmetry transformations that are commonly called *symmetry operations*.

If the body coincides with itself $n$ times under a rotation through an angle $2\pi$ about some axis, the axis is called an *n-fold (or nth-order) axis of symmetry* and is denoted by $C_n$ (Figure A2.5(a)). In this case the smallest angle of rotation under which the body can be made to coincide with itself is equal to $2\pi/n$. This rotation operation is denoted by the symbol $C_n$, the same as for the axis of rotation. A succession of $k$ rotations through $2\pi/n$, i.e. a rotation through $2\pi k/n$, is denoted by $C_n^k$. The operation $C_n$ when carried out $n$ times is obviously equivalent to the identity transformation. The latter is usually denoted by the symbol $E$:

$$C_n^n = E$$

The other possible symmetry element is a *plane of symmetry*, and the corresponding symmetry operation is a reflection of the body in the plane. The equilateral triangle with small tags at its vertices shown in Figure A2.5(a) possesses no planes of symmetry apart from the trivial plane of the triangle. The dumbbell in Figure A2.5(b) is obviously symmetric with respect to a reflection in the plane perpendicular to its connecting rod and which passes through its midpoint. A plane of symmetry is denoted in general by the symbol $\sigma$. A plane perpendicular to an axis $C_n$ is denoted by $\sigma_h$, and a plane passing through an axis $C_n$ is denoted by $\sigma_v$ (or by $\sigma_d$; see subsequent discussion). This notation is also used for the corresponding symmetry operations.

Clearly:

$$\sigma^2 = E$$

It may happen that a body can only be made to coincide with itself after the application of two symmetry operations in succession: a rotation through an angle $2\pi/n$ and a reflection in a plane perpendicular to the axis of rotation. This symmetry operation is called a *rotation–reflection* and is denoted by the symbol $S_n$. The order in which the rotation and reflection operations are carried out is unimportant:

$$S_n = \sigma_h C_n = C_n \sigma_h \tag{A2.197}$$

In Figure A2.5(c) two dumbbells are shown in a configuration where they are fixed at 90° with respect to each other. This configuration possesses a fourfold rotation–reflection axis, $S_4$.

A rotation–reflection axis only constitutes a new symmetry element if $n$ is even. If $n$ is odd, it follows from Equation (A2.197) that:

$$S_{2m+1}^2 = C_{2m+1}^2, \quad S_{2m+1}^{2m} = C_{2m+1}^{2m}, \quad S_{2m+1}^{2m+1} = \sigma_h \tag{A2.198}$$

i.e., there is an independent $(2m + 1)$-fold axis of symmetry and an independent plane of symmetry perpendicular to it. Note that if $n$ is even, a rotation–reflection axis $S_n$ is simultaneously a $C_{n/2}$-fold rotation axis.

A twofold rotation–reflection axis is equivalent to the presence of a *center of symmetry* in the body. This lies at the point where the axis $S_2$ and the plane $\sigma_h$ intersect. The operation $S_2$ constitutes an *inversion operation* through the center of symmetry, and is denoted by the symbol $I$:

$$I \equiv S_2 = C_2 \sigma_h \tag{A2.199}$$

Symmetry operations usually do not commute. The following operations only are an exception to this:

(a) Rotations about a given axis.

(b) Rotations through $\pi$ about mutually perpendicular axes (the result is equivalent to a rotation by $\pi$ about an axis perpendicular to the first two).

(c) A rotation and a reflection in a plane perpendicular to the axis of rotation [see Equation (A2.197)].

(d) Reflections in mutually perpendicular planes (the resulting operation is equivalent to a rotation by $\pi$ about the axis formed by the line of intersection of the two planes).

From (a) and (c) it follows that the operation of inversion commutes with any symmetry operation.

The set of symmetry operations for a given body forms the point group of the body. Groups possessing more than one axis with order $n > 2$ describe the symmetry of the regular polyhedra (Table A2.3). It is interesting to note that only five regular polyhedra can be constructed, and all these polyhedra were recognized by the ancient Greeks. They were discussed by Plato (427–347 BC).

The symmetry elements possessed by a molecule determine the particular point group to which it belongs. The point groups are therefore conveniently classified by the symmetry elements that generate the operations of a given group.

**Table A2.3** Regular polyhedra

| | Total number | | | Maximum order of axes $C_n$ |
|---|---|---|---|---|
| | faces | edges | vertices | |
| Tetrahedron | 4 | 6 | 4 | 3 |
| Cube | 6 | 12 | 8 | 3 |
| Octahedron | 8 | 12 | 6 | 4 |
| Dodecahedron | 12 | 30 | 20 | 3 |
| Icosahedron | 20 | 30 | 12 | 5 |

## Classification of point groups

There are several notations for point groups. We shall adhere to the notation due to Schönflies. We begin by classifying the simplest group, and proceed from these to the simplest groups with additional symmetry elements. Finally, we consider continuous point groups, picking out those to which it is possible for molecules to belong.

### *1. Discrete axial point groups*

These groups possess no more than one axis of symmetry, whose order is greater than twofold.[5]

$\mathbf{C}_n$ *Groups.* These possess a single symmetry element $C_n$, an $n$-fold axis of symmetry. A $\mathbf{C}_n$ group consists of $n$ operations, these being rotations through angles $2\pi k/n$ about the axis of symmetry. All the operations of the group commute, i.e. the group is Abelian. Every operation of the group therefore forms a class and the irreducible representations are all one dimensional.

A necessary condition for a molecule to belong to a $\mathbf{C}_n$ group is the absence of a plane of symmetry. The dichloro-substituted allene molecule (Figure A2.6(a)) is an example. The molecule $H_2C{=}CCl_2$ belongs to the group $\mathbf{C}_2$ and the molecule $H_3C{-}CCl_3$ to the group $\mathbf{C}_3$ if the terminal groups of atoms are slightly rotated with respect to each other (the equilibrium configurations of both molecules in fact possess higher symmetries).

$\mathbf{S}_{2n}$ *Groups.* These possess a single rotation–reflection axis, $S_{2n}$. An $\mathbf{S}_{2n}$ group is Abelian and consists of $2n$ rotation–reflection operations $S_{2n}^k$. As shown in the previous subsection, a rotation–reflection axis can only be of even order, since for an odd order the axis reduces to an ordinary symmetry axis of odd order and to a plane perpendicular to it. The operation $S_2$ is the inversion operation, so that the group $\mathbf{S}_2$ consists of just two elements, $E$ and $I$, and is often denoted by the symbol $\mathbf{C}_i$. The *trans*-form of the ClBr–HC–CHBrCl molecule (Figure A2.6(b)) serves as an example of a molecule which belongs to $\mathbf{S}_2$.

[5] The subscript $n$ in the notation for the point groups $\mathbf{C}_n, \mathbf{D}_n$ etc. denotes the order of the main symmetry axis. This should not be confused with the notation used for the linear groups $\mathbf{GL}_n$, $\mathbf{U}_n$, etc., where the subscript $n$ denotes the dimension of the vector space of the particular group.

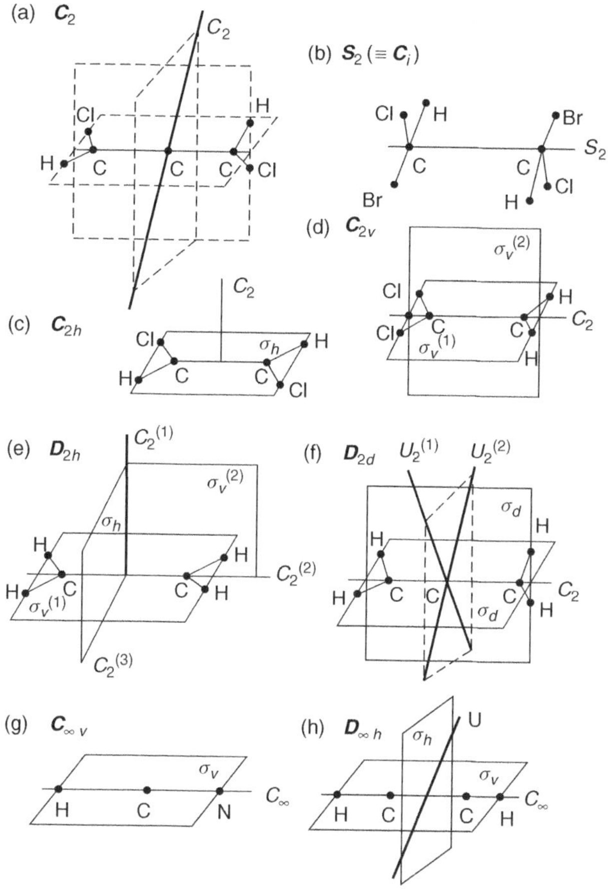


**Figure A2.6** Axial point groups

$\mathbf{C}_{nh}$ *Groups.* These possess an $n$-fold symmetry axis and a plane of symmetry perpendicular to it, $\sigma_h$. A $\mathbf{C}_{nh}$ group consists of $2n$ operations, $n$ rotations $C_n^k$, and $n$ rotation–reflection operations $C_n^k\sigma_h$. All the operations of the group commute and there are therefore $2n$ classes. The group $\mathbf{C}_{1h}$ contains just two elements, $E$ and $\sigma_h$; this group is usually denoted by $\mathbf{C}_s$. A group $\mathbf{C}_{nh}$ can obviously be written as a direct product of the groups $\mathbf{C}_n$ and $\mathbf{C}_s$, i.e. $\mathbf{C}_{nh} = \mathbf{C}_n \times \mathbf{C}_s$. All groups $\mathbf{C}_{2mh}$ possess a center of symmetry. The planar *trans*-isomers of the molecules $C_2H_2Cl_2$ (Figure A2.6(c)) and $C_6H_2Cl_2Br_2$ are examples of molecules which belong to the group $\mathbf{C}_{2h}$.

$\mathbf{C}_{nv}$ *Groups.* These possess an $n$-fold symmetry axis and $n$ planes of symmetry which pass through the axis. A $\mathbf{C}_{nv}$ group consists of $2n$ elements, $n$ rotations $C_n^k$ and $n$ reflections $\sigma_v$. The rotation and reflection operators do not commute with one another. The symmetry axis is bilateral due to the presence of planes $\sigma_v$, which pass through the axis, i.e., the rotations all belong to the same class. The $n$ reflections all belong to the same class only if $n$ is odd. If $n = 2m$, the reflection operations fall into two classes with $m$ elements in each. This is because rotations about an axis only bring alternating planes into coincidence with each other; adjacent planes cannot be made to coincide.

Many molecules belong to the $\mathbf{C}_{nv}$ point group. Thus $H_2O$, $H_2S$, $SO_2$, $NO_2$, *cis*-$C_2H_2Cl_2$ (Fig. A2.6(d)), $H_2CO$, etc. belong to $\mathbf{C}_{2v}$, and $NH_3$, $PCl_3$, $CH_3Cl$, etc. belong to $\mathbf{C}_{3v}$.

$\mathbf{D}_n$ *Groups.* These possess an $n$-fold symmetry axis, and $n$ twofold symmetry axes perpendicular to it; the twofold axes intersecting at an angle of $\pi/n$. A $\mathbf{D}_n$ group contains $2n$ operations: $n$ rotations $C_n^k$ about the vertical axis an $n$ rotations through $\pi$ about the horizontal axes, these usually being denoted by $U_2$. The operations of the group are divided into classes in the same way as for $\mathbf{C}_{nv}$ (the groups $\mathbf{D}_n$ and $\mathbf{C}_{nv}$ are isomorphic). The group $\mathbf{D}_2$ possesses three mutually perpendicular twofold axes and is denoted by the letter $\mathbf{V}$.

Molecules whose equilibrium configuration is characterized by the symmetry axes of the $\mathbf{D}_n$ group usually possess additional planes of symmetry, and so belong to point groups of higher symmetry than $\mathbf{D}_n$.

$\mathbf{D}_{nh}$ *Groups.* The axis system of the $\mathbf{D}_n$ group has added to it a horizontal plane of symmetry that passes through the $n$ axes $U_2$. This leads to the appearance of $n$ vertical planes of symmetry that pass through the $C_n$ axis. A $\mathbf{D}_{nh}$ group contains $4n$ operations: $n$ reflections $\sigma_v$ and $n$ rotation–reflection operations $C_n^k\sigma_h$ added to the $2n$ operations of $\mathbf{D}_n$. The group $\mathbf{D}_{nh}$ can be written as a direct product $\mathbf{D}_n \times \mathbf{C}_s$, since the reflection $\sigma_h$ commutes with all operations of $\mathbf{D}_n$.

The ethylene molecule in its equilibrium configuration (Figure A2.6(e)) serves as an example of a molecule which belongs to the group $\mathbf{D}_{2h}$. All planar $XY_3$ molecules belong to $\mathbf{D}_{3h}$, e.g. $BF_3$. In addition, cyclopropane, ethane (eclipsed conformation), 1.3.5-trichlorobenzene, etc. also belong to $\mathbf{D}_{3h}$. The symmetry group of the benzene molecule is $\mathbf{D}_{6h}$.

$\mathbf{D}_{nd}$ *Groups.* The axis system of the $\mathbf{D}_n$ group has added to it $n$ planes of symmetry, which pass through the $C_n$ axis and which bisect the angle between two adjacent twofold axes. A $\mathbf{D}_{nd}$ group contains $4n$ operations: $2n$ operations of $\mathbf{D}_n$, $n$ reflections $\sigma_d$, and $n$ rotation–reflection operations $\sigma_d U_2$. The rotation–reflection operations $\sigma_d U_2$ are equivalent to a rotation about the $C_n$ axis followed by a reflection in the plane perpendicular to $C_n$. The $C_n$ axis consequently becomes a rotation–reflection axis of twice the order, $S_{2n}$. If $n$ is odd, $\mathbf{D}_{2m+1,d} = \mathbf{D}_{2m+1} \times \mathbf{C}_i$.

The allene molecule, in which the planes of the two $CH_2$ groups are mutually perpendicular (Figure A2.6(f)), forms an example of a molecule possessing the symmetry of $\mathbf{D}_{2d}$. The ethane molecule in its staggered conformation (the angle between the $CH_3$ groups is $60^\circ$) belongs to $\mathbf{D}_{3d}$.

***2. Cubic point groups***

Point groups of this kind possess several symmetry axes, the orders of which are greater than twofold.

*The Group* **T**. If the group $\mathbf{D}_2 \equiv \mathbf{V}$ has added to it four oblique threefold axes, such that rotations about them carry the twofold axes into each other, the axis system of a tetrahedron is obtained. This system of axes is conveniently represented by drawing the threefold axes as the spatial diagonals of a cube and the twofold axes as passing through the centers of opposite faces of the cube (Figure A2.7(a)). Rotations about these axes constitute a group, which is denoted by the symbol **T**. The group **T** consists of 12 operations, which are divided into four classes: $E$, three rotations $C_2$, four rotations $C_3$ and four rotations $C_3^2$.

There is no molecule whose equilibrium configuration possesses the symmetry of the point group **T**. However, if the methyl groups in the tetramethylmethane molecule $C(CH_3)_4$ (which belongs to a group with higher symmetry, $\mathbf{T}_d$) are slightly rotated, the nonequilibrium configuration obtained possesses the symmetry of **T**.

*The Group* $\mathbf{T}_d$. The axis system of the group **T** has added to it six planes of symmetry, each of which passes through two of the threefold axes and through one of the twofold axes. The twofold axes thereby become fourfold rotation–reflection axes. The 24 operations of the group are divided into five classes: $E$, eight rotations $C_3$ and $C_3^2$; six reflections $\sigma_d$; six rotation–reflection operations $S_4$; $S_4^3$; and three rotations $C_2 \equiv S_4^2$.

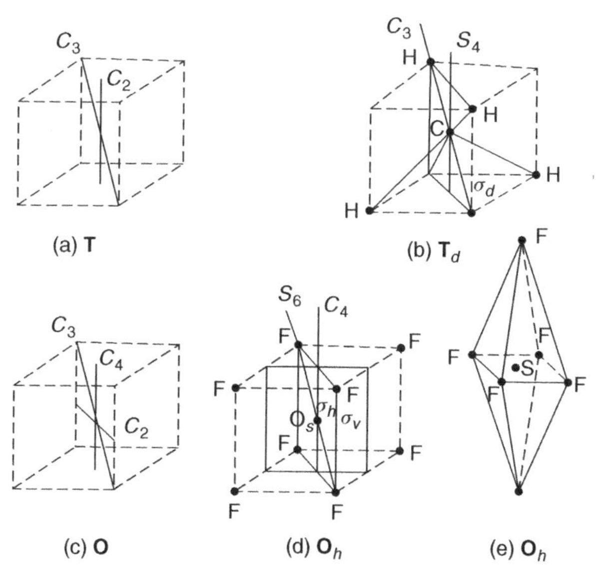


**Figure A2.7** Cubic point groups

All molecules which possess the symmetry of a tetrahedron belong to the group $\mathbf{T}_d$, e.g. $CH_4$ (Figure A2.7(b)), $CCl_4$, $P_4$, etc.

*The Group* $\mathbf{T}_h$. The axis system of the group **T** has added to it a center of symmetry. Three mutually perpendicular planes automatically appear as result, each of which passes through two of the twofold axes. Obviously, $\mathbf{T}_h = \mathbf{T} \times \mathbf{C}_i$. The point group $\mathbf{T}_h$ is apparently not realized as the symmetry group of a molecule.

*The Group* **O.** The symmetry axes of a cube form the symmetry elements of this group: three fourfold axes and four threefold axes (Figure A2.7(c)). The 24 operations of the group are divided into five classes: $E$, eight rotations $C_3$ and $C_3^2$, six rotations $C_4$ and $C_4^3$, three rotations $C_4^2$ about the fourfold axes and six rotations $C_2$ about the twofold axes.

Molecules which possess the axis system of the **O** group as a rule have additional planes of symmetry and so belong to a group with higher symmetry, $\mathbf{O}_h$.

*The Group* $\mathbf{O}_h$. The symmetry elements of this group are all the symmetry elements of a cube or of a regular octahedron. The symmetry axes of the group **O** have added to them six planes of symmetry $\sigma_h$ passing through opposite edges of the cube and three planes of symmetry $\sigma_v$ that pass through the center of the cube, parallel to its faces. The $C_3$ axes thereby become rotation–reflection axes $S_6$. The group $\mathbf{O}_h$ contains a center of symmetry and can be represented as the direct product $\mathbf{O}_h = \mathbf{O} \times \mathbf{C}_i$. The 48 operations of the group are divided into ten classes which are obtained from those of the group **O** by calculating the direct product $\mathbf{O} \times \mathbf{C}_i$.

Examples of molecules that possess the point symmetry of $\mathbf{O}_h$ are $OsF_8$ (Fig. A2.7(d)), $SF_6$ (Fig. A2.7(e)) and $UF_6$.

*The Group* **I.** This group contains 60 elements, all are pure rotations about the symmetry axes of a icosahedron or dodecahedron. There are six fivefold axes, 10 threefold axes, and 15 twofold axes. There are no molecules possessing the **I** symmetry because a highly symmetrical molecule usually has additional planes of symmetry.

*The Group* $\mathbf{I}_h$. The symmetry elements of this group are all the symmetry elements of a regular dodecahedron or icosahedron. The group $\mathbf{I}_h$ has a center of symmetry and can be represented as the direct product $\mathbf{I}_h = \mathbf{I} \times \mathbf{C}_i$. Its 120 elements are divided into 10 classes.

For a long time it was assumed that the icosahedral group was not important in physics and chemistry since it was not realized in nature. But, in the second half of the last century, it was revealed that the $B_{12}$ icosahedron is realized in the structures of many-boron compounds. For example, a stable icosahedral boron hydride ion $B_{12}H_{12}^{2-}$ was observed. The remarkable fullerene molecule, $C_{60}$, also possess the $\mathbf{I}_h$ symmetry; it has a truncated icosahedral structure (Figure A2.8).

This exhausts the finite point groups to which it is possible for molecules with rigid nuclear configurations to belong. By placing the groups considered in the order in which new symmetry elements are added, the following scheme is obtained:

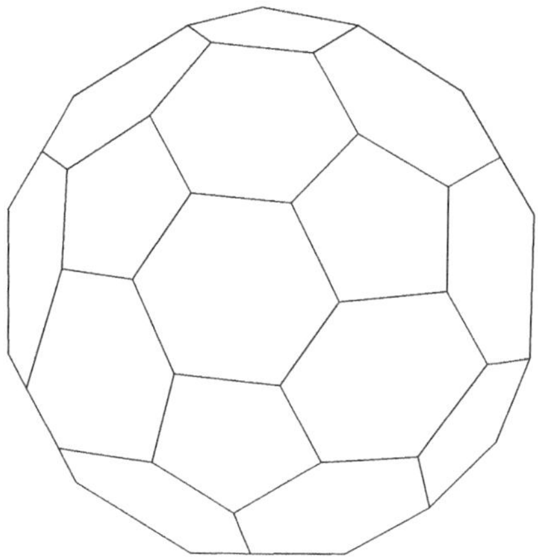

**Figure A2.8** The fullerene molecule; the point symmetry group of the truncated icosahedron (the $\mathbf{I}_h$ group)

$$
\begin{array}{ccccccccccc}
 & \mathbf{C}_n & & \rightarrow & \mathbf{D} & & \rightarrow (\mathbf{V}) \rightarrow & \mathbf{T} & \rightarrow & \mathbf{O} \rightarrow & \mathbf{I} \\
\swarrow & \downarrow & \searrow & & \swarrow & \searrow & & \swarrow \ \searrow & & \downarrow & \downarrow \\
\mathbf{S}_{2n} & \mathbf{C}_{nh} & \mathbf{C}_{nv} & & \mathbf{D}_{nh} & \mathbf{D}_{nd} & & \mathbf{T}_h \ \mathbf{T}_d & & \mathbf{O}_h & \mathbf{I}_h
\end{array}
$$

***3. Continuous point groups***

If in discrete axial groups, the order of the axis $C_n$ becomes infinite, the continuous axial point groups $\mathbf{C}_\infty$, $\mathbf{C}_{\infty h}$, $\mathbf{C}_{\infty v}$, $\mathbf{D}_\infty$ and $\mathbf{D}_{\infty h}$ are obtained (the group $\mathbf{D}_{\infty d}$ cannot be defined). All of these continuous groups are subgroups of the group of orthogonal transformations in three-dimensional space, $\mathbf{O}_3$. The groups $\mathbf{C}_\infty$, $\mathbf{C}_{\infty h}$ and $\mathbf{D}_\infty$ are not realized as symmetry groups for molecules.

*The Group* $\mathbf{O}_3$. The transformations of this group are combinations of rotations about any axis that passes through the origin of coordinates, and the inversion. The group $\mathbf{O}_3$ can be represented as the direct product of the three-dimensional rotation group and the group $\mathbf{C}_i$, $\mathbf{O}_3 = \mathbf{R}_3 \times \mathbf{C}_i$.

The classes of the group $\mathbf{O}_3$ can be obtained from those of the group $\mathbf{R}_3$. Besides the two classes with one element in each, $E$ and $I$, the $\mathbf{O}_3$ group contains two infinite sets of classes with a continuous set of elements in each class. These consist of rotations through an angle with absolute magnitude $|\phi|$ and a combination of rotations with the inversion. The group $\mathbf{O}_3$ is the symmetry point group of an atom.

*The Group* $\mathbf{C}_{\infty v}$. This group possesses a symmetry axis $C_\infty$ and an infinite set of planes $\sigma_v$ that pass through the axis. The group consists of a continuous set of classes, each of which contains two rotations, $C(\phi)$ and $C(-\phi)$, and of a single class of reflections containing an infinite set of reflection operations $\sigma_v$.

All linear molecules that are asymmetric about their midpoints have the symmetry of the group $\mathbf{C}_{\infty v}$, e.g. HD, NO, $N_2O$, HCN (Figure A2.6(g)).

*The Group* $\mathbf{D}_{\infty h}$. The symmetry elements of the group $\mathbf{C}_{\infty v}$ have added to them a center of symmetry. This leads to the appearance of a plane of symmetry $\sigma_h$ and of a continuous set of twofold axes $U_2$. The group $\mathbf{D}_{\infty h}$ can be represented as the direct product $\mathbf{D}_{\infty h} = \mathbf{C}_{\infty v} \times \mathbf{C}_i$. The classes of the group $\mathbf{D}_{\infty h}$ are obtained directly from those of $\mathbf{C}_{\infty v}$.

All linear molecules that are symmetric about their midpoint belong to $\mathbf{D}_{\infty h}$. This includes all diatomic molecules with identical nuclei, the molecules $C_2H_2$ (Fig. A2.6(h)), $C_2N_2$, $CO_2$, $C_3O_2$, etc.

The point groups that have been considered above consist of sets of rotation and reflection operations of the molecule as a whole. The molecule is regarded as a rigid configuration of atoms fixed in their equilibrium positions (*the equilibrium configuration*). However, there are a number of instances in which a molecule can possess several equilibrium configurations, each of which corresponds to an identical energy, and separated by a finite potential barrier. The molecule does not change its energy in making a transition from one configuration to another. In this way, new symmetry elements appear that enlarge the original point group. The symmetry groups of such 'nonrigid' molecules have been investigated by Longuet-Higgins [22]; see also References [23, 24].

Character tables for point groups are presented in most textbook on the group theory and its applications [7, 8, 11, 25]. The matrices of the irreducible representations of the most important point groups are given in the book [8], Appendix 2.

### A2.2.6 Irreducible tensor operators. Spherical tensors

#### Definition

In general, a set of $f_\alpha$ quantities $T_i^{(\alpha)}$ is called an *irreducible tensor* of a group of linear transformations if under the operations of the group, the $T_i^{(\alpha)}$ transform according to an irreducible representation $\mathbf{\Gamma}^{(\alpha)}$ of this group:

$$RT_i^{(\alpha)} = \sum_{k=1}^{f_\alpha} \mathbf{\Gamma}_{ki}^{(\alpha)}(R)\, T_k^{(\alpha)} \tag{A2.200}$$

From this definition it follows that any set of basis functions for an irreducible representation can be regarded as an irreducible tensor. Thus, the set of $2j+1$ spherical harmonic functions $Y_{JM}(\theta, \phi)$ forms an example of an irreducible tensor that belongs to the representation $\mathbf{D}^{(J)}$ of the group $\mathbf{R}_3$. The Cartesian components of an arbitrary vector $\mathbf{A}$ form a first rank tensor. However, their transformational properties under three-dimensional rotations are more complicated than those of the spherical components:

$$A_0 = A_z \quad A_{\pm 1} = \mp \frac{1}{\sqrt{2}}\left(A_x \pm i A_y\right) \tag{A2.201}$$

which constitute an irreducible tensor of the representation $\mathbf{D}^{(1)}$. When calculating matrix elements of vector quantities with basis functions of the group $\mathbf{R}_3$, it is therefore more convenient to use the spherical components of the vectors.

The addition of irreducible tensors is similar to the addition of Cartesian tensors (Section A2.1.6):

$$V_i^{(\alpha)} = T_i^{(\alpha)} + U_i^{(\alpha)} \tag{A2.202}$$

However, instead of tensor multiplication or contraction, the following operation is defined for irreducible tensors:

$$V_t^{(\tau)} = \sum_{i,k} T_i^{(\alpha)} U_k^{(\beta)} \langle \alpha i, \beta k \mid \tau t \rangle \tag{A2.203}$$

As a result one obtains a tensor that transforms according to an irreducible representation $\mathbf{\Gamma}^{(\tau)}$ that occurs in the decomposition of the direct product $\mathbf{\Gamma}^{(\alpha)} \times \mathbf{\Gamma}^{(\beta)}$. The coefficients in Equation (A2.203) are Clebsch–Gordan coefficients (A2.146).

A tensor that belongs to the unit (totally symmetric) representation of a group behaves as a scalar with respect to the operations of the particular group. The necessary and sufficient condition for the unit representation to occur in the decomposition of the direct product of two representations is that the representations are complex conjugates of each other (Equation (A2.143a)). Consequently, a scalar can never be formed from a product of two irreducible tensors that belong to two different representations.

In calculating the matrix elements of any operator it is important to know according to which irreducible representations the factors in the integrand transform. This knowledge enables, for example, nonzero matrix elements to be distinguished immediately and the group-theoretical selection rules to be obtained. It is therefore convenient to write the operator that occurs in the matrix element in the form of a sum of operators, each of which transforms according to definite irreducible representations of the group.

*An irreducible tensor operator* is defined (by analogy with an irreducible tensor) as a set of $f_\alpha$ quantities $T_i^{(\alpha)}$ whose transformation law is:

$$R^{-1} T_i^{(\alpha)} R = \sum_k \mathbf{\Gamma}_{ki}^{(\alpha)}(R)\, T_k^{(\alpha)} \tag{A2.204}$$

It is useful to note that irreducible tensor operators transform under the operations of a group in the same way as irreducible tensors. The difference between Equations (A2.204) and (A2.200) is due to the fact that operators in the new and old bases are related by Equation (A2.79).

The use of irreducible tensor operators facilitates the calculation of matrix elements considerably, since it is then possible to employ a whole series of useful relations first obtained by Wigner [17] and Racah [26, 27]. These relations rest upon the Wigner–Eckart theorem.

### The Wigner–Eckart theorem

Let us consider a matrix element of an irreducible tensor operator $T_t^{(\tau)}$:

$$\left\langle \alpha i \left| T_t^{(\tau)} \right| \beta k \right\rangle \tag{A2.205}$$

defined in terms of functions which transform according to the irreducible representations $\boldsymbol{\Gamma}^{(\alpha)}$ and $\boldsymbol{\Gamma}^{(\beta)}$ of the same group as the representation $\boldsymbol{\Gamma}^{(\tau)}$. We examine the transformational properties with respect to the operations of the group upon the function that results by allowing $T_t^{(\tau)}$ to operate upon the function $\psi_k^{(\beta)}$:

$$R\left(T_t^{(\tau)}\psi_k^{(\beta)}\right) = \left(R^{-1}T_t^{(\tau)}R\right)\left(R\psi_k^{\beta}\right) = \sum_{t',k'} \boldsymbol{\Gamma}_{t't}^{(\tau)}(R)\,\boldsymbol{\Gamma}_{k'k}^{(\beta)}(R)\left(T_{t'}^{(\tau)}\psi_{k'}^{(\beta)}\right)$$

The function $T_t^{(\tau)}\psi_k^{(\beta)}$ consequently transforms according to the direct product $\boldsymbol{\Gamma}^{(\tau)} \times \boldsymbol{\Gamma}^{(\beta)}$, and by using Clebsch–Gordan coefficients, it can be written in the form of a decomposition into basis functions for irreducible representations of the group [see Equation (A2.146a)]:

$$T_t^{(\tau)}\psi_k^{(\beta)} = \sum_{a,\mu,m} \Phi_m^{(a\mu)}(\tau\beta)\,\langle a\mu m \mid \tau t, \beta k\rangle \tag{A2.206}$$

In this equation the index $a$ distinguishes irreducible representations that are repeated, and the symbols $\tau$ and $\beta$ in the argument of the function $\Phi$ indicate that the form of $\Phi$ depends upon the basis functions in the direct product. Substituting Equation (A2.206) into (A2.205) we obtain:

$$\left\langle \alpha i \left| T_t^{(\tau)} \right| \beta k \right\rangle = \sum_{a,\mu,m} \langle \alpha\mu m \mid \tau t, \beta k\rangle \int \psi_i^{(\alpha)*}\Phi_m^{(a\mu)}(\tau\beta)\,dV \tag{A2.207}$$

The basis functions that transform according to different irreducible representations or according to different columns of the same representation are orthogonal to one another:

$$\int \psi_i^{(\alpha)*}\Phi_m^{(a\mu)}(\tau\beta)\,dV = \delta_{\alpha\mu}\delta_{im}A_a(\alpha, \tau, \beta) \tag{A2.208}$$

where the quantity $A_a(\alpha, \tau, \beta)$ is determined by the form of the functions $\psi_i^{(\alpha)}$ and $\Phi_m^{(a\mu)}(\tau\beta)$, but does not depend upon $i$ and $m$. Substituting Equation (A2.208) into Equation (A2.207), we obtain an analytical expression of the Wigner–Eckart theorem:[6]

$$\left\langle \alpha i \left| T_t^{(\tau)} \right| \beta k \right\rangle = \sum_{a} \langle a\alpha i \mid \tau t, \beta k\rangle\, A_a(\alpha, \tau, \beta) \tag{A2.209}$$

It follows from this equation that the Clebsch–Gordan coefficients completely determine the dependence of the matrix element upon the column numbers of the irreducible representations according to which the factors in the integrand transform. The presence of these coefficients allows one immediately to obtain the conditions under which a matrix element reduces to zero. These are as follows:

[6] This theorem was proved by Wigner [28] and by Eckart [29] for the three-dimensional rotational group. By writing it in the form (A2.209), Koster [30] subsequently extended the theorem to include an arbitrary finite group. Therefore it is more correct to call the generalized expression (A2.209) the Wigner–Eckart–Koster theorem.

the matrix element (A2.209) is zero whenever the decomposition of the direct product $\Gamma^{(\tau)} \times \Gamma^{(\beta)}$ into irreducible components does not contain the representation $\Gamma^{(\alpha)}$.

Systems with differing physical structures may possess identical symmetries. The matrix elements which occur in quantum-mechanical calculations on such systems differ only in the factor $A_a$, the Clebsch–Gordan coefficients being identical. Consequently, the Wigner–Eckart theorem enables the symmetry properties of a system that is being studied to be 'separated off' from the detailed physical structure.

The operators, which are usually employed in quantum mechanics, are symmetric with respect to all particles, i.e., they transform according to the unit representation $\mathbf{\Gamma}^{[N]}$ of the permutation group (see Section A2.2.3). Since $\mathbf{\Gamma}^{[N]} \times \mathbf{\Gamma}^{[\lambda]} = \mathbf{\Gamma}^{[\lambda]}$, the sum over $a$ in Equation (A2.209) reduces to a single term. As a result, the Wigner–Eckart theorem assumes the following form [8]:

$$\left\langle [\lambda_2] r_2 \left| T^{[N]} \right| [\lambda_1] r_1 \right\rangle = \langle [\lambda_2] r_2 \mid [N], [\lambda_1] r_1 \rangle \left\langle [\lambda_2] \left\| T^{[N]} \right\| [\lambda_1] \right\rangle = \delta_{\lambda_1 \lambda_2} \delta_{r_1 r_2} \left\langle [\lambda_1] \left\| T^{[N]} \right\| [\lambda_1] \right\rangle \quad \text{(A2.210)}$$

where the double bars in the matrix element denote its independence of the Young tableaux $r$ that enumerate the basis functions. Equation (A2.210) is the basis of the well-known quantum-mechanical selection rule according to which a perturbation described by a symmetric operator induces transitions only between states with the same permutational symmetry.

**Spherical tensors**

As we discussed above, the set of $(2l+1)$ spherical harmonics $Y_{lm}(\theta, \varphi)$ (see Equation (A2.185)) form an irreducible tensor which belongs to the representation of the group $\mathbf{R}_3$. Irreducible tensors of the group $\mathbf{R}_3$ are called *spherical tensors*.

In the case of the group $\mathbf{R}_3$, a scalar is characterized by a value of zero for the angular momentum, $J = 0$, and is constructed from two spherical tensors with the same $j$. In this case, the Clebsch–Gordan coefficients are given by:

$$\langle jm, j-m \mid 00 \rangle = (-1)^{j-m} (2j+1)^{-\frac{1}{2}} \quad \text{(A2.211)}$$

Substituting this expression into Equation (A2.203), we obtain the following scalar from two spherical tensors:

$$V^{(0)} = \frac{(-1)^j}{\sqrt{2j+1}} \sum_m (-1)^m T_m^{(j)} U_{-m}^{(j)} \quad \text{(A2.212)}$$

Expression (A2.212) without the multiplying factor in front of the sum is usually called the *scalar product of two spherical tensors* and is denote by:

$$\left(T^{(j)} \cdot U^{(j)}\right) = \sum_m (-1)^m T_m^{(j)} U_{-m}^{(j)} \quad \text{(A2.213)}$$

When $j = 1$, this expression coincides with the usual scalar product of two vectors expressed in spherical coordinates.

The Wigner–Eckart theorem (Equation (A2.209)) in the case of the group $\mathbf{R}_3$ reduces to a single term, because the decomposition of the direct product (Equation (A2.192)) does not give rise to repeated representations. Thus:

$$\left\langle jm \left| T_k^{(\kappa)} \right| j'm' \right\rangle = \langle jm \mid \kappa k, j'm' \rangle A\left(j, \kappa, j'\right) \tag{A2.214}$$

In place of the quantities $A\left(j, \kappa, j'\right)$, it is usual to introduce the so-called *reduced* matrix elements $\langle j \left\| T^{(\kappa)} \right\| j' \rangle$, which are related to the $A\left(j, \kappa, j'\right)$ by the equation:

$$A\left(j, \kappa, j'\right) = \frac{(-1)^{j+\kappa-j'}}{\sqrt{2j+1}} \langle j \left\| T^{(\kappa)} \right\| j' \rangle \tag{A2.215}$$

This leads to the following two equivalent statements of the Wigner–Eckart theorem written via the Clebsch–Gordan coefficients or the $3j$ symbols:

$$\begin{aligned}\left\langle jm \left| T_k^{(\kappa)} \right| j'm' \right\rangle &= (-1)^{j+\kappa-j'} \frac{\langle jm \mid \kappa k, j'm' \rangle}{(2j+1)^{\frac{1}{2}}} \langle j \left\| T^{(\kappa)} \right\| j' \rangle \\ &= (-1)^{j-m} \begin{pmatrix} j & \kappa & j' \\ -m & k & m' \end{pmatrix} \langle j \left\| T^{(\kappa)} \right\| j' \rangle \end{aligned} \tag{A2.216}$$

Since a reduced matrix element does not depend upon the values of the projections of the angular momenta $m$, $k$, and $m'$, it is sufficient from the point of view of the calculation to find the simplest matrix element $\left\langle jm \left| T_k^{(\kappa)} \right| j'm' \right\rangle$ and to determine $\langle j \left\| T^{(\kappa)} \right\| j' \rangle$ from Equation (A2.216). Let us calculate, for example, the reduced matrix element of the angular momentum operator $\mathbf{J}$. The spherical components of the angular momentum vector:

$$J_0 = J_z, \quad J_{\pm 1} = \mp\sqrt{\tfrac{1}{2}}\left(J_z \pm i J_y\right)$$

constitute a first rank irreducible spherical tensor. Since the $\psi_m^{(j)}$ are eigenfunctions of the operator $J_z$:

$$\left\langle jm \left| J_z \right| j'm' \right\rangle = \delta_{jj'}\delta_{mm'}m \tag{A2.217}$$

On the other hand, $J_z \equiv T_0^{(1)}$, and hence, according to Equation (A2.216):

$$\langle jm \left| J_z \right| jm \rangle = (-1)^{j-m} \begin{pmatrix} j & 1 & j \\ -m & 0 & m \end{pmatrix} \langle j \left\| J \right\| j \rangle \tag{A2.218}$$

The $3j$ symbol, which occurs in this equation, is equal to:

$$\begin{pmatrix} j & 1 & j \\ -m & 0 & m \end{pmatrix} = \frac{(-1)^{j-m} m}{\left[j(j+1)(2j+1)\right]^{\frac{1}{2}}} \tag{A2.219}$$

Taking into account Equations (A2.217) and (A2.218), we obtain:

$$\langle j \left\| J \right\| j' \rangle = \delta_{jj'} \left[j(j+1)(2j+1)\right]^{\frac{1}{2}} \tag{A2.220}$$

For a scalar operator, on has the equation:

$$\langle jm \left| T^{(0)} \right| j'm' \rangle = \left[ \delta_{jj'}\delta_{mm'} / (2j+1)^{\frac{1}{2}} \right] \langle j \left\| T^{(0)} \right\| j \rangle \tag{A2.221}$$

The matrix element of the scalar product of two spherical tensors and the corresponding reduced matrix element are given by the following formulae:

$$\begin{aligned} &\langle jm \left| \left( T^{(\kappa)} \cdot U^{(\kappa)} \right) \right| j'm' \rangle \\ &= \left[ \delta_{jj'}\delta_{mm'} / (2j+1) \right] \sum_{j''} (-1)^{j-j''} \langle j \left\| T^{(\kappa)} \right\| j'' \rangle \langle j'' \left\| U^{(\kappa)} \right\| j' \rangle \end{aligned} \tag{A2.222}$$

$$\begin{aligned} &\langle j \left\| \left( T^{(\kappa)} \cdot U^{(\kappa)} \right) \right\| j' \rangle \\ &= \left[ \delta_{jj'} / (2j+1)^{\frac{1}{2}} \right] \sum_{j''} (-1)^{j-j''} \langle j \left\| T^{(\kappa)} \right\| j'' \rangle \langle j'' \left\| U^{(\kappa)} \right\| j' \rangle \end{aligned} \tag{A2.223}$$

Many useful formulae with spherical tensors are represented in References [18–21, 26, 27].

# References


1. B.W. Lindgren, *Vector Calculus*, MacMillan, New York (1964).
2. D. Lovelock and H. Rund, *Tensors Differential Forms and Variational Principles*, Dover Publications, New York (1989).
3. D.A. Danielson, *Vectors and Tensors in Engineering and Physics*, Addison–Wesley, Redwood City, California (1992).
4. J.G. Simmonds, *A Brief on Tensor Analysis*, 2nd edn, Springer–Verlag, New York (1994).
5. G.B. Arfken and H.J. Weber, *Mathematical Methods for Physicists*, 4th edn, Academic Press, San Diego (1995).
6. F.D. Murnaghan, *The Theory of Group Representations*, John Hopkins Press, Baltimore, Maryland (1938).
7. M. Hamermesh, *Group Theory*, Addison–Wesley, Reading, Massachusetts (1962).
8. I.G. Kaplan, *Symmetry of Many-Electron Systems*, Academic Press, New York (1975).
9. I.G. Kaplan, in *Chemical Group Theory: Introduction and Fundamentals* (eds D. Bonchev and D.H. Rouvray), Gordon and Breach, London (1994), pp. 209–254.
10. L.S. Pontriagin, *Continuous Groups*, Princeton University Press, Princeton, New Jersey (1946).
11. L.D. Landau and E.M. Lifshitz, *Quantum Mechanics (Nonrelativistic Theory)*, 3rd edn, Pergamon Press, Oxford (1977).
12. M. Abramowitz and I.A. Stegun, *Handbook of Mathematical Functions*, Dover Publications, New York (1965).
13. G.S. Canright and S.M. Girvin, *Science* **247**, 1197 (1990).
14. F. Wilczek, *Phys. Rev. Lett.* **48**, 1144 (1982); **49**, 957 (1982).

15. I.G. Kaplan in *Fundamental World of Quantum Chemistry. A Tribute Volume to the Memory of Per-Olov Lowdin* (eds E.J. Brandas and E.S. Kryachko), Kluwer Academic Publishers, Dordrecht (2003), Vol. 1, pp. 183–220.
16. I.M. Gel'fand, R.A. Minlos and Z.Ya. Shapiro, *Representations of the Rotation and Lorentz Groups*, MacMillan, New York (1963).
17. E. Wigner, *Group Theory and its Application to the Quantum Mechanics of Atomic Spectra*, Academic Press, New York (1959).
18. M.E. Rose, *Elementary Theory of Angular Momentum*, John Wiley & Sons, Inc., New York (1957).
19. A.R. Edmonds, *Angular Momentum in Quantum Mechanics*, Princeton University Press, Princeton (1957).
20. A.P. Yutsis, I.B. Levinson and V.V. Vanagas, *The Theory of Angular Momentum*, Israel Program for Scientific Translations, Jerusalem (1962). Reproduced in *Handbook of Molecular Physics and Quantum Chemistry* (ed S. Wilson), John Wiley & Sons, Ltd, Chichester (2003), Vol. 1, Chapter 25.
21. D.M. Brink and G.R. Satchler, *Angular Momentum*, 3rd edn, Clarendon Press, Oxford (1993).
22. H.C. Longuet-Higgins, *Mol. Phys.* **6**, 445 (1963).
23. P.R. Bunker and P. Jensen, *Molecular Symmetry and Spectroscopy*, 2nd edn, NRC Research Press, Ottawa (1998).
24. S.F.A. Kettle, *J. Chem. Educ.* **79**, 258 (2002).
25. C.J. Bradley and A.P. Cracknell, *The Mathematical Theory of Symmetry in Solids: Representation Theory for Point Groups and Space Groups.* Clarendon Press, Oxford (1972).
26. U. Fano and G. Racah, *Irreducible Tensorial Sets*, Academic Press, New York (1959).
27. B.R. Judd, *Operator Techniques in Atomic Spectroscopy*, McGraw–Hill, New York (1963).
28. E. Wigner, *Zs. f. Phys.* **43**, 624 (1927).
29. C. Eckart, *Rev. Mod. Phys.* **2**, 305 (1930).
30. G.F. Koster, *Phys. Rev.* **109**, 227 (1958).

# Appendix 3
# Methods of Quantum-Mechanical Calculations of Many-Electron Systems

## A3.1 Adiabatic Approximation

The theory of intermolecular interactions is based on the quantum nature of the electron and nuclear motion that obey the Schrödinger equation for a system of interacting molecules. For a system of two interacting molecules, $A$ and $B$, containing $n_A$ and $n_B$ nuclei, and $N_A$ and $N_B$ electrons, respectively, the wave functions describing this interacting system must satisfy the stationary Schrödinger equation:

$$H\Psi\left(R', r'\right) = E\Psi\left(R', r'\right) \tag{A3.1}$$

where the set of $3\left(n_A + n_B\right)$ nuclear coordinates is denoted by $R'$ and the set of $3\left(N_A + N_B\right)$ electronic coordinates by $r'$, both relative to the laboratory coordinate frame. The nonrelativistic Hamiltonian of the system can be represented as a sum of the Hamiltonians $H_A$ and $H_B$ corresponding to the isolated molecules and the operator of the Coulomb interaction $V_{AB}$ between molecules:

$$H = H_A + H_B + V_{AB} \tag{A3.2}$$

$$\begin{aligned} H_A = &-\frac{\hbar^2}{2}\sum_{a=1}^{n_A}\frac{1}{M_a}\nabla_{R'_a}^2 - \frac{\hbar^2}{2m}\sum_{i=1}^{N_A}\nabla_{r'_i}^2 - \sum_{a=1}^{n_A}\sum_{i=1}^{N_A}\frac{Z_a e^2}{\left|\mathbf{r}'_i - \mathbf{R}'_a\right|} \\ &+\sum_{i<j}^{N_A}\frac{e^2}{\left|\mathbf{r}'_i - \mathbf{r}'_j\right|} + \sum_{a<b}^{n_A}\frac{Z_a Z_b e^2}{\left|\mathbf{R}'_a - \mathbf{R}'_b\right|} \end{aligned} \tag{A3.3}$$

where the first two terms are the kinetic energy operators of nuclei and electrons, respectively; the third term describes the attraction between electrons and nuclei, and the last two terms describe the electron–electron and nucleus–nucleus repulsion. The Hamiltonian $H_B$ has the same structure as in Equation (A3.3) and:

$$\begin{aligned} V_{AB} = &-\sum_{a=1}^{n_A}\sum_{j=1}^{N_B}\frac{Z_a e^2}{\left|\mathbf{r}'_j - \mathbf{R}'_a\right|} - \sum_{b=1}^{n_B}\sum_{i=1}^{N_A}\frac{Z_b e^2}{\left|\mathbf{r}'_i - \mathbf{R}'_b\right|} \\ &+\sum_{i=1}^{N_A}\sum_{j=1}^{N_B}\frac{e^2}{\left|\mathbf{r}'_i - \mathbf{r}'_j\right|} + \sum_{a=1}^{n_A}\sum_{b=1}^{n_B}\frac{Z_a Z_b e^2}{\left|\mathbf{R}'_a - \mathbf{R}'_b\right|} \end{aligned} \tag{A3.4}$$

*Intermolecular Interactions* I.G. Kaplan

The first two terms correspond to the Coulomb attraction of electrons of one molecule by the nuclei of the other molecule, and the last two terms correspond to the electron–electron and nucleus–nucleus repulsion between the molecules $A$ and $B$.

Equation (A3.1) can be solved only with the help of approximate methods. The ion $H_2^+$ is the only molecular system, for which the exact solution of the Schrödinger equation can be found. The fundamental simplification in the solution of the Schrödinger equation consists of the separation of the electronic and nuclear motions. Let us consider this problem in the case of two interacting atoms. It is convenient not to divide electrons between atoms $A$ and $B$ and enumerate them from 1 to $N = N_A + N_B$. The Hamiltonian (A3.2) is expressed as:

$$H = -\frac{\hbar^2}{2}\left(\frac{1}{M_a}\nabla^2_{R'_a} + \frac{1}{M_b}\nabla^2_{R'_b}\right) - \frac{\hbar^2}{2m}\sum_{i=1}^{N}\nabla^2_{r'_i} - \sum_{i=1}^{N}\left(\frac{Z_a e^2}{|\mathbf{r}'_i - \mathbf{R}'_a|} + \frac{Z_b e^2}{|\mathbf{r}'_i - \mathbf{R}'_b|}\right) + \sum_{i<j}^{N}\frac{e^2}{|\mathbf{r}'_i - \mathbf{r}'_j|} + \frac{Z_a Z_b e^2}{|\mathbf{R}'_a - \mathbf{R}'_b|} \tag{A3.5}$$

The interaction energy is a function only of the relative distances between the particles. Hence, it is natural to use the center-of-mass coordinate system, in which the translational motion of the whole system is separated. This procedure is ambiguous [1]. It is convenient to use the following set of relative coordinates [2–4]:

$$\mathbf{R} = \mathbf{R}'_a - \mathbf{R}'_{b'} \quad \mathbf{r}_i = \mathbf{r}'_i - \frac{M_a\mathbf{R}'_a + M_b\mathbf{R}'_b}{M_a + M_b} \tag{A3.6}$$

The detailed discussion of different choice of coordinate frames and separation of nuclear and electronic motion is given by Sutcliffe [5].

After separating the motion of the center of mass, the Hamiltonian of the relative motion has the form:

$$H_{rel} = H_e + K_R \tag{A3.7}$$

where $H_e$ is the Hamiltonian describing the electron motion at the clamped-nuclear approximation:

$$H_e = -\frac{\hbar^2}{2m}\sum_{i=1}^{N}\nabla^2_{r_i} - \sum_{i=1}^{N}\left(\frac{Z_a e^2}{r_{ai}} + \frac{Z_b e^2}{r_{bi}}\right) + \sum_{i<j}^{N}\frac{e^2}{r_{ij}} + \frac{Z_a Z_b e^2}{R} \tag{A3.8}$$

In Equation (A3.8) notations for distances between the electron $i$ and nucleus $A$, $r_{ai} = |\mathbf{r}_i - \mathbf{R}'_a|$, and between electrons $i$ and $j$, $r_{ij} = |\mathbf{r}_i - \mathbf{r}_j|$, are introduced.

The operator $K_R$ consists of the kinetic energy operator of the relative nuclear motion and the so-called *mass-polarization* term:

$$K_R = -\frac{\hbar^2}{2\mu}\nabla^2_R - \frac{\hbar^2}{2(M_a + M_b)}\left[\sum_i \nabla^2_{r_i} + 2\sum_{i<j}\nabla_{r_i}\nabla_{r_j}\right] \tag{A3.9}$$

where $\mu = M_a M_b/(M_a + M_b)$ is the reduced mass of the nuclei.

Electrons move considerably faster than nuclei due to their smaller masses. Hence, in a first approximation, the nuclei may be considered at rest and the Hamiltonian $H_e$ is considered as the clamped-nuclei Hamiltonian. Then, the wave function for the electron motion, $\Psi_n(r, R)$, depends on the distance $R$ between the nuclei as a parameter, and $r$ is the set of $3N$ coordinates of all the electrons relative to the center-of-mass frame. It satisfies the Schrödinger equation with the Hamiltonian (A3.8):

$$H_e \Psi_n(r, R) = E_n(R)\, \Psi_n(r, R) \tag{A3.10}$$

where the energy of the $nth$ electron state, $E_n(R)$, is a function of the distances between the nuclei.

The solution of the Schrödinger equation for the total Hamiltonian of the relative motion (Equation (A3.7)):

$$H_{rel} \Psi(r, R) = E \Psi(r, R) \tag{A3.11}$$

can be represented as a series, expanded over the complete set of the eigenfunctions $\{\Psi_n(r, R)\}$ of the Hamiltonian $H_e$:

$$\Psi(r, R) = \sum_n \chi_n(R)\, \Psi_n(r, R) \tag{A3.12}$$

where the coefficients $\chi(R)$ depend on the set $R$ (or the three coordinates of the relative motion of nuclei). Substituting Equation (A3.12) into Equation (A3.11), multiplying it by $\Psi_n^*(r, R)$, integrating over $r$, and taking the mutual orthogonality of the functions $\{\Psi_n(r, R)\}$ into account, gives the set of coupled equations:

$$\left[-\frac{\hbar^2}{2\mu}\nabla_R^2 + E_m(R) - E\right]\chi_m(R) = \sum_n W_{mn}(R)\, \chi_n(R) \quad m = 1, 2, \ldots \tag{A3.13}$$

where the operator $W_{mn}(R)$ has the form:

$$W_{mn}(R) = \int \Psi_m^*(r, R)\left[\frac{\hbar^2}{\mu}\nabla_R \Psi_n(r, R)\, \nabla_R - K_R \Psi_n(r, R)\right] dV_r \tag{A3.14}$$

The set of Equations (A3.13) is exact but it can be solved only via approximate methods.

The large difference in the masses of the nuclei and of the electrons determines that the term $W_{mn}(R)\, \chi_n(R)$ is small. The adiabatic approximation consists of the separation of the nuclear and electronic motions. It can be carried out in different ways. The *Born–Oppenheimer approximation* [6] corresponds to neglecting all of the coupling terms in the right-hand side of Equations (A3.13). Equations (A3.13) are then decomposed into a set of uncoupled Schrödinger equations:

$$\left[-\frac{\hbar^2}{2\mu}\nabla_R^2 + E_m(R)\right]\chi_{mv}(R) = E_{mv}\chi_{mv}(R) \tag{A3.15}$$

where the energy of the electronic motion in the $mth$ quantum state, $E_m(R)$, represents the potential energy for the nuclear motion.

Therefore, in the Born–Oppenheimer approach, the wave function of the system is reduced to a simple product:

$$\Psi_{mv}(r, R) = \Psi_m(r, R)\,\chi_{mv}(R) \tag{A3.16}$$

To each *mth* quantum state of electrons there corresponds a set of states of the nuclei, which are characterized by the quantum numbers $v$. The equilibrium nuclear configuration (in the case of two atoms, the equilibrium distance $R_0$) is determined by the minimum condition of the potential energy $E_m(R)$.

In molecular spectroscopy, if the displacements of the nuclei from their equilibrium positions are small, the *Condon adiabatic approximation* is often used. In this approximation the electronic wave function is evaluated only at the equilibrium internuclear distance $R_0$, and only the nuclear wave function depend upon the internuclear distance $R$:

$$\Psi_{mv}(r, R) = \Psi_m(r, R_0)\,\chi_{mv}(R) \tag{A3.17}$$

The *Born adiabatic approximation* [7, 8] corresponds to a separation of the set of Equations (A3.13), conserving the diagonal terms $W_{mm}(R)$. That is, Equation (A3.15) is replaced by the following one:

$$\left[-\frac{1}{2\mu}\nabla_R^2 + V_m(R)\right]\chi_{mv}(R) = E_{mv}\chi_{mv}(R) \tag{A3.18}$$

where the potential energy $V_m(R)$ becomes:

$$V_m(R) = E_m(R) - W_{mm}(R)$$

It is the *'best' adiabatic approximation.* The potential energy, $V_m(R)$, is called the *interatomic (intermolecular) potential* (see Section 1.3). The diagonal term $W_{mm}(R)$ is a correction to the potential energy, which arises from the coupling of the electron and nuclear motions. For precise calculations, this adiabatic correction has to be taken into account.

In Table A3.1, different contributions to the ground state binding energy of hydrogen are presented. The adiabatic correction $\Delta E^{ad} = 4.938\ \text{cm}^{-1}$ is larger than other correction, although it is only 0.01% of the Born–Oppenheimer value of the binding energy. The methods of calculating the adiabatic corrections for many-electron molecules have been elaborated in References [12–15].

**Table A3.1** Contributions to the dissociation energy in the ground state of $H_2$ (in $\text{cm}^{-1}$)

| | Born-Oppenheimer electronic energy [9] | Corrections | | | |
|---|---|---|---|---|---|
| | | Adiabatic [10] | Relativistic [11] | Radiative [11] | Nonadiabatic $V = 0, J = 0$ [9] |
| $R_0 = 1.4a_0$ | 38292.99 | 4.938 | − 0.517 | 0.205 | 0.499 |

According to Table A3.1, the precision of the adiabatic approximation for the ground state of the hydrogen molecule is very high, $\Delta E^{nonad} = 0.5\ \text{cm}^{-1}$ and $\Delta E^{nonad}/E_b \simeq 10^{-5}$. But the adiabatic approximation breaks down when there is a degeneracy or near-degeneracy of electronic states; the right-hand terms in Equation (A3.13) become large and cannot be neglected.

The broad scope of physical and physico–chemical problems arising in this case includes the Jahn–Teller effect [16, 17], inelastic collisions [18], nonadiabatic molecular dynamics [19–21], etc. (see the discussion in Section 1.3 and the references therein).

## A3.2 Variational Methods

### A3.2.1 Self-consistent field method

The first effective approximate method of solving the many-electron Schrödinger equation for atoms was proposed and developed by Hartree [22], without taking into account the Pauli principle, and by Fock [23], who employed the antisymmetric many-electron wave function. It is usually referred to as the *Hartree–Fock* (HF) or *self-consistent field* (SCF) method. It was generalized for molecule in the LCAO MO algebraic version by Roothaan [24, 25].

Within the framework of the HF approach, each electron is regarded as moving in some kind of 'self-consistent' field due to the other electrons and fixed nuclei. Each electron is characterized by a one-electron wave function, or simply, an *orbital.* Let us denote an orbital by $\varphi_m(\mathbf{r})$, where $m$ represents the set of quantum numbers describing a given one-electron state. Only two electrons can be described by a single orbital, due to the two possible orientations of the electron spin. They are characterized by the corresponding spin functions $\eta_\alpha\left(s_z = \frac{1}{2}\right)$ and $\eta_\beta\left(s_z = -\frac{1}{2}\right)$. The electrons in the orbital $\varphi_m$ are described by the two *spin-orbitals*:

$$\psi_{m\alpha}(x) = \varphi_m(\mathbf{r})\,\eta_\alpha(\xi) \qquad \psi_{m\beta}(x) = \varphi_m(\mathbf{r})\,\eta_\beta(\xi) \tag{A3.19}$$

where $x$ is a set of four electron coordinates, including the spin coordinate $\xi$.

If all the orbitals in the electronic configuration are doubly occupied, this is the so-called *closed shell,* with a total electron spin $S = 0$. The wave function, which satisfied the Pauli principle and describes a closed-shell configuration with $S = 0$, can be represented by the Slater determinant:

$$\begin{aligned}\Psi(x_1, x_2, \ldots, x_N) &= \frac{1}{\sqrt{N!}}\det\left|\psi_{1\alpha}\psi_{1\beta}\psi_{2\alpha}\ldots\psi_{\frac{N}{2}\beta}\right| \\ &= \frac{1}{\sqrt{N!}}\begin{vmatrix}\psi_{1\alpha}(1)\,\psi_{1\alpha}(2)\ldots\psi_{1\alpha}(N) \\ \psi_{1\beta}(1)\,\psi_{1\beta}(2)\ldots\psi_{1\beta}(N) \\ \psi_{2\alpha}(1)\,\psi_{2\alpha}(2)\ldots\psi_{2\alpha}(N) \\ \cdot\quad\cdot\quad\cdot\quad\cdot\quad\cdot\quad\cdot\quad\cdot \\ \psi_{\frac{N}{2}\beta}(1)\,\psi_{\frac{N}{2}\beta}(2)\ldots\psi_{\frac{N}{2}\beta}(N)\end{vmatrix}\end{aligned} \tag{A3.20}$$

where $\sqrt{N!}$ is the normalization factor assuming that the set $\{\psi_{n\sigma}\}$ is orthonormal, and $k$ in $\psi_{n\sigma}(k)$ designates four coordinates of the electron $k$.

Thus, the SCF method belongs to the one-electron approximation. The adiabatic Hamiltonian for an N-electron system contains the one-electron and two-electron terms, and in general can be represented as:[1]

$$\widehat{H} = \sum_{i=1}^{N} \widehat{f}_i + \sum_{i<j} g_{ij} \tag{A3.21}$$

where:

$$\widehat{f}_i = -\frac{\hbar^2}{2m}\nabla_i^2 - \sum_{a=1}^{n} \frac{Z_a e^2}{r_{ai}} \qquad g_{ij} = \frac{e^2}{r_{ij}} \tag{A3.22}$$

In the HF approach, the Hamiltonian (A3.21) must be transformed to the sum of one-electron Hamiltonians. This is achieved using the variational principle:

$$\delta E = \delta \langle \Psi | H | \Psi \rangle = 0 \tag{A3.23}$$

with the additional condition:

$$\langle \Psi \mid \Psi \rangle = 1 \tag{A3.24}$$

and restricting the trial functions to the determinants (A3.20). Since the Schrödinger equation follows directly from the variational principle in the form of Equations (A3.22) and (A3.23) for an arbitrary wave function $\Psi$ [26], the solution of the variational problem for the determinantal wave function should yield the best function of this class.

The expectation value of the energy in the state described by the determinantal function (A3.20), can be presented, as shown in various textbooks, via one-electron integrals, $f_n$, and two-electron *Coulomb*, $\alpha_{nm}$, and *exchange*, $\beta_{nm}$, integrals. Namely:

$$\begin{aligned} E &= \frac{1}{N!}\left\langle \det\left|\varphi_1^2\varphi_2^2\ldots\varphi_{\frac{N}{2}}^2\right| |H| \det\left|\varphi_1^2\varphi_2^2\ldots\varphi_{\frac{N}{2}}^2\right|\right\rangle \\ &= 2\sum_n f_n + \sum_{n,m}(2\alpha_{nm} - \beta_{nm}) \end{aligned} \tag{A3.25}$$

where:

$$f_n = \left\langle \varphi_n(i) \left| \widehat{f}_i \right| \varphi_n(i) \right\rangle \tag{A3.26}$$

$$\alpha_{nm} = \left\langle \varphi_n(i)\varphi_m(j) \left| g_{ij} \right| \varphi_n(i)\varphi_m(j) \right\rangle \tag{A3.27}$$

$$\beta_{nm} = \left\langle \varphi_n(i)\varphi_m(j) \left| g_{ij} \right| \varphi_n(j)\varphi_m(i) \right\rangle \tag{A3.28}$$

We consider the case when the Hamiltonian does not depend upon the spin interactions. That is why in Equation (A3.25) only orbital configuration is shown and all integrals in Equation (A3.26)–(A3.28) are expressed via orbitals $\varphi_n$.

---

[1] In this section, the operators that are not reduced to a simple multiplication are denoted by carets.

The derivation of the HF equations is simplified by introducing the Coulomb, $\widehat{\alpha}_m$, and exchange, $\widehat{\beta}_m$, operators [24], which are defined by the relations:

$$\widehat{\alpha}_m(i)\,\varphi_n(i) = \left[\int \varphi_m^*(j)\,\varphi_m(j)\,g_{ij}dV_j\right]\varphi_n(i)$$

$$\widehat{\beta}_m(i)\,\varphi_n(i) = \left[\int \varphi_m^*(j)\,\varphi_n(j)\,g_{ij}dV_j\right]\varphi_m(i) \qquad \text{(A3.29)}$$

Applying these operators, the two-electron Coulomb and exchange integrals may be represented as one-electron integrals:

$$\alpha_{nm} = \langle\varphi_n|\widehat{\alpha}_m|\varphi_n\rangle = \langle\varphi_m|\widehat{\alpha}_n|\varphi_m\rangle$$

$$\beta_{nm} = \langle\varphi_n|\widehat{\beta}_m|\varphi_n\rangle = \langle\varphi_m|\widehat{\beta}_n|\varphi_m\rangle \qquad \text{(A3.30)}$$

and Expression (A3.25) for the energy can be rewritten in the form:

$$E = 2\sum_n \langle\varphi_n|\widehat{f}|\varphi_n\rangle + \sum_{n,m}\langle\varphi_n|2\widehat{\alpha}_m - \widehat{\beta}_m|\varphi_n\rangle \qquad \text{(A3.31)}$$

The variational principle (A3.23) with the condition (A3.24) reduces to a search for an unconditional extreme of the functional:

$$J = \sum_n \left\langle\varphi_n\left|2\widehat{f} + \sum_m\left(2\widehat{\alpha}_m - \widehat{\beta}_m\right)\right|\varphi_n\right\rangle - \sum_{n,m}\varepsilon_{nm}\langle\varphi_n\mid\varphi_m\rangle \qquad \text{(A3.32)}$$

in which $\varepsilon_{nm}$ are the Lagrange multipliers. Equating to zero the first order variation of this functional and taking into account the linear independence of variations $\delta\varphi_n^*$ (or $\delta\varphi_n$), one easily finds that the orbitals that minimize the energy of a closed-shell electronic configuration must be solutions of the equations:

$$\widehat{h}^{HF}(i)\,\varphi_n(i) = \sum_m \varepsilon_{mn}\varphi_m(i) \qquad \text{(A3.33)}$$

with the *one-electron Hartree–Fock Hamiltonian*:

$$\widehat{h}^{HF}(i) = \widehat{f}(i) + \sum_m\left[2\widehat{\alpha}_m(i) - \widehat{\beta}_m(i)\right] \qquad \text{(A3.34)}$$

Since the set $\{\varphi_n\}$ is orthonormal, the Lagrange multipliers in Equation (A3.33) can be expressed as the matrix elements:

$$\varepsilon_{mn} = \langle\varphi_m|\widehat{h}^{HF}|\varphi_n\rangle \qquad \text{(A3.35)}$$

The one-electron Hamiltonian, $\widehat{h}^{HF}$, is invariant under any unitary transformation of the orbitals [24]; therefore, a unitary transformation that diagonalizes the Hermitian matrix $\varepsilon$ always exists. Thus, without any loss of generality, Equation (A3.33) can be rewritten as the eigenvalue equations of the operator $\widehat{h}^{HF}$ and we arrived at the famous *Hartree–Fock equations*[2]:

$$\widehat{h}^{HF}(i)\,\varphi_n(i) = \varepsilon_n \varphi_n(i) \qquad n = 1, 2, \ldots, N/2 \tag{A3.36}$$

As can be seen from Equation (A3.29), the operator $\widehat{h}^{HF}$ (Equation (A3.34)) itself depends upon the orbitals being sought. In the explicit form, Equation (A3.36) can be written as:

$$\left[-\frac{\hbar^2}{2m}\nabla_i^2 - \sum_{a=1}^{n}\frac{Z_a e^2}{r_{ai}} + \int \varphi_n^*(j)\frac{e^2}{r_{ij}}\varphi_n(i)\,dV_j \right.$$
$$\left. + 2\sum_{m\neq n}\int \varphi_m^*(j)\frac{e^2}{r_{ij}}\varphi_m(j)\,dV_j - \varepsilon_n\right]\varphi_n(i)$$
$$-\sum_{m\neq n}\left[\int \varphi_m^*(j)\frac{e^2}{r_{ij}}\varphi_n(j)\,dV_j\right]\varphi_m(i) = 0, \qquad n = 1, 2, \ldots, N/2. \tag{A3.37}$$

Thus, the HF equations belong to a class of nonlinear integro-differential equations of the third order in the unknown functions.

For atoms, the presence of central symmetry causes the variables in the HF equations to factorize. Integration over the angular variables reduces the problem to the HF equations for the radial function only. These cannot be solved analytically; the solution is found in the form of numerical tables. However, for molecules, factoring the variable is not possible and the numerical solution of equations becomes extremely complicated. The only practical way is to expand the unknown orbitals in a certain set of basis functions and reduce the problem to an algebraic one for the coefficients in this expansion [24].

Let $k$ denote the number of unknown orbitals $\varphi_n$ (for the N-electron closed-shell configuration $k = N/2$). In the expansions:

$$\varphi_n(i) = \sum_{q=1}^{\nu} c_{qn}\chi_q(i) \qquad \varphi_n^*(i) = \sum_{p=1}^{\nu} c_{pn}^*\chi_q^*(i) \tag{A3.38}$$

to preserve the linear independence of $k$ orbitals $\varphi_n$, the basis set $\{\chi_q\}$ must not be smaller than the set of $\varphi_n$, that is, $\nu \geq k$. Substituting the expansions (A3.38) into the HF Equations (A3.37), multiplying through by $\chi_p^*(i)$ and integrating over $dV_i$ while denoting:

$$s_{pq} = \langle\chi_p \mid \chi_q\rangle, \qquad f_{pq} = \langle\chi_p \left|\widehat{f}\right| \chi_q\rangle,$$
$$g_{pp',qq'} = \langle\chi_p(i)\,\chi_{p'}(j)\left|g_{ij}\right|\chi_q(i)\,\chi_{q'}(j)\rangle, \tag{A3.39}$$

[2] These equations describe the close-shell electron configuration with the total electron spin $S = 0$. The SCF equations for an arbitrary electron configuration with a definite spin $S$ taking into account the spin degeneracy were derived by the author and represented in the book [27], Chapter VIII.

gives, instead of $k$ integro-differential equations for orbitals, $\nu k$ nonlinear (third order) algebraic equations for the coefficients $c_{qn}$:

$$\sum_q c_{qn} f_{pq} + \sum_m \sum_{q,p',q'} \left[ 2c_{qn} c^*_{p'm} c_{q'm} g_{pp',qq'} - c_{qm} c^*_{p'm} c_{q'n} g_{pp',q'q} \right]$$
$$= \sum_q \varepsilon_n c_{qn} s_{pq}, \quad p = 1, 2, \ldots, \nu, \quad n = 1, 2, \ldots, k. \tag{A3.40}$$

In the algebraic representation, the orbital $\varphi_n$ is defined by $\nu$ coefficients $c_{qn}$ of the expansion (A3.38). The set $\{c_{qn}\}$ forms a vector $\mathbf{c}_n$ in the $\nu$-dimensional space and is represented as an one-column matrix (Equation (A2.70)); the conjugate matrix $\mathbf{c}^+$ is represented as an one-row matrix:

$$\mathbf{c}_n = \begin{pmatrix} c_{1n} \\ c_{2n} \\ \vdots \\ c_{\nu n} \end{pmatrix} \qquad \mathbf{c}_n^+ = \left( c^*_{1n} c^*_{2n} \ldots c^*_{\nu n} \right) \tag{A3.41}$$

The matrix analogue of operators $\widehat{\alpha}_m$ and $\widehat{\beta}_m$ (Equation (A3.29)) can be introduced as square $(\nu \times \nu)$ matrices, $\alpha_m$ and $\beta_m$, which are defined by matrix elements in the analogy of Equation (A3.29):

$$(\alpha_m \mathbf{c}_n)_{pn} = \sum_q \left( \sum_{p',q'} c^*_{p'm} g_{pp',qq'} c_{q'm} \right) c_{qn}$$
$$(\beta_m \mathbf{c}_n)_{pn} = \sum_q \left( \sum_{p',q'} c^*_{p'm} g_{pp',q'q} c_{q'n} \right) c_{qm} \tag{A3.42}$$

This allows to represent the Coulomb and exchange integrals in a form similar to that of Equation (A3.30):

$$\alpha_{nm} = \mathbf{c}_n^+ \alpha_m \mathbf{c}_n = \mathbf{c}_m^+ \alpha_n \mathbf{c}_m$$
$$\beta_{nm} = \mathbf{c}_n^+ \beta_m \mathbf{c}_n = \mathbf{c}_m^+ \beta_n \mathbf{c}_m \tag{A3.43}$$

If we introduce the square $(\nu \times \nu)$ matrix $\mathbf{h}^{HF}$ defined on the basis set $\{\chi_q\}$:

$$\mathbf{h}^{HF} = \mathbf{f} + \sum_{m=1}^{k} (2\alpha_m - \beta_m) \tag{A3.44}$$

and the square $(\nu \times \nu)$ matrix $\mathbf{S}$ of the overlap integrals $s_{pq}$, we obtain, instead of $\nu k$ algebraic Equations (A3.40) for the coefficients $c_{qn}$, $k$ equations for the columns $\mathbf{c}_n$:

$$\mathbf{h}^{HF} \mathbf{c}_n = \varepsilon \mathbf{S} \mathbf{c}_n \qquad n = 1, \ldots, k \tag{A3.45}$$

Comparison of this equation with Equation (A3.36) shows that to transfer the system of integro-differential Equations (A3.36) to a system of algebraic equations, the Hamiltonian operator in the former is replaced by the square matrix defined on the basis set used, the orbitals $\varphi_n$ by the column matrix $\mathbf{c}_n$ and the right-hand side multiplied by the matrix of overlap integrals.

The nonlinear algebraic system (Equation (A3.45)) is solved by the iterative procedure. Firstly, a starting set of coefficients, $\mathbf{c}_n^{(0)}$, is substituted into the matrix elements of $\mathbf{h}^{HF}$. Upon solving the linear algebraic system, a new set of coefficients, $\mathbf{c}_n^{(1)}$, is found that is once more substituted into the matrix $h^{HF}$; this process is continued until self-consistency is reached.

The SCF method discussed above is based on the one-determinantal wave function (Equation (A3.20)) with double occupied space orbitals, $\varphi_m^2$. So, it has a restriction that two electrons occupy one orbital and is called the *restricted Hartree–Fock* (RHF) method.

In real atomic and molecular systems, the motion of each electron is correlated with the motion of all the remaining electrons. It can be seen directly from the interaction term in the Hamiltonian. As a consequence of the Coulomb repulsion, the approach of two electrons is energetically unfavorable, since their interaction energy $e^2/r_{12}$ tends to infinity, as $r_{12}$ approaches to zero. It can be said that each electron is surrounded by a *'Coulomb hole'* with the respect to the other electrons.

Hence, the wave function must be constructed in such a manner that it prevents two electrons being located at the same point of space; and it is so for electrons with identically oriented spins. The Pauli principle forbids them to occupy the same orbital. This introduces a correlation in their motion, within the framework of the one-electron approximation. Therefore, for electrons with identically oriented spin, one can introduce, similar to the Coulomb hole, the term *'Fermi hole'*. On the other hand, the antisymmetrization of the wave function does not impose any restrictions on the motion of electrons with oppositely oriented spins. Both electrons can be characterized by the same orbital, as in Equation (A3.20).

The simplest way of eliminating this 'electron inequality' is to describe electrons with oppositely oriented spins by different orbitals also. This approach was developed in 1954 by Pople and Nesbet [28] and is designated as *unrestricted Hartree–Fock* (UHF) method. The wave function is still based on one determinant, but instead of Equation (A3.20) for the RHF wave function, the UHF wave function is represented by a determinant:

$$\Psi_0^{UHF} = \frac{1}{\sqrt{N!}}\det\left|\psi_{1\alpha}\psi_{2\beta}\psi_{3\alpha}\ldots\psi_{N\beta}\right| \tag{A3.46}$$

This function provides zero probability for all electrons of finding two electrons at the same point in space. The energy calculated with the function $\Psi_0^{UHF}$ is always lower than calculated with the function $\Psi_0^{RHF}$ due to the larger number of variational parameters. But $\Psi_0^{UHF}$ has one essential drawback, it is not the eigenfunction of the total spin square operator $\widehat{S}^2$:

$$\widehat{S}^2\Psi^{UHF} \neq S(S+1)\Psi^{UHF} \tag{A3.47}$$

$$\widehat{S}_z\Psi^{UHF} = S_z\Psi^{UHF} \tag{A3.48}$$

This means that $\Psi^{UHF}$ corresponds to a definite value $S_z$ but does not corresponds to a definite value of the total spin $S$. It can be represented as a superposition:

$$\Psi^{UHF} = \sum_S c_S \Psi_S \tag{A3.49}$$

where

$$\widehat{S}^2 \Psi_S = S(S+1)\Psi_S \tag{A3.50}$$

and the sum in Equation(A3.49), in general, includes all possible value of $S$ from $S = S_z$ up to $S_{\max}$ that may be realized in the given N-electron system. For instance, for negatively-charged molecular ions the main contribution must give the state with $S = \frac{1}{2}$, but other spin states also contribute and their contribution can be rather large. This situation is named *spin contamination*. To obtain the wave function corresponding to the proper value of $S$, the function (A3.49) found after solving the UHF equation should be projected on the proper spin state. Such kinds of procedures are embedded in the modern suites of program, although they do not always give a desired result.

## A3.2.2 Methods taking into account the electron correlation

### A3.2.2.1 $r_{12}$-dependent wave functions

The natural way of accounting of electron correlation is to insert the interelectronic distance $r_{ij}$ directly into the way function. For the simplest case of the two-electron helium atom, in the early days of quantum mechanics this was done by Hylleraas [29]. In the variational calculations on the ground state of helium, Hylleraas used a two-electron space function in the form of an infinite series:

$$\Phi(r_1, r_2) = e^{-\frac{r_1+r_2}{2}} \sum_{l,m,n=o}^{\infty} c_{lmn} (r_1 + r_2)^l (r_1 - r_2)^m r_{12}^n \tag{A3.51}$$

Eight terms of the series proved to be sufficient to obtain a good agreement with experimental data. The employment of the Hylleraas expansion for helium on a large scale resulted in a nanohartree accuracy $\left(1 \text{ nanohartree} = 0.00022 \text{ cm}^{-1}\right)$ [30, 31].

The first explicitly correlated wave function for molecules was employed for the hydrogen ($H_2$) molecule by James and Coolidge [32]. Following Hylleraas' successful calculation, they carried out a similar variational calculation on the ground state of the hydrogen molecule. The James–Coolidge function included the internuclear distance $R_{ab}$ as a parameter. When $R_{ab} = 0$, the function goes over into the Hylleraas function. A calculation with a 13-term function gave a value for the dissociation energy rather close to the experimental value. Further developments [33–36], with inclusion of large amount of terms (up to 883 [36]) in the variational function, results in an accuracy at nanohartree level (see Chapter 1, Table 1.1) that is unprecedented in molecular calculations.

For a long time it was accepted that the explicit $r_{ij}$-dependent variational functions could be used only for calculations of few-electron systems. James and Coolidge applied the Hylleraas-type function to calculate the ground state of the three-electron lithium atom [37]. In the 1970s this approach was applied in the high correlated calculations of four-electron atoms: Be [38] and $Li^-$ [39]. Clary and Handy employed the $r_{12}$-dependent wave functions in calculations of two-atomic molecules: $He_2$ [40] and LiH [41].

The main problem in extending the application of the explicitly correlated functions to many-atomic molecules was connected with computational difficulties of estimating cumbersome many-electron integrals. The $r_{ij}$-dependence of the variational function gives rise to three- and four-electron integrals, while in an one-electron basis, the variational problem includes only one- and two-electron integrals, in as much as Hamiltonian comprises only one- and two-electron operators.

Kutzelnigg, Klopper, and Noga developed so-called *linear R12 approach* [42–46]. In this approach, some completeness relationships are inserted that allow all many-electron integrals, except one- and two-electron integrals, to be eliminated from the final computation procedure. The $r_{12}$ distance dependence was directly incorporated into the standard Møller–Plesset (see Section A3.3.2) and Coupled Cluster (see Section A3.2.2.3) methods that permitted the precision of these methods to be increased considerably. The MP2-R12 [42, 43] and CC-R12 [44, 45] methods were elaborated and implemented in precise calculations of $Be_n$ and $Mg_n$ ($n = 2$–$4$) clusters [47], $H_2O$, $NH_3$ and $CH_4$ molecules [48], hydrogen-bonded complexes $(HF)_2$, $(HCl)_2$, $(H_2O)_2$ and some others [49]. The details of the linear R12 approach can be found in reviews [45, 46].

### A3.2.2.2 Configuration interaction

The two modifications of the SCF approach–RHF and UHF methods–belong to the one-electron approximation. Each electron is characterized by its one-electron wave function (spin orbital) and occupies some one-electron energy level. In the RHF approach it corresponds to the ground-state orbital configuration:

$$K_0 : \varphi_1^2 \varphi_2^2 \dots \varphi_{\frac{N}{2}}^2 \tag{A3.52}$$

in which all orbitals (one-electron energy levels) are double occupied.

In real systems of interacting electrons, the interaction mixed one-electron energy levels, the electrons lose their individuality. The simplest way to go beyond the SCF approach (or to take into account the electron correlation) is to apply a variational function as a linear combination of different orbital configurations constructed from $K_0$ (Equation (A3.52)) by transferring electrons on nonoccupied (virtual) orbitals. The number of solutions of the HF Equations (A3.45) is determined by the size $\nu$ of the basis set, $\nu$ is larger than the number of occupied orbitals $N/2$. Therefore, there always exist $\nu - N/2$ virtual orbitals, or unoccupied energy levels, on which electrons can be excited.

In the approach designated as the *configuration interaction (CI) method*, the variational function is represented as a linear combination of antisymmetric functions (determinants):

$$\Psi^{CI} = A_0\Psi^{HF}(K_0) + \sum_{n,e} A_n^e \Psi\left((K_0)_n^e\right) + \sum_{n,e}\sum_{m,i} A_{nm}^{ei}\Psi\left((K_0)_{nm}^{ei}\right) + \cdots \tag{A3.53}$$

where $(K_0)_n^e$ is a singly-excited configuration where an electron occupying the orbital $n$ in $K_0$ is excited to the virtual orbital $e$, $(K_0)_{nm}^{ei}$ is a doubly-excited configuration with two electrons excited to the virtual orbitals $e$ and $i$ from the orbitals $n$ and $m$, respectively. Expansion (A3.53) may include triply-, quadruply- and etc. excited configurations. It is convenient to write Equation (A3.53) in a more compact form:

$$\Psi^{CI} = \sum_{j=1}^{p} A_j \Psi\left(K_j\right) \tag{A3.53a}$$

where the coefficients $A_j$ are parameters that must be found by minimizing the variational functional:

$$\delta J = \delta\left\langle \Psi^{CI} \left| H - E \right| \Psi^{CI}\right\rangle = 0 \tag{A3.54}$$

Substituting Expansion (A3.53a) into Equation (A3.54) and taking into account that $\Psi(K_i)$ form an orthonormal set:

$$\left\langle \Psi(K_i) \mid \Psi\left(K_j\right)\right\rangle = \delta_{ij} \tag{A3.55}$$

one obtains the expression for $\delta J$:

$$\begin{aligned}\delta J &= \delta\left[\sum_{i,j} A_i^* A_j \left(H_{ij} - E\delta_{ij}\right)\right] \\ &= \sum_{i,j}\left(A_j\delta A_i^* + A_i^*\delta A_j\right)\left(H_{ij} - E\delta_{ij}\right) = 0\end{aligned} \tag{A3.56}$$

where:

$$H_{ij} = \left\langle \Psi(K_i) \left| H \right| \Psi\left(K_j\right)\right\rangle \tag{A3.57}$$

From Equation(3.56) it follows that $\delta J = 0$ at arbitrary and independent variations $\delta A_i^*$ and $\delta A_j$, if the coefficients $A_j$ fulfill the system of equations:

$$\sum_{j=1}^{p} A_j\left(H_{ij} - E\delta_{ij}\right) = 0 \qquad i = 1, 2, \ldots, p \tag{A3.58}$$

This homogeneous system of linear equations has a nonzero solution only if:

$$\det\left|H_{ij} - E\delta_{ij}\right| = 0$$

or:

$$\begin{vmatrix} H_{11}-E & H_{12} & \dots & H_{1p} \\ H_{21} & H_{22}-E & \dots & H_{2p} \\ \cdot\ \ \cdot & \cdot\ \ \cdot & \cdot & \cdot \\ H_{p1} & H_{p2} & \dots & H_{pp}-E \end{vmatrix} = 0 \tag{A3.59}$$

The algebraic Equation (A3.59) has $p$ roots. For each root $E^{(n)}$, it is necessary to solve Equation (A3.58) and find a set $\left\{A_j^{(n)}\right\}$ of $p$ coefficients. The ground state corresponds to the lowest energy value, which can be designated $E^{(1)} \equiv E_0$. It determines the set $\left\{A_j^{(1)}\right\}$ for the ground state wave function:

$$\Psi_0^{CI} = \sum_{j=1}^{p} A_j^{(1)} \Psi\left(K_j\right) \tag{A3.60}$$

Since the configuration Expansion (A3.60) is a convergent series, the calculation becomes more accurate as the number of configurations involved increases. In the full CI (FCI) calculation, all configurations that can be built up at a given basis set with size $\nu$ are taken into account. For the N-electron state with $S = 0$, the total number of configuration is equal to [50]:

$$p(FCI) = \frac{\nu!\,(\nu+1)!}{\left(\frac{N}{2}\right)!\,(N+1)!\,\left(\nu-\frac{N}{2}\right)!\,\left(\nu-\frac{N}{2}+1\right)!} \tag{A3.61}$$

For the relatively small ethane molecule ($C_2H_6$) and not large 6-31G** basis set ($\nu = 60,\ N = 18$), $n(\mathrm{FCI}) = 2.6 \times 10^{19}$.

Hence, the ethane molecule cannot be calculated at the FCI level with the 6-31G** basis set. However, it is worthwhile pointing out that in modern suites of programs, very effective methods are elaborated for finding only the lowest roofs of the secular equation for large-scale FCI, which includes up to $10^9$ configurations [51–53]. Mitrushenkov [53] calculated the ground state of water, applying the FCI function expanded over $7.2 \times 10^9$ determinants.

Since FCI is impossible except for small molecules and small basis sets, the reasonable limitation is to use only singly- and doubly-excited configurations, it is denoted as CI SD. In addition, the *frozen-core* approximation (where inner shells are not involved in the CI procedure) is often used.

An essential improvement in the convergence of the CI expansion is provided by configurations constructed from so-called *natural orbitals* (NO), that is, orbitals that diagonalize the density matrix of the first order [54]. An effective iterative procedure for the determination of NOs was developed by Bender and Davidson [55]. Employing NOs instead of the HF orbitals permits the size of the CI expansion to be decreased substantially without loss of precision.

Another way to reduce the size of the CI expansion in high-quality calculations is realized in the *multiconfiguration self-consistent field* (MC SCF) method. In this method, the coefficients $A_j$ in the CI expansion and the coefficients $c_{qn}$ in the orbital expansion are varied simultaneously. The main concepts of the MC SCF method were first discussed in the book by Frenkel [56]. He pointed out especially the importance of the linear independence of the variational conditions, which is imposed upon the configuration coefficients and upon the one-electron functions. The MC SCF method for molecular systems was first applied by Das and Wahl [57] for diatomic molecules and by Veillard and Clementi [58] for polyatomic molecules. Further improvements to this method [59, 60] have led to wide used of the MC SCF approach in the calculation of molecular systems.

An often-used modification of the MC SCF method is the *complete active space* SCF (CAS SCF) method [60], in which the orbitals are divided into *inactive* and *active* orbitals. Inactive orbitals usually belong to inner shells; they are kept doubly occupied in all configurations. The electrons, which do not occupy the inactive orbitals, are called active electrons. They are distributed among the active orbitals in all possible ways. In the N-electron system with k inactive orbitals, there are N–2k active electrons distributed between chosen set of active orbitals. Then, in this active configuration space, the FCI calculation is performed at the MC SCF level. At present, very effective procedures for improving the convergence in the MC SCF calculation are being elaborated [59–61]. Knowles and Werner [62] performed the CAS SCF calculation on the FeO molecule that included 178 916 configurations.

Another effective method in the CI family is so-called *multireference configuration interaction* (MRCI) method. It combines the direct CI and MC SCF methods. In the first stage, the MC SCF calculation is performed on a limited number of configurations chosen from physico–chemical considerations. As a result, one obtains configurations with optimized orbitals, which are much more precise than the configurations built with the HF orbitals. Then each of the configurations that is included in the optimized MC SCF function is used as a *reference configuration* (as the HF configuration, $K_0$, in the conventional CI method) to construct singly-excited, doubly-excited, etc. configurations that all are used in the direct CI (Equations (A3.58) and (A3.59)) calculation procedure.

#### A3.2.2.3 Coupled cluster method

This approach was first developed in nuclear physics in 1958–1960 by Coester and Kümmel [63, 64], see also Reference [65], but it was not implemented because at that time (and at present as well) the knowledge of nuclear forces was not sufficient to perform precise calculations and comparison with experiment. The *coupled cluster* (CC) approach was adopted for molecular calculations by Ciźek [66, 67] and then elaborated by many scientists (see reviews [68–71] and references therein). At present, it is one of the most popular approaches in *ab initio* calculations.

The main idea of the CC method is to present the unknown function as:

$$\Psi = e^{\widehat{T}}\Psi(K_0) \tag{A3.62}$$

where $\Psi(K_0)$ is the ground-state HF wave function and the exponential excitation operator is define by the Taylor-series expansion:

$$e^{\widehat{T}} = 1 + \widehat{T} + \frac{1}{2!}\widehat{T}^2 + \frac{1}{3!}\widehat{T}^3 + \cdots. \tag{A3.63}$$

The operator $\widehat{T}$ is called *cluster operator* and is represented as a sum:

$$\widehat{T} = \widehat{T}_1 + \widehat{T}_2 + \cdots + \widehat{T}_N \tag{A3.64}$$

where $\widehat{T}_k$ is the excitation operator that transfers $k$ electrons from occupied orbitals of $K_0$ to vacant orbitals in all possible ways; it converts the ground-state determinant into linear combinations of determinants corresponding to $k$ excited states. Namely:

$$\widehat{T}_1\Psi(K_0) = \sum_{e=\frac{N}{2}+1}^{\nu} \sum_{n=1}^{N/2} a_n^e \Psi\left((K_0)_n^e\right) \tag{A3.65}$$

$$\widehat{T}_2\Psi(K_0) = \sum_{\substack{e,i=\frac{N}{2}+1\\(e<i)}}^{\nu} \sum_{\substack{n,m\\(n<m)}}^{N/2} a_{nm}^{ei} \Psi\left((K_0)_{nm}^{ei}\right) \tag{A3.66}$$

and etc.
The coefficients $a_n^e$, $a_{nm}^{ei}$, etc. are found from the set of nonlinear algebraic equation by the iteration procedure [72].

Certainly, preserving all terms in the sum (Equation (A3.64)) is not realistic. If only single- and double-excitation operators, $\widehat{T}_1$ and $\widehat{T}_2$, are included, these are the CC *single and double* (CCSD) method [73, 74]. The next step is to include the triple and then quadruple excitation operators, $\widehat{T}_3$ and $\widehat{T}_4$. This has led to elaboration of the CCSDT [75–77] and CCSDTQ [78] methods. These methods produce very accurate results but are very time-consuming and therefore are feasible only for small molecules and not large basis sets. At present, the CCSD(T) method that includes the triple excitations noniteratively [79] is most widely used; an analysis of the CCSD(T) calculations is given in References [80, 81].

The CC methods considered in the preceding text are based on a single reference function $\Psi(K_0)$ (the HF closed-shell determinant); they are attributed to the *single-reference* CC (SR CC) approach. This approach is quite satisfactory for closed-shell systems in the ground-state geometry. However, for open-shell interacting systems, especially in the regions of chemical-bond breaking, the SR CC methods often give wrong results. For instance, the SR CCSD method leads to the completely wrong dissociation limit for $N_2$ and some other molecules [71, 82]. In many cases, the employment of the more precise SR CCSDT and SR CCSDTQ methods does not improve results; reliable results are provided by *multireference* CC (MR CC) methods.

The MR CC ansatz, rather convenient for implementing, was suggested by Jeziorski and Monkhorst [83]. It was realized by Kucharski and Bartlett [84] at the MR CCSD level. Mahapatra *et al.* [85] developed the MR CC scheme based on a reference function composed of determinants spanning a complete active space and calculated at the MC SCF level. The *reduced multireference* approach (RMR CCSD) was developed by Li and Paldus [86].

Another approach closely connected with the MR CC scheme is based on the Brillouin–Wigner perturbation theory and designated BW CC. It was developed by Hubać *et al.* [87–89]. Its application with *a posteriori* size-extensivity correction [88] see in Reference [89]. The connection between the BW CC and MR CC methods [83–85] was established by Pittner [90].

As shown by Piecuch and Kowalski [91–93], noniterative corrections to the energy obtained in SR CC methods can improve the results without addressing MR CC methods; for instance, they correctly describe the transition state with biradical character [94] that normally requires a multireference treatment. Inclusion of these noniterative corrections leads to renormalized CCSD(T) and CCSD(T,Q) methods.

As follows from the multitude of CC methods presented above, at present, CC theory is an actively developing area and it is impossible in this brief account to mention all new approaches.

### A3.2.2.4 Density functional theory approach

In this section the main ideas of the *density functional theory* (DFT), the *Kohn–Sham* (KS) equation and gradient, nonlocal corrections to it are presented. More detailed presentations are given in the monographs on this subject [95, 96].

Although, the wave function is a fundamental concept in quantum mechanics and determines all properties of an electronic system, it is not measured directly in experiment. All measured physical properties are expressed via bilinear products of wave functions. The expectation value of some physical observable, described in quantum mechanics by an operator $\widehat{L}$, is expressed as:

$$\overline{L} = \int \Psi^* (1, \ldots, N) \widehat{L} \Psi (1, \ldots, N) \, dV \tag{A3.67}$$

where numbers denote the four coordinates for each enumerated electron and the integration includes also a summation over spin variables.

All operators in quantum mechanics can be divided on two kinds: one- and two-particles operators. As a result, the matrix elements of the Hamiltonian are expressed via so-called *reduced one-* and *two-electron density matrices* [54, 97]. The reduced spinless one-electron density matrix is defined as[3]:

$$\rho \left(\mathbf{r}_1' \mid \mathbf{r}_1\right) = N \int \Psi^* \left(1', \ldots, N\right) \Psi (1, \ldots, N) \, dV^{(1)} \tag{A3.68}$$

where integration is performed over the configuration space of all electrons except the first; it includes also a summation over the spin-coordinates of all electrons.

Its diagonal element $\rho(\mathbf{r}_1)$ is determined by the probability density to find electron 1 at a point $\mathbf{r}_1$. Because of the indistinguishability of electrons in quantum

[3] In some publications, the expression (A3.68) is denoted as $n(\mathbf{r}_1' \mid \mathbf{r}_1)$ and by $\rho(\mathbf{r}_1' \mid \mathbf{r}_1)$ is denoted $-en(\mathbf{r}_1' \mid \mathbf{r}_1)$, which is called *charge density* matrix.

mechanics, the definition (A3.68) does not depend upon the electron number. The factor $N$ provides the natural result:

$$\int \rho(\mathbf{r})\, dV = N \qquad \text{(A3.69)}$$

If the N-electron wave function is represented as the double-filled determinant (Equation (A3.20)), the $\rho(\mathbf{r})$ in Equation (A3.68) is expressed as a simple sum over all occupied orbitals:

$$\rho(\mathbf{r}) = 2\sum_{n=1}^{N/2} \psi_n^*(\mathbf{r})\, \psi_n(\mathbf{r}) = 2\sum |\psi_n(\mathbf{r})|^2 \qquad \text{(A3.70)}$$

In Chapter 2, the electrostatic energy (Equation (2.6)) is directly represented via the electron densities $\rho^A$ and $\rho^B$ of interacting molecules. Hohenberg and Kohn [98] proved a general theorem that all properties of a molecule in its ground state, including the energy, are uniquely determined by the ground state electron density $\rho_o(\mathbf{r})$ that allows the ground-state energy to be represented as a function of $\rho_o$, designated as $E_o[\rho_o]$. They formulated an exact formal variational principle for $E_o[\rho_o]$, from which it follows that the exact ground state electron density minimizes $E_o[\rho_o]$, hence any trial variational $\rho_o$ gives the upper limit to the exact value of $E_o$.

The next step was made by Kohn and Sham [99]. They considered a system of independent electrons moving in some static potential $\nu(\mathbf{r})$, which include an external field $\upsilon_{ext}(\mathbf{r})$ and the electron-nuclear attraction:

$$\upsilon(\mathbf{r}) = \sum_a \frac{Z_a e}{|\mathbf{R}_a - \mathbf{r}|} + \upsilon_{ext}(\mathbf{r}) \qquad \text{(A3.71)}$$

and formulated the energy functional as a following sum:

$$E[\rho] = T[\rho] - \int e\upsilon(\mathbf{r})\,\rho(\mathbf{r})\, dV + \tfrac{1}{2}\iint \rho(\mathbf{r}_i)\,\rho(\mathbf{r}_j)\,\tfrac{e^2}{r_{ij}} dV_i dV_j + E_{XC}[\rho] \qquad \text{(A3.72)}$$

where $T[\rho]$ is the kinetic energy of a system of noninteracting electrons but with the same electron density as in the real system; the second term is the energy in the potential (A3.71); the third term is the energy of electron-electron repulsion corresponding to the Coulomb potential:

$$\Phi_{coul}(\mathbf{r}_i) = \tfrac{1}{2}\int \rho(\mathbf{r}_j)\,\tfrac{e^2}{r_{ij}} dV_j \qquad \text{(A3.73)}$$

the last term $E_{XC}[\rho]$ corresponds to the exchange–correlation energy, the expression of which has to be found. After assuming some analytical form for $E_{XC}$, the potential arising from $E_{XC}[\rho]$ is defined formally as the functional derivative:

$$\upsilon_{\text{xc}}(r) = \frac{\delta E_{XC}[\rho(\mathbf{r})]}{\delta\rho} \qquad \text{(A3.74)}$$

In contrast with the nonlocal exchange potential of the HF equation, the potential $\upsilon_{xc}(\mathbf{r})$ is local.

The Kohn–Sham (KS) equations can be obtained from the expression for $E[\rho]$ if one introduces some one-electron functions $\varphi^{KS}(\mathbf{r}_i)$ named *Kohn–Sham orbitals* and according to the Pauli principle constructs the Slater determinant (Equation (A3.20)). Then the electron density in the energy functional (A3.72) is expressed via $\varphi^{KS}(\mathbf{r})$:

$$\rho^{KS}(\mathbf{r}) = 2\sum_n \left|\varphi^{KS}(\mathbf{r})\right|^2 \tag{A3.75}$$

and for orbitals that minimize the $E[\rho]$, the following system of equations is valid (the *Kohn–Sham equations*) [99]:

$$\left[-\frac{\hbar^2}{2m}\nabla_i^2 + \upsilon(\mathbf{r}_i) + \Phi_{Coul}(\mathbf{r}_i) + \upsilon_{xc}(\mathbf{r}_i)\right]\varphi_n^{KS}(\mathbf{r}_i) = \varepsilon_n^{KS}\varphi_n^{KS}(\mathbf{r}_i), \quad n = 1, \ldots, N/2 \tag{A3.76}$$

or

$$\widehat{h}_i^{KS}\varphi_n^{KS}(\mathbf{r}_i) = \varepsilon_n^{KS}\varphi_n^{KS}(\mathbf{r}_i), \qquad n = 1, \ldots, N/2 \tag{A3.77}$$

where the expressions for potentials entering $\widehat{h}^{KS}$ are given by the Equations (A3.71), (A3.73) and (A3.74).

Let us denote, as in the HF equations:

$$\widehat{f}_i = -\frac{\hbar^2}{2m}\nabla_i^2 + \upsilon(\mathbf{r}_i) = -\frac{\hbar^2}{2m}\nabla_i^2 - \sum \frac{Z_a e^2}{|\mathbf{R}_a - \mathbf{r}_i|} + \upsilon_{ext} \tag{A3.78}$$

and rewrite Equation (A3.76):

$$\left[\widehat{f}_i + \tfrac{1}{2}\int \rho(\mathbf{r}_j)\frac{e^2}{r_{ij}}dV_j + \frac{\partial E_{XC}[\rho(r_i)]}{\partial \rho}\right]\varphi_n^{KS}(\mathbf{r}_i) = \varepsilon_n^{KS}\varphi_n^{KS}(\mathbf{r}_i), \quad n = 1, \ldots N/2 \tag{A3.79}$$

This is the one-electron Schrödinger equation in which the Hamiltonian itself depends upon orbitals being sought due to Equation (A3.75). For its solution, the KS orbitals are expanded, as in the HF case, over some basis set and nonlinear algebraic equations for the expansion coefficients are solved iteratively.

The main problem in DFT is how to find an explicit expression for the functional $E_{XC}[\rho]$; it is known only for the homogeneous electron gas [100]. Let us discuss the approximations used.

The first approximation was based on the homogeneous electron gas and named the *local density approximation* (LDA). If the electron density is a very slowly changing function, then $E_{XC}[\rho]$ can be presented as an integral:

$$E_{XC}[\rho] = \int \rho(\mathbf{r})\,\varepsilon_{xc}[\rho]\,dV \tag{A3.80}$$

where $\varepsilon_{xc}[\rho]$ is the exchange and correlation energy per electron in a homogeneous electron gas. It is separated on the exchange and correlation parts:

$$\varepsilon_{xc}^{LDA}[\rho] = \varepsilon_x[\rho] + \varepsilon_c[\rho] \tag{A3.81}$$

Exchange energy for an electron gas is equal to:

$$E_X^{LDA} = -\frac{4}{3}\left(\frac{3}{\pi}\right)^{\frac{1}{3}} \int \rho(\mathbf{r})^{\frac{4}{3}} \, dV \tag{A3.82}$$

The expression for the correlation part has been found by Vosko, Wilk and Nusair [101] and is denoted $E_c^{VWN}$. It is quite complex and we do not present it.

The LDA approach was first proposed for solids in the so-called *jellium model*, where electrons constitute a homogeneous electron gas. The jellium model for atoms and molecules is difficult to justify. Nevertheless, in many atomic and molecular applications, the LDA was working better than could be expected. From numerous calculations for atoms, it was estimated that the LDA underestimates the exchange energy by roughly 10% and overestimates the correlation energy by up to 100%. The same order of errors is found for diatomic molecules [102].

The spin-dependent DFT was developed by von Barth and Hedin [103], see also References [104, 105]. The KS equation was formulated in so-called *local spin-density approximation* (LSDA), where different orbitals (and different electron densities) for electrons with different spin projections are used, as in the UHF method. $E_X$ and $E_C$ contain two types of electron density, $\rho_\alpha$ and $\rho_\beta$:

$$E_X^{LSDA} = -\tfrac{4}{3}\left(\tfrac{3}{\pi}\right)^{\frac{1}{3}} \int \left[\rho_\alpha(\mathbf{r})^{\frac{4}{3}} + \rho_\beta(\mathbf{r})^{\frac{4}{3}}\right] dV \tag{A3.83}$$

$$E_C^{LSDA} = \int \left(\rho_\alpha + \rho_\beta\right) \varepsilon_c\left[\rho_\alpha, \rho_\beta\right] dV$$

The LSDA approach is capable of describing an electronic system in a magnetic field. But even in the absence of a magnetic field, it has many advantages in comparison with the LDA, for instance, it is appropriate for studying open-shell molecules. On the other hand, the LSDA approach, as the UHF method, does not correspond to the state with definite values for the total spin $S$ and corresponds to a state with a particular value of $S_z$ (compare Section A3.2.1, Equations (A3.47) and (A3.48)). It is important to stress that the LSDA equations do not describe space and spin multiplets because they correspond to a wave function presented as a single determinant [106, 107].

The LDA and LSDA approaches are based on the uniform-electron-gas model. At present, many density functionals appropriate for nonuniform electron distribution have been elaborated. They include the gradients of $\rho(\mathbf{r})$. These gradient-corrected functionals have been constructed for both exchange and correlated functionals in the LSDA approach and belong to the so-called *generalized gradient approximation* (GGA); the gradient corrections are often named *nonlocal corrections* (NL): In GGA:

$$E_X = E_X^{LSDA} + E_X^{NL} \tag{A3.84}$$

The gradient corrected functionals are included in different codes, among them is the Gaussian suite of programs. The first gradient-corrected exchange functionals were suggested by Perdew and Wang in 1986 (PW86) [108] and Becke in 1988 (B88 or B)[109]. Commonly used gradient-corrected correlation functionals include the Lee–Yang–Parr (LYP)[110] and Perdew–Wang (PW91) [111] functionals. Any exchange functional can be combined with any correlation functional. In the early 1990s, the combination of B88 with LYP or PW91 functionals, denoted as BLYP or BPW91 respectively, had been used.

In 1993, Becke [112] constructed the so-called *hybrid* functional as a semiempirical linear combination of the HF exchange energy designated as $E_X^{exact}$; B88 and PW91 gradient-corrected functionals based on the LSDA approach with the $\varepsilon_c^{LSDA}$ correlation energy density parameterized by Perdew and Wang [113]. This B3PW91 hybrid functional (where the number 3 indicates a three-parameter functional) and its modification, B3LYP, in which instead of PW91 the linear combination of LYP and VWN [101] correlation functionals is used, are implemented in various program suites and widely used.

In further improvements, Becke and Schmider [114–116] developed the ten-parameter hybrid functional. The systematic optimization based on least-square fitting to accurate thermochemical data led to very good agreement with the experimental data for the heat of formation of 148 molecules (so-called extended G2 test set [117]) with a 1.78 kcal/mol mean absolute error [116].

In the following years, many effective functionals were created. Among them (far from the complete selection) are:

MPWPW – created by Adamo and Barone [118], which modified the exchange functional proposed in 1991 by Perdew and Wang [111] and combined it with their correlation functional PW91 [113]. This functional showed quite good results for both covalent and noncovalent molecular systems.

B97-1/B97-2 – Becke's 97 hybrid functional [115] reparameterized by Hamprecht *et al.* [119]; the B97-1 functional became the best for the van der Waals complex calculations.

TPSS – nonempirical exchange-correlation functional constructed by Tao, Perdew, Staroverov and Scuseria [120] in the framework of the so-called *meta-generalized gradient approximation* (meta-GGA) and originating from the Perdew–Burke–Ernzerhof (PBE) functional [121]; it provides good results for molecules, solids and surfaces [122, 123].

X3LYP – developed by Xu and Goddard [124, 125], it gives a better performance than the popular B3LYP hybrid functional.

MPW1B95/MPWB1K – created by Truhlar and collaborators [126, 127] and based on the exchange MPW functional [111, 118] and Becke's correlation functional [114], it was named by authors [126, 127] *meta correlation functional* and labelled as B95. MPW1B95 was optimized against a representative atomization energy database for thermochemistry purposes and MPWB1K was optimized against the kinetic database.

In a recent review article [128], Zhao and Truhlar tested 44(!) DFT functionals using specially constructed databases. The comparative study of the best modern DFT functionals was carried out by the same authors in Reference [128a] where the best performance for organic complexes and clusters was demonstrated by their new PWB6K functional [128b].

From many excellent results obtained by the DFT approach, it should not be concluded that the DFT method may be put at a level with the best *ab initio* methods discussed in preceding sections. The great advantage of the DFT method is in its applicability for calculations of large systems for which *ab initio* methods are very expensive or limited by currently available computing power. However, the predictive power of semiempirical DFT methods depends upon the selected databases against which the parameters in the gradient functionals were adjusted. It is necessary to be very careful when applying DFT to new, unknown systems.

Effective DFT programs for excited states are still not available, although activity in this direction is rapidly developing [129–132]. On the other hand, the excited states are often degenerate and DFT cannot be, in principle, applied to degenerate and quasidegenerate states when the adiabatic approximation fails [133]. The application of DFT to van der Waals complexes meets serious problems (see Section 3.2.2.3).

Last but not least, DFT approaches cannot describe space and spin multiplets, a drawback that is connected with an inherent property of the electron density. Namely: it can be rigorously proved[4] that the expression for electron density does not depend upon the space and permutation symmetry of the quantum state. For instance, it does not depend upon the value of the total spin $S$ (the latter is uniquely connected with the permutation symmetry of the coordinate (space) wave function). The procedures developed for calculating the multiplet structure [106, 107, 135–138] do not follow from DFT principles and provide some *ad hoc* calculation recipes.

In spite of these drawbacks, the application of DFT calculation methods to the nondegenerate ground state of molecules and chemically-bound complexes demonstrates their high efficiency and precision, often comparable with *ab initio* MP2 calculations.

## A3.3 Perturbation Theory

The approximate solution of the Schrödinger equation:

$$H\Psi = E\Psi \tag{A3.85}$$

is considerably simplified, if the Hamiltonian can be expressed as a sum:

$$H = H_0 + V \tag{A3.86}$$

[4] The general proof is beyond the scope of this book and will be published elsewhere; the permutation symmetry case is considered in Reference [134].

where $H_0$ is the unperturbed Hamiltonian, the solution of which is assumed to be known:

$$H_0 \Psi_m^{(0)} = E_m^{(0)} \Psi_m^{(0)} \tag{A3.87}$$

and $V$ is a small correction called *perturbation*. The solution of the exact Schrödinger equation can be expressed in terms of the solutions of the unperturbed Equation (A3.87) as a series in powers of $V$. This expansion is not unique, different formalisms of the perturbation theory (PT) exist.

### A3.3.1 Rayleigh–Schrödinger perturbation theory

Let us expand the unknown function $\Psi$ over a complete set $\left\{\Psi_m^{(0)}\right\}$:

$$\Psi = \sum_m c_m \Psi_m^{(0)} \tag{A3.88}$$

and substitute this expansion into the Schrödinger Equation (A3.85) with the Hamiltonian (A3.86). Taking into account Equation (A3.87), the Schrödinger equation can be presented as:

$$\sum_m c_m \left(E - E_m^{(0)}\right) \Psi_m^{(0)} = \sum_m c_m V \Psi_m^{(0)} \tag{A3.89}$$

If Equation(A3.89) is multiplied by $\Psi_k^{(0)}$ and integrated over all space, then taking into account that the set $\left\{\Psi_m^{(0)}\right\}$ is orthonormal gives:

$$c_k \left(E - E_k^{(0)}\right) = \sum_m c_m V_{km} \tag{A3.90}$$

where $V_{km} = \left\langle \Psi_k^{(0)} \left| V \right| \Psi_m^{(0)} \right\rangle$.

Let us now present unknowns $E$ and $c_m$ as series over diminishing terms:

$$E = E^{(0)} + E^{(1)} + E^{(2)} + \cdots \tag{A3.91}$$

$$c_m = c_m^{(0)} + c_m^{(1)} + c_m^{(2)} + \cdots$$

where $E^{(1)}$ and $c_m^{(1)}$ are the same order as $V$, $E^{(2)}$ and $c_m^{(2)}$ are the same order as $V^2$, and so forth. We look for corrections to a state $n$. In this case, in Equation (A3.88) at the zero order, $c_n^{(0)} = 1$ and $c_m^{(0)} = 0$ for $m \neq n$. Substitution of Equations (A3.91) into Equation(A3.90) leads to the following equality:

$$\left(c_k^{(0)} + c_k^{(1)} + c_k^{(2)} + \cdots\right)\left[\left(E_n^{(0)} + E_n^{(1)} + E_n^{(2)} + \cdots\right) - E_k^{(0)}\right] \tag{A3.92}$$
$$= \sum_m \left(c_m^{(0)} + c_m^{(1)} + c_m^{(2)} + \cdots\right) V_{km}$$

Collecting in the left-hand and right-hand sides of the Equation(A3.92) the terms of the same order of magnitude, we obtain expressions for the energy and coefficients at this order. For instance, for the terms of the first order:

$$c_k^{(0)} E_n^{(1)} + c_k^{(1)} \left( E_n^{(0)} - E_k^{(0)} \right) = \sum_m c_m^{(0)} V_{km} = V_{kn} \tag{A3.93}$$

For $k = n$, the expression for $E_n^{(1)}$ is obtained:

$$E_n^{(1)} = V_{nn} = \left\langle \Psi_n^{(0)} \left| V \right| \Psi_n^{(0)} \right\rangle \tag{A3.94}$$

which is the expectation value of the perturbation operator in the unperturbed state. For $k \neq n$, we obtain the expression for all coefficients $c_k^{(1)}$, except for $c_n^{(1)}$ :

$$c_k^{(1)} = \frac{V_{kn}}{E_n^{(0)} - E_k^{(0)}} \qquad k \neq n \tag{A3.95}$$

Coefficient $c_n^{(1)}$ can be found from the normalization condition for:

$$\Psi_n = \Psi_n^{(0)} + \Psi_n^{(1)} \tag{A3.96}$$

that must be fulfilled with precision up to the first order. For this, we should put $c_n^{(1)} = 0$. Thus:

$$\Psi_n^{(1)} = \sum_m{}' \frac{V_{mn}}{E_n^{(0)} - E_m^{(0)}} \Psi_m^{(0)} \tag{A3.97}$$

the prime in this equation and later on designates that the sum does not contain terms, which turns denominators to zero; in Equation (A3.97) it means that $m \neq n$. The perturbation mixes $\Psi_n^{(0)}$ with all unperturbed states:

$$\Psi_n = \Psi_n^{(0)} + \sum_m{}' \frac{V_{mn}}{E_n^{(0)} - E_m^{(0)}} \Psi_m^{(0)} \tag{A3.98}$$

From the orthonormality of the set $\left\{ \Psi_m^{(0)} \right\}$, the so-called *intermediate normalization* condition is fulfilled:

$$\left\langle \Psi_n^{(0)} \middle| \Psi_n \right\rangle = 1 \tag{A3.99}$$

For the validity of PT, the evident condition must be fulfilled:

$$\left| \Psi_n^{(1)} \right| \ll \left| \Psi_n^{(0)} \right| \tag{A3.100}$$

or

$$\left| V_{mn} \right| \ll \left| E_n^{(0)} - E_m^{(0)} \right| \tag{A3.101}$$

For degenerate states, this inequality is not fulfilled and, in this case, a special approach is developed (see textbooks on quantum mechanics, for instance, Reference [26]).

Collecting terms of the second order in Equation (A3.92), we obtain the expressions for the energy and coefficients in the second order of PT. Below we represent the expressions for $E_n^{(2)}$ and $E_n^{(3)}$:

$$E_n^{(2)} = \sum_m{}' \frac{V_{nm} V_{mn}}{E_n^{(0)} - E_m^{(0)}} = \sum_m{}' \frac{|V_{mn}|^2}{E_n^{(0)} - E_m^{(0)}} \tag{A3.102}$$

the last equality in Equation (A3.102) follows from $V_{nm} = V_{mn}^*$ that is valid for Hermitian's operators. The expression for $E_n^{(2)}$ (A3.102) can be presented in a compact form via $\Psi_n^{(1)}$ (Equation (A3.97)):

$$E_n^{(2)} = \left\langle \Psi_n^{(0)} \left| V \right| \Psi_n^{(1)} \right\rangle \tag{A3.103}$$

The expression for $E_n^{(3)}$ is more complex:

$$E_n^{(3)} = \sum_k{}' \sum_m{}' \frac{V_{nm} V_{mk} V_{kn}}{\left(E_n^{(0)} - E_m^{(0)}\right)\left(E_n^{(0)} - E_k^{(0)}\right)} - V_{nm} \sum_m{}' \frac{|V_{mn}|^2}{\left(E_n^{(0)} - E_m^{(0)}\right)^2} \tag{A3.104}$$

## A3.3.2 Møller–Plesset perturbation theory

Møller and Plesset [139] proposed using the HF approximation as a zero-order and represent the Hamiltonian as:

$$H = \sum_{i=1}^{N} \widehat{h}^{HF}(i) + V \tag{A3.105}$$

The perturbation is then defined as a difference:

$$V = H - \sum_{i=1}^{N} \widehat{h}^{HF}(i) \tag{A3.106}$$

This kind of PT, named the *Møller–Plesset PT* (MPPT), is widely used in calculations of molecular systems; the application of MPPT to intermolecular interaction studies is discussed in Section 3.2.1.3.

The zero-order wave function is the HF N-electron wave function that can be found in the RHF or UHF approximation. More often, the UHF approximation is employed. In this case:

$$\Psi_0^{(0)} = \Psi_0^{UHF} = \frac{1}{\sqrt{N!}} \det \left| \psi_{1\alpha} \psi_{2\beta} \psi_{3\alpha} \ldots \psi_{N\beta} \right| \tag{A3.107}$$

if the ground state is considered. The expectation value of the energy in the state with the wave function (A3.107) is:

$$E_0^{UHF} = \left\langle \Psi_0^{UHF} \left| H \right| \Psi_0^{UHF} \right\rangle = \sum_{n=1}^{N} f_n + \tfrac{1}{2} \sum_{n,m} (\alpha_{nm} - \beta_{nm}) \tag{A3.108}$$

(compare this with Equation (A3.25) for double-filled $K_0$ configuration.)

Let us find the zero-order energy. It is defined as an eigenvalue of the Schrödinger equation:

$$H_0\Psi_0^{(0)} = \left[\sum_i \widehat{h}^{UHF}(i)\right]\Psi_0^{(0)} = E_0^{(0)}\Psi_0^{(0)} \tag{A3.109}$$

At the UHF level:

$$\begin{gathered}\widehat{h}^{UHF}(i)\,\psi_{n\sigma}(i) = \varepsilon_n\psi_{n\sigma}(i)\\ \varepsilon_n = f_n + \sum_m(\alpha_{nm} - \beta_{nm})\end{gathered} \tag{A3.110}$$

Using the definition of determinant (A2.24) gives:

$$\begin{aligned}&\sum_{i=1}^{N}\widehat{h}^{UHF}(i)\left(\frac{1}{\sqrt{N!}}\sum_P(-1)^p P\left(\psi_{1\alpha}(1)\,\psi_{2\beta}(2)\ldots\psi_{N\beta}(N)\right)\right)\\ &= \frac{1}{\sqrt{N!}}\sum_P(-1)^p P\left(\sum_{i=1}^{N}\widehat{h}^{UHF}(i)\,\psi_{1\alpha}(1)\,\psi_{2\beta}(2)\ldots\psi_{N\beta}(N)\right)\\ &= \left(\sum_{n=1}^{N}\varepsilon_n\right)\left(\frac{1}{\sqrt{N!}}\sum_P(-1)^p P\psi_{1\alpha}(1)\,\psi_{2\beta}(2)\ldots\psi_{N\beta}(N)\right)\\ &= \left(\sum_{n=1}^{N}\varepsilon_n\right)\Psi_0^{(0)}\end{aligned} \tag{A3.111}$$

From Equations (A3.109)–(A3.111) it follows that:

$$E_0^{(0)} = \sum_n\varepsilon_n = \sum_n\left[f_n + \sum_m(\alpha_{nm} - \beta_{nm})\right] \tag{A3.112}$$

Thus, $E_0^{UHF}$ does not equal $E_0^{(0)}$ :

$$E_0^{UHF} = E_0^{(0)} - \tfrac{1}{2}\sum_{n,m}(\alpha_{nm} - \beta_{nm}) \tag{A3.113}$$

The difference is due to the fact that in $E_0^{(0)}$ the interaction of each electron with itself is included.

The first order correction in MPPT is defined by Equation (A3.94):

$$\begin{aligned}E_0^{(1)} &= \left\langle\Psi_0^{UHF}\left|H - \sum_i\widehat{h}^{UHF}(i)\right|\Psi_0^{UHF}\right\rangle\\ &= \left\langle\Psi_0^{UHF}\left|H\right|\Psi_0^{UHF}\right\rangle - \left\langle\Psi_0^{UHF}\left|\sum_i\widehat{h}^{UHF}(i)\right|\Psi_0^{UHF}\right\rangle\\ &= E_0^{UHF} - \sum_n\varepsilon_n\end{aligned} \tag{A3.114}$$

Hence, up to the first order, the energy is:

$$E_0 = E_0^{(0)} + E_0^{(1)} = E_0^{UHF} \tag{A3.115}$$

where Equation (A3.112) is exploited. From Equation (A3.115) it follows that for an improvement of the UHF energy, the second-order correction must be calculated.

According to Equation (A3.102):

$$E_0^{(2)} = \sum_j{}' \frac{|V_{j0}|^2}{\left(E_0^{(0)} - E_j^{(0)}\right)^2} \tag{A3.116}$$

Let $K_1$ designate the UHF ground-state configuration described by determinant (A3.107). The excited-state functions $\Psi_j^{(0)}(K_j)$ can be chosen like in the CI method (Section A3.2.2.2), as determinants for each possible excited configuration: singly excited, $\Psi^{(0)}\left((K_1)_n^e\right)$, doubly excited, $\Psi^{(0)}\left((K_1)_{nm}^{ei}\right)$, etc. The matrix elements:

$$\begin{aligned} V_{j0} &= \left\langle \Psi_j^{(0)}(K_j) \middle| H - \sum_i \hat{h}^{UHF}(i) \middle| \Psi_0^{(0)}(K_1) \right\rangle \\ &= \left\langle \Psi_j^{(0)}(K_j) |H| \Psi_0^{(0)}(K_1) \right\rangle \end{aligned} \tag{A3.117}$$

since:

$$\left\langle \Psi_j^{(0)}(K_j) \middle| \sum_i \hat{h}^{UHF}(i) \middle| \Psi_0^{(0)}(K_1) \right\rangle$$

$$= \left(\sum_n \varepsilon_n\right) \left\langle \Psi_j^{(0)}(K_j) \middle| \Psi_0^{(0)}(K_1) \right\rangle = \left(\sum_n \varepsilon_n\right) \delta_{j0} = 0, \quad j \neq 0$$

It is easy to prove that:

$$\left\langle \Psi_j^{(0)}(K_j) |H| \Psi_0^{(0)}(K_1) \right\rangle \neq 0 \quad \text{only for } K_j = (K_1)_{nm}^{ei} \tag{A3.118}$$

The statement in Equation (A3.118) is valid in so far as the matrix element in Equation (A3.118) equals zero for all triply, quadruply, etc. excited configurations, because the Hamiltonian contains only single- and double-electron operators, so the excited configuration may not differ from the ground-state configuration on more than two orbitals. The singly-excited configurations are not mixed with the ground-state configuration due to the Brillouin theorem:

$$\left\langle \Psi^{(0)}\left((K_1)_n^e\right) |H| \Psi_0^{(0)}(K_1) \right\rangle = 0 \tag{A3.119}$$

(Its proof is given in Reference [140], Appendix 4).

Since in the matrix element in Equation (A3.118) the configurations differ by two orbitals, only the Coulomb interaction operator gives nonzero contribution. As a result, Equation (A3.116) becomes:

$$E_0^{(2)}\,(\text{MPPT}) = \sum_{n,m}{}' \sum_{e,i}{}' \frac{\left| \left\langle \Psi^{(0)}\left((K_1)_{nm}^{ei}\right) \left| \sum_{i<j} \frac{e^2}{r_{ij}} \right| \Psi^{(0)}\,(K_1) \right\rangle \right|^2}{\varepsilon_n + \varepsilon_m - \varepsilon_e - \varepsilon_i} \tag{A3.120}$$

The calculations taking into account the second order corrections:

$$E_0 = E_0^{(0)} + E_0^{(1)} + E_0^{(2)} = E_0^{HF} + E_0^{(2)}$$

are designated as MP2. The higher-order of MPPT corrections have been also elaborated [141]. The widely used MP4(SDTQ) method includes corrections up to the fourth order of MPPT with single, double, triple and quadrupole excitations. Its calculation results are close to CCSD(T) calculations.

### A3.3.3 Operator formalism and the Brillouin–Wigner perturbation theory

The perturbation series becomes more compact if the formalism of the projection operators is applied.

Let $|\Omega\rangle$ be a vector that belongs to an infinite Hilbert space and corresponds to a normalized many-electron function $\Omega$, so $\langle \Omega\,|\Omega\,\rangle = 1$. We introduce the operator[5]:

$$\widehat{P}_\Omega = |\Omega\rangle\,\langle\Omega| \tag{A3.121}$$

the action of which on an arbitrary vector $|\Phi\rangle$ selects its component along $|\Omega\rangle$. That is, using the Dirac notation:

$$\widehat{P}_\Omega\,|\Phi\rangle = |\Omega\rangle\,\langle \Omega\,|\Phi\,\rangle \tag{A3.122}$$

Therefore, the operator $\widehat{P}_\Omega$ is a *projection operator*, it projects a vectors $|\Phi\rangle$ onto a vector $|\Omega\rangle$. If the vectors $|\Omega\rangle$ and $|\Phi\rangle$ are orthogonal, i.e., $\langle \Omega\,|\Phi\,\rangle = 0$, it follows from Equation (A3.122) that $\widehat{P}_\Omega\,|\Phi\rangle = 0$. It is easy to prove directly that operator $\widehat{P}_\Omega$, as all projection operators, is idempotent, that is:

$$\widehat{P}_\Omega^2 = \widehat{P}_\Omega \tag{A3.123}$$

Let us introduce the complementary to the $\widehat{P}_\Omega$ operator:

$$\widehat{Q}_\Omega = 1 - \widehat{P}_\Omega \tag{A3.124}$$

that projects an arbitrary vector $|\Phi\rangle$ onto a manifold of vectors, which are orthogonal to $|\Omega\rangle$. In fact:

$$\left\langle \Omega \left| \widehat{Q}_\Omega \right| \Phi \right\rangle = \langle \Omega\,|\Phi\,\rangle - \langle \Omega\,|\Omega\,\rangle\,\langle \Omega\,|\Phi\,\rangle = 0 \tag{A3.125}$$

---

[5] In the case of unnormalized states, operator (A3.121) is replaced by $\widehat{P}_\Omega = \frac{|\Omega\rangle\langle\Omega|}{\langle\Omega|\Omega\,\rangle}$.

The unknown function in the Schrödinger Equation (A3.85) can be represented as a sum:

$$|\Psi_n\rangle = \left|\Psi_n^{(0)}\right\rangle + |\upsilon\rangle \tag{A3.126}$$

where $|\upsilon\rangle$ is determined by the perturbation and is orthogonal to the unperturbed function:

$$\left\langle\Psi_n^{(0)}\,|\upsilon\right\rangle = 0 \tag{A3.127}$$

This leads to the condition of intermediate normalization (Equation (A3.99)).

Let us denote the projection operator (A3.121) on the unperturbed state $\left|\Psi_n^{(0)}\right\rangle$ as $\widehat{P}_0$ and its complementary operator as $\widehat{Q}_0$, then from Equations (A3.125)–(A3.127), it follows that $\widehat{Q}_0\,|\Psi_n\rangle = |\upsilon\rangle$ or:

$$|\Psi_n\rangle = \left|\Psi_n^{(0)}\right\rangle + \widehat{Q}_0\,|\Psi_n\rangle \tag{A3.128}$$

Adding to the right- and left-hand side of the Schrödinger equation with the Hamiltonian (A3.86) a term $Z\Psi_n$, where $Z$ is an arbitrary number:

$$(Z + H_0 + V)\,|\Psi_n\rangle = (Z + E_n)\,|\Psi_n\rangle \tag{A3.129}$$

Equation (A3.129) can be written as:

$$(Z - H_0)\,|\Psi_n\rangle = (Z - E_n + V)\,|\Psi_n\rangle$$

Introducing the inverse operator $(Z - H_0)^{-1}$, we can express the equation for the wave function vector as:

$$|\Psi_n\rangle = (Z - H_0)^{-1}\,(Z - E_n + V)\,|\Psi_n\rangle \tag{A3.130}$$

To find the result of the action of the inverse operator $(Z - H_0)^{-1}$ on $|\Psi_n\rangle$, it is necessary to expand $|\Psi_n\rangle$ over the eigenvectors of the operator $H_0$:

$$|\Psi_n\rangle = \sum_n c_n \left|\Psi_0^{(0)}\right\rangle$$

Then, it can be shown that:

$$(Z - H_0)^{-1}\,|\Psi_n\rangle = \sum_n c_n \left(Z - E_n^{(0)}\right)^{-1} \left|\Psi_n^{(0)}\right\rangle \tag{A3.131}$$

since the action of the operator $(Z - H_0)$ on Equation (A3.131) leads to:

$$(Z - H_0)\,(Z - H_0)^{-1}\,|\Psi_n\rangle = \sum_n c_n \frac{(Z - H_0)}{\left(Z - E_n^{(0)}\right)} \left|\Psi_n^{(0)}\right\rangle = \sum_n c_n \left|\Psi_n^{(0)}\right\rangle = |\Psi_n\rangle$$

in a full agreement with the definition of the inverse operator.

Substituting Equation (A3.130) in the right-hand side of Equation (A3.128) we obtain:

$$|\Psi_n\rangle = \left|\Psi_n^{(0)}\right\rangle + \widehat{R}_0\,(Z)\,(Z - E_n + V)\,|\Psi_n\rangle \tag{A3.132}$$

where $\widehat{R}_o\,(Z)$ designates the so-called *resolvent operator*:

$$\widehat{R}_0\,(Z) = \widehat{Q}_0\,(Z - H_0)^{-1} = (Z - H_0)^{-1}\,\widehat{Q}_0 \tag{A3.133}$$

Equation (A3.132) is solved by an iteration procedure. At a first step, $|\Psi_n\rangle$ is substituted by the vector $\left|\Psi_n^{(0)}\right\rangle$, the obtained $\left|\Psi_n^{(1)}\right\rangle$ is then substituted for $|\Psi_n\rangle$ to obtain $\left|\Psi_n^{(2)}\right\rangle$, etc. This leads to the following series:

$$|\Psi_n\rangle = \sum_{k=0}^{\infty} \left\{\widehat{R}_0(Z)(Z - E_n + V)\right\}^k \left|\Psi_n^{(0)}\right\rangle \tag{A3.134}$$

To find the energy, we need to transform the Schrödinger equation:

$$(H_0 + V)|\Psi_n\rangle = E_n|\Psi_n\rangle \tag{A3.135}$$

selecting $E_n$. Multiplying Equation (A3.135) by $\left\langle\Psi_n^{(0)}\right|$ from the left gives:

$$\left\langle\Psi_n^{(0)}\left|H_0\right|\Psi_n\right\rangle + \left\langle\Psi_n^{(0)}\left|V\right|\Psi_n\right\rangle = E_n\left\langle\Psi_n^{(0)} \mid \Psi_n\right\rangle.$$

Acting with the operator $H_0$ on the $\left\langle\Psi_n^{(0)}\right|$ and taking into account the intermediate normalization condition (A3.99), one obtains the expression for the *energy level shift*:

$$E_n - E_n^{(0)} = \left\langle\Psi_n^{(0)}\left|V\right|\Psi_n\right\rangle \tag{A3.136}$$

Substituting Expansion (A3.134) for $|\Psi_n\rangle$ into Equation (A3.136), one finds the expansion for the energy shift:

$$E_n - E_n^{(0)} = \sum_{k=0}^{\infty}\left\langle\Psi_n^{(0)}\left|V\left\{\widehat{R}_0(Z)(Z - E_n + V)\right\}^k\right|\Psi_n^{(0)}\right\rangle \tag{A3.137}$$

Equations (A3.134) and (A3.137) allow the formulae of the *Rayleigh–Schrödinger* (RS) and *Brillouin–Wigner* (BW) perturbation theories to be obtained in all orders of PT. The RSPT is obtained if $Z = E_n^{(0)}$. It can be checked that for low orders of PT exactly the same expressions are obtained that were derived in Section A3.3.1 by the different method. The BWPT is follows from Equations (A3.134) and (A3.137) under the assumption that $Z = E_n$. This can now be carried out.

$$|\Psi_n\rangle = \sum_{k=0}^{\infty}\left\{\widehat{R}_0(E_n)V\right\}^k\left|\Psi_n^{(0)}\right\rangle \tag{A3.138}$$

$$E_n - E_n^{(0)} = \sum_{k=0}^{\infty}\left\langle\Psi_n^{(0)}\left|V\left\{\widehat{R}_0(E_n)V\right\}^k\right|\Psi_n^{(0)}\right\rangle \tag{A3.139}$$

Then we should take into account that the resolvent can be written as a series:

$$\widehat{R}_0(E_n) = \frac{\widehat{Q}_0}{E_n - H_0} = \sum_{m \neq n}\frac{\left|\Psi_m^{(0)}\right\rangle\left\langle\Psi_m^{(0)}\right|}{E_n - E_m^{(0)}} \tag{A3.140}$$

Taking into account Equation (A3.140), the first terms in the energy perturbation series can be written as:

$$E_n^{BW} = E_n^{(0)} + \left\langle \Psi_n^{(0)} \left| V \right| \Psi_n^{(0)} \right\rangle + \sum_{m \neq n} \frac{\left\langle \Psi_n^{(0)} \left| V \right| \Psi_m^{(0)} \right\rangle \left\langle \Psi_m^{(0)} \left| V \right| \Psi_n^{(0)} \right\rangle}{E_n - E_m^{(0)}} + \cdots. \tag{A3.141}$$

The first two terms in Equation (A3.141) are the same as in RSPT. Beginning from the second-order term, the energy $E_n$ that has to be evaluated enters the denominator, that is, the expression (A3.141) is the equation for the evaluation of $E_n$.

## A3.3.4 Variational perturbation theory

In this section, we consider briefly the basic statements of the variational perturbation theory. A more detailed presentation is given by Epstein [142].

Let us assume that the wave function $\Psi_n^{(0)}$ of the unperturbed system is known. However, evaluation of the correction function $\Psi_n^{(1)}$, for instance by Equation (A3.97), leads to some problems. In this case, evaluation of the correction function can be reduced to a variational problem.

Consider the ground state. The function $\Psi_0$ is assumed to depend not only upon space and spin coordinates, $x_i$, but also upon a set of real (or, in general, complex) parameters, $c_i$:

$$\Psi_0 = \Psi_0 \left( x_1, x_2, \ldots, x_N; c_1, c_2, \ldots, c_s \right) \tag{A3.142}$$

As we know, the Schrödinger equation follows from the variational principle (Equation (A3.23)) under the subsidiary condition (Equation (A3.24)). Then the energy:

$$\widetilde{E}_0 = \frac{\left\langle \Psi_0 \left| H \right| \Psi_0 \right\rangle}{\left\langle \Psi_0 \mid \Psi_0 \right\rangle} \tag{A3.143}$$

found with the approximate function (A3.142) will be the upper bound for the lowest eigenvalue, $E_0$, of the exact energy, or an equality:

$$\left| \left\langle \Psi_0 \left| H \right| \Psi_0 \right\rangle - E_0 \left\langle \Psi_0 \mid \Psi_0 \right\rangle \right| \geqslant 0 \tag{A3.144}$$

holds. Equation (A3.144) can be used to estimate $E_0^{(2)}$. Let us substitute the expansions for the wave function and energy in powers of a small parameter $\varkappa$:

$$\Psi_0 = \sum_{k=0}^{\infty} \varkappa^k \Psi_0^{(k)} \qquad E_0 = \sum_{k=0}^{\infty} \varkappa^k E_0^{(k)} \tag{A3.145}$$

into Equation (A3.144) with $H = H_0 + V$ and require that this inequality is satisfied for all values of $\varkappa$. Collecting like powers of $\varkappa^k$, is easy to show that the left-hand side of Equation (A3.144) is equal to zero for terms with $k = 0$ and 1. The terms with $\varkappa^2$ can be arranged as:

$$\left\{ J \left[ \Psi_0^{(1)} \right] - E_0^{(2)} \right\} \geqslant 0 \tag{A3.146}$$

where the functional:

$$J\left[\Psi_0^{(1)}\right]=\left\langle\Psi_0^{(1)}\left|H_0-E_0^{(0)}\right|\Psi_0^{(1)}\right\rangle+2\left\langle\Psi_0^{(0)}\left|V-E_0^{(1)}\right|\Psi_0^{(1)}\right\rangle \qquad \text{(A3.147)}$$

was first obtained by Hylleraas [143]. It is written for the real wave functions. If $\Psi_0^{(0)}$ is a real function and $\Psi_0^{(1)}$ and $V$ are complex, the functional $J\left[\Psi_0^{(1)}\right]$ becomes:

$$J\left[\Psi_0^{(1)}\right]=\left\langle\Psi_0^{(1)}\left|H_0-E_0^{(0)}\right|\Psi_0^{(1)}\right\rangle+2\,\mathrm{Re}\left\langle\Psi_0^{(0)}\left|V-E_0^{(1)}\right|\Psi_0^{(1)}\right\rangle \qquad \text{(A3.147a)}$$

On the other hand, substitution of the expansion (A3.145) into the Schrödinger equation and selection of terms of the same order up to $\varkappa^2$ leads to the following equations:

$$\left(H_0-E_n^{(0)}\right)\Psi_n^{(0)}=0 \qquad \text{(A3.148)}$$

$$\left(H_0-E_n^{(0)}\right)\Psi_n^{(1)}=\left(E_n^{(1)}-V\right)\Psi_n^{(0)} \qquad \text{(A3.149)}$$

$$\left(H_0-E_n^{(0)}\right)\Psi_n^{(2)}=\left(E_n^{(1)}-V\right)\Psi_n^{(1)}+E_n^{(2)}\Psi_n^{(0)} \qquad \text{(A3.150)}$$

The solution of these equations yields the perturbation corrections that coincide with those found by the similar method in Section A3.3.1.

If $\Psi_0^{(1)}$ is the exact correction to the wave function, from Equation (A3.146) it follows:

$$E_0^{(2)}=J\left[\Psi_0^{(1)}\right] \qquad \text{(A3.151)}$$

Substituting the first term in the functional (A3.147) and taking into account Equation (A3.149) leads to the formula (A3.103) for $E_n^{(2)}$. Otherwise, the functional $J\left[\Psi_0^{(1)}\right]$ yields an upper bound for $E_0^{(2)}$. Then it can be proved that Equation (A3.149) for the first-order correction follows directly from the variation of functional $J\left[\Psi_0^{(1)}\right]$ equated to zero.

In practice, the function $\Psi_n^{(1)}$ contains different variational parameters. As their number increases, the exact solution is more closely approached. This point is illustrated with the following example. Let the perturbation operator $V$ have the form:

$$V=\sum_{i=1}^{N}x_i \qquad \text{(A3.152)}$$

where the summation extends over all $N$ electrons of the system. In its simplest form, the first-order variational correction, $\Psi_0^{(1)}$, can be written as [144]:

$$\Psi_0^{(1)}=cV\Psi_0^{(0)} \qquad \text{(A3.153)}$$

where $c$ is considered as a variational parameter. A more sophisticated approximation involves more variational parameters and may be given in the form:

$$\Psi_0^{(1)}=\sum_{m=1}^{s}c_m V_m\Psi_0^{(0)} \qquad \text{(A3.154)}$$

where the operator $V_1$ coincides with Equation (A3.152) and the subsequent operators $V_2, V_3, \ldots$ contain all possible products of electron coordinates; the degree of the polynomial increases as it proceeds from $V_2$ to $V_3$, and so on.

Substituting the expression (A3.154) into the Hylleraas functional (A3.147) one finds $c_m$ on the basis of the variational principle. Namely, assuming $V_m$ and $c_m$ are real and taking the orthogonality of the function $\Psi_0^{(1)}$ and $\Psi_0^{(0)}$ into account, one obtains the equation:

$$\sum_n L_{mn} c_n = M_m \tag{A3.155}$$

where:

$$L_{mn} = \left\langle \Psi_0^{(0)} \middle| V_m \left( H_0 - E_0^{(0)} \right) V_n \middle| \Psi_0^{(0)} \right\rangle + \left\langle \Psi_0^{(0)} \middle| V_n \left( H_0 - E_0^{(0)} \right) V_m \middle| \Psi_0^{(0)} \right\rangle$$

$$M_m = -2\left\langle \Psi_0^{(0)} \left| V_1 V_m \right| \Psi_0^{(0)} \right\rangle$$

The matrix $\|L_{mn}\|$ can be simplified with the help of the relationship:

$$\left( H_0 - E_0^{(0)} \right) V_n \left| \Psi_0^{(0)} \right\rangle = [H_0, V_n] \left| \Psi_0^{(0)} \right\rangle$$

where $\Psi_0^{(0)}$ is assumed to be the exact eigenfunction of $H_0$. Evaluating the corresponding commutators, gives:

$$L_{mn} = \left\langle \Psi_0^{(0)} \middle| \sum_{1=1}^{N} (\nabla_i V_m)(\nabla_i V_n) \middle| \Psi_0^{(0)} \right\rangle \tag{A3.156}$$

The subsequent calculations require the explicit form of the ground-state wave function. In the case of molecules, $\Psi_0^{(0)}$ is assumed to be found by solution of the Hartree–Fock equations. It should be borne in mind that the function $\Psi_0^{(0)}$ was assumed to be exact in all the formulae given above. In a strict sense, the condition $J\left[\Psi_0^{(1)}\right] \geqslant E_0^{(2)}$ can be guaranteed only in that case. The problem of the approximation of $\Psi_0^{(0)}$ is discussed in detail in the review [145], where some examples of different variational principles are also presented.

## A3.3.5 Asymptotic expansions; Padé approximants

The proof of convergence of the perturbation expansion series for real systems is a rather complicated problem. In practice, one limits oneself by not very rigorous arguments such as, for instance, the smallness of $E_0^{(2)}$ and $E_0^{(1)}$ by comparison with $E_0^{(0)}$. In reality, for each term in the sum of $E_0^{(2)}$ (Equation (A3.102)) the inequality (A3.101) has to be fulfilled. For the ground state, in particular, the necessary condition is:

$$\left| \left\langle \Psi_1^{(0)} \left| V \right| \Psi_0^{(0)} \right\rangle \right| \ll \left| E_0^{(0)} - E_1^{(0)} \right| \tag{A3.157}$$

If this condition is violated, a *quasidegeneracy* is said to exist.

Condition (A3.157) may be written in a general form, which is valid when the matrix approach to the solution of the Schrödinger equation is applied. Let the Hamiltonian $H_0$ be represented by the square $(N \times N)$ diagonal matrix with the elements $E_0^{(0)} E_1^{(0)}, \ldots, E_{N-1}^{(0)}$ and the perturbation operator $V^{N-1}$ be given by the Hermitian symmetric matrix with vanishing diagonal elements. The so-called *norm* of matrix $\mathbf{v}$ designated by $|\mathbf{v}|$ is defined by the relationship [146]:

$$|\mathbf{v}|^2 = \mathrm{Sp}\left(\mathbf{v}^+\mathbf{v}\right)$$

It can be proved that the Rayleigh–Schrödinger series converges, if the inequality [146, 147]:

$$|\mathbf{v}| < \tfrac{1}{2}\left(E_1^{(0)} - E_0^{(0)}\right) \tag{A3.158}$$

is satisfied. To elucidate the meaning of this inequality, let us write the expression for the norm of the symmetric matrix ($\mathrm{v}_{ik} = \mathrm{v}_{ki}$) with vanishing diagonal elements:

$$|\mathbf{v}|^2 = \sum_{i,k} \mathrm{v}_{ik}\mathrm{v}_{ki} = \sum_{i,k} \mathrm{v}_{ki}^2 = 2\sum_{i<k} \mathrm{v}_{ki}^2 \tag{A3.159}$$

Therefore, the condition (A3.158) means that the square root of the sum of the square of all the matrix elements of the perturbation operator is smaller than half the difference between the energies of the unperturbed states.

If the criterion (A3.158) does not hold, the perturbation series converge, in must cases, asymptotically. According to Poincaré, *asymptotic series* are defined by the following features [148]:

a series $\sum_k c_k x^{-k}$ is said to be asymptotic, converging to a function $f(x)$ in the Poincaré sense, if for any positive integer $N$:

$$\lim_{x\to\infty} x^N \left[ f(x) - \sum_{k=0}^{N} c_k x^{-k} \right] = 0 \tag{A3.160}$$

even if for a certain $x$:

$$\lim_{N\to\infty} x^N \left[ f(x) - \sum_{k=0}^{N} c_k x^{-k} \right] = \infty \tag{A3.161}$$

As a rule, for each value of $x$, the asymptotic series diverges. Nevertheless, there exists an optimal number $N$, for which the function is appropriately represented by the series. For fixed $N$ and large $x$, the asymptotic series approaches the function $f(x)$ with any desired accuracy.

The Euler series is such an example. It has the form:

$$f(x) = 1 - 1!x + 2!x^2 - 3!x^3 + \cdots \tag{A3.162}$$

which can be related with the formal expansion of the function:

$$f(x) = \int_0^\infty \frac{e^{-t}}{1 + xt} dt \tag{A3.163}$$

in the Taylor series. The problem of approximating a divergent series by an analytical function can be solved with the help of the method of *Padé approximants* [149–151].

The idea of this method is the following. Let us assume that a given function $f(x)$ is represented by the Taylor series:

$$f(x) = \sum_k c_k x^k, \tag{A3.164}$$

which converges in a given interval. Cutting off this series at the *nth* term gives a polynomial, which approximates well this function when $x$ is near zero. The coefficients of this series, $c_k$, can be used to construct other approximate functions, which can represent the given function $f(x)$ at those values of $x$ where the Taylor series becomes divergent. In the Padé approach, $f(x)$ is approximated by the ratio of two polynomials, $P_m(x)$ and $Q_n(x)$, of degrees $m$ and $n$ respectively:

$$\begin{aligned} f(x) &\simeq \frac{P_m(x)}{Q_n(x)} \\ P_m(x) &= a_0 + a_1 x + a_2 x^2 + \cdots + a_m x^m \\ Q_n(x) &= 1 + b_1 x + b_2 x^2 + \cdots + b_n x^n \end{aligned} \tag{A3.165}$$

Coefficient $b_0$ is assumed to be unity. Therefore, the total number of coefficients $a_i$ and $b_i$ is equal to $m + n + 1$. The rational fraction $P_m(x)/Q_n(x)$ is denoted as $[m/n]$*th Padé approximant*, if the following condition holds:

$$f(x) - \frac{P_m(x)}{Q_n(x)} = 0\left(x^{n+m+1}\right) \tag{A3.166}$$

This condition permits the $(m + n + 1)$ numbers $a_i$ and $b_i$ to be expressed in terms of the $(m + n + 1)$ coefficients $c_k$ of the Taylor series (A3.164). The polynomials $P_m(x)$ and $Q_n(x)$ can be represented by the determinants [149]:[6]

$$P_m(x) = \begin{vmatrix} c_{m-n+1} & c_{m-n+2} & \cdots & c_{m+1} \\ c_{m-n+2} & c_{m-n+3} & \cdots & c_{m+2} \\ \vdots & \vdots & \vdots\vdots\vdots & \vdots \\ \sum_{j=n}^{m} c_{j-n} x^j & \sum_{j=n-1}^{m} c_{j-n+1} x^j & \cdots & \sum_{j=0}^{m} c_j x^j \end{vmatrix} \tag{A3.167}$$

[6] If $f(x)$ is a rational fraction, the Padé approximant for some $m$ and $n$ yields its exact value. In that case, the subsequent Padé approximants are not defined by the expressions (A3.167) and (A3.168), the latter vanish.

$$Q_n(x) = \begin{vmatrix} c_{m-n+1} & c_{m-n+2} & \cdots & c_{m+1} \\ c_{m-n+2} & c_{m-n+3} & \cdots & c_{m+2} \\ \vdots & \vdots & \vdots\vdots\vdots & \vdots \\ x^n & x^{n-1} & \cdots & 1 \end{vmatrix} \tag{A3.168}$$

In Equations (A3.167) and (A3.168) the coefficient $c_k$ vanishes if $j < 0$.

As an example of the Padé technique, let us consider the Euler series (A3.162) and construct the first approximants:

$$[1/1] = \frac{1+x}{1+2x}, \qquad [1/2] = \frac{1+3x}{1+4x+2x^2},$$

$$[2/2] = \frac{1+5x+2x^2}{1+6x+6x^2}, \qquad [1/3] = \frac{1+8x+11x^2}{1+3x+18x^2+6x^3}.$$

As proved in the theory of Padé approximants [149], the $[n/n]th$ approximant provides an upper estimate of the function being approximated, and the $[n-1/n]th$ approximant provides a lower estimate. For $x = 1$:

$$[1/1]_{x=1} = 0.6666, \qquad [1/2]_{x=1} = 0.5714,$$

$$[2/2]_{x=1} = 0.6154, \qquad [2/3]_{x=1} = 0.5882,$$

$$\vdots \qquad\qquad \vdots$$

$$[6/6]_{x=1} = 0.5968, \qquad [5/6]_{x=1} = 0.5960.$$

The exact value of the function (A3.163) at $x = 1$ is $f(1) = 0.5963$. We see that the interval between the upper and lower bounds is very small and contains the exact value.

The method of Padé approximants has been widely used to evaluate the dispersion coefficient, $C_n$, via the Casimir–Polder formula (Chapter 2, Equation (2.69)). The lower bound for $\alpha(i\omega)$ and, therefore, for the coefficients $C_n$ is obtained with the help of the $[n-1/n]th$ approximant. However, the $[n/n]th$ approximant does not give an upper bound for $C_n$, because it approaches a finite limit as $\omega \longrightarrow \infty$, and the integral over frequencies in Equation (2.69) becomes meaningless. The way of obtaining the upper bound for $C_6^{AB}$ is discussed in Reference [152], Chapter 2.

The generalization of the Padé method, which involves the simultaneous approximation of given function at some points, the multipoint Padé approximants, was studied in References [153, 154]. The method of Padé approximants had also been applied to the study of potential energy curves [155–158].

# References

1. D.W. Jepsen and J.O. Hirschfelder, *J. Chem. Phys.* **32**, 1323 (1960).
2. A. Dalgarno and R. McCarroll, *Proc. Roy. Soc.* (London) A**237**, 383 (1956); A**239**, 413 (1957).
3. A. Fröman, *J. Chem. Phys.* **36**, 1490 (1962).

4. J.O. Hirschfelder and W.J. Meath, *Adv. Chem. Phys.* **12**, 3 (1967).
5. B. Sutcliffe, in *Handbook of Molecular Physics and Quantum Chemistry*, S. Wilson (ed), John Wiley & Sons, Ltd, Chichester (2003), Vol. 1, Chapters 31 and 32.
6. M. Born and J.R. Oppenheimer, *Ann. der Phys.* (Leipzig) **84**, 457 (1927).
7. M. Born, *Göttingen Nachr. Acad. Wiss. Natl. Kl.* p.1 (1951).
8. M. Born and K. Huang, *Dynamical Theory of Crystal Lattice*, Oxford University Press, London (1956).
9. L. Wolniewicz, *J. Chem. Phys.* **103**, 1792 (1995).
10. W. Kołos and J. Rychlewski, *J. Chem. Phys.* **98**, 3960 (1993).
11. L. Wolniewicz, *J. Chem. Phys.* **99**, 1851 (1993).
12. H. Sellers and P. Pulay, *Chem. Phys. Lett.* **103**, 463 (1984).
13. B.H. Lengsfield and D.R. Yarkony, *J. Chem. Phys.* **84**, 348 (1986).
14. N.C. Handy, Y. Yamaguchi and H.F. Schaefer, III, *J. Chem. Phys.* **84**, 4481 (1986).
15. N.C. Handy and A.M. Lee, *Chem. Phys. Lett.* **252**, 425 (1996).
16. R. Engleman, *The Jahn–Teller Effect in Molecules and Crystals*, John Wiley & Sons, Inc., New York (1972).
17. I.B. Bersuker and V.Z. Polinger, *Vibronic Interactions in Molecules and Crystals*, Springer–Verlag, Berlin (1989).
18. A. Kuppermann, in *Dynamics of Molecules and Chemical Reactions*, R.E. Wayatt and J.Z.H. Zang (eds), Marcel Dekker, New York (1996), pp. 411–472.
19. H. Köppel and W. Domske, in *Encyclopedia of Computational Chemistry*, P.v.R. Schleyer (ed), John Wiley & Sons, Ltd, Chichester (1998), pp. 3166–3182.
20. D. Yarkony, *J. Phys. Chem.* A **105**, 6277 (2001).
21. A.J.C. Varandas, in *Fundamental World of Quantum Chemistry*, E.J. Brändas and E.S. Kryachko (eds), Kluwer Academic Publishers, Dordrecht (2003), Vol. 2, pp. 33–92.
22. D.R. Hartree, *Proc. Cambridge Phil. Soc.* **24**, 89, 111 (1928).
23. V.A. Fock, *Zs. f. Phys.* **61**, 126 (1930); *Ibid* **62**, 795 (1930).
24. C.C.J. Roothaan, *Rev. Mod. Phys.* **23**, 69 (1951).
25. C.C.J. Roothaan, *Rev. Mod. Phys.* **32**, 179 (1960).
26. L.D. Landau and E.M. Lifshitz, *Quantum Mechanics (Nonrelativistic Theory)*, Third Edition, Pergamon Press, Oxford (1977).
27. I.G. Kaplan, *Symmetry of Many-Electron Systems*, Academic Press, New York (1975).
28. J.A. Pople and R.K. Nesbet, *J. Chem. Phys.* **22**, 57 (1954).
29. E. Hylleraas, *Zs. f. Phys.* **54**, 347 (1929).
30. C.L. Pekeris, *Phys. Rev.* **112**, 1649 (1958); *Ibid.* **126,** 1470 (1962).
31. C. Schwarz, *Phys. Rev.* **128**, 1146 (1962).
32. N.M. James and A.S. Coolidge, *J. Chem. Phys.* **1**, 825 (1933).
33. W. Kołos and L. Wolniewicz, *J. Chem. Phys.* **43**, 2429 (1965).
34. W. Kołos and L. Wolniewicz, *Chem. Phys. Lett.* **24**, 457 (1974).
35. W. Kołos and J. Rychlewski, *J. Chem. Phys.* **98**, 3960 (1993).
36. L. Wolniewicz, *J. Chem. Phys.* **99**, 1851 (1993); **103**, 1792 (1995).
37. N.M. James and A.S. Coolidge, *Phys. Rev.* **49**, 688 (1936).
38. J.P. Perkins, *Phys. Rev.* A **8**, 700 (1973).
39. J.S. Sims, S.A. Hadstrom, D. Munch and C. Bunge, *Phys. Rev.* A **13**, 560 (1976).
40. D.C. Clary, *Mol. Phys.* **34**, 793 (1977).
41. D.C. Clary and N.C. Handy, *Chem. Phys. Lett.* **51**, 483 (1977).
42. W. Klopper and W. Kutzelnigg, *Chem. Phys. Lett.* **134**, 17 (1987).
43. W. Klopper, *Chem. Phys. Lett.* **186**, 583 (1991).

44. J. Noga and W. Kutzelnigg, *J. Chem. Phys.* **101**, 7738 (1994).
45. J. Noga, W. Klopper and W. Kutzelnigg, in *Recent Advances in Coupled-Cluster methods*, R.J. Bartlett (ed), *World Scientific*, Singapore (1997), pp. 1–48.
46. W. Klopper in *Encyclopedia of Computational Chemistry,* P.v.R. Schleyer (ed), John Wiley & Sons, Ltd, Chichester (1998), pp. 2352–2375.
47. W. Klopper and J. Almlöf, *J. Chem. Phys.* **99**, 5167 (1993).
48. H. Müller, W. Kutzelnigg and J. Noga, *Mol. Phys.* **92**, 535 (1997).
49. A. Halkier *et al.*, *J. Chem. Phys.* **111**, 9157 (1999).
50. S. Wilson, *Electronic Correlations in Molecules*, Oxford University Press, Oxford (1984).
51. C.B. Bauschlicher, Jr., S.R. Langhoff and P.R. Taylor, *Adv. Chem. Phys.* **77**, 103 (1990).
52. J. Olsen, P. Jørgensen and J. Simons, *Chem. Phys. Lett.* **169**, 463 (1990).
53. A.O. Mitrushenkov, *Chem. Phys. Lett.* **217**, 559 (1994).
54. E.R. Davidson, *Reduced Density Matrices in Quantum Chemistry*, Academic Press, New York (1976).
55. C.F. Bender and E.R. Davidson, *J. Phys. Chem.* **70**, 2675 (1966).
56. J. Frenkel, *Wave Mechanics*, Oxford University Press, London-New York (1934).
57. C. Das and A.C. Wahl, *J. Chem. Phys.* **44**, 87 (1966).
58. A. Veillard and E. Clementi, *Theor. Chim. Acta.* **7**, 133 (1967).
59. H.-J. Werner, *Adv. Chem. Phys.* **69**, 1 (1987).
60. B.O. Roos, *Adv. Chem. Phys.* **69**, 399 (1987).
61. R. Shepard, *Adv. Chem. Phys.* **69**, 63 (1987).
62. P.J. Knowles and H.-J. Werner, *Chem. Phys. Lett.* **115**, 259 (1985).
63. F. Coester, *Nucl. Phys.* **7**, 421 (1958).
64. F. Coester and H. Kümmel, *Nucl. Phys.* **17**, 477 (1960).
65. R.F. Bishop and H. Kümmel, *Physics Today* (March 1987), p. 52.
66. J. Čižek, *J. Chem. Phys.* **45**, 4256 (1966).
67. J. Čižek, *Adv. Chem. Phys.* **14**, 35 (1969).
68. R.J. Bartlett, *J. Phys. Chem.* **93**, 1697 (1986).
69. J. Paldus, in *Methods in Computational Molecular Physics*, NATO ASI Series, S. Wilson and J.H.F. Diercksen (eds), Plenum Press, New York (1992), pp. 99–194.
70. R.J. Bartlett, in *Modern Electronic Structure Theory*, D.R. Yarkony (ed), *World Scientific*, Singapore, Part II (1995), pp. 1047–1131.
71. J. Paldus and X. Li, *Adv. Chem. Phys.* **110**, 1 (1999).
72. P. Čarski and M. Urban, *Ab Initio Calculations*, Springer-Verlag, Berlin (1980).
73. R.J. Bartlett and G.D. Purvis, *Int. J. Quant. Chem.* **14**, 561 (1978).
74. J.A. Pople, R. Krishnan, H.B. Schlegel and J.S. Binkley, **14**, 545 (1978).
75. J. Noga and R.J. Bartlett, *J. Chem. Phys.* **86**, 7041 (1987).
76. G.E. Scuseria and H.F. Schaefer, III, *Chem. Phys. Lett.* **152**, 382 (1988).
77. J.D. Watts and R.J. Bartlett, *J. Chem. Phys.* **93**, 6104 (1990).
78. S.A. Kucharski and R.J. Bartlett, *J. Chem. Phys.* **97**, 4282 (1992).
79. K. Raghavachari, G.W. Trucks, J.A. Pople and M. Head-Gordon, *Chem. Phys. Lett.* **157**, 479 (1989).
80. K. Raghavachari and J.B. Anderson, *J. Phys. Chem.* **100**, 12960 (1996).
81. M. Head-Gordon, *J. Phys. Chem.* **100**, 13213 (1996).
82. X. Li and J. Paldus, *J. Chem. Phys.* **115**, 5759 (2001).
83. B. Jeziorski and H.J. Monkhorst, *Phys. Rev. A* **24**, 1668 (1981).

84. S.A. Kucharski and R.J. Bartlett, *J. Chem. Phys.* **95**, 8227 (1991).
85. U.S. Mahapatra, B. Datta and D. Mukherjee, *J. Chem. Phys.* **110**, 6171 (1999).
86. X. Li and J. Paldus, *J. Chem. Phys.* **110**, 2844 (1999).
87. I. Hubač *et al.*, in *Computational Chemistry. Reviews of Current Trends*, J. Leszczynski (ed), *World Scientific*, Singapore, Vol. 3 (1999), pp. 1–48.
88. I. Hubač and W. Wilson, *J. Phys. B* **33**, 365 (2000).
89. I. Hubač, J. Pittner and P. Čársky, *J. Chem. Phys.* **112**, 8779 (2000).
90. J. Pittner, *J. Chem. Phys.* **118**, 10876 (2003).
91. P. Piecuch and K. Kowalski, in *Computational Chemistry. Reviews of Current Trends*, J. Leszczynski (ed), *World Scientific*, Singapore, Vol. 5, 2000, pp. 1–105.
92. K. Kowalski and P. Piecuch, *J. Chem. Phys.* **113**, 18 (2000).
93. K. Kowalski and P. Piecuch, *Chem. Phys. Lett.* **344**, 165 (2001).
94. I. Özkan, A. Kinal and M. Balei, *J. Phys. Chem. A* **108**, 507 (2004).
95. R.G. Parr and W. Yang, *Density-Functional Theory of Atoms and Molecules*, Oxford University Press, New York (1989).
96. E.S. Kryachko and E.V. Ludena, *Energy Density Functional Theory of Many-Electron Systems*, Kluwer Academic Press, Dordrecht (1990).
97. R. McWeeny, *Rev. Mod. Phys.* **32**, 335 (1960).
98. P. Hohenberg and W. Kohn, *Phys. Rev.* **136**, B864 (1964).
99. W. Kohn and L.J. Sham, *Phys. Rev.* **140**, A1133 (1965).
100. D. Pines, *Elementary Excitations in Solids*, Benjamin Press, New York (1963).
101. S.H. Vosko, L. Wilk and M . Nusair, *Can. J. Phys.* **58**, 1200 (1980).
102. O.V. Gritsenko, P.R.T. Schipper and E.J. Baerends, *J. Chem. Phys.* **107**, 5007 (1997).
103. U. von Barth and L. Hedin, *J. Phys. C* **5**, 1629 (1972).
104. J.C. Stoddart and N.H. March, *Ann. Phys. (New York)* **64**, 174 (1971).
105. M.M. Pant and A.K. Rajagopal, *Solid State Commun.* **10**, 1157 (1972).
106. T. Ziegler, A. Rauk and E. Baerends, *Theor. Chim. Acta* **43**, 261 (1977).
107. T. Ziegler, *Chem. Rev.* **91**, 651 (1991).
108. J.P. Perdew and Y. Wang, *Phys. Rev. B* **33**, 8800 (1986).
109. A.D. Becke, *Phys. Rev. A* **38**, 3098 (1988).
110. C. Lee, W. Yang and R.G. Parr, *Phys. Rev. B* **37**, 785 (1988).
111. J.P. Perdew, in *Electronic Structure of Solids*, P. Ziesche and H. Eschrig (eds), Akademic Verlag, Berlin (1991).
112. A.D. Becke, *J. Chem. Phys.* **98**, 5648 (1993).
113. J.P. Perdew and Y. Wang, *Phys. Rev. B* **45**, 13244 (1992).
114. A.D. Becke, *J. Chem. Phys.* **104**, 1040 (1996).
115. A.D. Becke, *J. Chem. Phys.* **107**, 8554 (1997).
116. H.L. Schmider and A.D. Becke, *J. Chem. Phys.* **108**, 9624 (1998).
117. L.A. Curtiss, K. Raghavachari, P.C. Redfern and J.A. Pople, *J. Chem. Phys.* **106**, 1063 (1997).
118. C. Adamo and V. Barone, *J. Chem. Phys.* **108**, 664 (1998).
119. F.A. Hamprecht, A.J. Cohen, D.J. Tozer and N.C. Handy, *J. Chem. Phys.* **109**, 6264 (1998).
120. J. Tao, J.P. Perdew, V.N. Staroverov and G.E. Scuseria, *Phys. Rev. Lett.* **91**, 146401 (2003).
121. J.P. Perdew, K. Burke and M. Ernzerhof, *Phys. Rev. Lett.* **77**, 3865 (1996).
122. V.N. Staroverov, G.E. Scuseria, J. Tao and J.P. Perdew, *J. Chem. Phys.* **119**, 12129 (2003).

123. J.P. Perdew, J. Tao, V.N. Staroverov and G.E. Scuseria, *J. Chem. Phys.* **120**, 6898 (2004).
124. X. Xu and W.A. Goddard, III, *Proc. Natt. Acad. Sci. USA,* **101**, 2673 (2004).
125. X. Xu and W.A. Goddard, III, *J. Phys. Chem. A* **108**, 2305 (2004).
126. Y. Zhao, J. Lynch and D.G. Truhlar, *J. Phys. Chem. A* **108**, 2715, 4786 (2004).
127. Y. Zhao and D.G. Truhlar, *J. Phys. Chem. A* **108**, 6908 (2004).
128. Y. Zhao and D.G. Truhlar, *J. Chem. Theory Comput.* **1**, 415 (2005).
128a. Y. Zhao and D.G. Truhlar, *J. Phys. Chem. A* **109**, 6624 (2005).
128b. Y. Zhao and D.G. Truhlar, *J. Phys. Chem. A* **109**, 5656 (2005).
129. K. Burke and E.K.U. Gross, in *Density Functional: Theory and Applications,* D. Joubert (ed), Springer–Verlag, Berlin (1998), pp. 116–146.
130. M.E. Casida, C. Jamorski, K.C. Casida and D.R. Salahub, *J. Chem. Phys.* **108**, 5134 (1998).
131. M. Levy and A. Nagy, *Phys. Rev. Lett.* **83**, 4361 (1999).
132. A. Görling, *Phys. Rev. Lett.* **85**, 4229 (2000).
133. I.B. Bersuker, *J. Comput. Chem.* **18**, 260 (1997).
134. I.G. Kaplan, *Int. J. Quant. Chem.* **89**, 268 (2002).
135. U. von Barth, *Phys. Rev.* **20**, 1693 (1979).
136. O. Gunnarsson and R.O. Jones, *Physica Scr.* **21**, 394 (1980).
137. B.I. Dunlap, *Adv. Chem. Phys.* **69**, 287 (1987).
138. A. Nagy, *Phys. Reports* **298**, 1 (1998).
139. C. Møller and M.S. Plesset, *Phys. Rev.* **46**, 611 (1934).
140. J.C. Slater, *Electronic Structure of Molecules*, McGraw–Hill Book Company, New York (1963).
141. R. Krishnan and J.A. Pople, *Int. J. Quant. Chem.* **14**, 91 (1978).
142. S.T. Epstein, *The Variational Method in Quantum Chemistry*, Academic Press, New York (1974).
143. E. Hylleraas, *Zs. f. Phys.* **65**, 209 (1930).
144. J.G. Kirkwood, *Phys. Zs.* **33**, 57 (1931).
145. J.O. Hirschfelder, W. Byers Brown and S.T. Epstein, *Adv. Quant. Chem.* **1**, 255 (1964).
146. J. Killenbeck, *Rep. Progr. Phys.* **40**, 964 (1977).
147. T. Kato, *Perturbation Theory of Linear Operators*, Springer–Verlag, Berlin (1966).
148. E.T. Whittaker and G.N. Watson, *A Course of Modern Analysis*, Cambridge University Press, Cambridge (1927).
149. G.A. Baker, Jr., *Essentials of the Padé Approximants*, Academic Press, New York (1975).
150. *Padé Approximants and Their Applications*, P.R. Graves-Morris (ed), Academic Press, New York (1973).
151. J. Gilewicz, *Lecture Notes in Mathematics*, Springer–Verlag, Berlin (1973).
152. I.G. Kaplan, *Theory of Molecular Interactions*, Elsevier, Amsterdam (1986).
153. S.T. Epstein and M.F. Barnaley, *J. Math. Phys.* **14**, 314 (1973).
154. A.A. Nudelman and A.V. Tulub, *Teor. Mat. Fiz.* **39**, 359 (1979).
155. S.T. Epstein, *J. Chem. Phys.* **48**, 4716 (1968).
156. E. Brändas and D.A. Micha, *J. Math. Phys.* **13**, 155 (1972).
157. K.D. Jordan, J.L. Kinsey and R. Silbey, *J. Chem. Phys.* **61**, 911 (1974).
158. E.S. Kryachko, *Chem. Phys. Lett.* **116**, 411 (1985).

# Index



*Intermolecular Interactions* I.G. Kaplan

Printed and bound by CPI Group (UK) Ltd, Croydon, CR0 4YY
18/09/2026
14965009-0001